VISUAL DOUBLE STARS:
FORMATION, DYNAMICS AND EVOLUTIONARY TRACKS

ASTROPHYSICS AND SPACE SCIENCE LIBRARY

VOLUME 223

VISUAL DOUBLE STARS: FORMATION, DYNAMICS AND EVOLUTIONARY TRACKS

Edited by

J. A. DOCOBO

Astronomical Observatory "Ramon Maria Aller",
University of Santiago de Compostela, Spain

A. ELIPE

University of Zaragoza, Spain

and

H. McALISTER

Center for High Angular Resolution Astronomy,
Georgia State University,
Atlanta, GA, U.S.A.

KLUWER ACADEMIC PUBLISHERS

DORDRECHT / BOSTON / LONDON

A C.I.P. Catalogue record for this book is available from the Library of Congress

ISBN 0-7923-4793-5

Published by Kluwer Academic Publishers,
P.O. Box 17, 3300 AA Dordrecht, The Netherlands.

Sold and distributed in the U.S.A. and Canada
by Kluwer Academic Publishers,
101 Philip Drive, Norwell, MA 02061, U.S.A.

In all other countries, sold and distributed
by Kluwer Academic Publishers,
P.O. Box 322, 3300 AH Dordrecht, The Netherlands.

Printed on acid-free paper

Printed in the Netherlands

Table of Contents

INTRODUCTION

This workshop is devoted to Double stars. The general topics of the meeting were: formation, dynamics and evolutionary tracks. In accordance with the pure tradition of the *Saint James way*, "pilgrims" from all over the world come to meet together in Santiago. Although with a common interest (double stars), this meeting was a multidisciplinary one, since scientists with different backgrounds participated in it.

As a matter of fact, we think that this is the first workshop jointly supported by IAU Commissions 7 (Celestial mechanics) and 26 (Double and multiple stars). It is our belief that this meeting will be the origin of a more close relations and common research.

This meeting was held under the invitation of the University of Santiago de Compostela to commemorate its fifth centenary, and organized by the Astronomical Observatory named after its founder, Ramón M. Aller, who made significant contributions in the study of visual double stars, and was one of the pioneers who put the seeds of the present blossoming of Astronomy in Spain.

The Scientific Organizing Committee was formed by Drs. C. Allen, P. Couteau, J. A. Docobo, R. Dvorak, A. Elipe, S. Ferraz-Mello (co-chairman), H.A.McAlister, M. Valtonen, C.Worley (chairman) and H. Zinnecker.

The Local Organizing Committee was formed by Drs. J. A. Docobo (chairman), A. Elipe, J. F. Ling, C. Prieto and V. Tamazian. Thanks to the LOC, many of the participants discovered that Galicia is a nice green country and privileged *gastronomical* site. Local sponsors were, besides the University itself, the Xunta de Galicia and the Ministerio de Educación. To all of them, LOC and sponsors, our thanks.

The Proceedings include lectures and communications presented at the meeting. We are indebted to the referees who carefully reviewed the papers, resulting in many improvements in the form and content of the papers.

J. A. DOCOBO, A. ELIPE AND H. McALISTER

OPENING SESSION

Word of welcome by the Rector of the University of Santiago de Compostela, Professor Villanueva

On behalf of the University of Santiago de Compostela I would like to express our gratitude for your participation in the International Workshop *Visual Double Stars: Formation, Dynamics and Evolutionary Tracks* organized by the University of Santiago de Compostela.

The century-old tradition and Santiago's contribution to the formation of Europe have lately brought the city international recognition. Thus, Santiago has been declared Heritage of Humanity and the road to Santiago has been acknowledged as the most important cultural European route.

The Royal University of Santiago de Compostela, founded five centuries ago on the initiative of Lope Gómez de Marzoa, has taken profit of the dynamic use of the cultural and scientific heritage arisen from the *Way to Santiago* leading us to explore new motives and itineraries leading to the same final destination.

In this regard, we have developed a new network, called the Compostela Group of Universities integrated by 70 Universities from 17 different European countries. This initiative has as its main aim the development of cooperation agreements in the academic, cultural and scientific fields this understood as a social phenomenon that has contributed decisively to encounter the concept and the personality of Europe.

I would also deal with the coordination of the different activities in the field of the university international relations, the development of specific mechanisms for the immediate transmission of information, and the exchange and promotion of contacts among member universities and other interested universities, promoting social, educational and technological development with neighbouring countries, communities and regions.

Our five hundred years anniversary is the time to take stock of the tremendous scientific revolution that have occurred over the last decades and to discuss its impact on our future higher education schemes. This University has developed a great research capacity and potential in various fields, including Astronomy, first of all because the efforts of Prof. Ramón María Aller.

It is a great honour and privilege for me to welcome you to the University of Santiago de Compostela. I admire you for your high quality science and organizational capacity and last, but not least, for your warm and loveable personality.

Santiago and its University is well accustomed to receiving guests and offering hospitality. I am sure that you will find here an environment both lively and academic that will provide you with a vivid and memorable experience. Finally, I would like to give my most sincere thanks and congratulations to all the members of the Organizing Committee and its Chairman Prof. José Angel Docobo Durántez.

Darío Villanueva
Rector of the University of Santiago de Compostela

Word of welcome by the Chairman of the Local Organizing Committee, Professor Docobo

Dear Rector, Mr. Chancellor, Mr. Dean of Mathematics, Mr. Director of the Applied Mathematics Department, Mr. President of the Scientific Committee, Ladies and Gentlemen.

On behalf of the Organizing Committee of this International Workshop, it is a great honour for me to be able to welcome you to Santiago de Compostela. As many of you will already know, this age-old city has always had astronomical connotations: in Spanish , the name Camino de Santiago refers both to the pilgrims' way leading to the shrine of St. James in the cathedral, and to the Milky Way, which for centuries guided pilgrims here.

You will also know —if you have looked at your Workshop literature— that this event is one of the acts being held in celebration of the Fifth Centenary of the University of Santiago de Compostela under the apt motto "Gallaecia fulget" - Galicia shines- a motto taken from text carved in 1544 in the frieze of the cloister of this building. Although the region of Galicia, to which Santiago belongs, was later to become one of the more neglected and underdeveloped corners of Spain, in recent times it has begun to shine again in many fields, among them the field of higher education. This University, while proud of its ancient roots, is a dynamic modern institution that strives to meet the highest standards of European academic activity and pursues continual improvement.

By way of example, it is perhaps not out of place to mention here our modest aspirations as regards the University Observatory: though our astronomers enjoy a certain limited access to some of the internationally renowned astronomical facilities installed in recent years in other parts of Spain, our aim is, as it must be, to further the scientific capacity of our own Observatory, in readiness for the challenges of the 21st century, by acquiring new instruments and setting them up in some suitable location outside the city.

The University Observatory is named after its founder, the late Dr. Ramón-María Aller, the first Spanish member of the IAU Double Stars Commission. He would have been delighted to see the University hosting a scientific meeting such as this, which would have been impossible in his day. With very limited resources, Ramón-María Aller did much for research and teaching here, and for the prestige of this institution. He began his astronomical work privately, in his home town of Lalín, some 50 Km from Santiago; there, in the early years of the century, he built a private observatory and began to take astronomical measurements of surprising precision. On acquiring a Steinheil refractor in 1924 he began to concentrate on double stars, taking micrometric measurements and calculating orbits. In fact, the first double star orbit calculated in Spain, an orbit for the system STT77, was obtained by him in Lalín in 1935; it was subsequently published in Astronomische Nachrichten. When prevailed upon to teach at the University of Santiago, he devoted himself unsparingly to all facets of educational responsability at the same time as he continued his own research work. Despite the lack of suitable material in Spanish, he wrote two textbooks, Algoritmia and Introducción a la Astronomía, the latter of which was for many years a standard text in Spanish universities. His diligence, modesty and kindness were exemplary, and inspired all those who knew him and had the privilege of working with him. At his death in 1966 he left his library and astronomical apparatus to the University Observatory. Don Ramón's educational labours were not in vain: all the Spanish research groups currently working in the field of visual double stars have sprung directly or indirectly from his work here in Santiago. I should like to make special mention here of two of his former students. The first is Prof. Enrique Vidal Abascal, whose recent death has deprived us of a great geometer and a fervent supporter of the Observatory; Prof. Vidal Abascal developed extremely elegant methods for obtaining relative orbits from apparent orbits, and was the inventor of the "orbigraph", a unique mechanical device that automatically ensures compliance of (ρ, θ) curves with the areal law (it can be seen in the Observatory). The second of Don Ramón's students to have made a great contribution to Spanish astronomy is Prof. Rafael Cid Palacios, whose labours over many years are responsible for the high international reputation of the Space Mechanics Group of the University of Zaragoza. The work currently pursued here at the Ramón-María Aller Observatory owes much to both Prof. Vidal Abascal and Prof. Cid Palacios: in particular, it was Prof. Cid Palacios who developed the first analytical method for calculating the seven elements of a relative orbit from seven observational data without knowledge of the areal constant, a method on which I based the algorithm currently used to calculate double star orbits here in Santiago.

And now I must express my gratitude to those whose support of this Workshop has given me the opportunity to pay tribute to my mentors and voice my hopes for the future of the Observatory. My thanks go firstly to the University of Santiago, to the Department or Education of the Xunta de Galicia (Galicia's regional Government) and Spanish Ministry of Education, without whose financial contributions this event would not have been possible. Secondly, to the members of the IAU Celestial Mechanics and Double and Multiple Stars Commissions, who have acted as members of the Scientific Committee. And last, but not least, to all the other distinguished astronomers who have honoured us with their attendance, upon whose active participation the success of this Workshop depends, and who share, I am sure, my wish that the content of this meeting may serve, on the eve of a General Assembly of the IAU, to underline the importance of work on double and multiple stars for modern astronomy.

Once again, a heartily welcome to you all, and may you all enjoy these few days among us. Thank you.

José A. Docobo
Chairman of the LOC

Word of welcome by the Chairman of the Scientific Organizing Committee, Dr. Worley

I am happy to greet you in this beautiful city and hall, and trust that we will have a successful and informative scientific meeting.

Many years ago, as I began my scientific career in the study of double stars, the name of Ramón Aller was virtually the only one I knew belonging to a Spanish astronomer. And, indeed, there was little astronomical activity in the Spain of that time. But how circumstances have changed! Today there are in Spanish observatories of deservedly international repute, and a high level of competence. This has been a rather wonderful evolution in only few decades. May this conference contribute to the continued flowering of astronomy in this country.

Charles E. Worley
Chairman of the SOC
President IAU Commission 26

Word of welcome by the Chancellor for Territorial Policy, Public Works and Housing of the Xunta de Galicia, Mr. Cuiña

I would like to thank the Organizing Committee of The International Workshop on Double Stars and Celestial Mechanics and their Chairman for the honor they have bestowed upon me by inviting me to take part in today opening act.

I would like to offer my best wishes and gratitude to all those who organize and participate in this event.

I am truly touched, not only by the motive guiding this congregation but also by the link with these subjects that an admirable and dear friend and neighbor from my home town had: *Don Ramón María Aller Ulloa,* who in 1948 was elected member of Commission 26 of the International Astronomical Union. Please allow me to dedicate, in this opening act, a few words to this illustrious figure.

I will not make any reference to his contributions to astronomy, a field which I know little about and about which important contributions have already been made, such as those made by his admirers Enrique Vidal Abascal or the Chairman of the Organizing Committee of this Workshop and Director of the Observatory of Santiago, Dr. José Angel Docobo.

My purpose is to speak about his vital, especially social, importance just as professor Vidal Abascal said, "Don Ramón's masterpiece was his life".

This son of Lalín has received many acknowledgments, distinctions and honors in his lifetime including membership of academies, councils and scientific societies at a national and international level; important teaching and cultural positions; different social medals of recognition and streets named after him in towns and cities all over Galicia. But I am sure that none of this would fill him with more pride than to be able to live in the hearts and memory of his people for whom he felt so much affection.

For eleven years I have had the honor of being Mayor of Lalín, a town in Central Galicia on its way to becoming a city, where the name of Ramón María Aller Ulloa maintains - was able to demonstrate that big plants with deep roots can also grow in small pots.

Without his love of the land and its people, few renowned figures would have been able to reach the profound feelings of a village as Don Ramón did.

Hardly any area of knowledge was unknown to him and people marvelled at his capacity to apply his deep knowledge of exact sciences to the most varied fields.

His whole life was dedicated to pastoral work - characteristic of his condition as a priest - which included study and teaching. His home was a centre for investigation, assessment, layman and religious teaching, activities which are still remembered by the older members of the community. This is because many of the community members, from all walks of life, acquired theoretical knowledge which Don Ramón applied to the fields of architecture, mechanics, carpentry or watch-making. He taught in an unselfish constant manner, always supported by his great humility which lead him to say to those students with learning difficulties: "it is not that you are not intelligent, otherwise you would not be coming here to learn, it is me who is not intelligent because I am unable to explain well to you." His teaching was not only theoretical, as what he taught emerged as tangible everyday realities. Don Ramón had the great gift of drawing and designing, he has left his mark on many houses, churches, musical instruments and images of saints either by directing the building of his observatory or repairing all sorts of mechanical apparatus which belonged to him or to others, always thinking of how he could make them work better. It was precisely his search for perfection, which he applied to celestial mechanics, that lead him to propose a series of modifications in the manufacturing of one of his first theodolites by a German company. These were welcomed with admiration but he did not receive any compensation. As he would later tell with habitual humor and irony, "they liked them...and they didn't want to charge for them!." His link with the society in which he lived made him love, cultivate and dignify the Galician language, confined mainly in those days to rural areas. He, who dominated ten languages, contributed to the elimination of barriers and prejudices that limited the use of Galician - a language in existence for eight centuries and with a golden literary epoch - to folkloric and popular events by using it in various articles thus becoming one of the propagators of scientific prose in Galician. His open mindedness, his universal knowledge did not stop him appreciating in a just measure all the value enclosed in the most immediate reality. It was perhaps that interest and that deep knowledge of things around him that led him to understand more clearly other more abstract concepts. Maybe it was his need to know that didn't stop in this world and by analyzing space he believed he could discover new horizons that were distant from Earth which, as a good christian, he believed as transitory. Studies of his works coincide in confirming that he was one of the most important astronomical authorities of the century. This is due to his observations, his books, his work, the apparatus he invented or the school he founded. However, the most concise and exact summary that could define his life can be read on the monument that the town of Lalín dedicated to him: "Science, humility and virtue".

His life, dedicated to research, gave him scientific recognition but it was also his humility, his modesty and his dedication to society that brought him closer to sanctity and make him worthy of the category of exemplary life.

I know that all you present at this Workshop will be able to understand and forgive this small personal tribute to an astronomer who, as with his other neighbors, has also left in me a mark of admiration and recognition.

I would like this speech to serve also to appreciate and propagate the ethical and moral value, further from the purely scientific, that scientific men like those present have. I, as member of the Galician Government, would like to take this opportunity to wish you a fruitful day and a happy stay in our land.

I also hope that this Workshop will contribute to bring this exciting world of astronomy closer to Galician citizens, so that - as my fellow countryman said in one of his publications - they will be able to read the book of the sky simultaneously , "open to all as the most beautiful."

José Cuiña Crespo
Chancellor for Territorial Policy,
Public Works and Housing, Xunta de Galicia

VISUAL DOUBLE STARS : FORMATION, DYNAMICS AND EVOLUTIONARY TRACKS

A WORKSHOP ON DOUBLE STARS AND CELESTIAL MECHANICS ORGANIZED BY THE ASTRONOMICAL OBSERVATORY "RAMON MARIA ALLER" COMMEMORATING OF THE 5TH CENTENARY OF THE UNIVERSITY OF SANTIAGO DE COMPOSTELA (GALICIA, SPAIN)

THIS WORKSHOP HAS BEEN SUPPORTED BY THE FOLLOWING SPANISH INSTITUTIONS:

UNIVERSIDADE DE SANTIAGO DE COMPOSTELA

XUNTA DE GALICIA - REGIONAL GOVERNMENT OF GALICIA

D.G.I.C.Y.T. - SPANISH MINISTRY OF EDUCATION

PHOTO 1. The presidential table photograph

The presidential table during the Opening Act which took place in the Salon Noble, Fonseca Palace, at the University of Santiago de Compostela (USC)

From left to right: Prof.Dr. A. Bermúdez de Castro, Head of the Applied Mathematics Department (USC); Dr.C.E. Worley, President of the IAU Commission 26, SOC Chairman; Prof.Dr. D. Villanueva, Rector of the University of Santiago de Compostela; Mr.X. Cuiña, Chancellor for Territorial Policy, Public Works and Housing of the Xunta de Galicia; Prof.Dr. E. Macias, Dean of the Mathematics Faculty (USC), Prof.Dr. J.A. Docobo, Director of the Astronomical Observatory "R.M.Aller" (USC), LOC Chairman

MEETING PHOTOGRAPH

1. Mrs.Nunes
2. M.Nunes
3. Mrs.Ferraz-Mello
4. H.Zinnecker
5. P.Magdalena
6. J.Ling
7. J.Ferraz
8. E.Vinuales
9. C.Calvo
10. Mrs.Elipe
11. A.Elipe
12. J.Elipe
13. J.A.Docobo
14. Mrs.Docobo
15. T.Kalvuridis
16. D.Benest
17. C.Allen
18. P.Gruener
19. A.Costa do Campo
20. Mrs.Fernandes
21. J.Fernandes
22. S.Ferraz-Mello
23. C.Hummel
24. M.Moure
25. C.Martin
26. P.Couteau
27. R.Argyle
28. Mrs.Argyle
29. A.Harpaz
30. Mrs.Harpaz
31. Mrs.van Altena
32. R.de la Fuente
33. M.Herrera
34. C.Alvarez
35. V.Tamazian
36. W. van Altena
37. E.Grosheva
38. P.Lampens
39. E.Oblak
40. J.Dommanget
41. A.Kisselev
42. H.Abt
43. S.Soderhjelm
44. Mrs.Baba
45. N.Baba
46. Y.Balega
47. Mrs.Balega
48. P.Virelizier
49. A.Abad
50. C.Abad
51. R.Dvorak
52. P.Abad
53. N.Shakht
54. J.L.Halbwachs
55. M.Bate
56. J.Seimenis
57. M.Valtonen
58. H.Beust
59. O.Pols
60. Y.Cuypers
61. I.Bonnell
62. C.Worley
63. O.Franz
64. Mrs.Franz
65. S.Mikkola
66. Mrs.Mikkola
67. Mrs.Abad
68. M.Lara
69. C.Burger
70. I.Aparicio
71. L.Floria
72. C.Prieto
73. Mrs.McAlister
74. D.M.Jasinta
75. M.McAlister
76. E.Horch
77. H.McAlister
78. Mrs.Oblak

LIST OF PARTICIPANTS

Armenia	
Vakhtang Tamazian	Yerevan State University
Austria	
Christian Burger	Institut fur Astronomie, Wien
Rudolf Dvorak	Institut fur Astronomie, Wien
Belgium	
Jean Dommanget	Observatoire Royal de Belgique
Yan Cuypers	Observatoire Royal de Belgique
Patricia Lampens	Observatoire Royal de Belgique
Brazil	
Sylvio Ferraz-Mello	Instituto Astronomico e Geofisico, Sao Paolo
Finland	
Seppo Mikkola	Tuorla Observatory, University of Turku
Mauri Valtonen	Tuorla Observatory, University of Turku
France	
Daniel Benest	Observatoire de la Cote d'Azur
Herve Beust	Observatoire de Grenoble
Paul Couteau	Observatoire de la Cote d'Azur
Jean-Louis Halbwachs	Observatoire de Strasbourg
Christian Martin	Observatoire de la Cote d'Azur
Eduard Oblak	Observatoire de Besancon
Philippe Virelizier	Observatoire de Strasbourg
Germany	
Matthew Bate	Max Planck Insitut fur Astronomie, Heidelberg
Kurt Wagner	Institut fur Astronomie und Astrophysik, Tubinge
Hans Zinnecker	Astrophysikalisches Institut, Potsdam
Greece	
Telemachus Kalvuridis	National Tecnical University of Athens
John Seimenis	University of Aegean
Israel	
Amos Harpaz	Oranim University of Haifa
Indonesia	
Dini Maria Dewi Jasinta	Bosscha Observatory
Japan	

Naoshi Baba	Hokkaido University

Mexico

Christina Allen	Instituto de Astronomia, UNAM
Miguel Herrera	Instituto de Astronomia, UNAM
Arcadio Poveda	Instituto de Astronomia, UNAM

Portugal

Alfredina Costa do Campo	Observatorio Astronomico de Lisboa
Manuel Nunes	Observatorio Astronomico de Lisboa
Joao-Manuel Fernandes	Centro de Astrofisica da Universidade do Oporto

Russia

Yuri Balega	Special Astrophysical Observatory
Ildiko Balega	Special Astrophysical Observatory
Elena Grosheva	Pulkovo Observatory
Kisselev Alexei	Pulkovo Observatory
Natalya Shakht	Pulkovo Observatory

Spain

Alberto Abad	Universidad de Zaragoza
Carlos Alvarez	Universidade de Santiago de Compostela
Ignacio Aparicio	Universidad de Valladolid
Carmen Calvo	Universidad de Zaragoza
Raul de la Fuente Marcos	Universidad Complutense, Madrid
Jose-Angel Docobo	Observatorio Astronomico "Ramon Maria Aller"
Antonio Elipe	Universidad de Zaragoza
Luis Floria	Universidad de Valladolid
Alvaro Gimenez	Instituto Nacional de Tecnica Aerospacial
Martin Lara	Real Instituto y Observatorio de la Armada
Josefina Faen Ling	Observatorio Astronomico "Ramon Maria Aller"
Pilar Magdalena	Universidade de Santiago de Compostela
Jose Luis Martin	Universidade de Santiago de Compostela
Jorge Nunez	Universitat de Barcelona
Cristina Prieto	Universidade de Vigo
Ederlinda Vinuales	Universidad de Zaragoza
Ana Ulla	Instituto de Astrofisica de Canarias - LAEFF

Sweden

Staffan Soderhjelm	Lund Observatory

United Kingdom

Robert Argyle	Royal Greenwich Observatory
Ian Bonnell	Cambridge University Institute of Astronomy
Onno Pols	Cambridge University Institute of Astronomy

USA

Helmut Abt	Kitt Peak National Observatory
Otto Franz	Lowell Observatory
Elliott Horch	Yale University Observatory
Christian Hummel	US Naval Observatory
Harold McAlister	Georgia State University, CHARA
William van Altena	Yale University Observatory
Wolszczan Alex	Pennsylvania State University
Charles Worley	US Naval Observatory

CONCLUDING REMARKS

My impression is that this has been a very successful meeting scientifically. It is notable as the first collaboration of Commission 7 with Commission 26. This is important for the future understanding of binary and multiple star formation and dynamical evolution; subjects where we are only now beginning to receive the first inklings of the pertinent processes.

Our hosts have provided us with a wholly satisfactory venue for the meeting, and we are grateful for their efficiency and consideration, which has made this a very enjoyable occurrence. May our efforts contribute to the future success of Dr. Docobo and his colleagues here at the University of Santiago de Compostela

Charles E. Worley
President
IAU Commission 26

The current policy of the Commission 7 (Celestial Mechanics) of the International Astronomical Union is to strongly support initiatives leading to stimulate investigations of all astronomical problems akin with Celestial Mechanics.

The role played by Celestial Mechanics in the construction of reference frames and ephemerides, as well as in the unraveling of the dynamics of the Solar System, is well known. The contributions of Celestial Mechanics to other branches of Astronomy and Astrophysics is less known. However, one may cite that in the last decade, Celestial Mechanics contributed to the understanding of the dynamics of the Hamiltonian systems arising in Cosmology, the stability of the dust cloud around β Pic, the putative planets around pulsar 1257+12, the chaotic motion of stars in elliptic galaxies the dynamics of planetary formation.

In what concerns visual double stars, Celestial Mechanics has not been much farther than classical methods of orbit determination. However, multiple systems offer many possibilities of study and I was glad to see, in this conference, a new generation of astronomical investigations related to Celestial Mechanics.

I do remember the first meeting on N-body gravitational systems which I attended, in the sixties, when many simulations were shown. At the end of each presentation, Victor

Szebehely, one of the most distinguished names of our time's Celestial Mechanics, was asking: 'Did you regularize your equations?' The invariable answer was 'No!'. In fact, the theories of regularization in three-dimension collisions were recent and not yet known of astronomers. Today, the information seems to diffuse more quickly. I just quote, as an example of the instances in which modern techniques and facts of Celestial Mechanics were present in this meeting, the accurate study of third-body perturbations and, in particular, the coupling of eccentricity and inclination known as Kozai resonance and whose importance in Solar System Dynamics was just recently discovered.

Multidisciplinary meetings have a large importance and should be strongly encouraged and supported by the astronomical community.

Sylvio Ferraz-Mello
President
IAU Commission 7

SECTION I

OBSERVATIONS AND THEIR RESULTS

TWENTY YEARS OF SPECKLE INTERFEROMETRY

H.A. McALISTER
CHARA, Georgia State University, Atlanta, Georgia, USA

1. Introduction

It is an honor for me to be the first speaker at this symposium celebrating the five hundredth anniversary of the Universidad de Santiago de Compostela as well as recognizing the contributions to astronomy made by the past and present staff members of the Observatorio Astronómico "Ramón Maria Aller." I congratulate Dr. Docobo and his co-organizers for putting together such an interesting meeting and wish them the very best in their efforts to advance astronomical research and education in this historic and lovely part of the world.

My contribution to this gathering will be a summary of my thoughts and experiences following two decades of activity in the field of binary star speckle interferometry. I have seen this technique advance from being a novel and experimental concept with relatively few observational results to its present status as the premier means for measuring the orbital motions of resolved binary star systems.

At the Flagstaff double stars meeting in 1981 (McAlister 1983), I presumptuously appropriated a title from W.S. Finsen (1971) with only five years of experience behind me. I now find that I have as many years in my field as Finsen did at the time he wrote his evaluation of the technique of visual interferometry. I corresponded with Finsen in the last few years prior to his death in 1979, and I believe that some of his philosophy about the observation of binary stars and the utilization of new and potentially powerful techniques rubbed off on me. No doubt I have come to some of the same conclusions after twenty years as did Finsen in his 1971 retrospective.

2. Speckle Interferometry and Binary Stars

Antoine Labeyrie (1970) brilliantly realized that the fine structure in highly magnified images of stars was easily exploitable for its diffraction limited information content. He showed that a straightforward power spectrum analysis (and its Fourier equivalent, the autocorrelation) could yield such information as diameters of supergiant stars and relative positions of the components of close visual binaries. Labeyrie's speckle interferometry quickly came into a kind of vogue, and several promising avenues were developed that would hopefully lead to the recovery of phase information and the full reconstruction of diffraction limited images of binary stars. The supergiant Betelgeuse became the favored target for such imaging attempts.

J. A. Docobo et al. (eds.), Visual Double Stars: Formation, Dynamics and Evolutionary Tracks, 3–8.

While binary stars were frequently used as demonstration objects for these exercises, those who employed them often took little care in calibrating their data for scale and orientation. I know of at least one case where the calibration was based upon a separation and position angle for a binary taken from the Yale Bright Star Catalogue, and the actual separation of the "calibration" object was half the BSC value at the epoch of the speckle observation. To their defense, many of these early imaging attempts were aimed at developing algorithms and not really intended to produce useful astronomical results. Double stars were merely demonstration objects towards the really interesting goal of imaging the surfaces of supergiants. Of course, much of the early speckle work was supported by agencies interested in imaging objects in Earth orbit. In those instances, nothing could be farther from their goal than was double star astronomy.

Now, nearly 25 years after Labeyrie's inspiration, double star research is clearly the prime beneficiary of speckle interferometry. The long sought images of the surface of Betelgeuse remain elusive, but hundreds of bright new and close binaries have been discovered and thousands of known pairs have been measured with accuracies not anticipated prior to 1970.

3. An Observing Program

The first task following my extraordinarily good fortune in being employed as a post-doc at Kitt Peak National Observatory in 1975 to use a speckle camera developed by Roger Lynds was to develop an observing list for an exploratory program of double star research. The 4-meter Mayall telescope on Kitt Peak would resolve binaries down to about 35 milliarcseconds (mas), and I realized that this offered the extremely interesting potential of resolving spectroscopic binaries. Prior to speckle interferometry, the ability to combine the complementary angular data from visual resolution with the linear data from radial velocity curves was limited to a few famous cases, most notably Capella whose resolution by visual interferometry at the 100-inch Hooker telescope in the 1920's was a landmark in twentieth century astronomy.

An analysis of Alan Batten's "Sixth Catalogue of the Orbital Elements of Spectroscopic Binary Stars" became my first paper in the field (McAlister 1976) and provided a list of 73 systems that were promising targets for resolution. Some of these objects had already been resolved by the very first speckle practitioners, and I succeeded in resolving three more on my first observing run. This list, which merits updating, remains useful in providing target objects at the new generation of large reflectors (if time for binaries is allocated on these instruments) and for long-baseline interferometers (where the distinction between spectroscopic and visual binaries will become obsolete at last).

The "Index Catalogue of Double Stars", now succeeded by the "Washington Double Star Catalogue," was the important source for close visual binaries. These objects, which had been discovered over the past century or more by extremely skilled visual observers, would form the majority of the speckle candidates. Other potentially resolvable systems were taken from lists of occultation binaries, composite spectrum stars and stars with suspected variable radial velocities. These all proved to be productive sources for newly resolved pairs.

This observing list, which continues to evolve and expand since its inception, has been taken to a variety of large telescopes during the twenty years of this program. These telescopes include the Kitt Peak 4-m Mayall; the 72-inch Perkins located on Anderson Mesa, Arizona; the Cerro Tololo 4-m Blanco; the 3.6-m Canada-France-Hawaii Telescope on Mauna Kea; and the 120-inch Shane reflector of the Lick Observatory. We now are routinely using the beautifully renovated 100-inch Hooker telescope on Mt. Wilson.

4. Current Status

With my colleagues William Hartkopf and Brian Mason, I am preparing a detailed review article for publication elsewhere that will summarize the status and major contributions of binary star speckle interferometry. So I will only include here a brief summary table describing the extent to which speckle interferometry has been applied to binaries. This table is immediately obsolete as new measures are being produced by several groups around the world each year. For a continuously updated record, I refer the reader to CHARA's "Catalog of Interferometric Measurements of Binary Stars" maintained by William Hartkopf and accessible on-line at http://www.chara.gsu.edu.

TABLE 1. Speckle Measures of Binary Stars

Category	Measured	Unresolved	Mean Sep. (arcsec)
CHARA Photographic	2,779	1,653	0.22
CHARA ICCD	10,597	2,991	0.29
Other Interferometric	5,741	1,794	0.46
Occultation	831	n/a	0.09
TOTAL	19,948	6,438	0.32

5. Lessons Learned

One of the first lessons I learned was that speckle interferometry was fantastically efficient at the telescope. One minute of data accumulation was usually more than sufficient to produce a detection with useful signal-to-noise ratio. On a long winter's night, and with a patient and understanding telescope operator, more than 200 stars can be observed. The observing program quickly grew to more than 2,000 objects, and surveys were initiated as background programs to fill in observing runs.

I have described in two popular articles my observing experiences over the years with speckle interferometry (McAlister 1977, 1996), and so I will refrain from repeating those experiences here. One point worth making for which I have been criticized on occasion is the reason for not using seeing calibration objects.

Labeyrie's method calls for dividing the power spectrum of a science target star by the power spectrum of a point source to deconvolve the low frequency seeing background which biases the measurement of angular separation at best and hides the detection at worst. For this to work well, the instantaneous point-spread function must not change from the calibration to the science object. This requires stationary seeing conditions and the utilization of a calibration star close in position and similar in brightness to the science target. I developed an opinion early on that these conditions were never met in reality, and that the PSF was inevitably changed from target to calibrator. I chose instead to compensate for the seeing bias by smoothing and subtracting from the power spectrum or autocorrelation, and we now routinely use a boxcar subtraction from the latter function.

We have never seen any evidence that this approach has produced any systematic effects in the data. It also has the happy consequence of employing most of the observing hours for binaries rather than for calibrators.

The Georgia State/CHARA speckle program has been successful, I believe, because we have emphasized both quantity and quality in our measurements. I continue to be surprised when I see a speckle paper with only a few dozen measurements resulting from several nights at the telescope. There really are no valid reasons for not observing many dozens if not several hundred stars per night. Binary star speckle interferometry is not a field for dilettantism.

Furthermore, what the field needs is more observers with long-term commitments. A few scattered measures over several years really makes a small contribution. Orbits must be measured over the long term, and sufficient measures must be available to convince the orbit computer of their accuracy and reliability. One should start a speckle program with the intent to keep at it for decades.

I should also emphasize that large telescopes are not required in order to produce useful and important results. Nature has provided us with binaries occupying a continuum in angular separation, and telescopes of even less than one meter aperture can measure large numbers of pairs with great accuracy. With this in mind, Charles Worley has initiated an extremely productive speckle effort on the 26-inch USNO refractor in Washington and has retired his micrometer in favor of a speckle camera.

There is more room in the field for productive and committed speckle observers. The southern sky remains a particularly rich arena for new entries. A southern program like the USNO's new effort would encounter a wealth of close visual binaries discovered by Herschel, van den Bos, Finsen, and Rossiter. The motions of these southern pairs continues to go on unmeasured after decades of neglect.

While the acquisition of speckle observations is fast and easy, the reduction of the massively acquired data is very time consuming. To avoid a stultifying program, a small group of speckle observers, sharing a common philosophy to maintain quality control, is required. Our present speckle group consists of two faculty members, one post-doc and one graduate student. A small team like this is really quite essential.

In my "five year" review (McAlister 1983), I emphasized the need for careful calibration for scale and orientation. I must continue to stress that this is absolutely basic in binary star speckle interferometry. A double slit mask or course grating located at the entrance pupil of the telescope remains the most effective means for spatially calibrating the data. While this mask can also provide a zero-point for position angle measurements, orientation is also (and more easily) established by trailing a star across the speckle camera detector using low magnification. Modern detectors employing intensified CCDs must also be calibrated for flat field and bias response of the CCD/intensifier combination.

The importance of calibration cannot be over stressed. A significant fraction of the early speckle data deserve zero weight in orbit solutions because of the lack of attention to calibration.

In that first "lessons" paper, I also spoke of the goal of extracting differential photometry from speckle observations. My optimism at the time has never been rewarded as "speckle photometry" continues to be an elusive and unproven goal. Photometric calibration of the speckle process is very difficult, and although such techniques as triple correlation give lovely images of binaries, there is no convincing evidence that these images possess reliable photometric information. This continues to be an important goal, however, and CHARA graduate student Lewis Roberts is mobilizing various speckle imaging algorithms for a comparison with aperture masking and adaptive optics to see what might be done with speckle photometry.

With the increasing accessibility to adaptive optics systems, one might question the need for further activity in speckle interferometry. However, in comparison with adaptive optics, speckle methods have higher resolution and fainter limiting magnitude. Speckle is also quite easy and inexpensive to implement compared with adaptive optics. Our own program is evolving into dual use – speckle for binary star astrometry and adaptive optics for photometry. We have been fortunate to use the adaptive optics system developed by the Mount Wilson Institute at the 100-inch telescope and hope to have continued access to that instrument to determine Δm's for a large number of stars on our program.

The field of high resolution imaging is about to see a revolution through the upcoming routine operation of several long-baseline interferometers. Our own "CHARA Array" on Mt. Wilson will incorporate binaries as a prime component of its observing program. With limiting resolutions some 100 times greater than speckle interferometry at 4-m class telescopes, these instruments will resolve the majority of the spectroscopic systems.

Does this make speckle interferometry a doomed field. By no means. Many, if not most, of the binaries observed by telescope arrays will have suffered mass exchange at some time in their past, and their placement on the empirical mass-luminosity relation will be clouded by major evolutionary effects in the binary. While such effects make these systems very, very interesting, a clean mass-luminosity relation will continue to benefit from determining masses of fully detached binaries. Advances in the accuracy of radial velocity measurements and the anticipation of improved parallaxes from the Hipparcos spacecraft means that many visual/speckle binaries will give up their mass/luminosity secrets when their orbital motions are sufficiently well determined.

If we have the available people power, CHARA will continue its speckle work indefinitely into the future with the goal of measuring the orbital motions of binaries in the separation regime of 1.0 arcsec to 0.2 mas. These objects will possess periods in the range of centuries to hours. The best is yet to come!

6. Acknowledgments

My twenty years in this field have been made possible through the kind and continuous support of the National Science Foundation through a series of grants to Georgia State University, the most current of which is NSF Grant AST-9416994. Critical support from the Air Force Office of Scientific Research during the early 1980's gave CHARA the computing resources it needed to compete productively with other speckle groups around the world, and I thank Dr. Henry Radoski at AFOSR for his interest in our efforts. Georgia State University has contributed significantly to this effort through partial salary support for several post-doctoral research scientists and especially through the Chancellor's Initiative Fund administered by our Vice President for Research and Sponsored Programs. I have benefited from the collaboration with several dozen scientists over the years. I particularly want to thank my CHARA colleague William Hartkopf for his dedication to binaries since joining me in 1981 and to note that in recognition of his efforts he officially took over management of our speckle program in 1995.

References

Finsen, W.S. "Twenty Years of Double Star Interferometry and its Lessons," (1971) *Astrophysics and Space*

Science, **11**, p. 13.
Labeyrie, A. "Attainment of Diffraction-Limited Resolution in Large Telescopes by Fourier Analysing Speckle Patterns in Star Images," (1970) *Astronomy and Astrophysics*, **6**, p. 85.
McAlister, H.A. "Spectroscopic Binaries as a Source for Astrometric and Speckle Interferometric Studies," (1976) *Publications of the Astronomical Society of the Pacific*, **88**, p. 317.
McAlister, H.A. "Binary Star Speckle Interferometry," (1977) *Sky and Telescope*, **53**, p. 346.
McAlister, H.A. "Five Years of Double Star Interferometry and Its Lessons," (1983), Proceedings of IAU Colloquium No. 62 on Current Techniques in Double and Multiple Star Research, *Lowell Observatory Bulletin*, **9**, p. 125.
McAlister, H.A. "Twenty Years of Seeing Double," (1996) *Sky and Telescope*, **92**, p. 28.

LES BINAIRES VISUELLES SERRÉES

P. COUTEAU
O.C.A. Observatoire de Nice. B.P. 4229
F-06304 Nice Cedex 4
France

1. RAPPEL HISTORIQUE

C'est Claude Ptolémée qui signale la première étoile double, ν Sagittarii, dans son Almageste qui reproduit le catalogue d'Hipparque paru en 127 avant notre ére. Après l'invention de la lunette d'approche, Galilée signale la binarité de Mizar (14") en 1611, et le Père Richaud celle de α Centauri en 1689. Castor, γ Virginis et 61 Cygni son découverts au milieu du XVIIIème siècle, ainsi qu'une centaine "d'étoiles à compagnons" cataloguées par Christian Meyer en 1777.

William Herschel (1738-1822) le premier se demande si les étoiles doubles ne sont pas, en fait, de véritables systèmes stellaires, dont il porte le nombre connu à près d'un millier.

Mais la véritable histoire des étoiles doubles commence le 25 février 1825, lorsque Wilhem Struve (1793-1864) entame sa prospection de binaires avec le "Grand réfracteur" de 24cm construit par Fraunhofer. C'est la première lunette astronomique moderne, munie d'un micromètre à fils et d'une monture équatoriale entraînée par une horloge. Struve découvre 2.600 couples et en catalogue plus de 3.000; ce sont le permières mesures modernes qui constituent le "Catalogus Novus", ou Catalogue de Dorpat, paru en 1827. Quinze ans plus tard, Otto, le fils de Wilhem publie son "Nouveau Catalogue d'Etoiles doubles" à Saint Petersbourg; ce sont les fameuses $O\Sigma$ découvertes au tout nouveau 38cm. L'ensemble des Σ (W. Struve) et des $O\Sigma$, totalisant 3.626 paires constitue pour plus d'un siècle la bible des observateurs d'étoiles doubles.

Très vite des mouvements orbitaux se dessinent, ce qui engage les astronomes à les observer. Une foule d'observateurs apparaît, surtout anglais. Les plus célèbres son le Révérend W.R. Dawes (1799-1868) et Jhon Herschel (1792-1883), le fils de William. En Italie, E. Dembowski (1812-1883) installe sa lunette dialytique sur les rivages du lac de Côme, il passe ses nuits entre son équatorial et le catalogue de Dorpat.

Mais un constructeur de génie fait son entrée su la scène astronomique: Alvan Clark. Il commence par fabriquer de petits objectifs de 15 à 20 cm, des merveilles, qui seront entre les mains de Dawes et surtout du jeune S. W. Burnham (1838-1921) de Chicago. Burnham pense que les Struve n'ont pas épuisé les "moissons d'or des champs célestes d'étoiles

J. A. Docobo et al. (eds.), Visual Double Stars: Formation, Dynamics and Evolutionary Tracks, 9–13.

doubles", comme aimait le répéter Sir James South. Avec un modeste 6 pouces, installé dans son jardin, il découvre 451 binaires dont β Delphini de 27 ans de période. Dembowski s'acharne à la mesure de ces nouvelles binaires et Flammarion (1878) répertorie les couples en mouvement.

La reprise des prospections lancée par Burnham va trouver son grand élan avec Aitken et Hussey. De 1900 à 1910, ils découvrent 4.700 binaires dont beaucoup à la grande lunette de Lick. Ce qui est fondamental, c'est l'importante moisson de couples serrés, qui n'étaient découverts jusque-là qu'épisodiquement. Enfin des centaines de binaires serrées livreront leurs orbites à l'horizon des années 50.

D'autres prospections ont lieu à la même époque, en particulier par Jonckheere et Espin récoltant 6.000 binaires d'écartement voisin d'une à deux secondes. Un peu plus tard l'hémisphère sud est exploré par Innes, van den Bos, Finsen et Rossiter rassemblant plus de 10.000 paires.

A partir de 1920, la prospection fait place à l'observation des couples connus avec principalement P. Baize et G. van Biesbroeck. Vers 1970, à la suite des découvertes de Kuiper, 4.000 couples sont engrangés par P. Muller, P. Couteau et W. D. Heintz, démontrant que le ciel des étoiles doubles est plus riche que certains ne le pensent.

D'autant plus que l'interférométrie se banalise à partir de 1975, sous l'impulsion de Labeyrie, et permet l'accès au pouvoir séparateur des grands télescopes. C'est l'équipe du CHARA (Center for High Angular Resolution Astronomy) avec le télescope Mayal de 4m à Kitt Peak qui réalise la soudure tant attendue entre les couples visuels et les binaires spectroscopiques. Quelques centaines de ces dernières sont résolues, révolutionnant ainsi les bases de la relation Masse-Luminosité.

2. QU'EST-CE QU'UN COUPLE VISUEL SERRE?

Le cualificatif de serré est relatif à l'instrument. Une binaire visuelle est serrée lorsque la séparation des composantes est inférieure au pouvoir séparateur ρ. $\rho = 12/D$ (D ouverture en centimètres). Le pouvoir séparateur n'est pas une limite. Une étoile double à composantes d'égal éclat, suffisamment lumineuse, paraît absolument ronde lorsque la séparation descend endessous du demi-pouvoir séparateur (0.5ρ). Soit 0"12 pour un objectif de 50cm, 0"16 pour les remarquables 15 pouces du siècle dernier (38cm) et, théoriquement, 0" 06 pour la grande lunette de Yerkes.

Dawes et Burnham, avec les ouvertures modestes dont ils disposaient (6 à 8 pouces), décelaient la binarité jusqu'à 40% du pouvoir séparateur. La performance est moins bonne à mesure qu'augmente la taille des objectifs. Baize s'arrête à 55% du pouvoir séparateur avec 15 pouces, tandis qu'Aitken et Hussey ne descendent pas sous le pouvoir séparateur avec la grande lunette de Lick (36 pouces). Les couples les plus serrés découverts à ce dernier instrument sont A 2909 (0"137), A 883 (0"139), Hu 1247 (0"153). On aurait dû s'attendre à déceler des couples deux fois plus serrés, mais l'oeil est de plus en plus mal à l'aise derrière les images fragiles, agitées, cassées que donnent les grands objectifs.

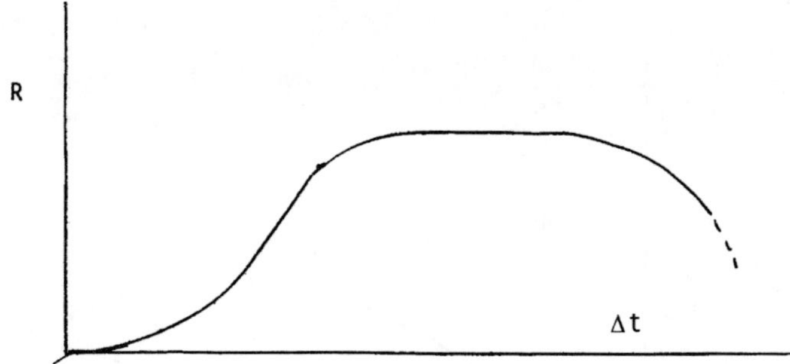

Figure 1. Rendement Orbital normalisé

L'interférométrie des tavelures et le compositage des images CCD permettent de s'affranchir des dégradations de la turbulence et de restituer l'image d'Airy, sans toutefois "descendre" en-dessous du pouvoir séparateur.

Le télescope Mayall de 4m, utilisé régulièrement à la détection et à la mesure des binaires, atteint les "records" actuels à ouverture pleine. La mesure la plus serrée concerne COU 1145 (0"023), deux fois plus serrés que Capella. Toutefois il est rare que les observations descendent sous 0"03, ce qui correspond au pouvoir séparateur du télescope, elles deviennent fréquentes à partir de 0"034. Le télescope de Zelenchuk (6m) ne fait pas mieux.

Il en résulte que la notion de binaire visuelle serrée évolue avec le temps. Cette évolution suit parallèlement les progrès de la dimension des lunettes, puis des télescopes. L'interférométrie accède plus facilement aux binaires serrées, mais ne va pas plus loin que l'oeil, sauf en ce qui concerne les couples à grande différence d'éclat.

3. Bilan des prospections

L'intérête des prospections est de fournir des orbites à moyen et long terme. Mais elles n'ont pas toutes le même intérêt. Par exemple, Jonckheere et Espin ne dévoile qu'une orbite sur leurs 6.000 couples, tandis que Kuiper en récolte une quinzaine avec un apport d'une centaine d'étoiles doubles.

Introduisons la notion de Rendement Orbiral Normalisé

$$R = \frac{\text{Nombre orbites}}{\text{N. de découv.}} \times \frac{1000}{\Delta t}$$

Δt étant le temps écoulé depuis la fin de la prospection, le rapport R donne le nombre d'orbites annuels pour mille couples. A la fin d'une prospection, ce rendement est nulle, il ne démarre qu'au bout de quelques années, puis prend de l'ampleur, pour se stabiliser pendant assez longtemps, enfin el diminue quand les orbites possibles se sont manifestées (fig. 1)

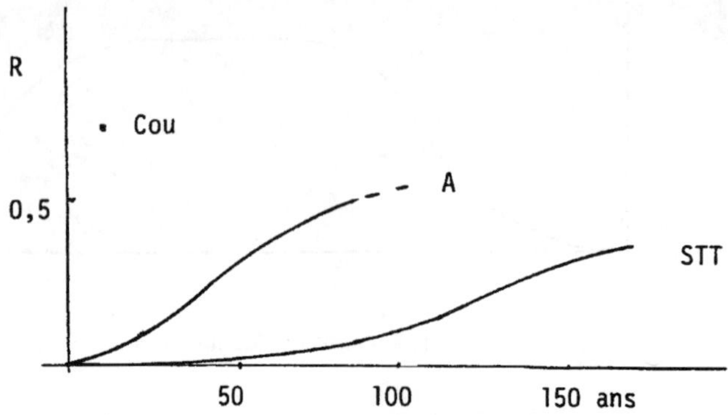

Figure 2. Rendement Orbital nNormalisé por quelques observateurs

Il est instructif d'étudier ce rendement pour quelques observateurs à travers l'histoire des étoiles doubles. Formons le tableau I dans lequel nous trouvons en première colonne le sigle de l'observateur, l'ouverture utilisée en seconde colonne, le nombre de couples découverts, le recul en années en 1995. En cinquième colonne le nombre d'orbites connues en 1934 suivi entre parenthèses du Rendement Orbital Normalisé, en sixième colonne les mêmes paramètres pour 1995.

TABLEAU 1

Observateurs	Ouverture (cm)	N. bin.	Δt	1934	1995
STF	24	2.600	170	51 (0.17)	137 (0.31)
STT	38	514	150	18 (0.37)	73 (0.94)
Ho	48	622	100	4 (0.16)	24 (0.39)
Bu	6 à 91	1.300	110	23 (0.44)	104 (0.80)
A	30 à 91	3.000	85	7 (0.10)	210 (0.80)
Rst	68	5.600	50	–	27 (0.10)
B	66	2.800	65	–	23 (0.13)
Fin	66	550	50	–	28 (1.02)
Cou	50	2.700	10	–	23 (0.80)

Les anciennes prospections de 1825 à 1910 affichent des rendements à peu près égaux pour O. Struve, Burnham, Aitken et Hussey qui n'a pas été mentionné car très voisin de son "alter ego" Aitken. Remarquons la prospection remarquable de Otto Struve faite avec le tout nouveau 38cm, et recensant des couples lumineux un peu difficile pour l'instrument de 24cm de son père William. Burnham, avec des réfracteurs de petite dimension, a eu la chance de passer avant ses élèves Aitken et Hussey, il montre un étonnant sens de la prospection.

Rossiter et van den Bos, ainsi que Innes ont déployé leur talent dans l'hémisphère sud où ils ont recueilli "le tout venant", ce qui conduit à un résultat long à se manifester. Finsen fait preuve d'une remarquable perspicacité avec son interféromètre, de même que Kuiper qui se limite à de sûrs critères de proximité. Les prospections modernes manquent de recul, mais elles sont prometteuses ayant révélé des binaires difficiles bien cachées au fond des ciels perturbés par la turbulence.

Despuis la banalisation de l'interférométrie, les prospections s'orientent vers les couples serrés. Leur recensement es très en retard, de plus, leur recherche est faite davantage par critères astrophysiques cas par cas que par porspection systématique. Le CHARA s'engage dans la voie à suivre par ses visites systématiques d'étoiles brillantes (HR). L'an 2000 s'ouvre sous la perspective de parallaxes nombreuses et précises et d'orbites à courtes périodes associées aux binaires serrées de parallaxes connues par Hipparcos.

IMPROVING ASTROMETRIC MEASUREMENTS USING IMAGE RECONSTRUCTION

ALBERT PRADES
Escola Universitària Politècnica de Barcelona
Universitat Politècnica de Catalunya
Barcelona, Spain

AND

JORGE NÚÑEZ
Departament d'Astronomia i Meteorologia
Universitat de Barcelona and
Observatorio Fabra. Barcelona, Spain

Abstract.

The improvement of astrometric measurements by the Maximum Likelihood Image Reconstruction method (MLE) is investigated. It is shown that the reconstruction process may provide benefits in centering precision for both ground- and space-based astrometric images. A gain of a factor of two in astrometric precision can be obtained. It is also shown that the reconstruction process does not introduce any systematic bias (in the case of isolated stellar images) in position, as previous studies seemed to indicate.

1. Introduction

During the past years a substantial amount of work has been done in image reconstruction directed towards optical astronomy, mainly after the discovery of the spherical aberration on the Hubble Space Telescope in 1990. Since 1988 our group has been working on the development of statistically based algorithms for Image Reconstruction (Núñez and Llacer 1991, 1993a, 1993b, 1994, 1996). Since 1992 our group has been working in collaboration with Yale University on the appliaction of the reconstruction methods to optical astrometry. The first studies (Girard et al. 1994, 1995) showed the

J. A. Docobo et al. (eds.), Visual Double Stars: Formation, Dynamics and Evolutionary Tracks, 15–25.

feasibility of the reconstruction of HST Planetary Camera astrometric images. However, in these studies an astrometrically unacceptable systematic bias seemed to be present in the reconstructed images. Here we discuss both the improvement of the astrometric precision after the reconstruction and the problem of the systematic bias. We also address the case of blended double star images.

2. The Maximum Likelihood Algorithm

2.1. NOTATION AND IMAGE MODEL

The notation used in this paper is the following:

Let $p_j, j = 1, \cdots D$ be the projection (measured) data; $a_i, i = 1, \cdots B$ the emission density in the image (parameters to be estimated); f_{ji} the Point Spread Function (PSF) or probability that an emission in pixel i in the source be detected at detector j; $b_j, j = 1, \cdots D$ the background in the data; $n_j, j = 1, \cdots D$ the readout noise in the data (Gaussian); $C_j, j = 1, \cdots D$ the detector gain corrections (flatfield); $f'_{ji} = \frac{f_{ji}}{C_j}$ the corrected PSF; $q_i = \sum_{j=1}^{D} f'_{ji}$ the total detection probability for an emission from pixel i; $h_j = \sum_{i=1}^{B} f'_{ji} a_i + b_j$ the forward projection or blurring operation.

Let $\mathbf{p}, \mathbf{a}, \mathbf{f}, \mathbf{b}, \mathbf{n}, \mathbf{C}, \mathbf{f}', \mathbf{q}$, and $\mathbf{h}$ be the corresponding arrays.

We shall work with the following imaging model: an object emits light with an intensity given by a spatial distribution **a**. The light is focused by the optical system over a detector array consisting of individual, discrete, independent detectors. Each detector has a different quantum efficiency characterized by a gain correction distribution **C**. A certain background radiation **b**, coming mainly from the sky but also from sources inside the detector, is detected along with the spatial distribution **a**. We assume that the detection process is Poisson distributed. Finally, the detector is read by an electronic process which adds a Gaussian readout noise **n** with zero mean and known standard deviation σ. The imaging equation corresponding to this model is:

$$\mathbf{f}' \mathbf{a} + \mathbf{b} + \mathbf{n} = \mathbf{p} \ . \tag{1}$$

Equation (1) in discrete form becomes:

$$\sum_{i=1}^{B} \frac{f_{ji}}{C_j} a_i + b_j + n_j = p_j \quad j = 1, \cdots, D \ .$$

Most imaging systems are described by equation (1), particularly those based on Charge Coupled Device (CCD) cameras, and Image Pulse Counting Systems (IPCS).

2.2. MAXIMUM LIKELIHOOD ALGORITHM

Here we use the Maximum Likelihood Algorithm (MLE) stopped after a certain number of iterations (Núẽz and Llacer, 1993a). We are also working on Maximum a Posteriori (Bayesian) approaches but in this paper we prefer to use the more standard MLE method to demonstrate that image reconstruction is useful for Astrometry.

The conditional probability $\mathbf{P(p|a)}$ describes the noise in the data and its possible object dependence. It is fully specified in the problem by the likelihood function. This is the function to be maximized. As indicated above, we have two processes: the first to form the image on the detector array and the second to read the detector. Taking both processes into account, the compound likelihood is (Núñez and Llacer 1993a):

$$\mathbf{L} = \mathbf{P(p|a)} = \prod_{j=1}^{D} \sum_{k=0}^{\infty} \frac{1}{\sqrt{2\pi}\sigma} e^{-\frac{(k-p_j)^2}{2\sigma^2}} e^{-h_j} \frac{(h_j)^k}{k!}$$

and its logarithm is:

$$\log \mathbf{L} = \sum_{j=1}^{D} \left[-\log(\sqrt{2\pi}\sigma) - h_j + \log \sum_{k=0}^{\infty} \left(e^{-\frac{(k-p_j)^2}{2\sigma^2}} \frac{(h_j)^k}{k!} \right) \right] . \quad (2)$$

If the process were pure Poisson (no readout noise), the logarithm of the likelihood would be the classical expression:

$$\log \mathbf{L} = \sum_{j=1}^{D} [-h_j + p_j \log h_j - \log(p_j!)] . \quad (3)$$

The function to be maximized is:

$$F = \sum_{j=1}^{D} \left[-\log(\sqrt{2\pi}\sigma) - h_j + \log \sum_{k=0}^{\infty} \left(e^{-\frac{(k-p_j)^2}{2\sigma^2}} \frac{(h_j)^k}{k!} \right) \right] -$$

$$- \mu \left(\sum_{i=1}^{B} q_i a_i - \sum_{j=1}^{D} p_j + \sum_{j=1}^{D} b_j \right) , \quad (4)$$

where μ is a Lagrange multiplier for the conservation of counts.

To obtain the maximum of (4), we set $\partial F/\partial a_i = 0$ and apply the method of successive substitutions, which affords us greater flexibility than other methods and results in rapidly converging algorithms. The Maximum Likelihood Estimator algorithm (MLE) is given by the iterative formula:

$$a_i^{(k+1)} = K a_i^{(k)} \left[\frac{1}{q_i} \sum_{j=1}^{D} \frac{f_{ji} p_j'}{\sum_{l=1}^{B} f_{jl} a_l^{(k)} + C_j b_j} \right]^n \qquad i = 1, \cdots, B. \tag{5}$$

where

$$p_j' = \frac{\sum_{k=0}^{\infty} \left(k\, e^{-\frac{(k-p_j)^2}{2\sigma^2}} \frac{(h_j)^k}{k!} \right)}{\sum_{k=0}^{\infty} \left(e^{-\frac{(k-p_j)^2}{2\sigma^2}} \frac{(h_j)^k}{k!} \right)} . \tag{6}$$

In (5) k is the index of the iteration, K is a constant to preserve the energy in the form $\sum_{i=1}^{B} q_i a_i = \sum_{j=1}^{D} p_j$, computed at the end of each iteration and n is a constant to accelerate convergence up to approximately three times ($n = 3$). Constant n does not affect the point to which the algorithm converges.

The iterative algorithm (5), (6) has a number of desirable characteristics: It solves the cases of both pure Poisson data and Poisson data with Gaussian readout noise. The algorithm maintains the positivity of the solution; it is easy to implement; it includes flatfield corrections; it removes background and can be accelerated to be faster than the Richardson-Lucy algorithm (Lucy, 1974). The main loop (projection and backprojection) of the algorithm is similar in nature to the Expectation Maximization algorithm. The algorithm can be applied to a large number of imaging situations, including CCD and Pulse-counting cameras both in the presence and in the absence of background.

We note that by its definition, p_j' $j = 1, \cdots, D$ is always positive, while the original data p_j $j = 1, \cdots, D$ can be negative. In the absence of readout noise or when it is negligible, $\sigma \to 0$ in (6). Then the exponentials are dominant at $k = p_j$ and $p_j' \to p_j$.

In the case of no background and no readout noise, algorithm (5) becomes:

$$a_i^{(k+1)} = a_i^{(k)} \left[\frac{1}{q_i} \sum_{j=1}^{D} \frac{f_{ji} p_j}{\sum_{l=1}^{B} f_{jl} a_l^{(k)}} \right]^n \qquad i = 1, \cdots, B\ . \tag{7}$$

For $n = 1$ and disregarding the gain (flatfield) corrections ($q_i = 1$), (7) is identical to the Richardson-Lucy algorithm.

2.3. STOPPING THE ALGORITHM

It is well known that the MLE solution is an unconstrained "classical" solution of the ill-posed Fredholm integral equation (1). The result is that the iterative algorithms (5) or (7) produce solutions that are highly unstable, with high peaks and deep valleys. It is necessary to stop the process before reaching convergence. To stop the algorithm we use feasibility and cross-validation tests.

The feasibility tests stop the algorithm when the first n moments of the distribution are consistent with the Poisson/Gauss hypothesis (Llacer, Veklerov, and Núñez 1989).

The cross-validation tests computes the likelihood of the solution with respect to an alternative set of data (cross-likelihood). The stopping point is given by the maximum of the cross-likelihood against the iteration number (Núẽz and Llacer, 1993a).

3. Reconstruction of Astrometric Frames

3.1. GENERAL APPEARANCE

Image reconstruction of astrometric frames makes substantial improvements in qualitative appearance and recovery of faint star images in terms of the number of stars that can be detected. As an example, we present here the reconstruction of two astrometric frames: one obtained from space and the another from the ground. All images for these examples are shown in a linear grey scale, except where noted.

In the first example, we demonstrate the reconstruction of an astrometric image obtained with the PC camera of the aberrated Hubble Space Telescope by W. Van Altena (Yale). Figure 1 (left) shows the raw data to be reconstructed. Figure 1 (right) shows the reconstruction after 30 iterations. Although by technical constraints it is difficult to include very detailed images in this paper, the improvement in both appearance and number of stars seen is evident.

In the second example, we reconstructed a 512x512 astrometric image obtained from the ground. The image was obtained by Yves Requieme at Bordeaux Observatory using a CCD camera working in scanning mode attached to the Bordeaux Meridian Circle. Figure 2 (left) shows the raw data. Figure 2 (right) shows the reconstruction after 20 iterations. Again the improvement of the reconstructed image with respect to the raw data is evident. The improvement is better seen in the plots of Figure 3. The left part of Figure 3 shows a three dimensional plot of a detail of the raw image displayed in Figure 2 (left). The detail corresponds to the close cluster located at bottom right. Figure 3 (right) shows the same region after reconstruction (20 iterations).

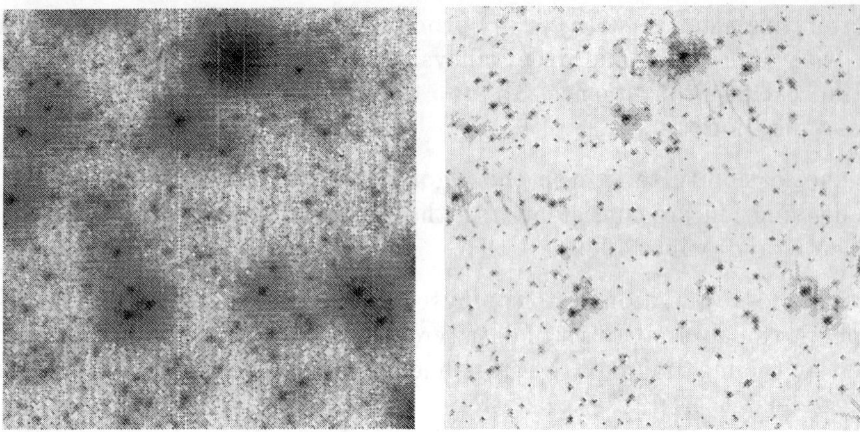

Figure 1. Left: Raw image of a HST PC astrometric frame. Right: Reconstructed image after 30 iterations.

Figure 2. Left: Raw image of a ground based CCD astrometric image. Right: Reconstructed image after 20 iterations.

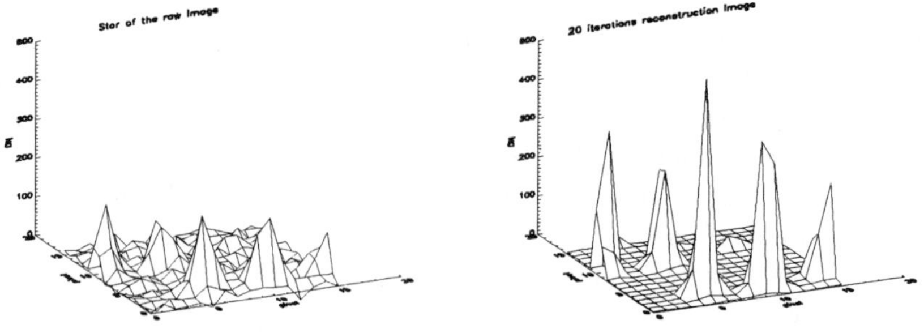

Figure 3. Left: Three dimensional plot of a detail of the raw image displayed in Figure 2 (left). Right: Same after reconstruction with 20 iterations.

3.2. ASTROMETRIC PRECISION OF THE RECONSTRUCTED IMAGES

In a previous paper (Girard et. al, 1995), the astrometry group at Yale University in collaboration with the authors performed a series of tests to assess whether the image reconstruction can retain or improve the astrometric information to be extracted from an image. The tests involved two HST-PC frames of NGC6752 (40s and 500s exposure respectively). Both images were reconstructed with 100 iterations. Given the space-variant character of the PSF of the HST, Girard et al. used three different PSF centered at pixels (200,600), (400,400) and (600,200) respectively. The reconstructed intensity profiles were centered, and the long exposure positions were transformed into those of the short exposure to determine the unit weight measuring error. The results are given in Table 1.

TABLE 1. Positional Accuracy (long- to short-exposure) from Girard et al. (1995)

Image	σ_x (mas)	σ_y (mas)
Raw	1.4	2.1
PSF(200,600)	1.4	1.2
PSF(400,400)	1.3	1.0
PSF(600,200)	0.8	0.8

The results shown in Table 1 suggest that 1 mas positional precision can be obtained using reconstructed images. The accuracy is, however, only 2 mas using the raw images.

To test the improvements in positional precision, we carried out a second experiment: we used a two-dimensional Gaussian fitting to obtain the positions of the stars in both the raw image and the reconstruction of the Meridian Circle CCD image (Figure 2). Table 2 shows the means of the standard deviation in X and Y coordinates of the fitted Gaussian profiles.

The results of Table 2 show that after reconstruction, the standard deviation of the Gaussian-fitted stellar intensity profiles are about half those obtained using the raw data. Given that the σ of the fitted stellar profile is directly related to the seeing present during the exposure, the results show that the reconstructed image is equivalent to an image obtained with a seeing that is half of the real.

TABLE 2. Standard Deviations of the Gaussian Fitted Profiles

Image	σ_x (pixels)	σ_y (pixels)
Raw	0.51	0.67
Reconstructed (20 its.)	0.24	0.34
Reconstructed (100 its.)	0.16	0.25

1 pixel = 1.65 arcsec

4. ABSENCE OF BIAS

In the first studies of the use of image reconstruction in Astrometry, Girard et al. (1995) noticed indications that, in the reconstructed images of the HST PC, the derived image centers were biased toward the center of the brightest pixel. This effect, if confirmed, can be due to several factors. One possible cause could be the use by computational constraints of a constant PSF to reconstruct images with a known space-variant PSF. Another cause could be the reconstruction algorithm itself.

To investigate this bias further, we carried out two new tests using computed generated data.

In the first experiment, we simulated the observation of one star in ideal conditions. We generated 1000 Poisson realizations of a Gaussian profile. We used the same position (out of the center of the pixel) to generate the 1000 realizations. We, thus, simulated 1000 independent observations (each with different Poisson noise) of the same star located at the same point. This is our raw data. Then, we reconstructed each one of the 1000 raw stellar images using the MLE algorithm. Finally the intensity profiles of the 1000 raw and the 1000 reconstructed stellar images were centered using a two-dimensional Gaussian profile.

The effect of the Poisson noise is that the centering process gives for each of the 1000 raw stars a position that is slightly different from the true value. Thus, there is a certain scatter of the positions obtained arround the expected point. In Figure 4 (left) we plot the distribution of the differences between the positions of the 1000 raw stars and the true value. The units of the plot are pixels. Note that, as expected, almost all the positions are uniformly distributed inside a radius of 0.05 pixels centered on the true position.

Figure 4 (right) shows the distribution of the differences between the positions of the same 1000 stars after the reconstruction process and the true position. Note that, again, almost all the positions are uniformly distributed arround the true position. Note also that the the points are inside

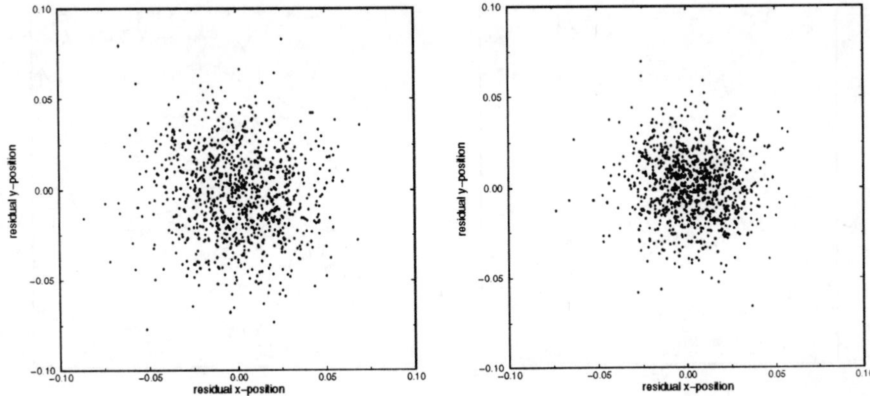

Figure 4. Left: Distribution of the positions of 1000 raw stars with respect to the expected point. Right: Distribution of the positions of the same 1000 stars after reconstruction.

a radius that is still smaller than in the raw data. Thus, the scatter of the positions obtained after the reconstruction is smaller than in the raw data.

If there were a systematic bias of the derived image centers toward the center of the pixel after the reconstruction, the cluster of points of Figure 4 (right) would be not cetered at the point (0,0). Besides, other bias would not give a uniform distribution of the positions arround the origin. Thus, there is no evidence of any sytematic bias of the positions obtained after the reconstruction with regard to the ones obtained using the raw data.

To investigate this absence of bias further, we carried out a second experiment consisting of generating 500 Poisson realizations of a Gaussian profile but, in this case, we generated each realization not at the same point but randomly distributed across the pixel. Figure 5 (left) shows the distribution of the positions used for generating the stars. Then, we reconstructed each one of the frames using the MLE algorithm and obtained the positions of the stars using the centering algorithm. Again, the differences between the true positions and the positions obtained after the reconstruction are less than 0.05 pixels in each coordinate, showing no bias. To be sure that there is no systematic bias, we computed the directions of the shifts between the positions obtained using the raw and the reconstructed stars respectively. Figure 5 (right) shows the directions of the shifts. In the plot, the magnitude of all the shifts has been artificially set to the same value to emphasize the absence of sytematic trends. The real values of the shifts are, however, smaller (less than 0.05 pixels in any case).

Again, the result of the experiment is that there is no evidence of any systematic bias due to the reconstruction process.

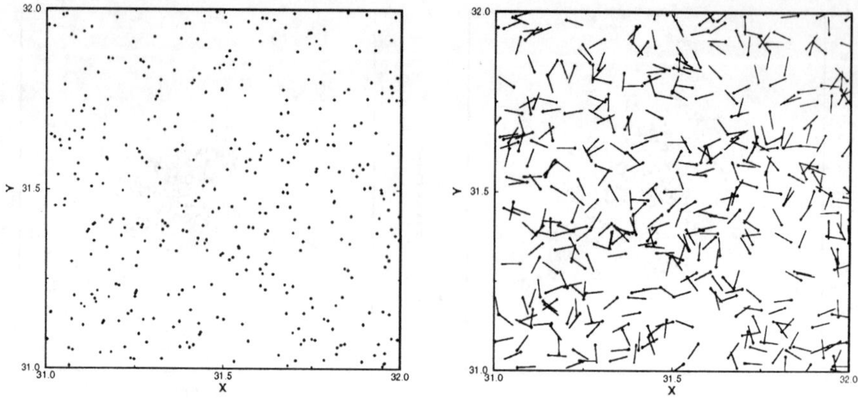

Figure 5. Left: Positions of 500 raw stars randomly distributed inside a pixel. Right: Directions of the shifts between the raw and reconstructed data. In the plot, the magnitude of the shifts has been artificially set to the same value.

5. BLENDED IMAGES

Another important problem is to determine the effect of the reconstruction process in the case of blended images of stars. To this end, we performed a series of tests consisting of generating pairs of stars with decreasing separation. The centering algorithm was modified to fit both stars simultaneously instead of centering each one separately. We used two two-dimensional elliptic Gaussian functions.

The pairs of blended star images were reconstructed and the centering algorithm was applied to both the raw data and the reconstructed images. The positions obtained using the raw data show no bias with respect to the true values. However, the positions obtained using the reconstructed data seem to show a bias consisting of a shift towards the centroid of the blended image. This result is only provisional because we have performed only preliminary tests. We are working to confirm or rule out the presence of this bias in the reconstruction of the blended images. If the bias is confirmed, the next step would be to assess whether it is it is due to the nature of the problem or to the reconstruction algorithm itself.

Thus, at this point we issue a warning on the use of the reconstruction techniques for astrometry if blended star images are involved.

6. CONCLUSIONS

The effect of image reconstruction on the astrometric accuracy of space- and ground-based images has been investigated. Image reconstruction provides a gain in centering precision for both ground-based and space-based astrometric images. The results show that, after reconstruction, the fitted stellar intensity profiles (Gaussian functions) have standard deviations that are about a half those obtained using the raw data. Thus, the accuracy obtained with image reconstruction can be improved by a factor of two with regard to the raw data. Moreover, image reconstruction may provide benefits in terms of the number of stars that can be detected and measured.

The result of our experiments using isolated stellar images do not reveal any systematic bias due to the reconstruction process. We should, however, indicate that in the case of blended double stars images, a systematic shift toward the optical centroid of the system can be present. We are investigating this effect.

7. ACKNOWLEDGEMENTS

This work was supported in part by the the DGICYT M.E.C. (Spain) under grants no. PB94-0905 and PB95-1031. Partial support was also obtained from the D.G.U. Generalitat de Catalunya under fellowship 1995BEAI200053 and from the Gaspar de Portola Catalan Studies Program of the University of California and Generalitat de Catalunya.

References

Girard, T.M., van Altena, W.F., Núñez J., Benedict, G.F., Duncombe, R.L., Hemenway, P.D., Jeffreys, W.H., MCArthur, B., Cartney, J.Mc, Nelan, E., Shelus, P.J., Story, D., Whipple, A.L., Franz, O.G., Wasserman L.H. and Frederick, L.W. (1994), Astrometry with the HST Planetary Camera. *IAU Symp. 166*, 101-106

Girard, T.M., Y. Li, Y., W.F. van Altena, W.F., J. Núñez, J. and y A. Prades, A. (1995) Astrometry with Reconstructed HST Planetary Camera (WF/PC 1) Images, *International Journal of Imaging Systems and Technology (Special Issue on Image Reconstruction and Restoration in Astronomy)* **6**, n. 4, 395-400

Lucy, L.B. (1974). An iterative technique for the rectification of observed distributions. *Astron. J.* **79**, 745-759.

Llacer, J., Veklerov, E., and Núñez, J. (1989). Statistically Based Image Reconstruction Algorithms for Emission Tomography. *International Journal of Image Systems and Technology*, **1**, 132-148

Núñez, J., and Llacer, J. (1991). Maximum Likelihood estimator and Bayesian reconstruction algorithms with likelihood cross-validation. *Astron. Soc. Pacific Conf. Ser.* **25**, 210-214.

Núñez, J., and Llacer, J. (1993a). A general Bayesian image reconstruction algorithm with enropy prior. Preliminary application to HST data. *Publ. Astron. Soc. Pacific* **105**, 1192-1208.

Núñez, J. and Llacer, J. (1993b). HST image reconstruction with variable resolution. *The restoration of HST Images and Spectra II* (R.J. Hanisch and R.L. White eds.). STScI: Baltimore, 123-130.

Núñez, J. and Llacer, J. (1994), A Bayesian Algorithm for Image Reconstruction with Variable Hyperparameter. In *Fifth Int. Meeting on Bayesian Statistics*, (J.M. Bernardo ,J. Berger, A.P. Dawid and A. Smith. eds.). Oxford Univ. Press. 713-722.

Núñez, J. and Llacer, J. (1996), Image Reconstruction with Variable Resolution Using Gaussian Invariant Functions in a Segmentation Process. *Vistas in Astron.*, **40**.

OPTICAL SPECTRA OF SOME VISUAL BINARIES WITH VARIABLE COMPONENT

V.S. TAMAZIAN, J.A. DOCOBO
Astronomical Observatory "Ramón María Aller"
P.O. Box 197, Santiago de Compostela, Spain

AND

V.H. CHAVUSHYAN, V.V. VLASYUK
Special Astrophysical Observatory RAS
Nizhnij Arkhyz, Karachai-Cherkessia, 357147, Russia

Abstract. The slit spectra of 11 components in 6 visual binary stars (UU Psc, YZ Cas, RS Tri, BX And, COU 14, IL Cep) with variable components are studied. MK classification were made for observed stars, for the first time for all secondary components except those of COU 14 and IL Cep. No significant variations have been detected in the spectrum of COU 14. The visual companion of Herbig Be star IL Cep is possibly an Am star.

1. Introduction

The study of variable stars as a whole forms one of the major branches of stellar astronomy providing us with additional parameters (time scales, amplitudes etc.) which are not available for non-variable stars.

When a variable star is a component of visual binary (VB) it becomes of special interest because we are dealing with the physical system of two stars with equal initial age and chemical composition but which are apparently at present at different stages of their evolution.

Are we sure that the VB system of two stars undergoing their evolution process does not suffer concomitant orbit evolution?

Does its present orbit keep the traces of the physical evolution of the components?

Do any significant differences exist in astrophysical and orbital parameters between visual binaries with variable components (hereafter VBVC) and other, much more numerous VBs whose components do not exhibit light variations?

These and many other questions concerning the origin and evolutionary tracks of VB stars still remain unclear despite a permanently voiced (Dommanget, 1988; this Proceedings) lack of information on their basic astrophysical parameters in general, and for those with variable components in particular.

Despite many efforts to collect spectral data on VBs by Meisel (1968, and references therein for earlier articles), Murphy (1969), Bouige (1974), Levato (1975), Lutz and Lutz

J. A. Docobo et al. (eds.), Visual Double Stars: Formation, Dynamics and Evolutionary Tracks, 27–33.

(1977), Abt (1981; 1985), Gahm et al.,(1983), Lindroos (1985) and others, we still lack information on such important parameter as their MK spectral class.

Abt (1985) clearly demonstrated the lack of spectral information even for relatively bright stars from ADS catalogue (Aitken, 1932) as well as many reasons to obtain their MK types through classification based on slit spectra, namely to provide spectroscopic distances; to learn whether or not the system is physical; to identify stars of unusual interest etc. Important statistical studies can also be performed when homogeneous astrophysical data sample are collected.

Among the articles dedicated to certain kinds of variables of particular interest is an exhaustive study of eclipsing binaries among VB systems published recently by Chambliss (1992) who presented an extensive overview of previous works on VBVC. The lack of information is noted by Chambliss for many stars investigated in his article.

Fekel (1981) demonstrated another approach to the problem by studying double and multiple stellar systems around 34 variable stars as did Gershberg (1985) who considered different aspects of duplicity among flare stars.

It is noteworthy that most radiostars are variables forming physical pairs for which both astrophysical and orbital parameters are also strongly needed (Debarbat and Chollet, 1988). One of the stars included in the present study (UU Psc) appears on the short list of such stars given by these authors in the cited article.

Different aspects of duplicity among variable stars are extensively reviewed by Heintz (1978), Petit (1989) and Sterken and Jaschek (1996).

The most recent published catalogue of VBVC is that of Proust et al. (1981) which contains 300 entries including all the stars from previous similar catalogues published by Plaut (1934; 1940), Baize (1962) and Petrova (1963). The same procedure as used by Proust et al.(1981) has been adopted by P.Lampens (1992) to elaborate a modified and updated version with the newest catalogues of both variable (GCVS, Kholopov et al., 1982-1985; NSV, Kukarkin et al., 1982) and components of visual double and multiple stars (Dommanget, 1989). Its latest version has been kindly provided to the authors by P.Lampens (1996) and contains 444 entries including all the stars from the catalogue of Proust et al., (1981).

Simple countings show that amongst the main components of VBVC in both catalogues of Proust and Lampens the percentage of stars unclassified in the MK system is 20peculiar and hardly classified stars such as WR-C-N, Of etc. The lack of spectral data becomes evident for secondary components over 90unclassified in the MK system in both catalogues.

THus, a lot of work must be done to study the basic astrophysical parameters of many primaries and the vast majority of secondaries amongst VBVC, thus obtaining more observational data on these systems is of certain interest.

Once gathered, such data should then be used for the comparative study of their astrophysical and orbital properties both to improve our knowledge of masses, radii, luminosity etc. as well as to learn more on the structure and evolution of stars and stellar systems.

The main aim of our ongoing study is to collect a sample of VBVC with well determined astrophysical and orbital data which will allow us furthermore to perform statistical analysis and a comparative study of visual binaries with components belonging to different kinds of variables.

As a first step this paper presents the MK classification of 10 components in 5 VBVC from Proust's catalogue (eclipsing binaries UU Psc, YZ Cas, RS Tri, BX And and Herbig Be star IL Cep) based on the slit spectra obtained for the first time for their visual companions. Close visual binary COU 14 = 13 Peg = HR 8344 (Couteau, 1993) with well determined orbit (Hartkopf et al., 1989) have been observed in order to check its suspected variability first noted by Couteau (1960) but which had still not been confirmed with certainty.

The spectra of the main components has been obtained with a twofold purpose to complement already known data on these variable stars largely investigated before and to compare our MK types with those published by other authors.

Both the MK classification and brief comments on some individual stars are given whereas detailed results of the spectroscopic study are in preparation and will be published elsewhere in the near future.

2. Observations

In Table 1 main data on observations are given.

All observations were made in November, 1995 on 6m telescope of Russian Special Astrophysical Observatory (Long-slit spectrograph installed in the prime focus + CCD 580x380pix., as described in Afanasiev et al., (1995), resolution 3-4 Å as well as on 1.22m telescope of Asiago Astrophysical Observatory in Italy (prism spectrograph camera VI + CCD 512x512pix., resolution 3 Å at Hgamma) and on 1.82m telescope of Astronomical Observatory of Padova at mount Ekar station in Asiago (grating spectrograph Boller and Chivens with CCD TH 7882, resolution 5-6 A).

Spectral range 3750 - 5200 Å was covered by two exposures in the range 3750 - 4550 and 4500 - 5200 Å.

All necessary corrections have been made (bias, dark current etc.) using standard IRAF package. Aperture extraction has been performed following the optimal method implemented to IRAF by Horne (1986).

Twenty standard lines of Fe and Ne taken from Munari and Tomov's (1995) have been used for the wavelength correction of spectra obtained on 1.22m telescope. More details on 6m telescope observations and data reduction may be found in Vlasyuk (1993).

The stars from Stone's list (1977) have been observed as standards.

All data on observations are presented in Table 1 where they are given as follows : Column 1 : Bayern name of star Columns 2,3: Right ascension and declination for epoch 2000.0 Column 4 : HD or BD number Column 5 : ADS number from Aitken's (1932) catalogue Column 6 : Observation date Column 7 : Telescope aperture Column 8 : Exposure time (sec) and spectral range: R1 - 3750-4550 Å R2 - 4500-5200 Å

3. MK classification

All spectra obtained for each star on different telescopes have been carefully compared in order only to identify the spectral lines definitely detected in the spectrum. Only relatively strong representative lines indicated by Jaschek and Jaschek (1987) have been considered for classification purposes. We used "An atlas of stellar tracings" by Goy et al. (1995) which covers all spectral classes and luminosity types being the most complete available atlas of spectra tracings including those of fundamental standards in the MK system.

The results are presented in Table 2 where data are given as follows : Column 1 Bayern name Column 2 Visual V magnitude at minimum (GCVS) for variable (main) component. For secondary stars they are taken from Proust et al.(1981) Column 3 Type of variability Column 4 Separation (in arcsec) taken from Proust et al. (1981) Column 5 MK types as defined by authors Column 6 Other MK types and corresponding references

TABLE 1. Main observational data.

Name	Coordinates (2000.0) alpha	delta	HD, BD	ADS	Date	Tele-scope	Exp.-Range (sec)
UU Psc	00 14 59	+08 49 17	1061	191 A	18.11.95	1.22m	200-R1; 500-R2
				191 A	29.11.95	6m	15-R1; 15-R2
				191 B	18.11.95	1.22m	900-R1; 1200-R2
				191 B	29.11.95	6m	180-R1; 180-R2
YZ Cas	00 45 39	+74 59 18	4161	624 A	19.11.95	1.22m	600-R1;
				624 A	30.11.95	6m	120-R1; 120-R2
				624 B	19.11.95	1.22m	3600-R1;
RS Tri	01 34 49	+29 35 21	+28 258	1236 A	29.11.95	6m	60-R1; 300-R2
				1236 B	29.11.95	6m	300-R1; 1200-R2
BX And	02 09 04	+40 47 39	13078	1671 A	29.11.95	6m	600-R1; 600-R2
				1671 B	29.11.95	6m	600-R1;
COU 14	21 50 08	+17 17 11	207652	-	15.11.95	1.82m	20-R1; 30-R2
				-	19.11.95	1.22m	600-R1; 1200-R2
				-	29.11.95	6m	30-R1; 30-R2
IL Cep	22 53 15	+62 08 45	216629	16341 A	29.11.95	6m	30-R1; 60-R2
				16341 B	29.11.95	6m	60-R1; 180-R2

TABLE 2. MK classification of observed stars.

Name	V	Var. type	Sep. (sec.)	MK classification: This paper	MK classification: Other classifications
UU Psc	6.05	Algol	11".6	F0 IV-V	F0IV-F4V (Cowley and Fraquelli, 1974) F0 V (Levato, 1975)
Second.	7.7			F5 V	—
YZ Cas	6.12	Algol	36.1	A2 IV-V	A2 IV (Abt and Bidelman, 1969) A1 V (Hill et al., 1975)
Second.	9.7			G0 V	—
RS Tri	11.00	Algol	5.0	A4 V	A5 V (Lampens, 1996)
Second.	11.2			F6 V	—
BX And	9.57	W UMa	19.6	F5 V	F2 V var. (Hill et al., 1975)
Second.	10.7			G5 IV-V	—
COU 14	5.32	?	0.3	F2 IV	F2 III (Nassau and van Albada,1947) F2 III-IV (Harlan, 1969)
IL Cep	9.61	HerbigBe	7.0	B3 IVe	B2 IV-Vne (Garrison, 1970)
Second.	11.7			Am	

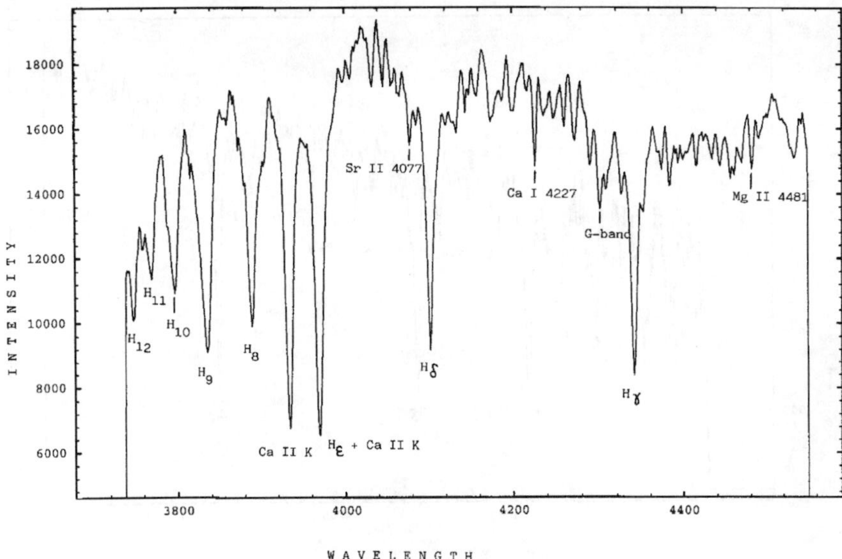

Figure 1.

The components separation of COU 14 has been taken from Catalogue of Couteau (1993). This star appears as NSV 13891 in NSV catalogue (Kukarkin et al., 1982) but its variability type is not yet known. Note that its apparent magnitude 6.1 is taken from Fernie (1981) and possibly corresponds to its minimum.

From data presented in Table 2 one can see that our classification differs by 3 subclasses only for BX And which might be explained by the variable character of its spectra as noted by Hill et al. (1975) while in all other cases classification remains well concordant and does not exceed one subclass. As regards luminosity classes, they always remain within the usual error of 0.6 class (Jaschek and Jaschek, 1987) and even coincide for BX And and RS Tri.

Thus, comparison with previously published MK types for main components shows that our classification agrees fairly closely with the others.

4. Some comments on COU 14 and IL Cep

Fig.1 represents the composite spectra of COU 14 as typical slit width was 1-2 arcsec whilst separation between components is much smaller.

Comparison of spectra obtained for COU 14 at different dates (see Table 1) shows no significant changes in its spectrum.

As there was a 4 and 10 day interval between observations and taking into account that Breger (1969) certainly discarded short-term (3-6 hours) variability one may suppose that we are possibly dealing with long-term variability of COU 14 due to its secondary component.

The unique spectral class to the visual companion of IL Cep cannot be assigned because of a relatively weak Ca II K line which does not correspond to the strength of hydrogen and

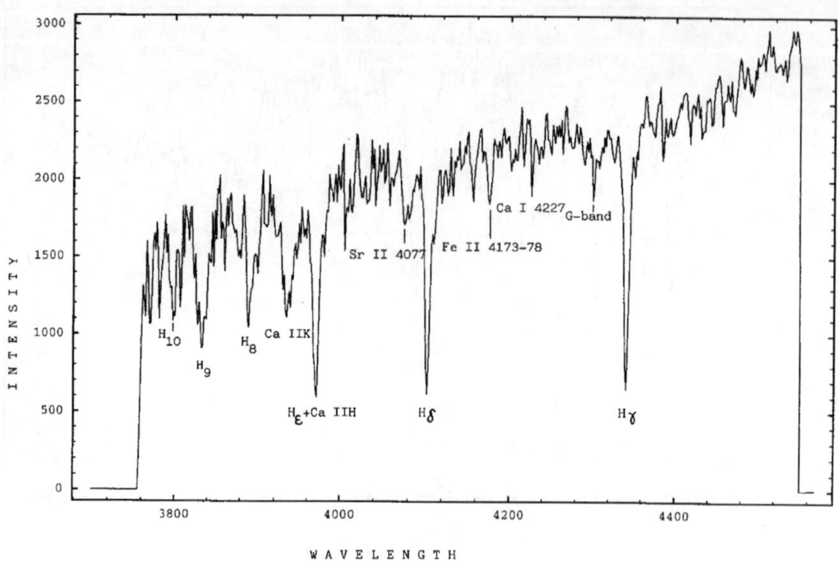

Figure 2.

metallic lines (Fig. 2). We are possibly dealing with an Am star estimated provisionally as K/H/M=A3/F0/F3.

Notice that both Ca II K and Sr II lambda 4077 are unusually wide, the latter having a clearly visible emission core. A detailed spectroscopic study of this star will be presented separately.

Acknowledgments

The authors would like to thank Prof. C. Jaschek for many helpful comments. One of the authors (V.S.T.) thanks Prof. R. Barbon and Drs. L. Buzon and A. Niedzielski for their support during observations at Asiago.

References

1. Abt, H.A., Bidelman W.: 1969, *Astrophys. J.*, **158,** 1091
2. Abt, H.A.: 1981, *Ap. J. Suppl.*, **45**, 437
3. Abt, H.A.: 1985, *Ap. J. Suppl.*, **59**, 95
4. Afanasiev, V.L., Burenkov, A.N., Vlasyuk, V.V., Drabek, S.V.: 1995, *SAO Technical report* **No. 234**
5. Aitken, R.G.: 1932, *New General Catalogue of Double Stars Within 120deg. of the North Pole* (Washington, DC, Carnegie Institute)
6. Baize, P.: 1962, *J. Observateurs,* **45**, 117
7. Bouige, R.: 1974, *Vistas Astron.*, **16**, 117
8. Breger, M.: 1969, *Ap. J. Suppl.*, **19,** 79
9. Chambliss, C.: 1992, *PASP,* **104**, 663
10. Couteau, P.: 1960, *J. Observateurs,* **43**, 1
11. Couteau, P.: 1993, *Catalogue de 2700 Etoiles Doubles COU,* Second Edition, Cote D'Azur Observatory, Dept.
12. Fresnel Cowley, A., Fraquelli D.: 1974, *PASP,* **86**, 70

13. Dommanget, J.: 1988, *Astrophys. and Space Sci.*, **142**, 5
14. Dommanget, J.: 1989, *Catalogue of Components of Double and Multiple Stars* Contrib. Van Vleck Obs., 8, 77
15. Debarbat, S., Chollet F.: 1988, *Astrophys. and Space Sci.* , **142**, 61
16. Fekel, F.C.: 1981, *Ap. J.* , **246**, 879
17. Fernie, J.D.: 1976, *J. R. Astron. Soc. Can.* , **70**, 77
18. Gahm, G.F., Ahlin P., Lindroos K.P.: 1983, *Astron. and Astrophys.*, **51**, 143
19. Garrison, R.F.: 1970, *Astron.J.*, **75**, 1001
20. Goy, G., Jaschek, M. and Jaschek, C.: 1995, *An atlas of stellar tracings*, Geneva Observatory
21. Harlan, E.A.: 1969, *Astron.J.*, **74**, 916
22. Hartkopf, W.I., McAlister, H.A., Franz, O.G.: 1989, *Astron. J.*, **98**, 104
23. Horn, K.: 1986, *PASP*, **98**, 609
24. Heintz, W.D.: 1978 , *Double stars* , D.Reidel Publishing Co., Dordrecht, Holland
25. Hill, G., Hilditch, R.W., Younger, F., Fisher, W.A.: 1975, *Mem. R. Astron. Soc.*, **79**, 131
26. Jaschek, C., Jaschek, M.: 1995, *The bevavior of chemical elements in stars*, Cambridge Univ. Press, Cambridge
27. Jaschek, C., Jaschek, M.: 1987, *The Classification of Stars*, Cambridge Univ. Press, Cambridge
28. Kholopov, P.N., Samus, N.N., Frolov, M.S., Goranskij, M.P., Gorynya, M.P., Kireeva, N.N., Kukarkina, N.P., Kurochkin, N.E., Medvedeva, G.I., Perova, N.B., Shugarov, S.Yu.: 1982-1985, *General Catalogue of Variable Stars (GCVS)*, vol. I-III, Nauka Publ. House, Moscow, 4th ed.
29. Kukarkin, B.N., Kholopov, P.N., Artiukhina, N.M., Federovich, V.P., Frolov, M.S., Goranskij, M.P., Gorynya, M.P., Karitskaya, E.A., Kireeva, N.N., Kukarkina, N.P., Kurochkin, N.E., Medvedeva, G.I., Perova, N.B., Ponomareva, G.A., Samus, N.N., Shugarov, S.Yu.: 1982, *New Catalogue of Suspected Variable Stars (NSV)*, Nauka Publ. House, Moscow
30. Lampens, P.: 1992, in *Variable Star Research: An international perspective*, J.R. Percy, J.A.Mattei and C.Sterken (eds.), Cambridge University Press, Cambridge, p.60
31. Lampens, P.: 1996 (private communication)
32. Levato, H.: 1975, *Astron. and Astrophys.*, **19**, 91
33. Lindroos, K.P.: 1985, *Astron. and Astrophys. Suppl.*, **60**, 183
34. Lutz, T.E., Lutz, J.H.: 1977, *Astron. J.*, **82**, 431
35. Meisel, D.D.: 1968, *Astron. J.*, **73**, 350
36. Murphy, R.E.: 1969, *Astron. J.*, **74**, 1082
37. Nassau, J.J., van Albaba, G.B.: 1947, *Ap. J.*, **106**, 20
38. Petit, M.: 1987, *Variable stars*, J.Wiley&Sons Publishing,
39. Petrova, C.A.: 1963, *Perem. zvezdy*, **14**, 357
40. Plaut, L.: 1934, *Bull. Astron. Inst. Netherlands*, **227**, 181
41. Plaut, L.: 1940, *Bull. Astron. Inst. Netherlands*, **232**, 121
42. Proust, D., Ochsenbein F., Pettersen B.R.: 1981, *Astron. and Astrophys. Suppl.*, **44**, 179
43. Sterken, C., Jaschek, C.: 1996, (eds.) *Light curves of variable stars*, Cambridge Univ. Press, Cambridge
44. Stone, R.P.S.: 1977, *Ap. J.*, **218**, 767
45. Tomov, T.V., Munari, U.: 1995 , *Padova and Asiago Obs. Techn. Rep.* **No.9**
46. Vlasyuk, V.V.: 1993, *Bull. Spec. Astrophys. Obs.*, **36**, 107

REDUCTION OF CCD OBSERVATIONS OF DOUBLE AND MULTIPLE STARS

J. CUYPERS
Koninklijke Sterrenwacht van België,
Ringlaan 3, B-1180 Brussels, Belgium

Abstract. The method used at the Royal Observatory to extract photometric and astrometric information from CCD images of visual double stars with a separation in the range 1.5 to 20 arcsec is described. Since standard methods are not applicable, a short history of the technique selection process is given. Finally direct profile fitting was used and the adopted point spread function is a modified Moffat profile, corrected for ellipticity of the isophotes. The problems related to the technique are described, some results are given and preliminary errors are estimated.

1. Introduction

In recent years the Royal Observatory was involved in several observational programmes of wide visual double stars (Argue et al., 1992; Oblak et al., 1992; Sinachopoulos et al., 1995). Besides classical photometry also CCD imaging of double stars was done (see Figure 1 for an example).

Extracting accurate photometric and astrometric information of bright visual double stars on CCD images is not straightforward. When bright stars are involved, mostly none or very few additional stars are recorded on the image and, as a consequence, crowded field packages for CCD reduction cannot be used in a proper way. It is also not clear whether a profile derived from stars further away from the components will give the accurate information on the double or multiple star contained in the CCD image.

If there is overlap between the images of the components (even under good seeing conditions stars separated by 10 arcsec and more still show some overlap in 1 m telescopes), a classical approach (aperture photometry and centering routines) cannot be used.

J. A. Docobo et al. (eds.), Visual Double Stars: Formation, Dynamics and Evolutionary Tracks, 35–41.

Consequently, direct profile fitting was used, since this is a method were the information can be extracted sufficiently accurately.

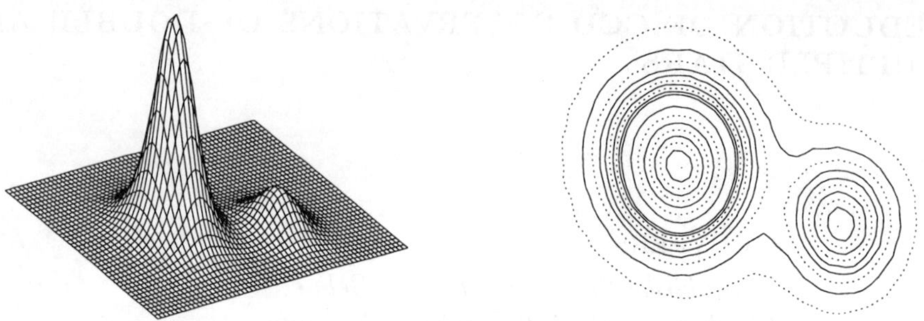

Figure 1. Example of a double star image on CCD (3D view and contour plot): HIC 24366 ($\rho = 7.8"$, $magA = 10.5$, $\Delta mag = 1.7$)

2. Choosing a profile

After some experiments with one-dimensional projections of the images (inspired by Rakos et al., 1982 and Sinachopoulos, 1988) a two-dimensional approach was adopted. The point spread function (PSF) is given as a function of the two spatial coordinates x and y.

Several profiles were tested to fit to an image of a single star. For the fitting process a least-squares approach was used, where minimization was done with a modified Gauss-Newton algorithm with numerical derivatives (as implemented in the ESO-MIDAS configuration with use of the numerical subroutines of the NAG library).

A simple Gaussian

$$f(x, y) = h_A e^{-br^2},$$

or Lorentz-like profiles

$$f(x, y) = \frac{h_A}{(1 + (r_A b)^2)^t} \qquad \text{(with } t \text{ fixed at 1, 2, ...)}$$

where

$$r_A = \sqrt{(x - x_A)^2 + (y - y_A)^2}$$

and the unknowns are x_A, y_A, h_A and b, were clearly not sufficient to describe the core and the wings of the images accurately.

Therefore, a Moffat profile (Moffat, 1969) was used, written as:

$$f(x, y) = \frac{h_A}{(1 + (r_A b)^2)^q}$$

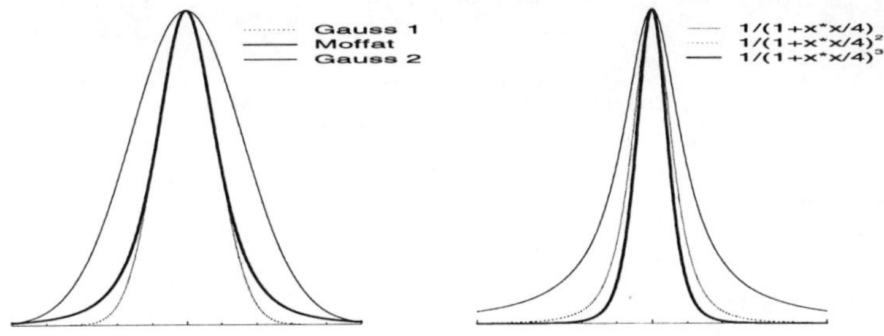

Figure 2. Gaussian and Moffat profiles compared

The extra unknown parameter q makes it possible to fit the core and the wings as well (see Figure 2). In all profiles (x_A, y_A) gives the central position and h_A is the height at that position. Remark that b is related to the Full Width at Half Maximum : $FWHM = 2\sqrt{\{\sqrt[q]{2} - 1\}}/b$. In this modified form the fitting proces was usually more "stable", then when written otherwise. Other profiles, such as two-dimensional analogues of the Franz profile (Franz, 1973) and those suggested for use with the one-dimensional area scanner (Rakos et al., 1982), were also tested. An additional unknown describing the dependency on distance in the exponent should make it possible to fit even better the observed profiles, but a lot of numerical instabilities, especially in the derivatives, are introduced. Moreover, the important quantities to be derived (height and central position of the profile) did not change significantly compared to the results obtained with the Moffat profile (E. Ruymaekers, unpublished annex thesis, 1994).

In a lot of cases it was necessary to use elliptical isophotes. This was done by replacing r_A with the expression:

$$r_A = \sqrt{(x - x_A)^2 + e(x - x_A)(y - y_A) + (1 + f)(y - y_A)^2}$$

with two additional unknowns e and f.

3. Fitting multiple components

When a small number of stars is present on the CCD frame a sum of profiles with identical shapes has to be fitted. The two components model will be:

$$f(x, y) = \frac{h_A}{(1 + (r_A b)^2)^q} + \frac{h_B}{(1 + (r_B b)^2)^q} + sky,$$

where A and B refer to the two components and b, q, h and r are defined as above. From the results of the fit, one can derive the differential position and magnitude of B with respect to A.

For N components this becomes:

$$f(x,y) = \sum_{i=1}^{N} \frac{h_i}{(1+(r_i b)^2)^q} + sky,$$

where, as above,

$$r_i = \sqrt{(x-x_i)^2 + e(x-x_i)(y-y_i) + (1+f)(y-y_i)^2}$$

A few tests (Lampens and Seggewiss, 1995) indicated that at least ten components can be fitted with this method.

4. The programme in the MIDAS environment

The fitting proces was done within the ESO-MIDAS environment. This was done for reasons of availability when the observations started (most sets of data resulted from ESO) and tradition. The treatment of the "raw" data could be done in the same environment as well.

The CCD images were bias subtracted and flat-fielded with MIDAS commands. Since almost all the stars observed so far had very bright components, and, therefore, only very short exposure times were used, a simple sky subtraction, based on the mean of the three smallest values of small squares in the corners of the images was used. In the subsequent analysis it was assumed that the sky background was equal to zero.

Initial values for the parameters could easily be estimated: the parameter b is directly related to the seeing, q is usually in the range 2 to 3. For e and f (if used), zero is the starting value. A good guess for the position of the main component are the coordinates of the maximum of the frame and the position of the other component(s) can be estimated directly on the frame (mouse clicking) or from an astrometric list (e.g. Hipparcos Input Catalogue). Special attention is necessary when the magnitude differences between the components have different signs in different filters.

5. Tests and estimates of errors in the astrometry

For synthetic data the fitting routines perfectly reproduced the profiles. On single, isolated, well observed stars, remaining residuals were only slightly above the photon noise. It is clear that when the isophotes do not have a perfect elliptical shape, residuals will be larger. In those cases the largest residuals where five to ten times the photon noise, but the influence on the derived photometric and astrometric parameters, was very small.

To evaluate the internal errors we compared the results of observations of the same double star, on images taken one after another on one night,

TABLE 1. Differential data of 1 night

HIC Nr	Filter	$\rho('')$	$\theta(°)$	Δmag	Date
24366	V	7.787	106.687	1.725	1991 10 16 (1)
24366	V	7.775	106.657	1.725	1991 10 16 (2)
24366	V	7.778	106.601	1.718	1991 10 16 (3)
24366	V	7.803	106.860	1.712	1991 10 16 (4)
24366	V	7.782	106.711	1.724	1991 10 16 (5)
24366	V	7.782	106.671	1.718	1991 10 16 (6)

on different nights and in different seasons (see Tables 1 to 3 for a few examples). The agreement is excellent. Internal errors on separations are only a few milliarcseconds.

TABLE 2. Differential data of consecutive nights

HIC Nr	F.	$\rho('')$	Stdρ	$\theta(°)$	Stdθ	Δmag	StdΔ	Date
25870	V	4.813	0.005	326.17	0.02	0.618	0.006	91.10.21
25870	V	4.814	0.005	326.24	0.02	0.624	0.001	91.10.22

The true (relative) apparent separation ρ between A and B, is function of the scale on the CCD image. If the same camera is used with the same instrument and filter, the scale will not change. The position angle θ changes with every dismount and remount of the camera. When observing a star near the equator in a "non-tracking" mode, a trace on the CCD is marked, but the precision of the extracted orientation is not much better than 0.4°, as repeated observations indicated.

In the range of expected precision (0.01-0.001 arcsecs in ρ and 0.1° in θ), there was not sufficient comparison material available. This will be possible when the definitive Hipparcos results are released. Comparisons with a few preliminary Hipparcos data, indicate that the scale factor can be obtained with a 0.05% precision and orientation with a 0.2° precision on a telescope of the 1m class, even in non-optimal conditions (badly focused images, no excellent seeing, ...). Remark that for large separations (>15 arcsec), the PSF is not necessarily identical for the different components and, therefore, an extra error will be introduced. At the very small separations (less than the $FWHM$) the quality of the results will degrade as well.

TABLE 3. Differential data in different seasons (no correction for scale and orientation)

HIC Nr	F.	$\rho('')$	Stdρ	$\theta(°)$	Stdθ	Δmag	StdΔ	Date
25436	I	12.005	0.005	251.34	0.01	0.704	0.004	91.10.23
25436	I	12.003	0.012	249.59	0.17	0.703	0.001	92.02.21

6. Towards standard magnitudes

The difference in magnitude between the components can be derived from the ratio of the heights of the fitted profile, as described above. It is difficult to give a reliable estimate of the errors on the differential magnitudes, since there exists almost no comparison material. Internal errors are very small, usually less than 0.005 mag (see Tables 1 to 3 for some examples).

If a total intensity can be extracted and the extinction is available, the total instrumental magnitude can be calculated. When the transformation to a standard system is known (e.g. from measurements of standard stars), there is sufficient information to calculate the standard magnitudes and colours of each component.

For the observing campaigns considered one tried to measure as many (single) standard stars as feasible and to use them as well for the calculation of the extinction. By combining the standard measurements per season, but with a different (calculated) extinction coefficient for each night, a reliable colour transformation could be done. Typical errors are around 0.01 magnitude (see Table 4). Combining all errors involved in the reduction proces of the double or multiple stars one arrives at an error of less than 0.03 mag in the standard magnitudes and colours of each component. For a detailed comparison of the magnitudes with other sources we refer to the paper presented by Lampens et al. at this meeting.

TABLE 4. Standard magnitudes in different seasons

HIC Nr	Filter	Mag A	Mag B	Δ Mag	Date
25436	V	8.260	9.102	0.842	91.10.23
25436	V	8.262	9.092	0.830	92.02.21
25436	I	7.765	8.469	0.704	91.10.23
25436	I	7.764	8.464	0.699	92.02.21

7. Conclusions

The procedure described here is adequate to extract photometric and astrometric information of double and multiple stars on CCD images. In view of the limited capacity of the telescopes used (< 1 m), the relatively few measurements taken of each star system and the non optimized CCD camera and acquisition system, a large quality improvement is still possible. Nevertheless, this procedure has produced data of unprecedented accuracy on individual components of a large group of "intermediate" visual double stars.

Acknowledgements

Many colleagues contributed with ideas and discussions to the described reduction procedure. I thank them all. Part of this research was carried out with the help of the project "Service Centres and Research Networks", financed by the Belgian Federal Scientific Services (DWTC/SSTC).

References

Argue, A.N., Bunclark, P.S., Irwin, M.J., Lampens, P., Sinachopoulos, D. and Wayman, P.A., 1992, 'Double Star CCD astrometry and photometry', Mon. Not. R. Astron. Soc. **259**, 563-568.

Franz, O.G., 1973, J. Royal Ast. Soc. Canada **67**, 81

Lampens, P., Seggewiss, W., 1995, reported at the *Astronomische Gesellschaft Herbsttagung*, 20 september 1995, Bonn.

Oblak, E., Argue, A.N., Brosche, P., Cuypers, J., Dommanget, J., Duquennoy, A., Froeschlé, M., Grenon, M., Halbwachs, J.L., Jasniewicz, G., Lampens, P., Mermilliod, J.C., Mignard, F., Sinachopoulos, D., Seggewiss, W. and Van Dessel, E., 1992, 'The European Network of Laboratories: Visual Double Stars', IAU Coll. 135, ASP Conference Series **32**, eds. H.A. McAlister and W.I. Hartkopf, 454-456.

Moffat, A.F.J, 1982, Astron. Astrophys. **3**, 455-461.

Rakos, K.D., Albrecht, R., Jenkner, H., Kreidl, T., Michalke, R., Oberlechner, D., Santos, E., Schermann, A., Schnell, A., Weiss, W., 1982, Astron. Astrophys. Suppl. Series **47**, 221-235.

Sinachopoulos, D., 1988 Astron. Astrophys. Suppl. Series **76**, 189.

Sinachopoulos, D., Cuypers, J., Lampens, P., Oblak, E., Van Dessel, E., 1995, Astron. Astrophys. Suppl. Series **112**, 291-297.

MEASUREMENT PRECISION OF THE YALE-SAN JUAN SPECKLE INTERFEROMETRY PROGRAM

E. P. HORCH, T. M. GIRARD, W. F. VAN ALTENA AND R. D. MEYER
Department of Astronomy, Yale University
P.O. Box 208101, New Haven, CT 06520-8101 USA

C. E. LOPEZ
Observatorio Astronómico "Félix Aguilar"
Av. Benavidez 8175 Oeste, 5407 Marquesado, San Juan, Argentina

AND

O. G. FRANZ
Lowell Observatory
1400 West Mars Hill Road, Flagstaff, AZ 86001 USA

Abstract. We present an update on our progress in taking speckle observations of double stars from the Southern Hemisphere. The work here includes a measurement precision study, where we compare some of our measures to ephemeris positions of binaries with very well-determined orbits.

1. Introduction

In the talk that opened this meeting, Dr. McAlister showed that there is still a large disparity in the number of speckle measures of southern double stars compared with northern ones. As late as 1988 there were almost no speckle measures of double stars south of $-30°$ declination (Hartkopf 1992). Since that time, the situation has improved somewhat with the publication of three large sets of position angle and separation measures from data taken at the Cerro Tololo 4-m telescope (McAlister, Hartkopf & Franz 1990, Hartkopf *et al.* 1993, Hartkopf *et al.* 1996), but a large imbalance favoring the Northern Hemisphere objects still persists today.

By way of example, Figure 1 shows a situation typical of the kind of data that exist for many of the far southern double stars today. This object,

J. A. Docobo et al. (eds.), Visual Double Stars: Formation, Dynamics and Evolutionary Tracks, 43–53.

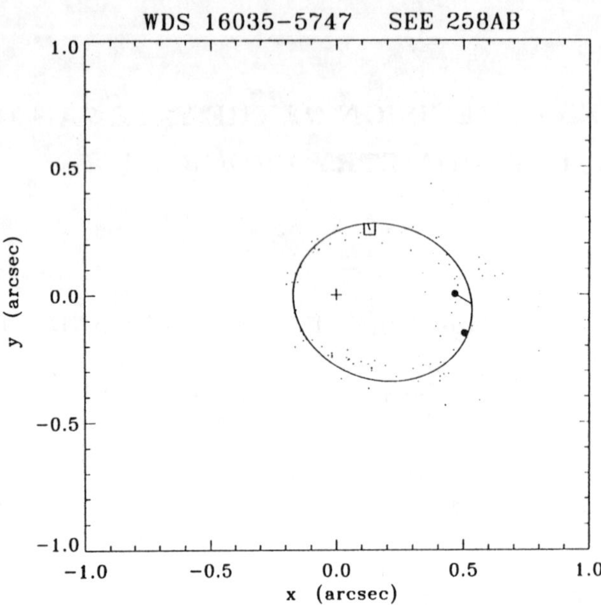

Figure 1. Measures and orbit of See 258 AB.

See 258 AB, has visual data dating back to the turn of the century, which are represented by dots in the figure. van den Bos published an orbit of See 258 AB in 1961 that, according to *The Fourth Catalog of Orbits of Visual Binary Stars* of Worley and Heintz (1983), is a "grade 1" or "definitive" visual orbit with semi-major axis of 0.366 arc seconds and period of 26.93 years. (The curve in the figure is the orbit of van den Bos.) Also plotted in the figure are all published speckle measures of See 258 as of this writing, a grand total of three, two from the 1970's plotted with filled circles (Morgan *et al.* 1978, Bonneau *et al.* 1980) and a third measure enclosed in a box, which is our recently published measure (Horch *et al.* 1996). Despite the fact that the system is easily resolvable with speckle interferometry even with small telescopes, there is virtually no data on this well-known system with the speckle technique.

Of course the primary motivation for collecting high-quality orbital data on visual binary stars is deriving good stellar masses, and in this regard it is often not so important to derive better orbits because if the parallax of the system is not well-known, parallax error dominates the error in the estimate of the mass. However, in the near future, Hipparcos parallaxes will be made available for a large number of the well-known visual binaries, and then there will be many cases where the error from the semi-major axis estimate will be much more significant. In addition, with smaller and smaller radial velocities being measured every year, there are many "visual" systems that can be observed by means of spectroscopy through at least part of the orbit. These facts make it very important to obtain the best possible astrometry of these systems in order to get the best mass estimates. Due mainly to the more than 20 years of speckle data produced by the Center for High-Angular Resolution Astronomy (CHARA) under the direction of Dr. McAlister, there is a long baseline of high-precision data on all the important systems observable from the Northern Hemisphere, but until recently the southern systems have been neglected. This is why we have started a long-term speckle observing program in the Southern Hemisphere.

2. Update and Current Status

We began taking speckle observations from El Leoncito, Argentina (latitude −31° 48′) in July of 1994. There are actually two observatories at that location, Carlos U. Cesco Observatory, which is jointly run by the National University of San Juan (Argentina) and Yale Southern Observatory, and the Complejo Astronómico El Leoncito (CASLEO), which is the national observing facility of Argentina. At Cesco Observatory, we use a 76-cm reflector, and at CASLEO, there is a 2.1-m telescope.

Since the beginning of the project we have been using a multi-anode microchannel array (MAMA) detector to record speckle patterns. We have the camera on loan from J. G. Timothy of the University of New Brunswick, Canada. Because this device has a bialkali photocathode, the quantum efficiency in the visible is low compared with most other speckle cameras being used today and there is virtually no detector response redder than about 6000Å. We are hoping to replace the MAMA with an intensified-CCD with a red-extended S-20 photocathode, but at least through the end of 1996 we will continue to use the MAMA.

Absolute scale calibration and orientation measurements are made at least once per observing run with a full-aperture slit mask on the 76-cm telescope. We derive a secondary scale by allowing selected bright stars to drift across the detector with the telescope tracking off. The declination of the star and the diurnal rate are then used to derive the plate scale. While we currently use the aperture mask for primary scale calibration, we have found that the zero point in the position angle is better determined by the drift scans. Up until now, we have not had an aperture mask for scale calibration at the 2.1-m telescope, so we rely on drift scans and observing some stars in common with the 76-cm telescope to fix the scale and orientation there. An aperture mask for the 2.1-m telescope has been constructed, however, and we will begin using it on the upcoming run in October 1996.

As of this writing, the program is now two years old and we have made 1503 observations of 701 double stars. We have observed a total of 81 nights at the 76-cm telescope and 23 nights at the 2.1-m telescope. Eight nights have already been awarded to us at the 2.1-m telescope in the latter half of 1996.

3. Measurement Precision Study

Since our goal is to provide high-quality astrometric data of Southern visual binaries, we have recently completed a measurement precision study to assess our progress in this regard. The idea of the study is straight-forward: we measure position angles and separations of binaries with well-determined orbits, then compare our results to the predicted ephemeris position of the object at the epoch of observation. Originally, we thought that using binaries with grade 1 visual orbits from the orbit catalog of Worley and Heintz (1983) would be sufficient for studying our measurement precision and scale calibration. In this case, we would have many objects from our database to use in such a study. However, after finding many examples like Figure 1 where the visual orbit, although in some sense "definitive," was of insufficient quality to judge the speckle observations, and many other cases in the literature where speckle data has led to a substantial revision in some orbital elements of the binary, we decided against using these objects in our study. Instead, we turned to the much smaller sample of binaries with orbits determined with the highly-weighted inclusion of speckle data. Though there are few such objects that are observable from the Southern Hemisphere, they have some of the highest quality "visual" orbits that exist today.

Between February 1995 and March 1996 we collected 37 observations of 8 objects fitting the above criteria. Those observed with the 76-cm telescope are shown in Table 1 while those observed at the 2.1-m telescope are shown in Table 2. In addition to the orbital elements of these systems, the tables give the *Washington Visual Double Star Catalog* (WDS) number of each object (Worley & Douglass 1984). After taking speckle data with our system, we used a weighted least-squares fitting approach to fit the fringe pattern of the binary power spectra. This is the same technique described in Horch *et al.* (1996), and it yields the position angle and separation based on the orientation and spacing of the fringes.

One of the most obvious things we noticed about the precision of our measures for the 37 observations in the sample is that it is strongly dependent on the seeing conditions. Figures 2 and 3 show the separation and position angle residuals as a function of seeing full width at half maximum

TABLE 1. Orbital Elements of the Cesco Objects

Parameter	Bu 101 [1]	Sp 1 AB [2]	A 2768 [1]	StF 1728 AB [1]
WDS (α,δ 2000)	07518-1354	08468+0625	10427+0335	13100+1732
P (yr)	23.34	15.0507	80.56	25.804
	±0.17	±0.0064	±0.30	±0.055
a ($''$)	0.573	0.2547	0.3778	0.6684
	±0.010	±0.0009	±0.0014	±0.0013
i (°)	79.68	50.01	145.92	90.06
	±0.06	±0.27	±0.78	±0.05
Ω (°)	102.5	107.99	56.8	192.34
	±1.6	±0.35	±1.9	±0.24
T_o	1962.381	1991.247	1976.674	1963.468
	±0.039	±0.005	±0.030	±0.021
e	0.735	0.6558	0.546	0.497
	±0.016	±0.0018	±0.001	±0.012
ω (°)	71.4	266.10	355.3	101.08
	±1.6	±0.27	±1.9	±0.24

TABLE 2. Orbital Elements of the CASLEO Objects

Parameter	Bu 1163 [2]	StF 2597 [2]	Stt 535 AB [2]	Ho 296 AB [1]
WDS (α,δ 2000)	01243-0655	19553-0644	21145+1000	22409+1433
P (yr)	16.114	425.	5.6998	20.83
	±0.024	±22.	±0.0023	±0.15
a ($''$)	0.1974	1.085	0.2313	0.2907
	±0.0015	±0.024	±0.0005	±0.0002
i (°)	116.1	103.05	99.57	140.12
	±2.9	±0.84	±0.18	±0.02
Ω (°)	30.38	264.03	203.68	252.37
	±0.70	±0.77	±0.12	±0.23
T_o	1988.860	1974.28	1987.166	1983.557
	±0.014	±0.23	±0.010	±0.004
e	0.927	0.9414	0.4386	0.738
	±0.012	±0.0020	±0.0027	±0.001
ω (°)	350.5	327.9	7.02	23.28
	±1.2	±1.7	±0.64	±0.23

[1] Orbital elements from Hartkopf, McAlister & Franz (1989).
[2] Orbital elements from Hartkopf, Mason & McAlister (1996).

(FWHM). When the seeing is better than about 2 arc seconds, the residuals in both coordinates cluster more tightly together than the data with seeing worse than 2 arc seconds, indicating better precision when the seeing is good. There is one data point in the position angle plot which has seeing of about 1.3 arc seconds but a large residual of about $-3°$. This is an observation of Sp 1 AB, which at the epoch of observation had a separation of 0.271 arc seconds, fairly close to the diffraction limit of the 76-cm telescope. For such a small separation, a small deviation in the x- or y-coordinate leads to a substantial difference in position angle. In addition, the magnitude difference is 1.5, which makes the measure additionally challenging near the diffraction limit. The same residual data are plotted in Figures 4 and 5 in rectilinear coordinates (right ascension and declination residuals).

Although the data are few, we can make some preliminary statements about absolute scale calibration at the 76-cm telescope. If we confine the discussion to the data taken under seeing conditions better than 2 arc seconds, there seems to be no reason to suspect a serious problem in the zero point of our position angle calibration; the average of the θ-residuals differs from zero only by 0.1°. The average of the separation residuals is -5.6 milliarcseconds (mas), which is potentially significant. We will continue to monitor this both with future scale calibrations and binary observations.

We can also compute the standard deviations (sigmas) of position angle, separation, x and y residuals, to get an idea of the measuring precision so far on the project. These are shown in Table 3 along with the results discussed in the previous paragraph. Our position angle residuals have a sigma of about 1.2°, and the separation residuals appear at this point to have a sigma of about 7.5 milliarcseconds (mas). These numbers can be compared with those quoted by the CHARA group for their observations at the 4-m telescope at Kitt Peak, which are 1.01° in position angle and 3.5 mas in separation (Hartkopf, McAlister & Franz 1989). There are probably two reasons for the difference in precision between the two programs: 1) the size of the telescope aperture (4-m for CHARA versus 76-cm for the work here), and 2) our current detector has lower quantum efficiency than most other speckle cameras as discussed in the previous section.

Since there are only six observations from the 2.1-m telescope in this study, it is too soon to say more than there appears to be no improvement in the measuring precision at the larger telescope, as evident from Figures 2 through 5. In fact, it may be that it is worse, for reasons that are not yet entirely clear. We will continue to take more data of these and other binaries with well-determined orbits to get a better picture of the measuring precision at both telescopes.

4. Work on Systematic Errors

There are three components that contribute to the residuals discussed in the previous section. First, there are random or accidental errors, whose distribution is controlled by photon statistics and the speckle process; there

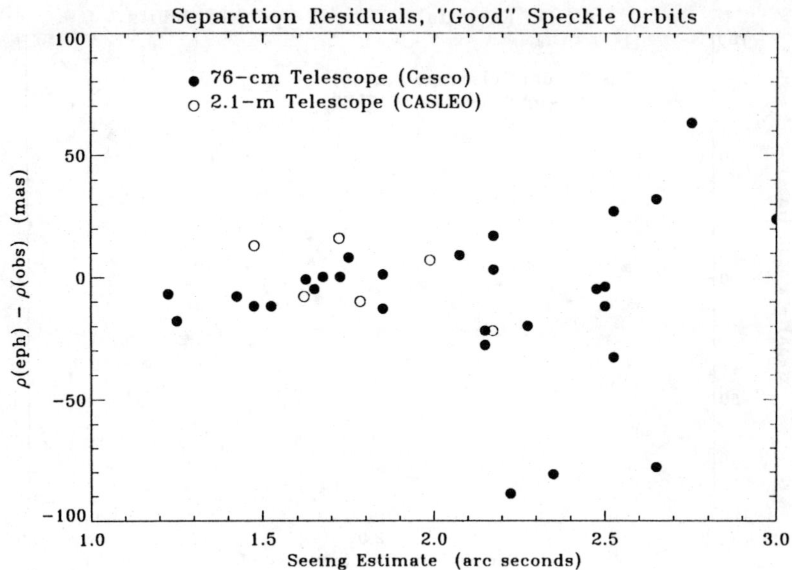

Figure 2. Separation residuals as a function of seeing FWHM.

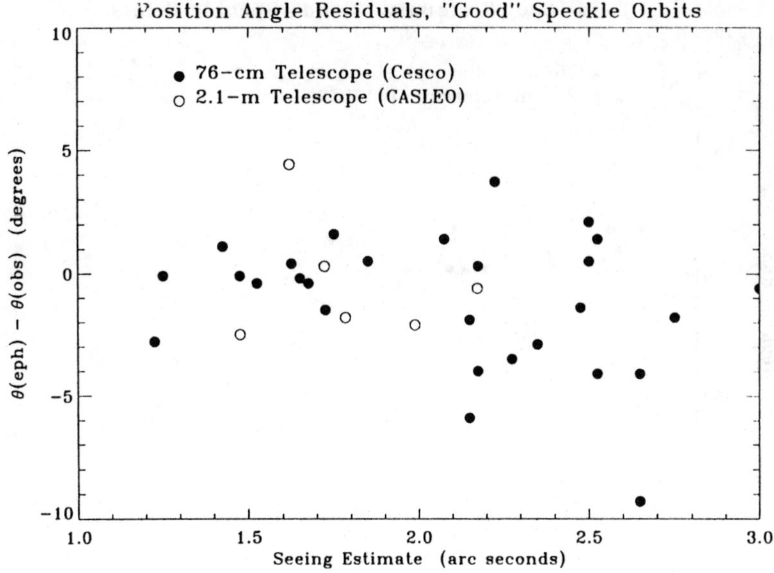

Figure 3. Position angle residuals as a function of seeing FWHM.

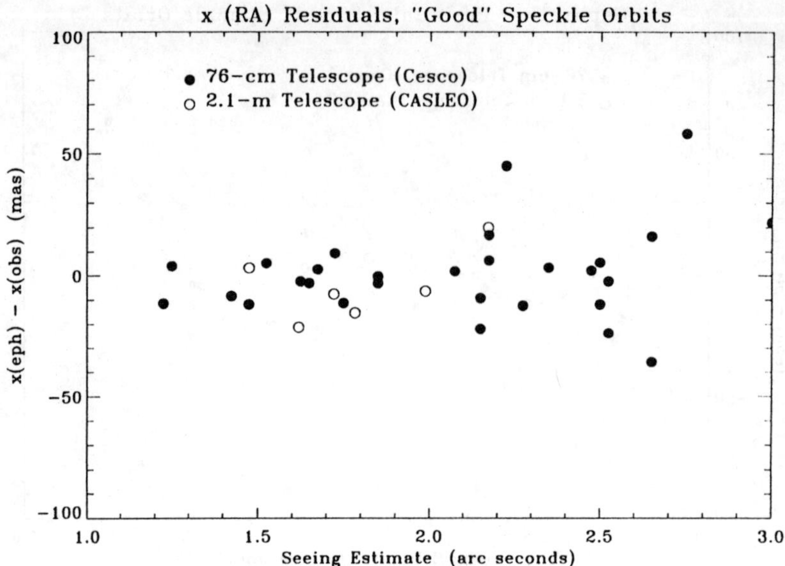

Figure 4. Right ascension residuals as a function of seeing FWHM.

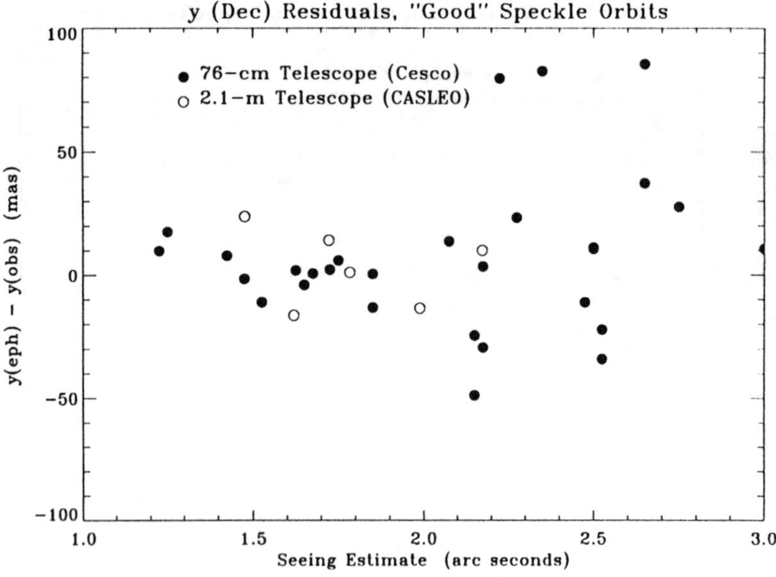

Figure 5. Declination residuals as a function of seeing FWHM.

TABLE 3. Summary of Residuals, 76-cm Telescope (Cesco)

Seeing $\leq 2''.0$	
$\overline{\Delta\rho} = -5.6 \pm 2.2$ mas	$\sigma_\rho = 7.5 \pm 1.5$ mas
$\overline{\Delta\theta} = -0.12° \pm 0.33°$	$\sigma_\theta = 1.16° \pm 0.24°$
$\overline{\Delta x} = -2.7 \pm 2.0$ mas	$\sigma_x = 7.1 \pm 1.4$ mas
$\overline{\Delta y} = +1.3 \pm 2.5$ mas	$\sigma_y = 8.6 \pm 1.8$ mas

are systematic errors induced by the telescope and camera system or the analysis routines; finally there are errors in the orbits themselves, although we have selected the objects so that these errors are as small as possible. The analysis presented in the previous section does not correct for systematic errors. We are continuing to examine the possibility of systematic error generated by change of focus, wavelength of observation, and optical field angle distortion. So far, we have found no evidence for any measurable change of scale as a function of telescope focus or wavelength, but we have found evidence for a small amount of field angle distortion caused by the magnification element in the speckle camera.

Most speckle cameras have a very small field of view, a few arc seconds on a side at most, but with our speckle camera we have a much larger field of view, about 15 × 60 arc seconds at the 76-cm telescope. Even with this large field, we oversample the diffraction-limited point spread function of the telescope by a factor of about 2.7. The large size is also convenient for acquiring targets since we do not have a flexure or atmospheric dispersion model for making corrections to the positions of stars at the 76-cm telescope. We try with each observation to place the star in approximately the same location on the detector, but with such a large field, the position can easily vary by a couple of arc seconds. This makes us especially susceptible to field angle distortion.

We have been able to begin to map out the distortion in our camera both with drift scan star trails and the aperture mask. When using the aperture mask, the slits create a diffraction pattern on the image plane. By moving the pattern around over the full field of view and measuring the change in the derived scale, we can measure the distortion. This work is in progress. The star trails also provide information about distortion, because if distortion exists, the trails will not be straight lines and/or the velocity of the star will not appear to be constant across the detector. Figure 6 shows a long-axis residual plot of many star trails after a best fit line has been subtracted from each trail. The systematic deviation from zero is an indication of distortion. The level of the distortion appears to be small compared to the typical accidental errors of our double star observations,

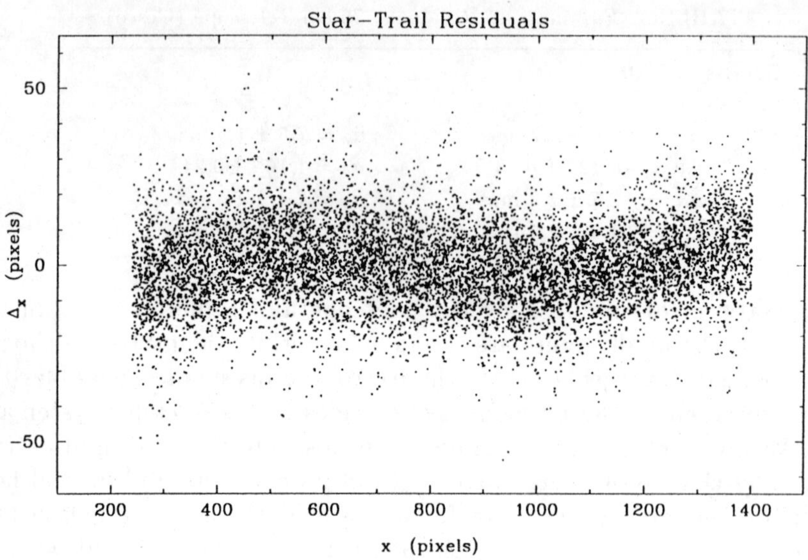

Figure 6. Star trail long-axis residuals.

but nonetheless we are currently trying to model and remove it from our double star measures. We hope that once we do this, the standard deviation values in Table 3 will decrease a little.

5. Conclusions

The Yale-San Juan speckle interferometry program has taken 1500 speckle observations of double stars in the last two years. In our initial measurement precision study, it appears that for data taken under seeing conditions better than 2 arc seconds, the standard deviation in separation residuals is about 7.5 mas, and the standard deviation in position angles is about 1.2° when we compare our measures from the 76-cm telescope to ephemeris positions of binaries with very well-determined orbits.

We would like to thank J. G. Timothy for the loan of the MAMA detector system, and CASLEO for granting us observing time. This work is funded in part by NSF grant AST93-12148.

References

Bonneau, D., Blazit, A., Foy, R. & Labeyrie, A. 1980, A&AS, 42 185
Hartkopf, W. I., McAlister, H. A. & Franz, O. G. 1989, AJ, 98, 1014

Hartkopf, W. I., Mason, B. D., Barry, D. J., McAlister, H. A., Bagnuolo, W. G. & Prieto, C. M. 1993, AJ, 106, 352

Hartkopf, W. I., Mason, B. D. & McAlister, H. A. 1996, AJ, 111, 370

Hartkopf, W. I., Mason, B. D., McAlister, H. A., Turner, N. H., Barry, D. J., Franz, O. G. & Prieto, C. M. 1996, AJ, 111, 936

Horch, E. P., Dinescu, D. I., Girard, T. M., van Altena, W. F., López, C. E. & Franz, O. G. 1996, AJ, 111, 1681

McAlister, H. A., Hartkopf, W. I. & Franz, O. G. 1990, AJ, 99, 965

McAlister, H. A., these proceedings

Morgan, B. L., Beddoes, D. R., Scaddan, R. J. & Dainty, J. C. 1978, MNRAS, 183, 701

van den Bos, W. H. 1961, Union Obs. Circ. 6, 377

Worley, C. E. & Douglass, G. G. 1984, Washington Visual Double Star Catalog (U.S. Naval Observatory, Washington, D.C.)

Worley, C. E. & Heintz, W. D. 1983, Fourth Catalog of Orbits of Visual Binary Stars, Pub. USNO, 24, part 7

IMAGES OF MIZAR A FROM LONG BASELINE OPTICAL INTERFEROMETRY

C.A. HUMMEL

Universities Space Research Association
NRL/USNO Optical Interferometer Project
c/o US Naval Observatory - AD5
3450 Massachusetts Avenue NW
Washington DC 20392, USA

1. Introduction

In long baseline interferometry, a single large aperture is replaced by a comparatively small selection of its subapertures, which is physically realized as an array of independent telescopes or siderostats. For each subaperture pair (baseline), the mutual coherence, or fringe visibility, of the two light beams is measured by correlation or superposition, giving a component of the Fourier transform of the object brightness distribution (van Cittert-Zernike theorem). Thus, by measuring amplitude and phase of the visibility over as many baselines as possible, the image can be reconstructed.

A long baseline interferometer is an adaptive optics system, since the wavefronts received at each subaperture have to be aligned before combination. Once this is accomplished, the array can be spread out over a large area, giving much higher angular resolution capabilities than a single aperture with limited extent. Due to the diluted aperture, however, the dynamic range of the map and its fidelity to complex structure are limited.

Optical long baseline interferometry is following in the footsteps of radio interferometry on very long baselines (VLBI), a technique which has become well established over the last three decades. We report on recent successes in imaging the spectroscopic binary Mizar A with the Navy Prototype Optical Interferometer (NPOI).

J. A. Docobo et al. (eds.), Visual Double Stars: Formation, Dynamics and Evolutionary Tracks, 55–61.

2. The Navy Prototype Optical Interferometer

The NPOI is a joint project of the U.S. Naval Research Laboratory and the U.S. Naval Observatory, in association with the Lowell Observatory. It is under construction on Anderson Mesa, near Flagstaff, Arizona. The imaging array features six siderostats (35 cm clear aperture) which are operated simultaneously. They can be mounted on any one of an array of concrete piers in a Y configuration. Each arm of the Y will eventually be 250 m in length, providing baseline lengths of up to 437 m. The maximum resolution will be about 200 microarcseconds, while the faintest stars observable will be of about 8th magnitude. For a more detailed description of the NPOI with respect to binary star research, see Armstrong (1996).

The observations reported here were made with the siderostats of the astrometric array, with baselines up to 38 m. (Astrometric siderostats feature a complex laser metrology system to monitor motions of the pivot point. They are used for the determination of stellar positions with milliarcsecond precision. A more complete description of the astrometric array is given by Hutter [1995].)

The beam of light from any siderostat travels through evacuated pipes into a laboratory, which houses the delay lines and the beam combiner. The former are used to compensate for the geometric path length differences between the beams; the latter is used to form interference patterns for any combination of incoming beams. The interference patterns are dispersed using prisms and then focused onto 32-channel optical fiber arrays, which are connected to avalanche photodiodes for fringe detection. The latter is achieved through modulation of the delay with a triangle wave and synchronous detection of the intensity fluctuations in the detectors. The wavelength range is 440 nm to 850 nm.

The fringe detection algorithm determines the residual group delay of the fringe packet, which is then sent to a servo which controls a small mirror mounted on a piezoelectric stack on the cart in the delay line tank. This servo aims to track the rapid movements of the fringe packet due to atmospheric turbulence. At the same time, some light diverted by a beamsplitter from the beam is focused on a quad cell detector in order to determine the offset of the stellar image from the center of the field. This signal is sent to a servo controlling another mirror mounted on piezoelectric stacks next to the siderostat in order to track the rapid movements of the star. Fringe and star tracking thus comprise all elements of an adaptive optics system for subapertures not larger than the seeing cell size (about 10 cm to 20 cm in the optical).

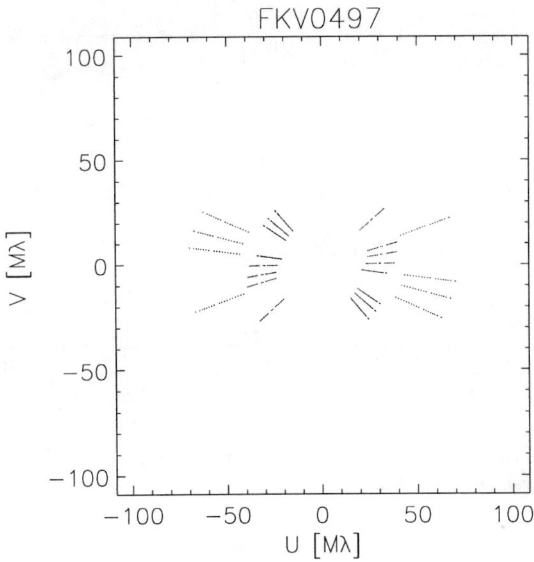

Figure 1. Aperture coverage for the binary Mizar A (the number 497 in the FK5) on May 1st, 1996.

3. Observations

We observed Mizar A (ζ^1 Ursae Majoris) seven times during May and June of 1996. Between three and four scans of about two minutes duration (i.e. on-source integrations with the entire array) were recorded each night. We used the center (C), east (E), and west (W) siderostats of the astrometric array, giving three baselines (EW: 38 m, CE: 18 m, CW: 22 m) simultaneously. In Fig.1, we show the achieved aperture coverage for the night of May 1st as a typical example. Each point in the aperture coverage plot corresponds to a visibility measurement for a specific time, baseline, and channel. A point's position (coordinates u and v) relative to the origin corresponds to the baseline orientation and projected length, measured in units of the wavelength of the channel. (For each point with visibility V, there is one on the opposite side relative to the origin with the visibility equal to the complex conjugate of V.) On May 1st, four scans were obtained at different times, so that earth rotation would change the orientation of the baselines. In the plot, each radial line corresponds to a scan on a baseline, with the uv-points spreading out along a radial line due to the broad bandwidth.

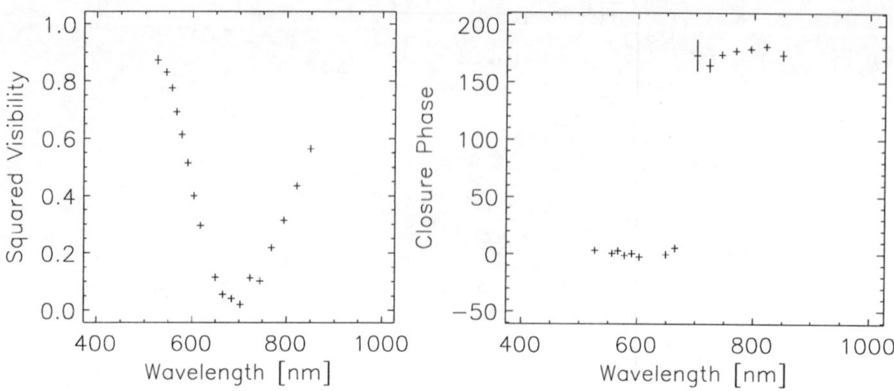

Figure 2. Squared visiblity amplitude and closure phase on the EW baseline vs wavelength. Data is the first scan on May 1st.

4. Data reduction

The calibration of the observed average visibility amplitudes and closure phases is based on unresolved single stars. For these sources, the amplitude must equal unity and the closure phase zero. The deviation of the measured (uncalibrated) visibilities from the expected values can be written as a function of atmospheric and instrumental variables, the coefficients of which can be determined from the calibration scans and then applied to the data of the program sources. For the maps presented here, a single calibration factor for the visibility amplitudes was derived for each channel, night, and baseline independently; more sophisticated methods are being developed currently, based on procedures used for the Mark III interferometer on Mt Wilson, California (Mozurkewich *et al.* 1991). For the closure phases, a quadratic polynomial with wavelength was fitted to the calibrator data. In Fig.2 we show squared visibility amplitude of the first scan of May 1st on the EW baseline, as a function of wavelength. The quasi-sinusoidal variation of the amplitude with wavelength (baseline length) shows that the binary is clearly resolved on this baseline.

The closure phase for the same scan is also plotted as a function of wavelength in Fig.2. The jump in phase from zero degrees to 180 degrees takes place around 700 nm, where the squared visibility amplitude has a minimum close to zero. The 180 degree change in phase indicates that the two components of the binary are of equal brightness.

5. Imaging

The complex-valued (amplitude and phase) visibility function is the Fourier transform of the source's brightness distribution. The original image, therefore, can be obtained by back-transformation. However, phases in ground-based interferometry are severely disturbed due to propagation effects in the turbulent atmosphere. Because the phase noise is additive at each telescope and the visibility phases are phase differences between telescopes, the phase noise cancels when the visibility phases are added up along a closed loop in the aperture plane. This closure phase, the sum of the visibility phases around a loop, is thus free of atmospheric phase noise, and is equal to the intrinsic source closure phase.

The fraction of phase information lost by forming closure phases is $(n-2)/n$, where n is the number of subapertures. Imaging software developed for VLBI employs phase self-calibration methods in which the current image itself is used to complete the phase information with an iterative procedure in a way that converges toward the true image.

To make use of existing software, the NPOI calibrated visibility data of the binary were written into a FITS file, a format which can be read by the AIPS software for imaging of aperture synthesis data. The squareroot was taken of the squared visibility amplitude data, and the closure phase data were arbitrarily assigned to the phase of the EW baseline, the phase on the other baselines being set to zero. This procedure preserves the closure phase, while the standard phase self-calibration algorithm developed for the imaging of interferometric data affected by atmospheric phase noise would redistribute the phase among the three baselines in the course of mapping. Due to the negligible color difference between the two components in Mizar A, we were able to combine the data of all channels without corrections being made to the source structure as a function of wavelength. (The invalidity of this assumption in most cases, as well as significant orbital motion during the observations of the short-period systems, will make imaging these sources a challenge.) We show the resulting map for the May 1st observations in Fig.3, and a map for May 29th in Fig.4.

We have produced seven maps of the binary for May 1-4, May 29th, June 1st and 4th. We fitted Gaussian components to the maps to determine the relative positions. These are plotted with an orbit published for Mizar A by Hummel *et al.* (1995) in Fig.5. The size of the uncertainty ellipses corresponds to one-fifth of the synthesized beam FWHM, a standard criterion used in VLBI. The agreement with the Mark III orbit is good, but shows slight systematic differences which could be used to improve the orbital solution.

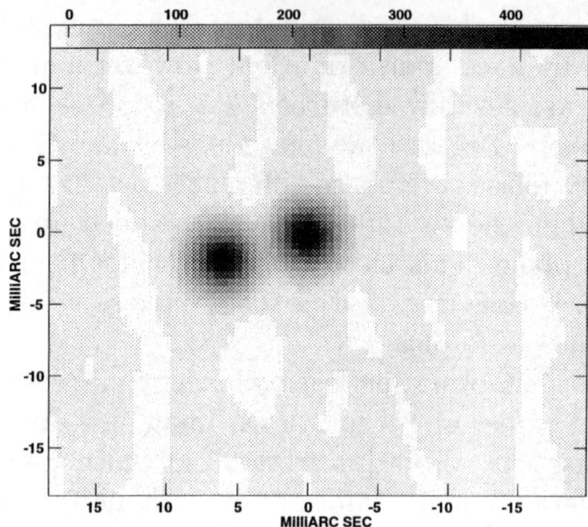

Figure 3. Image of Mizar A on May 1st, 1996. The flux density scale given at the top is in units of thousandths of the total flux density.

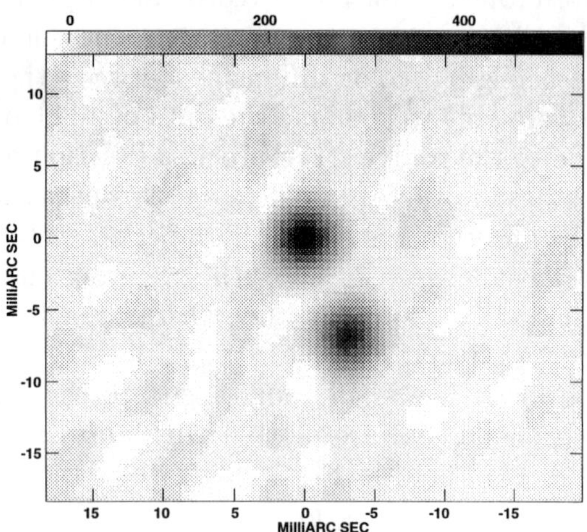

Figure 4. Image of Mizar A on May 29th, 1996.

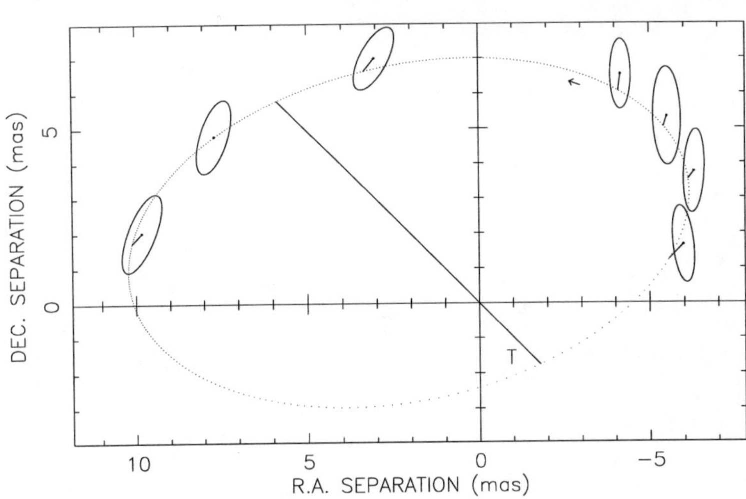

Figure 5. Binary positions (and uncertainty ellipses) from fits to images made with the NPOI of Mizar A for seven epochs superposed on the orbit by Hummel *et al.* (1995) from observations with the Mark III optical interferometer. The measured positions are connected with lines to the predicted model positions. The line marked with a T (the periastron) is the major axis of the orbit. The spacing of the dots corresponds to $P/360$, where $P = 20.54$d. A little arrow indicates the sense of the orbital motion.

Acknowledgements

The author acknowledges the great efforts of his colleagues in the Optical Interferometer Project in building the Optical Interferometer and performing the observations.

References

Armstrong, J.T. (1996), in Proceedings of the 3rd Pacific Rim Conference on Recent Developments in Binary Star Research, ed. K. Leung and B. Soonthornthum (ASP Conference Series, in press)

Hummel, C.A., Armstrong, J.T., Buscher, D.F., Mozurkewich, D., Quirrenbach, A., Vivekanand, M. (1995), *Astronomical Journal*, **110**, 376

Hutter, D.J., Johnston K.J., and Mozurkewich, D. (1995) in IAU Symposium 166, Astronomical and Astrophysical Objectives of Sub-Milliarcsecond Optical Astrometry, ed. E. Høg and P.K. Seidelmann (Kluwer, Dordrecht), p. 23

Mozurkewich, D., Johnston, K.J., Simon, R.S., Bowers, P.F., Gaume, R., Hutter, D.J., Colavita, M.M., Shao, M., and Pan, X.P. (1991), *Astronomical Journal*, **101**, 2207

OBSERVATIONS OF BE AND B BINARY STARS WITH THE SPECKLE SPECTROSCOPIC METHOD

N. BABA
Department of Applied Physics, Hokkaido University
Sapporo 060, Japan

S. KUWAMURA
Institute of Physical and Chemical Research, RIKEN
Wako 351-01, Japan

Y. NORIMOTO
Okayama Astrophysical Observatory, NAO
Kamogata 719-02, Japan

R. HIRATA
Department of Astronomy, Kyoto University
Kyoto 606-01, Japan

S. ISOBE
National Astronomical Observatory
Mitaka 181, Japan

AND

S. CUEVAS AND A. RUELAS-MAYORGA
Instituto de Astronomia, UNAM
70-264 México, D.F., México

Abstract. Speckle spectroscopy is a method to obtain objective spectra of stellar objects with high spatial resolution. We conducted speckle spectroscopic observations of binary stars at the Okayama Astrophysical Observatory in Japan and at the San Pedro Mártir Observatory in Mexico. Spectra of primary and secondary stars are spatially separated and the spectral distribution around Hα line is examined. We report observational results for some Be and B binary stars.

1. Introduction

Atmospheric turbulence hampers high resolution imaging of stellar objects from the ground. Stellar speckle interferometry (Labeyrie 1970) was proposed to get image information with diffraction-limited spatial resolution of a telescope and has been intensively used for binary star observations (McAlister and Hartkopf 1988; McAlister 1996). Weigelt (1981) and Beckers (1982) applied the speckle technique to high-spatial resolution spectroscopy.

J. A. Docobo et al. (eds.), Visual Double Stars: Formation, Dynamics and Evolutionary Tracks, 63–71.

Speckle spectroscopy is a spectroscopic method to obtain the objective spectrum of a stellar object with high-spatial resolution. Baba *et al.* (1988) proposed a speckle spectroscopic method based on the shift-and-add method (Bates and Cady 1980) for wideband spectroscopy (Kuwamura *et al.* 1993a, 1993b). In our method a speckle image and its dispersed specklegram of a stellar object are simultaneously detected, and then the object and the spectrum are reconstructed with high spatial resolution from the ensemble of the data frames. We currently use a cross-correlation method (Hege 1989) for narrow band speckle spectroscopy (Kuwamura *et al.* 1993c).

Application of speckle spectroscopy to binary star observations enables us to examine each spectrum of the primary and the secondary stars. Since speckle spectroscopy separates spatially each spectrum of a binary star, it is unnecessary to suppose the spectral type *etc.* in the reconstruction process. Binary stars with angular separation greater than not seeing-limited but diffraction-limited resolution of a telescope are targets of our observations.

Baba *et al.* (1994a, 1996) reported preliminary results of speckle spectroscopic observations of Be binary stars. In this paper observational results of Be and B binary stars conducted in 1995 are described.

2. Observations

We conducted speckle spectroscopic observations of Be and B binary stars at the San Pedro Mártir Observatory in Mexico and at the Okayama Astrophysical Observatory in Japan. The 2.12 m and the 1.88 m telescopes were used in these observations at SPM and OAO, respectively. Two separate but rather similar speckle spectroscopic cameras were used in these observations, and their schematic illustration is shown in Fig. 1.

A beam from a telescope is collimated by lens L1 and divided into two beams by beamsplitter BS. The transmitted beam passes through an interference filter F and is focused by lens L2. A speckle image at a specified wavelength of F is detected by an intensified CCD-TV camera, ICCD-TV1. The beam reflected by the BS is dispersed by a holographic blazed reflection grating (600 grooves/mm). The dispersed specklegram is focused by lens L3 and detected by ICCD-TV2. The focal length of lens L3 is the same as L2.

In order to capture a speckle image and its dispersed specklegram at the same instant, two ICCD-TV cameras are synchronized by a video controller, which also synthesizes two video frames. The data frames are recorded on S-VHS videotape with a normal video rate of 1/30 s. An interlacing operation at the data reduction stage makes the effective exposure time 1/60 s.

Our speckle spectroscopic camera is mounted to the Cassegrain rotator of a telescope. The position angle of our camera is usually fixed so as to align two stars perpendicular to the dispersion direction. A data frame of ϕAnd (see Sect. 4.2) is shown in Fig. 2. A speckle image at $\lambda = 643$ nm ($\delta\lambda = 13$ nm) is shown at the bottom and its dispersed specklegram at the top. Two video frames, namely a speckle image and its dispersed specklegram are synthesized to one data frame as shown in Fig. 2. The spectral region covers nearly from 630 nm (right-hand side in Fig. 2) to 690 nm (left-hand side).

3. Data reduction

A simple and fast method for image and spectral reconstruction is based on the shift-and-add (SAA) technique. From the SAA image reconstruction the angular separation,

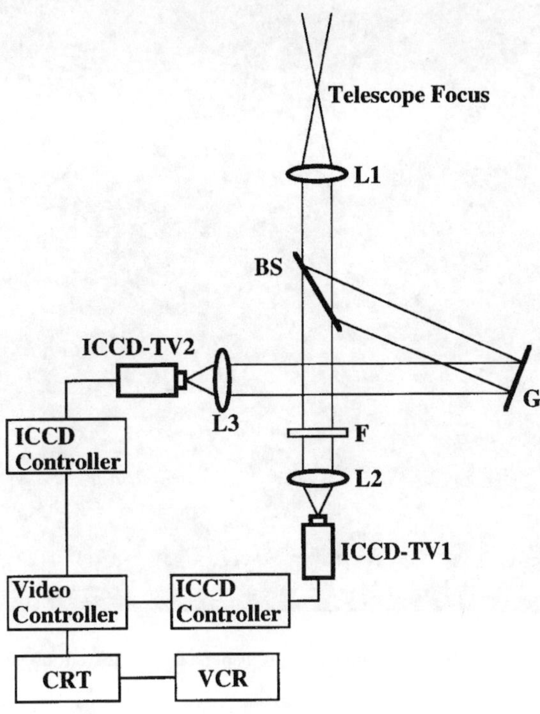

Figure 1. Speckle spectroscopic camera

position angle and intensity ratio of a binary star can be estimated. The reconstructed spectra by SAA shows characteristic features of each spectrum of a binary star.

A more sophisticated reconstruction of spectra is conducted by the cross-correlation method. Our final spectral results are obtained through cross-correlation reduction.

3.1. SHIFT-AND-ADD OPERATION

The SAA is mainly used to reconstruct an image of a binary star. The brightest speckle pair in a speckle image is found and then the whole frame, which contains the speckle image and its dispersed specklegram, is shifted so as to make the brightest speckle pair in the speckle image center on a reconstruction frame. The brightest speckle pair in each specklegram is found by matched filtering (Ribak 1986). A suitable matched filter for a binary star is iteratively constructed. Because a speckle image and its dispersed specklegram are detected at the same instant and there is the one-to-one positional relationship between the speckle image and the dispersed specklegram, the spectra of the binary star are also reconstructed. Although the SAA operation results in a large background component, the subtraction of the background component from the SAA reconstruction can be easily done (Kuwamura *et al.* 1993b).

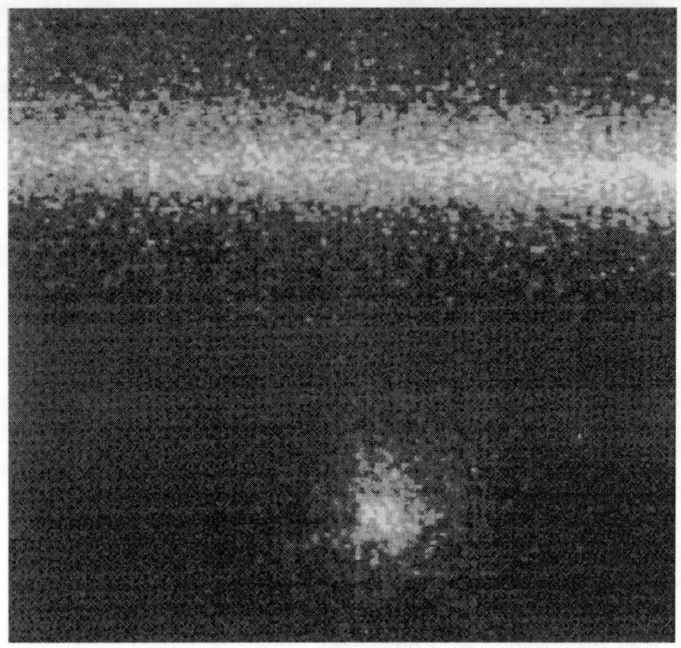

Figure 2. Speckle image and its dispersed specklegram of ϕAnd

For example, Fig. 3 shows the SAA reconstructed image and the accompaningly resulted spectra of ϕAnd from 2100 data frames. The SAA reconstructed spectra are very useful for the initial inspection of the spectra. ¿From Fig. 3 one can readily notice that the reconstructed spectrum of the primary star exhibits an Hα emission line.

3.2. CROSS-CORRELATION METHOD

A speckle image of a binary star contains information about the instantaneous point-spread function (PSF) of telescope-atmospheric system. Therefore, it is possible to deconvolve the dispersed specklegram with the PSF.

A speckle image can be written as

$$i(x, y; \lambda_0) = o(x, y; \lambda_0) * h(x, y; \lambda_0), \tag{1}$$

where $o(x, y; \lambda_0)$ is the intensity distribution of an object at wavelength λ_0, $h(x, y; \lambda_0)$ is the PSF and $*$ denotes convolution. On the other hand the dispersed specklegram at the same instant is written as

$$d(x, y; \lambda) = s(x, y; \lambda) * h(x, y; \lambda), \tag{2}$$

where $s(x, y; \lambda)$ is the spectral distribution of an object. The PSF, $h(x, y; \lambda)$ varies with λ, but in case of narrowband spectroscopy around λ_0, it can be assumed that $h(x, y; \lambda) \simeq h(x, y; \lambda_0)$. The Fourier transforms of the speckle image and its dispersed specklegram

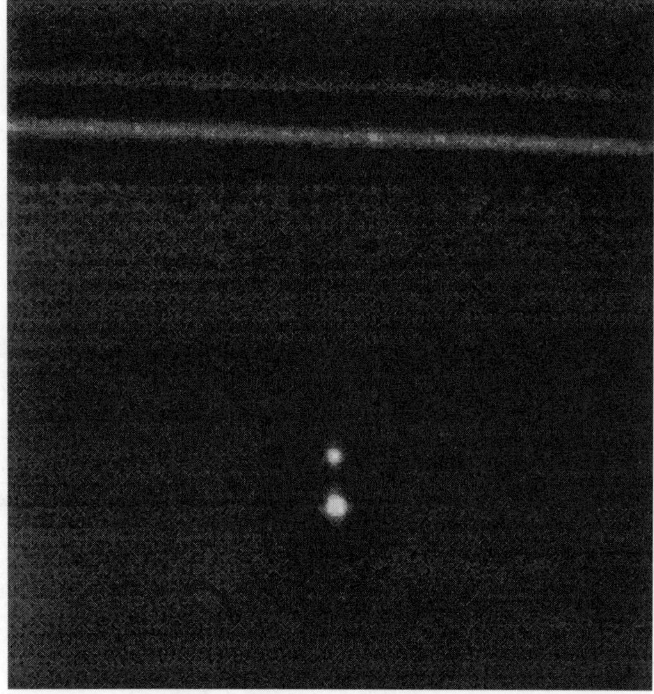

Figure 3. SAA reconstruction of ϕAnd

with respect to spatial coordinates are

$$I(u, v; \lambda_0) = O(u, v; \lambda_0)H(u, v; \lambda_0), \tag{3}$$

and

$$D(u, v; \lambda) = S(u, v; \lambda)H(u, v; \lambda_0), \tag{4}$$

where uppercase letters denote the 2-D Fourier transforms of the corresponding lowercase letter functions. From Eqs. (3) and (4), the spectrum of the object is reconstructed as

$$s(x, y; \lambda) = FT^{-1}\left\{\frac{< DI^* >}{< |I|^2 >}O\right\}, \tag{5}$$

where FT^{-1} means to take inverse Fourier transformation, the angle brackets denote the ensemble average over data frames and the asterisk indicates a complex conjugate. The basic operation in the above equation is to take the cross-correlation between a speckle image and its dispersed specklegram. In the spectral reconstruction step it is necessary to input the Fourier transform of the object distribution. In the case of a binary star this information is derived from the three parameters and the SAA reconstructed image can supply them.

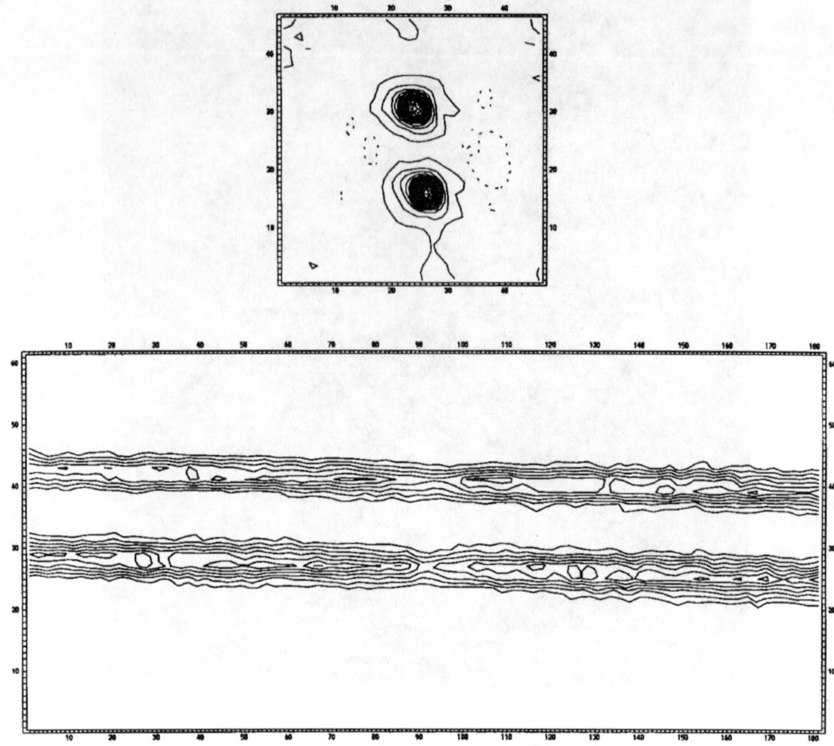

Figure 4. Reconstructed image and spectra of λCas

4. Results

The following are SAA reconstructed images and cross-correlation results on the spectra of λCas, ϕAnd and λCyg. λCas is catalogued as a B star and the others as Be stars in the Bright Star Catalogue (Hoffleit 1982).

4.1. λ CAS

The data frames of this star were recorded at the Okayama Astrophysical Observatory on December 9, 1995. In this reconstruction 2089 data frames were used. The speckle images were taken with an interference filter of $\lambda = 643$ nm ($\Delta\lambda = 13$ nm). The angular separation between the primary and the secondary stars (at the bottom and the top in the reconstructed image in Fig. 4, respectively) is $0.42''$. The spectra range from 635 nm (right) to 675 nm (left). As can be seen both spectra exhibit Hα absorption lines. ¿From the other spectroscopic observations it is known that the spectrum of this star exhibits broad lines. It will be interesting to examine whether one or both of the stars have broad lines.

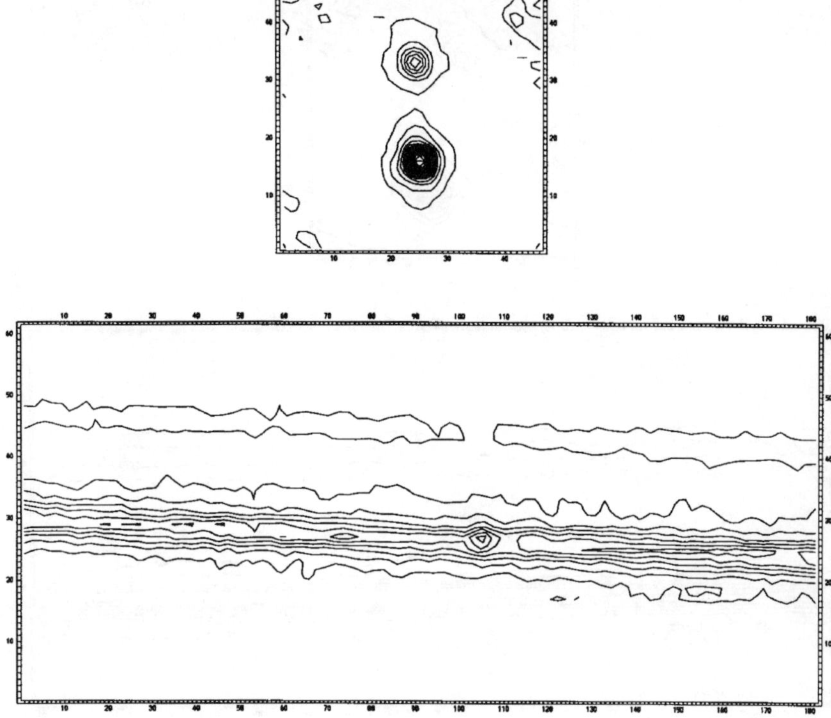

Figure 5. Reconstructed image and spectra of ϕAnd

4.2. ϕ AND

The data frames were obtained at the Okayama Astrophysical Observatory on December 10, 1995. We used 2100 frames in the data reduction. ¿From the SAA reconstructed image (λ = 643 nm, $\Delta\lambda$ = 13 nm) as shown in Fig. 3, the angular separation is measured as 0.53″. The spectra in Fig. 5 range from 632 nm (right) to 672 nm (left). It is noticed that the primary star shows an Hα emission line while the secondary has an Hα absorption line. This result confirms our earlier one (Baba *et al.* 1994a). Balega showed a result of their speckle spectroscopic observations of ϕAnd after his talk in this Workshop and their result was very similar to ours.

4.3. λ CYG

The data frames were collected at the San Pedro Mártir Observatory on August 11, 1995. These data were obtained with the 2.1m telescope before the introduction of the active support system. We reconstructed the image at λ = 643 nm ($\Delta\lambda$ = 13 nm) and the spectra from 2120 frames as shown in Fig. 6. The measured angular separation is 0.89″. The spectral range for the primary star is from 637 nm (right) to 677 nm (left). From the contour plot of the spectra both stars have Hα absorption lines. It should be noticed

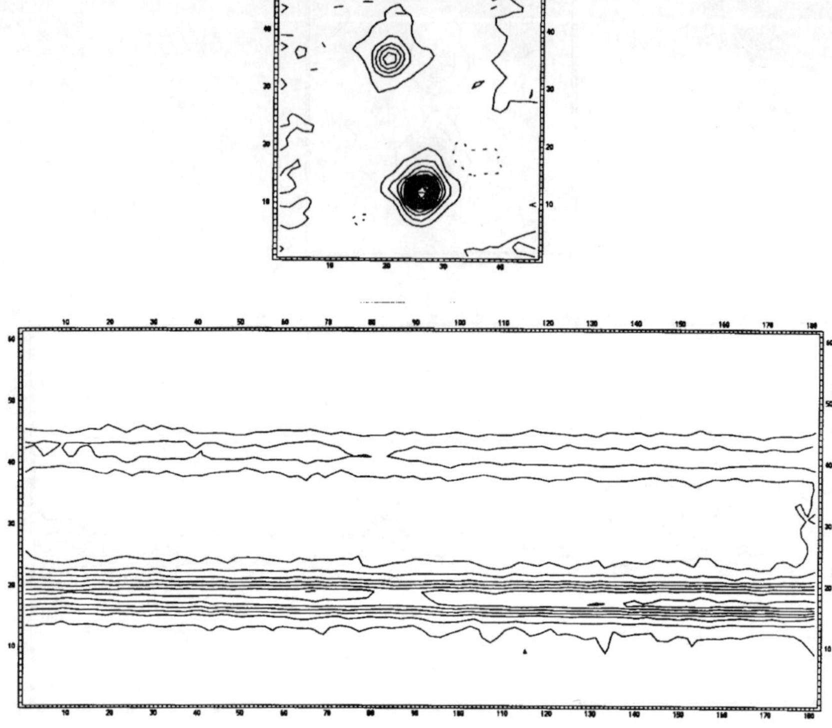

Figure 6. Reconstructed image and spectra of λCyg

that the positions of the Hα lines in the primary and the secondary stars are shifted in accordance with the disposition of the image. Our earlier result on this star (Baba *et al.* 1996) showed that the Hα absorption line of the primary star was nearly filled-in. Further monitoring observations of this star will be necessary. McAlister and Fekel (1980) found that the primary star is also binary. Therefore, the spectrum of the primary star reconstructed here is blended spectra of two stars. Observations with a larger telescope than 3 m will be able to separate the spectra of the primary star.

5. Conclusion

We examined one B binary star (λCas) and two Be binary stars (ϕAnd and λCyg) with the speckle spectroscopic method. To exemplify the usefulnes of speckle spectroscopy the spectrum of each star around Hα line was shown. It will be important to analyze each component spectrum in binary and multiple stars to investigate characteristics of such stars.

Adaptive optics is a very efficient technique for high-spatial resolution imaging and spectroscopy from the ground. Contrary to speckle interferometry, adaptive optics enables us long exposure observations and is usable for faint objects. However, at the moment, perfect wavefront correction seems hard with large telescopes in the visible region. If the adaptive optics system would not be perfect, spectra of stellar objects would be mixed up and distinction of each spectrum of binary or multiple stars would be difficult. On the other hand, speckle spectroscopy can be also applied to 8-10m telescopes in the visible region.

References

Baba, N., Kuwamura, S., Miura, N. and Norimoto, Y. (1994a) Imaging Spectroscopy with High-Spatial Resolution, *ApJL*, **431**, L111-L114

Baba, N., Kuwamura, S. and Norimoto, Y. (1994b) Stellar Speckle Camera for Spectroscopy, *Appl. Opt.*, **33**, 6662-6666

Baba, N., Kuwamura, S., Norimoto, Y. and Cuevas, S. (1996) Speckle Spectroscopic Observations of Be Binary Stars, in *Recent Development of Binary Star Researches*, Leung, K., ed. (Astron. Soc. Pacific, San Francisco, in press)

Baba, N., Tabata, M. and Murata, K. (1988) Wideband Speckle Spectroscopy Based on the Shift-And-Add Method, *Opt. Lett.*, **13**, 616-618

Bates, R. H. T. and Cady, F. M. (1980) Toward True Imaging by Wideband Speckle Interferometry, *Opt. Commu.*, **32**, 365-369

Beckers, J. M. (1982) Differential Speckle Interferometry, *Opt. Acta*, **29**, 361-362

Hege, E. K. (1989) First Order Imaging Methods: An Introduction, in *Diffraction- Limited Imaging with Very Large Telescopes*, Alloin, D. M. and Mariotti, J. -M., eds. (Kluwer, Dordrecht), pp. 141-155

Hoffleit, D. (1982) *The Bright Star Catalogue, 4th ed.* (Yale Univ. Obs., New Haven)

Kuwamura, S., Baba, N., Miura, N. and Hege, E. K. (1993a) Stellar Spectra Reconstruction from Speckle Spectroscopic Data, *AJ*, **105**, 665-671

Kuwamura, S., Baba, N., Miura, N., Noguchi, M., Norimoto, Y. and Isobe, S. (1993b) Preliminary Observational Results of Wideband Speckle Spectroscopy, *Proc. ESO Conf. on High-Resolution Imaging by Interferometry II*, Beckers, J. M. and Merkle, F., eds. (ESO, Garching), pp. 461-469

Kuwamura, S., Baba, N., Miura, N. and Norimoto, Y. (1993c) Stellar Spectra Reconstruction from Speckle Spectroscopic Data. II., *AJ*, **106**, 2532-2539

Labeyrie, A. (1970) Attainment of Diffraction Limited Resolution in Large Telescope by Fourier Analyzing Speckle Pattern in Star Images, *A&A*, **6**, 85-87

McAlister, H. A. (1996) Twenty Years of Seeing Double, *S&T*, **Vol. 92 no. 5**, pp. 28-35

McAlister, H. A. and Fekel, F. C. (1980) Speckle Interferometric Measurements of Binary Stars. V., *ApJS*, **43**, 327-337

McAlister, H. A. and Hartkopf, W. I. (1988) *Second Catalog of Interferometric Measurement of Binary Stars* (Georgia State University, Atlanta)

Ribak, E. (1986) Astronomical Imaging by Filtered Weighted-Shift-And-Add Technique, *J. Opt. Soc. Am. A*, **3**, 2069-2076

Weigelt, G. (1981) Speckle Interferometry, Speckle Holography, Speckle Spectroscopy, and Reconstruction of High-Resolution Images from Space Telescope, *Proc. ESO Conf. on the Scientific Importance of High Angular Resolution at Infrared and Optical Wavelengths*, Kjaer, K., ed. (ESO, Garching), pp.95-114

SPECKLE MASKING IMAGING OF THE SPECTROSCOPIC BINARIES GLIESE 150.2 AND 41 DRACONIS

I.BALEGA AND Y.BALEGA
Special Astrophysical Observatory
Zelenchukskaya region, Karachai-Cherkesia, 357147 Russia

AND

H.FALCKE, R.OSTERBART, M.SCHÖLLER AND G.WEIGELT
Max-Planck-Institut für Radioastronomie
Auf dem Hügel 69, 53121 Bonn, Germany

1. Introduction

Despite a history of micrometric observations dating back more than 150 years, and many decades of astrometric and interferometric study of low-mass main sequence binary stars in the solar neighborhood, their masses and luminosities are known with much lower accuracies than for massive ($1.5 \leq \mathcal{M}/\mathcal{M}_{\odot} \leq 10$) stars of early spectral types. The latest "conservative" list of 45 binaries with normal components, whose masses and radii are measured with an error $\leq 2\%$, includes only 3 objects of spectral type G, one pair of the type M, and no K-type binaries (Andersen, 1991). Only 3 systems from this list are within 50 pc from the Sun. This is explained by the observation selection effect which confines the list of stars with fundamental masses and luminosities to double-lined eclipsing binaries. The HIPPARCOS astrometric mission will not change the situation significantly because its parallax errors will be too high to satisfy the 2% accuracy criterion even for stars within 25 pc from the Sun. Therefore, we decided to focus our attention on spectroscopic pairs recently discovered by cross-correlation spectroscopy. For nearby solar-type stars the most extensive surveys of radial velocities have been done by Duquennoy and Mayor (1991) and Tokovinin (1988). From Tokovinin's list, a few spectroscopic binaries with fast orbital motion were first resolved by speckle interferometric observations at the 6-

J. A. Docobo et al. (eds.), Visual Double Stars: Formation, Dynamics and Evolutionary Tracks, 73–78.

m telescope of the Special Astrophysical Observatory (SAO) in Zelenchuk. All of them are interesting new candidates for precise mass determination.

Here we present the first results for two binaries included in our observational program in 1993: the nearby (π=49 mas), 7.7 magnitude, K2 dwarf star Gliese 150.2 and the F7V star 41 Dra. The 41 Dra binary is a member of the hierarchical quadruple system ADS 11061AB=40/41 Dra. Both 40 and 41 Dra pairs are SB2 (Tokovinin, 1995). For Gl 150.2 and 41 Dra we present diffraction-limited images obtained by the speckle-masking method. The images allow us to determine not only the geometry of the binaries, but also their exact maginitudes. The 41 Dra system could become one of the most interesting multiples for overall study by means of different methods.

2. Observations and image reconstructions

Gl 150.2 and 41 Dra were first resolved at the 6-m telescope in 1993 using the SAO speckle camera (Balega *et al.*, 1994). Speckle interferograms were recorded through 605/24 nm and 667/20 nm interference filters with 20 ms exposure time under 1-1.5″ seeing. Later, the measurements were continued using the Max-Planck-Institut für Radioastronomie speckle camera (Baier and Weigelt, 1983) which provides higher dynamic range and better astrometric accuracy. The new detector of this camera is a Thomson 512^2 pixels CCD optically coupled to a 3-stage electrostatic image intensifier. The new data were collected through a 656/30 nm red filter, with the magnification 40× corresponding to an image scale of 4.97 mas/pixel. In addition, in 1996 we recorded 41 Dra speckle images in the K-band (filter 2191/411 nm) with a 256^2 pixels NICMOS-3 camera. For infrared observations the exposure time of the speckle interferograms was 150 ms, and the scale in the image plane was 33.2 mas/pixel. A system of Digital Signal Processors was used for real-time speckle-masking computations, so we could see the evolution of the restored image during observing. Simultaneously, the data were recorded on Exabyte streamers for future detailed analysis.

Diffraction-limited images in the visible and in the infrared were reconstructed from speckle interferograms by the speckle-masking method (Weigelt,1977; Lohmann *et al.*, 1983). The following processing steps were applied to the speckle data:

1. Subtraction of the detector average dark current and division by the flatfield for each speckle interferogram;
2. Calculation of the average power spectrum of all speckle data;
3. Subtraction of detector noise bias terms from the average power spectrum;
4. Calculation of the average bispectrum of all speckle interferograms;
5. Subtraction of detector noise bias terms from the average bispectrum;

6. Compensation of the speckle interferometry transfer function in the bias-compensated average power spectrum to obtain the Fourier modulus;
7. Retrieval of Fourier phase from the bias-compensated average bispectrum;
8. Reconstruction of the diffraction-limited image from the object modulus and phase.

The object Fourier phase was reconstructed from the bias-compensated average bispectrum using the phase recursion method (Lohmann *et al.*, 1983). The bispectrum of each frame consisted of $\approx 31.7 \times 10^6$ elements (maximum length of bispectrum vectors: u=50 px, v=95 px = diffraction cut-off frequency). For correct recursive phase reconstruction each bispectrum element was weighted with its SNR.

3. Results

During the observations of Gl 150.2 in 1993 we suspected the existence of three components in the system at the following position angles and separations: $\theta_{AB} = 107°$, $\rho_{AB} = 186$ mas; $\theta_{AC} = 134°$, $\rho_{AC} = 45$ mas. The contrast of the secondary peaks in the correlation was rather low because of the large magnitude difference between the stars in Gl 150.2.

TABLE 1. Interferometric measurements of Gl 150.2 and 41 Dra

Binary	Epoch	$\theta°$	ρ mas	Filter $\lambda/\Delta\lambda$, nm	Δm
Gl 150.2AB	1993.8418	107	186	667/20	2.2
	1994.7130	102.2	249	656/30	2.49
Gl 150.2AC(?)	1993.8418	134	45	667/20	1
41 Dra	1993.3492	320	102	605/24	0.2
	1993.7646	328	93	605/24	0.3
	1993.8437	327	92	605/24	0.3
	1994.7129	296.0	25	656/30	0.38
	1995.7757	316.9	101	850/30	–
	1996.2667	321.9	113	2191/411	0.47

The bispectrum image reconstruction of Gl 150.2AB through the red 656/30 nm filter is shown in Fig.1a. The measurements of the geometry and magnitude difference are reported in Table 1. No evidence for the third star was found in 1994. The upper limit for the separation of suspected companion is about 10 ms for an equal brightness pair. Orbital motion of such a close secondary would cause radial velocity variations with a period of 1 yr and an amplitude of $\sim 10\,\mathrm{km/s}$. Such variations have never been reported for Gl 150.2, so we conclude that the suspected triple structure of the system is spurious. A year after the first resolution, the angular distance

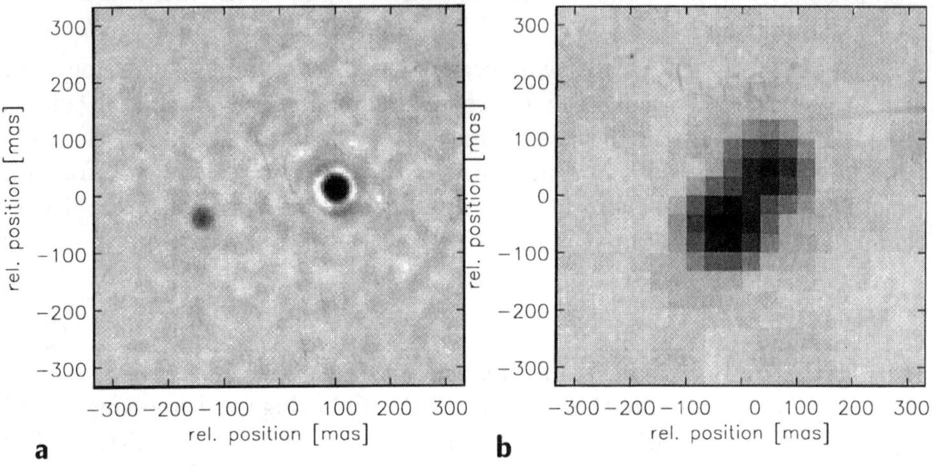

Figure 1. (a) Diffraction-limited 656/30 nm image of spectroscopic binary Gl 150.2AB reconstructed by the speckle masking method from 1600 speckle interferograms. (b) Image reconstruction of 41 Dra from 533 infrared speckle interferograms recorded through a 2191/411 nm filter. The scale and orientation are identical in both images: each panel covers $0.7'' \times 0.7''$, north is up and east is to the left.

in Gl 150.2AB system increased by 63 mas thus confirming the identity of spectroscopic and interferometric pairs. From the estimated magnitude difference $\triangle m = 2.49$ we obtain for the second star $m_R = 10.3$. The absolute magnitude for the main K2V star must be close to M_V=6.4 (Allen, 1973), corresponding to $m - M$=1.4 and π_{ph}=52 mas. The absolute magnitude for the secondary is then M_V=8.9 and its spectral type is M0V. Approximate masses of the A and B components are $0.7\mathcal{M}_\odot$ and $0.5\mathcal{M}_\odot$ respectively, and their orbital period is ~10 yrs. In the following years, speckle measurements combined with radial velocity data and HIPPARCOS parallaxes will provide high precision masses and luminosities for Gl 150.2AB.

Speckle measurements of 41 Dra are given in Table 1. In 1994.7129, 40 days after the periastron passage, the weaker star of the binary was observed in the NW direction at ρ=25 mas - the separation equal to the diffraction limit of the 6-m telescope at 650 nm. In Fig.1b it is shown one of the speckle masking images of 41 Dra, recorded through the 2191/411 nm filter where the diffraction limit of the telescope is 75 mas.

As was recently shown by Tokovinin (1995), all four components of 40/41 Dra = ADS 11061AB system have very similar spectral types and masses. The wide pair ADS 11061Aab shows the highest eccentricity among known spectroscopic binaries. By combining P, T, e elements from the spectroscopic orbit with our 6 speckle measurements in different filters, we can define the parameters of the visual orbit. This was done with the help of the Monet method (1979) giving us the following:

$P^* = 3.4147$ *yrs*, $T^* = 1994.5988$, $e^* = 0.9754$,
$a'' = 70$ *mas*, $i^\circ = 50$, $\Omega^\circ = 358$, $\omega^\circ = 130$.

The corresponding ellipse of the relative motion is shown in Fig.2. From the known values for $(\mathcal{M}_{Aa} + \mathcal{M}_{Ab}) \sin^3 i = 1.100\mathcal{M}_\odot$ and $K_{Ab}/K_{Aa} = 1.07$ (Tokovinin, 1995), we can now estimate the masses of the Aa and Ab components:
$\mathcal{M}_{Aa} = 1.26\mathcal{M}_\odot$, $\mathcal{M}_{Ab} = 1.18\mathcal{M}_\odot$.
The masses are found with an uncertainty of about 15%. Taking into account the bolometric correction for an F7V star, the dynamic parallax of the system is π_{dyn}=21 mas, which is in agreement with the published trigonometric parallax (Turon *et al.*, 1992). It is interesting to note that exactly the same parallax has been reported during this Workshop for the wide system ADS 11061AB by A.Kiselev, as a result of astrometry at the Pulkovo observatory. The orbital parallax for the 41 Dra, found by combination of interferometric and spectroscopic orbit parameters, is $\pi_{orb} = 23 \pm 3$ mas.

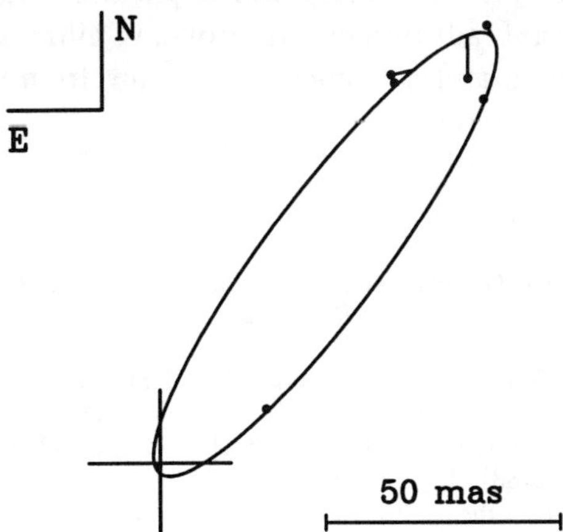

Figure 2. The relative interferometric orbit for 41 Dra=ADS 11061Aab. Speckle measurements with the 6-m telescope are connected to their predicted positions on the orbit.

Close to the periastron passage the two components of 41 Dra are separated by only 1–2 mas, or 5–10 stellar radii. Short-term tidal frictions near the periastron could cause the evolution of orbit eccentricity, period, and semi-major axis. The quadruple system ADS 11061AB as a whole is therefore a critical test object for the verification of theories of multiple stars origin and evolution.

4. Conclusion

We obtained the first diffraction-limited images of the nearby spectroscopic binaries Gl 150.2 and 41 Dra. For 41 Dra, speckle observations were made both at visible wavelengths and in the K-band. This pair with K2V and M0V components shows orbital motion with a period of about 10 years. Because of a significant magnitude difference between the two stars of Gl 150.2, it can be observed only as SB1. For this system, precise parallaxes and new speckle measurements at different wavelengths can improve our knowledge about the physical characteristics of the components.

The pair 41 Dra is an interesting case of a relatively short-period SB2 with an eccentricity close to 1, and it is at the same time a member of a quadruple system. The preliminary combined spectroscopic-interferometric orbit is derived for 41 Dra, yielding the masses $1.26\mathcal{M}_\odot$ and $1.18\mathcal{M}_\odot$ for its components. The dynamic parallax of the system is $\pi_{dyn} = 21$ mas.

All the astrometric and photometric measurements for the two stars were made with the speckle-masking method. The use of Digital Signal Processors for bispectrum computations allows diffraction-limited image reconstruction in real time. For binary star application, the speckle-masking method has significant advantages. It provides diffraction-limited image with high photometric and astrometric accuracy from 400 nm to the infrared.

References

Allen, C.W. Astrophysical Quantities. Univ. of London. Atlone Press, 1973

Andersen, J. (1991) *Astron. Astropys. Rev.* **3**, 91

Balega, I.I. et al. (1994) *Astron. Astrophys. Suppl.Ser.* **105**, 503

Baier, G. and Weigelt, G. (1983) *Astron. Astrophys.* **121**, 137

Duquennoy A. and Mayor M. (1991) *Astron. Astrophys.* **248**, 485

Hofmann, K.-H. and Weigelt, G. (1986) *Astron. Astrophys.* **167**, L15

Lohmann, A.W. et al. (1983) *Appl. Optics* **22**, 4028

Monet, D.G. (1979) *Astrophys. J.* **234**, 275

Tokovinin, A.A. (1988) *Astrophysics* **28**, 173

Tokovinin, A.A. (1995) *Astronomy Lett.* **21**, 286

Turon, C. et al. (1992) The HIPPARCOS Input Catalogue, ESA SP-1136, **4**

Weigelt, G. (1977) *Opt. Commun.* **21**, 55

OBSERVATIONS OF DOUBLE STARS FROM THE VENEZUELAN NATIONAL ASTRONOMICAL OBSERVATORY

C. ABAD AND F. DELLA PRUGNA
Centro de Investigaciones de Astronomía, CIDA
Apdo. P. 264, Mérida 5101-A, Venezuela

Abstract. The Venezuelan National Astronomical Observatory was dedicated in 1975. Located in Los Andes at 8^0 47' north and at an elevation of 3600 mt., almost the entire sky is visible in one year. At present three telescopes are in working order: the 1-mt. f/3 Schmidt telescope, the 1-mt. f/21 Coudé Reflector and the Zeiss 0.65-mt. f/16 Refractor. Short spells of double star observations have been carried out since the creation of the Centro de Investigaciones de Astronomía CIDA to date, using the Zeiss 0.65-mt. Refractor. Early measurements were reported by MacConnell (1978, 1984) and Valbousquet (1980). Since 1993, CIDA joined the project "*Determinación de Medidas Astrométricas y Parámetros Astrofísicos de Interés de Estrellas Dobles y Múltiples*". This is a comprehensive project headed by the "Ramón M^a. Aller" Astronomical Observatory of Santiago de Compostela University, involving also Madrid, Nice and CIDA Observatories. The latter is especially suited for observations of southern double stars. The use of CCD at the 1-mt. CIDA Reflector significantly increased the number of astrometrically observed binaries. At the same time, several methods were developed to reduce CCD frames of double stars (Abad and Della Prugna, 1995). In this poster, we show the telescopes and some graphical examples of the astrometric methods employed.

1. Telescopes and Equipment used during observations

The CIDA Zeiss 65-cm. f/16 Refractor is especially suited for observations of double stars thanks to its fine optics and the large filar micrometer, here shown at the focus. A close-up of the large Zeiss filar micrometer shows the dial used to read out separations coupled to the moving wire frame. This dial has been added recently to supersede the old reading mechanism which relies on the engraved drum used to advance the screw and frame. The CIDA Zeiss 1-mt f/21 Coude Reflector and a digital camera , based on a Thomson TH-7883 CCD chip (shown at the focus), are routinely used to observe visual binaries. Usually, seven or more exposures are taken onto the same frame shifting the double along the East-West direction. This procedure gives at the same time the correct orientation of the sky onto the frame and a reasonable number of images to determine the Position Angle and Separation of the pair.

J. A. Docobo et al. (eds.), Visual Double Stars: Formation, Dynamics and Evolutionary Tracks, 79–83.

2. Astrometric Techniques used to reduce CCD frames taken at the CIDA 1-mt. f/21 Coudè Reflector

CCD observations of visual binaries are carried out using the 1-mt. Reflector and a digital camera that uses a Thomson TH-7883 CCD chip. The camera shutter allows a multiple exposures mode by which the orientation of the sky onto the frame is determined using the telescope's Right Ascension offset (Fig.1). To determine the separation of the pair, we employ one of three different methods, depending on the degree of images overlap.

2.1. METHOD 1.

When the images of the pair are well resolved, their centroids can be determined individually. Thanks to the high linearity of CCDs to light and the proximity of the pair, we can presume a common Point Spread Function (PSF). Modeling of this PSF will give the centroids positions. Nevertheless, in practice we find that a simple radial Gaussian function fit, or similar algorithms, is quite appropriate to determine the centroids and therefore the separation of the pair if the scale factor is known.

2.2. METHOD 2.

When the images of the components overlap, their centroids must be determined simultaneously, since any individual fit is prone to introduce systematic errors that depend on seeing conditions and especially on the magnitude difference of the pair. In this case, the image can be represented with enough accuracy by the sum of two radial gaussian functions. The non linear aspect of the parameters to be determined calls for good initial values in the iterative process. The convergence is eventually limited by the severe overlap of images. If this occurs, the following method is recommended.

2.3. METHOD 3.

When the pair shows a heavy overlap there is little hint of duplicity. To cope with this situation, we devised a simple method to determine the separation of the components. The original image of the pair is duplicated and rotated 180^0 using the pixel with maximum ADU (Analog to Digital Units) as pivot. Next, the original and rotated images are registered and subtracted. Registering is carried out to "suppress" the image of the brighter component (A), leaving only the images of the companion (B). Appropriate criteria used to suppress A are: the symmetry of the B component's images left on the frame, or the proportion of the A and B images. Finally, the separation of the pair is given by half the distance between the two images of the B component. Using this procedure, we were able to measure visual binaries down to 0.5 arcsecs.

Acknowledgements

Financial support for this project has been from Xunta de Galicia under XUGA 24301B92 and from D.G.I.C.Y.T. under PB92-1074, both coordinated by J.A. Docobo.

References

Abad, C. and Della Prugna, F.: 1995, CCD Measurements of Visual Binaries, *Astronomy & Astrophysics Supplement Series*, **111**, 229.

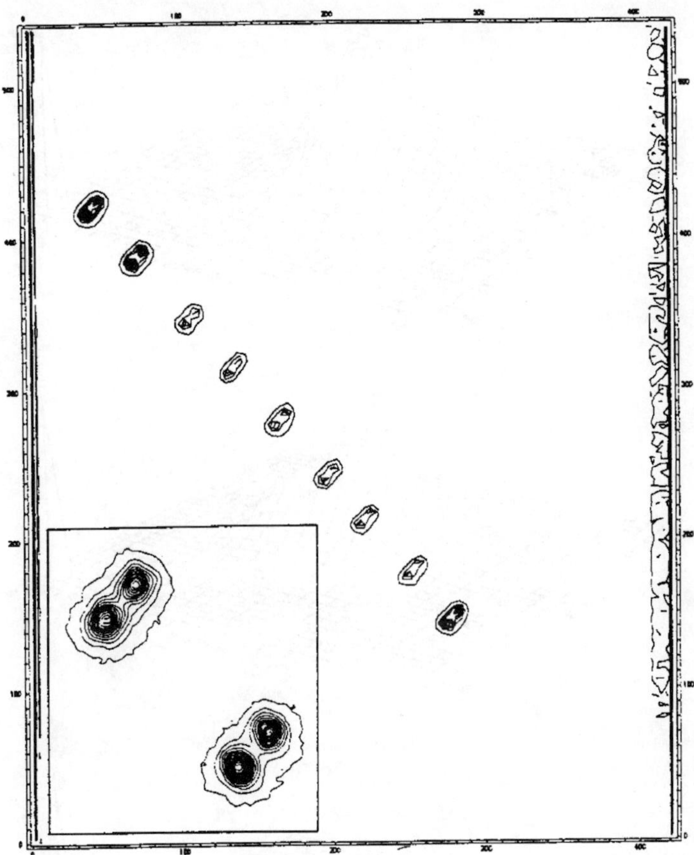

Figure 1. Plot of isophotes of a typical CCD frame showing the series of multiple exposures, spaced along the East-West direction, of the double star HU 139 AB-C. The inset shows details of two of these images

Abad, C., Della Prugna, F. and García, L.: 1995, An Image Processing Method to Extend CCD Measurements of Visual Binaries to Closer Pairs, *Astronomy & Astrophysics Supplement Series*, submitted.

MacConnell, D. J.: 1978, Micrometer Observations of Double Stars I, *The Astronomical Journal*, **84**, 436.

MacConnell, D. J.: 1984, Micrometer Observations of Double Stars II, *The Astronomical Journal*, **89**, 876.

Valbousquet, A.: 1980, Measures d'Ètoiles Doubles a Mérida, Venezuela, *Astronomy & Astrophysics Supplement Series*, **40**, 347

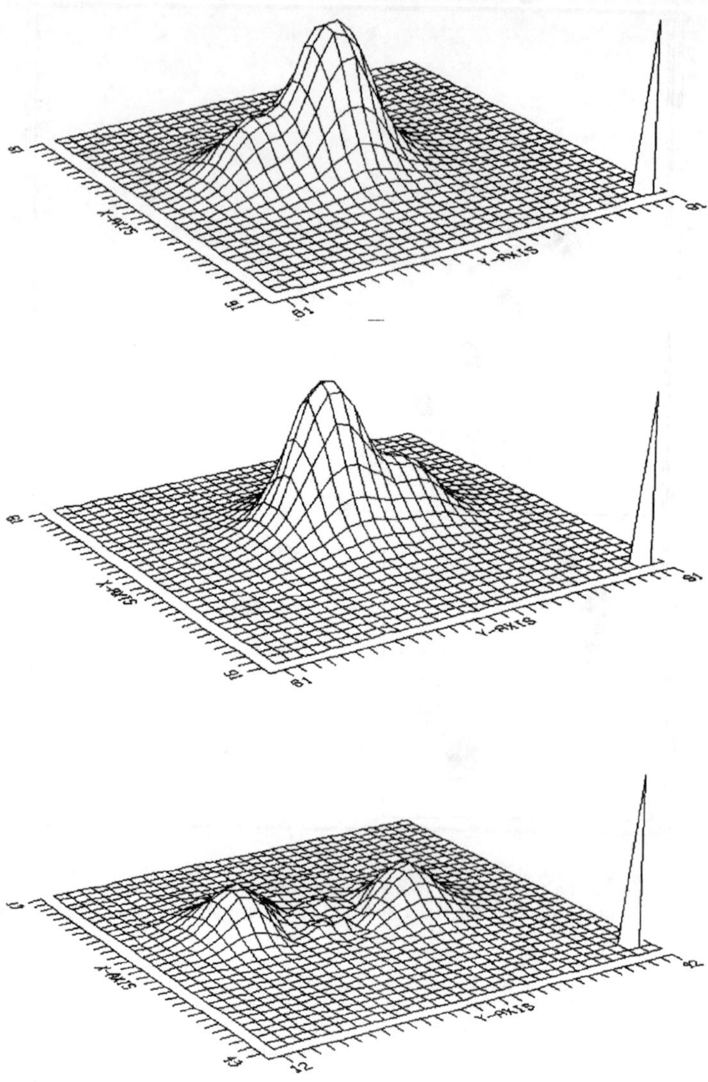

Figure 2. To show how Method 3 works, we selected a double star with slightly overlapped images. This pair can be easily handled by Method 2 but is used here to better explain the procedure of Method 3. The upper graph *(a)* shows a surface plot of the double star image. The middle graph *(b)* is the duplicated image of the original rotated 180 degrees. The lower graph *(c)* shows the resulting image (absolute values) after registering and subtraction of the original and the rotated image. The bright star A has been almost suppressed and only the images of the B component are left. Separation is half the distance between these

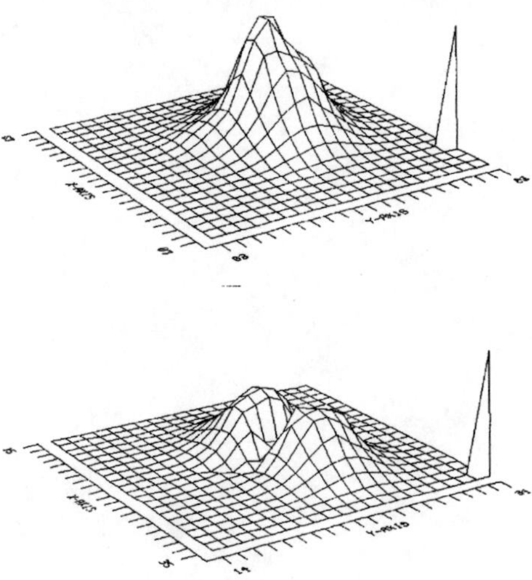

Figure 3. This example shows the power of Method 3. The upper graph shows a surface plot of a double star with heavily overlapped images. The lower graph shows the resulting image after processing the frame using Method 3. Images of the B component indicate a separation slightly over two pixels, or 0.5 arcsecs.

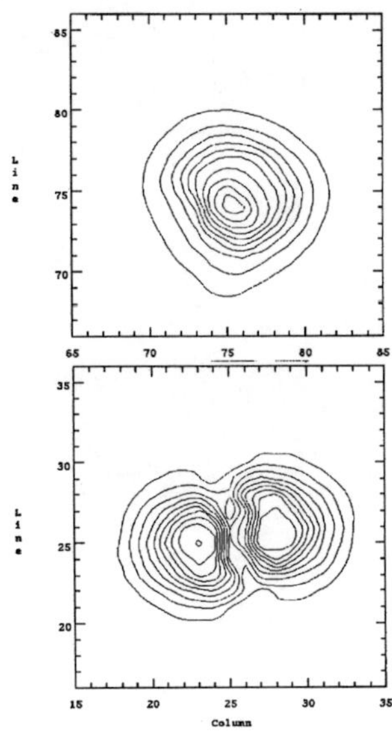

Figure 4. The same example shown in *Figure 3.* as isophotal plots. Note that in the original image there is almost no hint of duplicity.

PHYSICAL AND OPTICAL VISUAL DOUBLE STARS OF THE NORTH-POLAR AREA AS OBTAINED BY PHOTOGRAPHIC OBSERVATIONS AT PULKOVO

E.A.GROSHEVA
Main Astronomical Observatory of RAS
196140, 64 Pulkovskoe sh., St.Petersburg, Russia

The subject of this paper is the results of photografic observations of the visual double stars of the North-polar area. These observations are a part of the Pulkovo's program for the systematic observations of the components relative positions of the visual double stars situated in the North-polar area [2]. A northern location of Pulkovo Observatory gives favourable conditions for observations of stars in the North-polar area, which cannot be considered exhaustively investigated so far. Within this part of the program the main attention has been paid to the physical double stars situated close to the Sun, which are the most perspective for a determination of their orbits by the method of apparent motions parameters [1].

The selection of physical double stars has been done on the basis of comparison of the component's proper motions. A modern observations by 26"-refractor were compared with positions from catalogue "Carte de Ciel" [4]. The proper motions were obtained for components of 50 star pair of the North-polar area, using the following methods:

1. comparing equatorial coordinates were calculated with catalogue PPM as reference one. The considerable difference of epoches ($80 - 90 years$) allowed to obtain proper motions with precision $\pm 0.005''/year$;
2. in cases, when not enough reference stars near investigating double star, the proper motions of components were obtained by the method of homographic coordinates [1,3]. The proper motions calculated by this way are relative.

The results of these methods are presented in Table 1. The comparison of these results showed that a difference of the component's proper motions in both cases is equal, of cause, in the ranges of random errors, although, the proper motioms themselves obtained by different ways can be unequal in consequence of various influences of the systematic errors due to motions of reference stars.

Tables 2,3 contains the data for visual double stars. Tables 4,5 contains the proper motions of component of the double stars. The optical pairs are marked by 'o!'; the relative proper motions are marked by '*'. There are 7 physical pairs situated close to the Sun among these double stars. These are AC 81 181/2; AC 79 2968/69; AC 76 4078/9; AC 85 1539/40; AC 86 1020/21; AC 76 5185/86; AC 76 5490/91.

J. A. Docobo et al. (eds.), Visual Double Stars: Formation, Dynamics and Evolutionary Tracks, 85–88.

TABLE 1. The comparison of two methods of obtaining a relative proper motions of the double star components.

AC		by method of homographic coordinates		by method of the position comparison		ΔT (*years*)
		μ_α	μ_δ	μ_α	μ_δ	
	A	-0.0238	-0.0300	0.0020	-0.0133	
76 7042/3	B	-0.0232	-0.0247	0.0026	-0.0081	92.2
	(B-A)	0.0006	0.0053	0.0006	0.0052	
	A	0.0189	-0.0520	0.0227	-0.0645	
81 182/2	B	0.0150	-0.0554	0.0187	-0.0678	84.8
	(B-A)	-0.0039	-0.0034	-0.0040	-0.0033	
	A	-0.0019	0.0014	0.0065	-0.0512	
79 2142/3	B	0.0114	-0.0493	-0.0023	0.0005	88.8
	(B-A)	-0.0133	0.0507	-0.0088	0.0517	
	A	0.0153	-0.0060	0.0196	-0.0122	
81 161/2	B	0.0101	-0.0070	0.0142	-0.0131	84.8
	(B-A)	-0.0052	-0.0010	-0.0054	-0.0009	

References

1. A.A.Kiselev. (1978) The theoretical foundation of photographic astrometry. **pp. 237–255**, Moscow, Nauka, (in Russian).

2. A.A.Kiselev *et al.*. (1988) The catalogue of relative positions and motions of 200 visual double stars as obtained by $26''$-refractor at Pulkovo. **pp. 3–12**, St.Peterburg, Nauka,(in Russian).

3. A.A.Kiselev.(1968) The calculation of the star's proper motions with catalogue *"Carte de Ciel"*. *Izvestia GAO*, **Vol. no. 183**, **pp. 105–117**, St.Peterburg,(in Russian).

4. Astrografic Catalogue, 1900.0, Greenwich section. Edinburg,1908.

TABLE 2. The double stars of North-polar area.

N	AC	ADS	α_{2000}	δ_{2000}		m	ρ	Θ
1	73 158/9	-	0 27.8	73 59		11.0,10.5	10.8	2
2	73 129/30	-	0 28.9	73 23		10.9,10.7	9.4	238
3	81 161/2	830	1 2.3	81 52	AB	var,11.2	13.6	63.2
4		-			AC	var,12.2	17.7	342.1
5	81 181/2	-	1 3.1	82 2		11.5,11.7	6.6	12.2
6	81 156/7	-	0 54.0	81 55		10.0,11.4	13.3	306.1
7	81 124/5	-	0 53.7	81 39		11.8,12.2	3.3	349.6
8	75 465/6	-	3 34.7	75 28		10.8,11.8	12.6	131.6
9	72 1847/8	-	3 41.2	73 18		10.0,10.2	13.1	203
10	82 874/5	-	4 57.9	82 27		11.4,11.2	6.8	359.2
11	79 2043/4	-	4 59.6	80 8		9.9,10.0	7.8	202
12	72 3008/9	-	5 41.5	72 14		11.2,11.2	4.3	323.3
13	79 2142/3	-	5 45.3	79 5		11.0,11.5	4.8	286
14	72 3491/2	5701	7 4.9	72 40		8.6,10.1	3.7	256
15	79 2968/9/61	6646	8 16.5	79 30	AB	7.6, 7.9	20.7	14.9
16		-			AC	7.6, 12.1	65.9	125.4
17	79 2927/8	-	8 20.7	79 6		10.8,11.1	2.9	96.6
18	87 847/8	-	9 21.3	87 1		11.5,11.3	6.5	279.9
19	70 4336/7	-	9 42.8	70 2		10.9,11.4	88.6	253.1
20	69 4012/3	7565	9 55.1	68 56		9.5, 9.5	8.9	271.5
21	76 4078/9	-	10 11.3	76 22		9.1, 9.5	23.8	187.4
22	87 867/8	-	10 30.2	87 27		12.6,12.6	13.7	215.7
23	79 3594/5	-	10 38.9	78 38		9.6,10.0	12.6	233.8
24	74 4014/5	-	11 1.2	73 40		9.4, 9.9	8.9	248.9
25	74 4074/5/6	8100	11 15.5	73 28	AB	7.0, 7.5	56.7	98.8
26		-			AC	7.0,11.0	6.4	314.6
27	79 3914/5	-	11 51.9	78 56		10.6,11.0	10.6	89.9
28	71 5765/6	-	12 1.9	70 56		8.2,12.5	5.4	232.2
29	89 120/1	8604	12 28.4	88 40		8.6,10.0	8.4	325.8
30	74 4969/70	-	14 17.4	73 44		10.7,10.9	6.9	56.3
31	70 6103/4/5	9275	14 26.5	70 19	AB	9.0,10.6	21.60	49.1
32		-			AC	9.0,11.0	15.16	35.6
33	83 2000/1	-	14 30.6	83 8		10.3,10.6	3.4	89.9
34	85 1509/10	9509	14 57.6	85 30		9.3, 9.4	3.4	91.0
35	86 1018	-	14 51.8	85 42		12.0,12.2	8.4	336.8
36	85 1539/40	-	14 57.6	85 29		9.5,11.5	8.3	100.6
37	86 1024	-	14 59.4	85 48		11.8,12.0	18.6	350.4
38	86 1025	-	15 4.2	85 47		11.8,12.0	5.8	46.6
39	86 1020/1	-	15 7.8	85 44		11.5,11.2	23.7	241.6
40	71 6644	9460	14 56.8	70 50		8.3, 8.3	3.7	159.1
41	76 5185/6	-	15 2.6	76 34		11.2,10.6	5.1	246.8
42	80 3643/4	-	15 4.9	79 42		12.4,12.6	5.8	173.9
43	80 4757/8	-	15 3.3	79 17		11.5,12.0	8.6	256.7
44	80 4784/5	-	15 0.6	79 33		12.0,12.1	36.7	338.1
45	73 5919/20	-	15 56.7	73 14		11.0,11.4	11.2	140.2
46	76 5514/5	-	15 57.2	76 29		9.5,10.0	23.6	285.4
47	76 5490/1	-	16 0.0	76 15		11.2,11.9	26.5	239.1
48	79 5084/5	-	16 30.4	79 30		10.6,10.7	3.5	92.6
49	75 6652/3	11072	18 3.8	75 48	AB	7.0,10.0	5.7	245.1
50		-			AC	7.0, 9.5	22.9	352.5
51	76 7042/3	-	19 16.1	76 46		11.4,11.4	3.8	147.7
52	80 5313/4	-	21 6.1	81 20		11.2,11.5	10.03	121.4

TABLE 3. The proper motions of component.

N	AC	epoch	$\mu_\alpha(A)$	$\mu_\beta(A)$	$\mu_\alpha(B)$	$\mu_\delta(B)$	$\Delta_{(B-A)}\mu_\alpha$	$\Delta_{(B-A)}\mu_\delta$	
1	73 158/9	1993	0.0049	0.0205	0.0016	0.0198	-0.0033	-0.0007	*
2	73 129/30	1993	0.0032	-0.0081	0.0016	-0.0096	-0.0016	-0.0015	*
3	81 161/2	1983	0.0196	-0.0122	0.0142	-0.0131	-0.0054	-0.0009	
4	81 161/2(AC)	1983	-	-	-	-	-	-	
5	81 181/2	1983	0.0227	-0.0645	0.0187	-0.0678	-0.0040	-0.0033	
6	81 156/7	1983	0.0081	-0.0138	-0.0283	-0.0054	-0.0368	0.0084	o!
7	81 124/5	1983	-0.0012	-0.0038	0.0004	0.0144	0.0016	0.0182	
8	75 465/6	1993	0.0047	-0.0209	0.0238	-0.0219	0.0191	-0.0010	
9	72 1847/8	1988	-0.0147	0.0214	-0.0188	0.0311	-0.0041	0.0097	*
10	82 874/5	1992	-0.0220	0.0073	-0.0214	0.0165	0.0006	0.0092	
11	79 2043/4	1994	0.0166	-0.0113	0.0161	-0.0093	-0.0005	0.0020	*
12	72 3008/9	1986	0.0020	0.0310	0.0000	0.0360	-0.0020	0.0050	
13	79 2142/3	1988	0.0065	-0.0512	-0.0023	0.0005	-0.0088	0.0517	*,o!
14	72 3491/2	1993	-0.0087	-0.0329	-0.0094	-0.0377	-0.0007	-0.0048	*
15	79 2968/9(AB)	1987	-0.0339	-0.0678	-0.0373	-0.0688	-0.0034	-0.0010	
16	79 2968/61(AC)	1987	-0.0339	-0.0678	0.0027	-0.0350	0.0366	0.0328	o!
17	79 2927/8	1987	-0.0177	0.0004	-0.0137	0.0064	0.0040	0.0060	
18	87 847/8	1982	-0.0290	-0.0290	-0.0260	-0.0280	0.0030	0.0010	
19	70 4336/7	1990	-0.6525	-0.3070	-0.6497	-0.3031	0.0028	0.0039	*
20	69 4012/3	1993	-0.0495	-0.0591	-0.0513	-0.0615	-0.0018	-0.0024	*
21	76 4078/9	1900	-0.0594	-0.0474	-0.0581	-0.0491	0.0013	-0.0019	
22	87 867/8	1986	-0.013	-0.037	-0.0170	-0.0270	-0.0040	0.0100	
23	79 3594/5	1988	-0.0615	-0.0994	0.0081	-0.0071	0.0696	0.0923	o!
24	74 4014/5	1980	-0.0081	-0.0297	-0.0037	-0.0290	0.0044	0.0007	*
25	74 4074/5/6(AB)	1990	-0.3973	0.0806	0.0111	-0.0013	0.4084	-0.0819	*,o!
26	74 4074/5/6(AC)	1990	-0.3973	0.0806	-0.3747	0.0888	0.0226	0.0082	
27	79 3914/5	1975	0.0477	0.0124	0.0548	0.0156	0.0071	0.0032	*
28	71 5765/6	1986	-0.0129	0.014	-0.0194	0.0081	-0.0065	-0.0060	
29	89 120/1	1993	-0.0182	-0.0102	-0.0234	-0.0105	-0.0052	0.0003	
30	74 4969/70	1987	-0.0191	0.0210	-0.0207	0.0249	-0.0016	0.0039	*
31	70 6103/4/5(AB)	1990	-0.1731	0.1209	-0.1195	0.0708	0.0536	-0.0501	*,o!
32	70 6103/4/5(AC)	1990	-0.1731	0.1209	-0.1743	0.1200	-0.0012	-0.0009	
33	83 2000/1	1992	0.1099	-0.0632	0.0142	-0.0299	-0.0957	0.0332	o!
34	85 1509/10	1982	0.0147	0.0159	0.0134	0.0153	-0.0013	-0.0006	
35	86 1018	1993	-0.0058	-0.0033	-0.0097	-0.0218	-0.0039	-0.0185	o!
36	85 1539/40	1993	-0.1111	0.0712	-0.1129	0.0697	-0.0018	-0.0015	
37	86 1024	1993	0.0118	-0.0185	0.0007	-0.0062	-0.0111	0.0123	o!
38	86 1025	1993	0.0048	0.0006	0.0248	-0.0214	0.0200	-0.0220	o!
39	86 1020/1	1993	-0.1172	0.0487	-0.1072	0.0494	0.0100	0.0007	
40	71 6644	1913	0.0244	-0.0090	0.0236	-0.0103	-0.0008	-0.0013	*
41	76 5185/6	1993	-0.0367	-0.0312	-0.0345	-0.0328	0.0022	-0.0016	
42	80 3643/4	1988	-0.0050	-0.0236	0.0019	-0.0227	0.0069	0.0009	
43	80 4757/8	1988	-0.0007	-0.0086	-0.0058	-0.0094	-0.0051	-0.0008	
44	80 4784/5	1988	-0.0220	0.0187	-0.0025	-0.0049	0.0195	-0.0236	o!
45	73 5919/20	1992	0.0029	0.0424	0.0012	0.0440	-0.0017	0.0016	*
46	76 5514/5	1900	-0.0121	-0.0066	0.0142	-0.0415	0.0263	-0.0349	o!
47	76 5490/1	1982	0.0622	-0.0545	0.0419	-0.0169	-0.0203	0.0375	
48	79 5084/5	1990	0.0077	-0.0116	0.0137	-0.0090	0.0060	0.0026	
49	75 6652/3(AB)	1983	-0.0252	0.0319	-0.0218	0.0462	0.0034	0.0143	
50	75 6652/3(AC)	1983	-0.0252	0.0319	-0.0134	0.0388	0.0118	0.0069	
51	76 7042/3	1987	0.0054	-0.0128	0.0061	-0.0077	0.0007	0.0051	
52	80 5313/4	1983	-0.0103	-0.0122	-0.0026	0.0012	0.0077	0.0134	

A FEW NEW WIDE PAIRS

M. ODENKIRCHEN AND P. BROSCHE
Sternwarte der Universität Bonn
Auf dem Hügel 71, D - 53121 Bonn, Germany

Abstract. We report on a small number of wide visual binaries and common proper motion pairs, which have been measured in the course of our work for the Hipparcos extragalactic link. The nature of these pairs is analysed by means of a combination of relative position, proper motion and two-colour-photometry.

1. Introduction

The sample described here originates from a photographic proper motion study in two fields related to the bright galaxies M 51 and M 81 (Odenkirchen 1996; Odenkirchen & Brosche 1995, 1996). Each field covers about $(2°)^2$ and about 350 measured stars. The limiting magnitudes are $15^{\mathrm{m}}.5$ and $14^{\mathrm{m}}.5$ and the galactic latitudes $+69°$ and $+42°$ respectively. Although the sample is small, it is statistically well defined. The search for visual pairs is complete in the angular distance interval from $4''$ to $20''$. The search for common proper motion pairs is complete with respect to a more complex set of selection criteria (see section 2). The astrometric parameters which result from our measurements are presented in Tab. 1. For the reason of brevity we do not report individual errors. However, proper motions are generally accurate to 1 mas/y or better and the errors in relative position are $\leq 0''.1$. Photometric and spectroscopic informations (partly from external sources) and a number of derived quantities are given in Tab. 2. From the 16 pairs in those tables, 11 are new in the sense that they do not occur in double star catalogues. For the other ones, at least our proper motions are an essential new piece of information, since they are of high accuracy. Therefore we take the liberty to use the word "new" in the headline for all our pairs.

J. A. Docobo et al. (eds.), Visual Double Stars: Formation, Dynamics and Evolutionary Tracks, 89–94.

TABLE 1. Position and proper motions of wide pairs from photographic measurements

No.	Name	Position		Proper Motion		Relative Position		
	BD, ADS or other	α_{2000} [h m s]	δ_{2000} [° ′ ″]	μ [mas/a]	μ'	ρ [″]	θ [°]	t
1A	+47 2056	132331.32	+471230.7	−21.7	6.1	12.53	48.1	1977
B		132332.24	+471239.0					
2A	8910 A	132745.73	+474532.2	5.9	−38.1	14.92	345.4	1970
B	8910 B	132745.36	+474546.7	5.7	−38.3			
3A	+48 2124	133046.98	+475936.8	−16.5	−3.0	13.88	331.2	1940
B		133046.36	+475949.2	−10.5	1.3			
4A		133225.64	+471703.1	−11.9	−24.0	4.00	239.2	1994
B		133225.29	+471701.2	−28.0	9.2			
5A	8945 A	133334.00	+464748.4	6.4	9.3	8.85	342.7	1970
B	8945 B	133333.75	+464756.9	7.1	10.9			
6A		95439.81	+683637.5	20.8	−16.6	5.40	114.5	1969
B		95440.70	+683635.2	19.7	−18.1			
7A	7565 A	95502.69	+685622.0	−62.6	−62.3	8.92	272.5	1960
B	7565 B	95501.01	+685622.4	−63.5	−61.7			
8A		95746.54	+682929.0	−26.4	−18.5	14.81	31.4	1963
B		95748.09	+682942.2	−4.3	−2.4			
9A		95905.06	+684726.2	−1.7	38.9	18.80	0.8	1966
B		95905.09	+684745.1	−5.2	41.9			
10A		100028.73	+694251.5	−0.3	−1.0	17.30	260.9	1962
B		100025.54	+694249.1	12.5	6.8			
11A		100132.48	+683635.3	−28.7	−3.6	5.43	94.7	1967
B		100133.50	+683634.8	−24.5	−4.4			
12A	+48 2129	133344.13	+480053.7	−70.0	9.7	123.9	134.9	1943
B	+48 2130	133352.88	+475926.1	−69.6	7.7			
13A	Gl 360	94234.84	+700202.0	−671.0	−269.3	88.7	77.2	1965
B	Gl 362	94251.73	+700221.9	−671.1	−264.9			
14A		95202.46	+685108.1	−4.5	−12.2	4459.	97.1	1963
B		100537.63	+684202.1	−4.3	−12.0			
15A	7611 AB	100256.39	+684708.8	−29.8	−23.3	675.5	298.6	1960
B		100106.91	+685232.2	−32.2	−21.6			

μ, μ' = proper motion components $\mu_\alpha \cos\delta$, μ_δ
ρ, θ = Mean separation and position angle between B and A (in J2000.0 system)
t = Mean epoch of the measurements of ρ, θ

TABLE 2. Photometry, spectral type and values of some derived parameters

No.	Photometry		SpT	Derived Quantities							
	B	B−V		d	$\rho \cdot d$	$\Delta\mu$	Δv_t	$\overline{v_t}$	M	β	n
	[mag]			[pc]	[AU]	[mas/a]	[km/s]		[M$_\odot$]		
1A	10.48	0.14	A2	490	6100			53	2.5		u
B	14.38										
2A	10.58	0.8	K0	61	775	0.3	0.1	10	0.8	-1.0	p
B	10.59	0.8	K0						0.8		
3A	10.52	1.39	K0	45	625	7.4	1.6	3.6	0.8	0.2	p
B	14.15										
4A	14.01					37.0					o
B	14.32										
5A	9.84	1.02	K0III	430	3700	1.7	3.4	24	1.1	0.8	u
B	14.26										
6A	13.58	0.94		164	890	1.9	1.5	21	0.8	0.1	p
B	13.62	0.94							0.8		
7A	11.30	0.72	G	105	930	1.1	0.5	44	0.9	-0.4	p
B	11.36	0.73							0.9		
8A	13.18	0.63		370	5500	27.3	48	40	1.0	1.9	o
B	14.71	0.63		750					1.0		
9A	10.65	0.71		85	1600	4.6	1.9	16	0.9	0.3	p
B	13.43	1.38							0.5		
10A	13.08	1.12		85		15.0					o
B	13.74	0.34		900							
11A	13.95	0.48		780	4200	4.3	16	100	1.2	1.5	o
B	15.84	0.97									
12A	10.23	0.55	G0	120	15000	2.0	1.0	40	1.0	0.5	p
B	10.53	0.53	G0	140					1.0		
13A	12.02	1.46	M3	11	1100	4.4	0.3	41	0.3	-0.4	p
B	12.68	1.49	M4						0.3		
14A	11.50	0.57		200	$8.9\,10^5$	0.3	0.3	12	1.0	0.8	u
B	14.16	1.00		200					0.7		
15A	8.39	0.51	F5	60	40500	2.9	0.8	11	1.4	0.6	p
B	11.73	1.08		50					0.7		

$\Delta\mu$ = proper motion difference (total amount) between B and A
$\overline{v_t}, \Delta v_t$ = Mean and difference in tangential space velocity
M = estimated mass
β = $\log(\Delta v_t / v_{orb})$ logarithm of ratio of tangential to orbital velocity (see text)
n = remark on nature: o = optical, p = physical, u = undecided

2. Methods and Criteria

The central question with regard to wide pairs is that for their nature: Are they physical or optical? The most elementary criterion for selecting pairs, which tend to be physical, consists in a reasonable upper limit for the angular separation. Figure 1 shows, how the number N of arbitrarily chosen pairs in both fields increases with angular separation ρ. For large ρ we observe the square law, which is valid for a random distribution of stars with the given surface density (dashed line in Fig. 1), but for $\rho < 20''$ we find significantly more pairs than expected for the random case. Therefore we took up in our sample all pairs of stars with $\rho \leq 20''$ (nos. 1 to 11 in the tables). The lower limit of ρ is given by the resolution limit of our photographic observations and lies around $4''$. Of course, not all of these statistically selected pairs are physical. Therefore further selection criteria on the basis of proper motion and photometry must be applied.

Physical pairs with $\rho > 20''$ exist also, but they are outnumbered by random occurences and hence do not show up in configuration space. One may think that such pairs could be identified by an analogous statistical criterion in proper motion space. However, this does not work satisfactorily because a) the influence of non-negligible measuring errors and b) the inhomogeneous distribution of stars in proper motion space come into play. Therefore one needs to add some physical information on the system: Two stars are probably physically associated to each other if they are close to each other in space and if also the size of their relative space motion is comparable to the expected orbital velocity. This kind of criterion requires estimates for a number of physical parameters, which can be obtained by combining astrometric and photometric measurements. They are listed in the right part of Tab. 2 and were derived in the following way:

1. From colour B−V and apparent magnitude V a distance modulus and hence a photometric distance was derived using a main sequence colour-magnitude relation. Giants were identified and excluded from the sample by means of a 'reduced proper motion' diagram (for an explanation see for example Evans & Irwin 1992), since they would yield largely erroneous distance estimates. Anyway, their true distances would be too large to obtain reasonable results on the question of physical association.
2. Using the distance estimate d the angular separation was converted into a spatial separation $\rho \cdot d$ and the relative proper motion into a relative tangential space velocity Δv_T. From the colour (or spectral type) an estimate of the stellar mass M was obtained and an expectation value for the orbital velocity of the system calculated in the form $v_{orb} = 2\pi[(\mathsf{M}_1 + \mathsf{M}_2)/(\rho \cdot d)]^{1/2}$. In an earlier investigation of a much larger but otherwise similar sample of wide binaries Brosche et al. (1992a,b) showed by analysing the ratio

$\lambda = \Delta v_t / v_{orb}$ that pairs with $\beta = \log \lambda > 0.8$ are probably optical while with decreasing values of β pairs become more likely physical. Thus the selection criteria for 'motion pairs' were set up such that all pairs were accepted which a) have components that appear as main sequence stars, b) have distance moduli that differ by not more than 1 mag (in accordance with a photometric error of about $0\overset{m}{.}15$ in B−V) and c) have $\beta \leq 0.8$. The search resulted in four additional pairs which are noted in the tables as nos. 12 to 15. Of course, the above given set of criteria is appropriate not only for selecting pairs at large angular separations, but also for segregating physical from optical pairs among the candidates with 'small' angular separation ($\rho \leq 20''$).

3. Discussion of Results

Among the 11 pairs with $\rho < 20''$ our criteria identify 5 to be physical and 4 to be optical. No. 10 appears to be optical because of largely disparate distances whereas no. 4 (for which we have no information on colour) is recognized as an optical pair because of large deviations in the proper motions which result in a large relative space velocity even if the stars are assumed to be very red and thus nearby. No. 5 is a borderline case falling into the intermediate zone between physical and optical pairs. However, this may only be an effect of the combination of proper motion error and relatively large distance. For the moment we regard it an undecided case. The nature of pair no. 1 remains unclear since proper motion and colour of the second component are still unknown. Among the 4 additional pairs with larger angular separation no. 13 is definitely a physically bound system whereas nos. 12 and 15 are formally identified as physical pairs ($\beta < 0.8$), but have rather large spatial separations. They may be physical pairs in the sense of a common origin of the components and may undergo a process of dissolution. In the case of no. 14 it is clear that with a separation of the order of several parsec the components are not directly associated to each other, but the hypothesis of a common origin might apply here as well. Further observations, in particular measurements of the radial velocities are necessary in order to clear up the physical state of these three pairs.

Summarizing our work, the analysis of astrometric and photometric data on 15 'new' pairs has shown that the existence of wide physical binaries far beyond the solar neighborhood can be proven not only statistically, but *individually* on the basis of precise proper motions and photometry. While in current catalogues precise proper motions are mostly lacking, future astrometric satellite surveys with limiting magnitude 15 or fainter will bring a wealth of such measurements and hence provide a profound basis for the individual study of wide binaries.

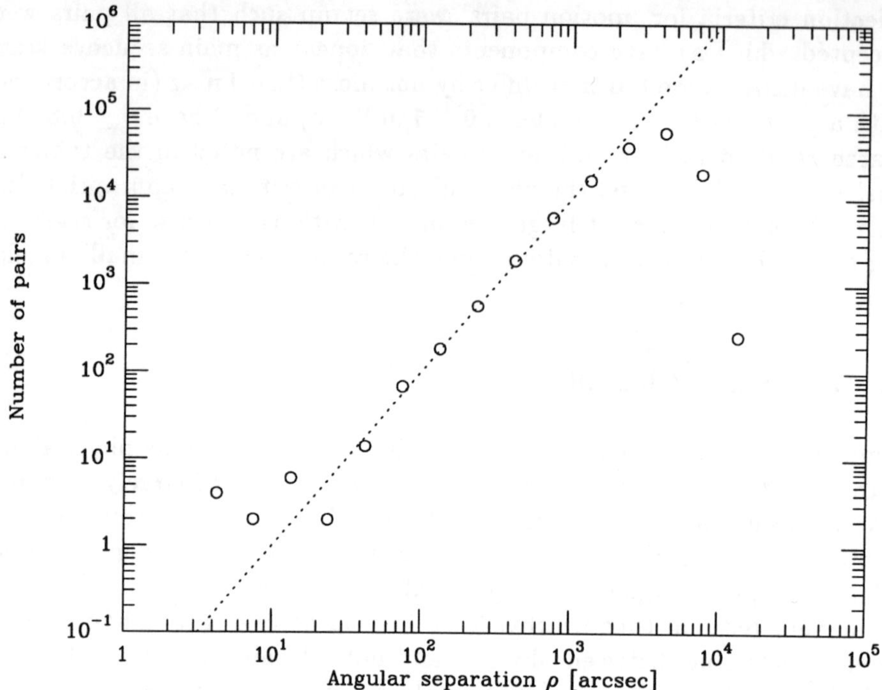

Figure 1. Number of stellar pairs as a function of angular separation. Open circles: Number of pairs within consecutive intervals of constant width 0.25 in $\log \rho$ plotted against ρ in logarithmic scale. Dashed line: Square-law corresponding to the observed surface density and a random distribution of stars. The *excess* in the number of pairs in the lowest bins reveals the existence of physical pairs. The decrease in number at the upper end of the ρ axis is due to the limited size of the fields.

References

Brosche P., Denis-Karafistan A.I. and Sinachopulos D. (1992a) *Astron. Astrophys.* **235**, 113

Brosche P., Denis-Karafistan A.I. and Denis C. (1992b) *Astron. Nachr.* **313**, 341

Brosche P., Odenkirchen M. and Tucholke H.-J. (1995) *Astron. Nachr.* **316**, 35

Brosche P. and Sinachopoulos D. (1988) *Astrophys. Space Sci.* **142**, 255

Evans D.W. and Irwin M.J. (1992) *Mon. Not. R. Astron. Soc.* **255**, 521

Odenkirchen M. and Brosche P. (1995) *Astron. Astrophys.* **302**, 915

Odenkirchen M. (1996) Ph.D. thesis, University of Bonn

Odenkirchen M., Brosche P., Börngen F., Meusinger H. and Ziener R. (1996) *Astron. Astrophys.* (submitted)

THE PULKOVO PROGRAMME OF PHOTOGRAPHIC OBSERVATIONS OF SPECTROSCOPIC BINARY STARS AS THE STARS WITH INVISIBLE SATELLITES.

O.V. KIYAEVA
Pulkovo observatory

Abstract. The investigation of stars with invisible satellites on the basis of photographic observations with Pulkovo 26-inch refractor is one of the traditional problems in the Pulkovo observatory. Now 11 spectroscopic binary stars with known orbit are included in the program of photographic observations of stars with invisible satellites. Among them there is a control star ADS 15600 for which a speckle-interferometric orbit agreeing with spectroscopic orbit has been already obtained (McAlister,1980). The aim of this program is to investigate the methods of determining a photocentric orbit for the star with a dark component comparing results with control stars. For other stars —to complete the spectral orbit, that is to determine inclination and longitude of ascending node. The observations are carried out since 1993. The first results of treatment for ADS 15600 don't contradict to the control orbit. We obtain also that photocentre is placed closer to the less massive component.

Long-term series of relative positions of wide visual double stars ($\rho > 3''$, $P >$200 years) describing a short arc of a visual orbit are accumulated in the Pulkovo observatory with 26-inch refractor (F=10.4m, D=65cm).

Periodic fluctuations relative to the orbital motion are found out for some stars [1–3]. These fluctuations can be caused by the presence of an invisible satellite. These fluctuations are of the order of errors of photographic observations. It considerably complicates the task of a photocentre orbit determination.

Eleven stars from the catalogue of spectroscopic binary stars orbits SPB8 [4] were included in the programme of photographic observations in 1993 for to test the algorithms of photocentre orbit determination. Having obtained a visual astrometric orbit we complement the spectroscopic orbit and establish orbit orientation in space which cannot be obtained from radial velocities observations.

If the spectroscopic binary star is not a component of a wide visual double star, we obtain photocentre coordinates relative to the system of reference stars from long-term photographic observations, and we study the deviations from rectilinear movement.

The stars on our programme have the following numbers according to the catalogue SPB8 [4]: 244, 442 (ADS 5983A), 473, 491, 648 (ADS 8035), 807 (ADS 9173 A), 947 (ADS 10345), 1215 (ADS 13554 A), 1218, 1242, 1350 (ADS 15600 A).

J. A. Docobo et al. (eds.), Visual Double Stars: Formation, Dynamics and Evolutionary Tracks, 95–98.

The best control example is spectroscopic binary star SPB 1350=ADS 15600 A=ξ Cephei A ($\alpha = 22^h02^m\!.2, \delta = 64°23'$ for equinox 1950.0, $m = 4.6$, $SP = A_m$, $\pi_{tr} = 0''\!.030$) which has also a visual orbit determined from speckle interferometry.

The combination of spectroscopic and speckle interferometric orbits yields parallax and mass of each component. Orbital elements of each component relative to the centre of mass are the same apart from semi-major axes ($a_1/a_2 = M_2/M_1$) and longitudes of periastron ($\omega_1 = \omega_2 + 180°$).

A visual orbit of the photocentre relative to the centre of mass is obtained as a result of photographic observations treatment. All elements of this orbit, except a semi-major axis (α), should be the same that the elements of the orbit for one of the components. It is possible that the geometrical place of the photocentre is far from a geometrical place of a main component. Comparing the values of semi-major axes α, a_1 and a_2, a difference between magnitudes of components $\Delta m'$ can be determined by the following formula:

$$\Delta m' = m_2 - m_1 = 2.5 \log\left(\frac{1}{\beta} - 1\right), \qquad \text{where} \qquad \beta = \frac{I_2}{I_1 + I_2} = \frac{a_1 \pm \alpha}{a_1 + a_2}. \quad (1)$$

The top mark is used if the photocentre is located closer to a less-massive component relative to the centre of mass of a system; the bottom one if it is located closer to a more-massive component. Obviously there is a point to determine $\Delta m'$ if both stars are comparable on brightness and if the value β is determined enough confidently.

The double-lined spectroscopic orbit of a star ADS 15600 Aa is derived by Drs. C.R.Vickers and C.D. Scarfe [5]. According to this orbit, we have the mass ratio $M_1/M_2 = 2.8 \pm 0.2$. They obtained also a magnitude difference $\Delta B = 0.55 \pm 0.2$ and $\Delta V = 0.3$ by analyzing the spectra. Dr. Harold A. McAlister obtained the speckle interferometric orbit [6].

The orbits agree well, but if we base on the spectroscopic orbit and use the value $i = 72°$ derived from speckle interferometric orbit we obtain the following values for the masses of the components $M_1 = 1.0 M_\odot$, $M_2 = 0.36 M_\odot$. If we base on the speckle interferometric orbit and trigonometric parallax we obtain $M_1 + M_2 = 2.79 M_\odot$, that corresponds better to spectral types. Therefore, we have calculated the control orbit of each component relative to the centre of mass on the basis of McAlister's orbit, including $M_1/M_2 = 2.8$. The Pulkovo 26-inch refractor has photo-visual object-lens, its system is close to V. If it is assumed $\Delta m = 0.3$, we come to the conclusion that the photocentre must be located closer to the less-massive component relative to the centre of mass.

The Pulkovo series of photographic observations of a wide visual double star ADS 15600 AB contains 42 plates, but 37 plates are obtained for 3 seasons 1993-95 and only 5 plates - for a period 1983-89. To exclude extra systematic errors observations were carried out near meridian ($< \pm 30^m$) and we used only one type of plates. Plates are measured with the measuring machine "Ascorecord". There are 15–20 exposures on each plate.

To obtain the photocentre orbit we used a usual method. The following systems of 42 equations are solved by the least-squares method:

$$\begin{aligned} x_i &= x_o + \dot{x}_o(t_i - t_o) + AX_i + FY_i, \\ y_i &= y_o + \dot{y}_o(t_i - t_o) + BX_i + GY_i, \end{aligned}$$

where A, B, F, G are the Thiele-Innes elements,

$$X_i = \cos(E_i) - e \qquad andqquadY_i = \sqrt{1 - e^2} sin(E_i)$$

The orbital coordinates X_i, Y_i depend on the dynamical parameters of an orbit r, t and E, which are defined from spectroscopic orbit.

It is enough to present the wide pair orbital motion by linear formulas, because the observed arc of this orbit is very small due to the long orbital period (3800 years [7]).

The results are presented in Table 1. Parameters P, T, e have been taken from the spectroscopic orbit for the solution I and from the speckle interferometric orbit for the solution II.

TABLE 1. Comparison of orbits for ADS 15600.

	Solution I	Solution II	Control orbit of components A (McA)	a	Spectroscopic orbit of components A (V,S)	a	The best limit errors *
P	2.22 yr	2.25	2.254 ± 5		2.220 ± 6		–
T	1993.23	1993.51	1993.51 ± 4		1993.23 ± 27		±0.02yr
e	0.46	0.59	0.59 ± 1		0.46 ± 1		±0.2
a	–	–	0.019″ ± 2	0.054 ± 2	0.47(AU) ± 2	1.32 ± 5	–
α	0.013″ ± 9	0.010 ± 10	–	–	–	–	±0.002″
i	70° ±53	75 ±34	72 ±1		–		±5°
ω	238° ±42	301 ±44	89 ±1	269 ± 1	106 ± 5	286 ± 5	±12°
Ω	106° ±42	110 ± 44	94 ±1		–		±6°
$\Delta\ m'$	0.25 ±70	0.44 ±75	–		–		–

∗ The values of these errors were obtained by analyzing the artificial examples.

Despite large errors, it is possible to affirm the following:

1) The values i and Ω, independently derived, agree with the control orbit.

2) Comparing the values of ω for the photocentre orbit and for the orbits of components, we prove directly that the photocentre is closer to the less-massive star relative to the centre of mass of a system.

3) Values $\Delta m'$ do not contradict the result of Vickers and Scarfe.

4) The large errors of calculated parameters are expected. The main reasons are: the series of observations is short for such work and the astrometric effect on Y is too small for this star.

We have investigated another algorithm [8] for photocentre orbit determination. This method is useful in case the astrometric effect is more essential in one direction than in others. It must be effective for this star, but now we cannot use this method because the series of observations is not yet sufficient.

The observations of ADS 15600 are being continued. In 3-4 years we plan to obtain more precise orbit, to determine independently dynamic parameters P, T, e and to compare them with control values.

We calculated the orbital elements errors which can be obtained for this star under ideal conditions of observations. These values are calculated by analyzing artificial examples when mean accidental error is equal to $0.01''$. They are presented in the last column of Table 1.

As a conclusion we can say following.

The accomplishment of this programme is being planned for some next years.

The majority of stars included in this programme consists of single-lined spectroscopic binaries. Completing these orbits with the parameters i and Ω, we can find a low limit for parallax from comparison α (in arcsec) and a (in Astronomical unit), as well as evaluate the mass of the secondary component, as it is usually done for dark satellites.

We plan to compare different methods of photocentre orbit determinations. The methods can be used for CCD observations of any double objects.

Thus we have opportunity to improve the technique of determination of a photocentre orbit and to receive new results for elected stars.

The author is very grateful to Drs. Alexey A. Kiselev and Andrey A. Toko-vinin for discussion, useful advises and their attention to this study.

References

1. A.A.Kisselev,O.V.Kiyaeva,N.A.Shakht (1992) Astrometry with a long-foci telescope at Pulkovo, *in "Problems of the study of the Universe", St.Petersburg*, **Vol. no. 13**,p. 142–164 (in Russian).
2. O.V.Kiyaeva (1992) The astrometric study of ADS 48 - a nearby double star with a probable invisible satellite, *ASP*, **Vol. no.32**,p. 330–332.
3. O.V.Kiyaeva, N.A.Shakht (1992) The study of ADS 5983 (δGeminorum) motion, *ASP*,**Vol. no.32**,p. 349–351.
4. A.Batten, J.Fletcher, D.MacCarthy (1991) Eighth catalogue of the orbital elements of spectroscopic binary systems.
5. C.R.Vickers, C.D.Scarfe (1976) A spectroscopic study of the triple system ξCephei, *PASP*,**Vol. no.88**,p. 944–948.
6. H.A.McAlister (1980) Speckle interferometry of the spectroscopic binary 17 ξCephei A, *Astrophys. J.*,**Vol.no. 236**,p. 522–525.
7. G.Zeller (1965) *Ann. Sternwarte Wien*,**Vol.no 26**,p. 112.
8. O.V.Kiyaeva (1995) The half-automatic algorithm for a determination of the photocentric orbit of the star which has an invisible satellite with a known period, *in "Computer Methods of Celestial Mechanics-95", ITA RAN, St.Petersburg*,p. 125-126 (in Russian).

THE AUTOMATIC MACHINE "FANTASY" EMPLOYMENT FOR THE MEASUREMENTS OF THE STARS WITH DARK COMPANIONS

N.A.SHAKHT, E.V.POLYAKOV AND V.B.RAFALSKY
Pulkovo Observatory, 196140, Saint-Petersburg, Russia
e-mail: shakht@gao.spb.su

keywords: measuring machine, stars, dark components

1. Measuring machine "Fantasy"

The measuring laboratory of the Pulkovo Observatory is equipped with "Fantasy" which is an universal measuring machine with wide capabilities. The machine base is a massive cast iron table. Over its polished surface the carriage is moved by two line electric motors in an air bearing. The carriage position is determined by a laser interferometer. The scanning system consists of a Cathode Ray Tube (CRT) and a measuring photo multiplier (PMP) with a plate in-between. The scanning beam trajectory is specified by the program. If after an analysis the current image can be classified as the star image then its diameter and the image center coordinates will be measured.

Table 1. Some Characteristics of "Fantasy".

Positioning system

carriage movement field	400 × 400 mm
carriage positioning time	not longer 4 s
error of positioning	less 1 micron
carriage position measurement accuracy	0.32 micron
rate of carriage movement	330 mm/s

J. A. Docobo et al. (eds.), Visual Double Stars: Formation, Dynamics and Evolutionary Tracks, 99–106.

Scanning system

working field of CRT	$4 \times 4mm$
pixel scanning matrix dimensions	4096×4096
pixel size	1×1 micron
aperture of scanning beam	$2 - 3$ microns
pixel sampling time	40 microsec ($25Kpx/s$)
dynamics	256 levels ($0 - 2D$)

2. Measurements of Plates

This automatic complex has been put into practice for measurements of a set of plates of Pulkovo 26-inch refractor with the star Gliese 623

(AC 48° 1595/1589), [$R.A.$ = $16^h22^m.6$, $Decl.$ = $+48^o28'$ (1950.0) $m_v = 10.3$ sp $dM3$, $\pi = 0.''138$]. (Gliese, 1969)

It is known that this star is a double system whose second component Gliese 623 B is a dark object with a lower stellar mass (0.08 solar mass) which is close to substellar one , see, for instance, (Marcy *et al.*, 1989). The set of observations at Pulkovo in 1979 – 1994 has been measured earlier by means of semiautomatic machine "Ascorecord" and some results of this measurement with determination of proper motion, parallax and the preliminary photocentric orbit have been presented (Shakht, 1995).

A list of 89 individual relative positions in 1979 – 1994 and the residuals reflected the orbital motion Gliese 623 A under the influence of a dark component with a mass of 0.09 ± 0.03 solar mass has been given (Shakht, 1996). Now these plates have been used as a test for the comparison of visual bisections with automatic ones.

The observations of this star have been made at Pulkovo since 1979 yr to present by means of 26-inch refractor ($D = 65cm$, $F = 10.4m$, $M = 19.''86$ in $1mm$, the field for the plates, $13cm \times 18cm$ is $40' \times 1^o$).

The set of 1979 – 1995 consists of 100 plates with average 6 exposures on each plate.

All of these plates have been measured by means of "Fantasy" during some days together with the process of the preparation. The total duration of the measuring consisted of about 10 hours.

Then the data of a measurement have been processed by means of special computer programme with comparison of each individual relative position of the central star for individual exposure with the "mean" position on the plate . This method allows one to control all exposures and cut off some bad ones by means of adopted criteria. This technique is applied at Pulkovo for treatment of parallactic serie (Kisselev, 1982) and it is useful for more impartial estimation of the weights for individual plates.

The data about reference stars are given in Table 2. X, Y are the distances of stars with respect to the main star, μ_x, μ_y are relative proper motions determined in this reference system.

Table 2. The reference stars.

N^o	m_{vis}	X(mm)	Y(mm)	μ_x	μ_y
1	10.5	-23.94	+14.26	+0."008	-0."001
2	9.0	-16.84	-54.88	+0.014	-0.003
3	11.0	+11.67	-40.08	-0.017	-0.004
4	10.2	+24.71	+52.94	-0.008	-0.001
5	9.1	+31.83	-22.86	+0.008	+0.006
6	11.7	+15.92	+33.69	+0.001	-0.003
7	11.3	-31.95	+30.44	-0.029	+0.001
8	11.1	-24.64	+27.25	+0.024	-0.017

The reference stars have been chosen according to Palomar Atlas close to the magnitude and color of the object.

They were used on the plates of the set 1979 – 1990 when the observations were carried out with a rectangular shutter $13cm \times 18cm$. Since 1991 due to some problems with maintenance of the telescope it has been substituted for a new petal-shutter, and the working field of the plate has decreased and thus we can measure only 5 stars $N^o N^o 1, 3, 5, 6, 8$. Then we used two systems of the reference stars for comparison: the system I with 8 stars for set 1979-1990 and the system II with 5 stars in which all of set 1979-1995 has been treated.

The comparison of the accuracy between automatic complex "Fantasy" and Ascorecord for the system I is shown in the Table 3, where the mean square errors of one exposure σ_x, σ_y in microns are given. N is a number of plates.

Table 3. The errors of one exposure of the object.

	σ_x	σ_y	N
automatic	1.4	1.4	75
visual	1.7	1.9	72

We have attemted to measure some plates by means of "Fantasy" which could not be measured with "Ascorecord" because of poor quality and 50% of them have given satisfactory results.

The error of the mean position on one plate with average number of the exposures 5.7 is 0.6 microns for "Fantasy" and 0.9 microns for "Ascorecord" and it is 0."012 and 0."018 in the scale of the 26-inch refractor.

Then each plate has been related to the standard plate which has been chosen in the middle of interval of observations at the moment 1987.3. The following formulae have been used:

$$x_i - x_{i(st)} = ax_i + by_i + \Delta X$$
$$y_i - y_{i(st)} = a'x_i + b'y_i + \Delta Y \qquad (1)$$

where x_i, y_i – measured coordinates of each reference star on the current plate, $x_{i(st)}$, $y_{i(st)}$ – corresponding coordinates on the standard plate, a, a', b, b' – constants of the plate, $\Delta X, \Delta Y$ – position of the object on each plate with respect to the standard one.

The residuals obtained after calculation of the constants of the plate gives the mean square error of the reduction σ_{xy}^*.

This error equals :

$$\sigma_{xy}^* = (\sigma_m^2 + \sigma_\mu^2(\Delta t)^2)^{1/2}$$

where σ_m is the error of the measurements of the reference stars σ_μ – the error which depends on the proper motions of the reference stars , Δt – the interval of time between the standard and current plate.

The errors of measurements of one exposure for the "mean" star can be estimated on the basis of observations which are away from the standard plate no more than 1 year with the purpose of precluding the influence of motions of reference stars.

These errors obtained on the basis of automatic and visual measurements are given in micron in the table 4.

Table 4. The errors of one exposure for reference stars.

	σ_{mx}	σ_{my}
automatic	1.7	2.1
visual	2.1	2.4

Then the following equations have been solved:

$$\Delta X_j^{'} = C_x + \mu_x(t_j - t_o)$$

$$\Delta Y_j^{'} = C_y + \mu_y(t_j - t_o) \qquad (2)$$

where $\Delta X_j^{'}, \Delta Y_j^{'}$ are the "geliocentric" positions of the object on each plate eliminating the influence of the parallax motion, μ_x, μ_y – are the proper motion of the object and C_x, C_y – are constants which depends on the errors of the standard plate.

The following values of the proper motion have been obtained:

$\mu_x = 1.''1476 \pm 0.''0012$; $\mu_y = -0.4457 \pm 0.''0012$(m.e.).

The error of the unit weight obtained in consequence of the solution of equations (2) σ_o is 0."038, whereas that based on the visual measurements was equal to 0."044.

It should be noted that this error does'not characterize the accuracy of the set because it also contains the orbital motion of Gliese 623A around the baricenter of the system.

The mean yearly residuals Rx , Ry with their mean errors ϵ_x, ϵ_y obtained after the excluding proper motion and parallax and confirmed the orbital motion of Gliese 623A are given in the Table 5 in milliarcsec. N is the number of plates.

Table 5. The mean annual residuals with their mean errors.

DAte	*N*	R_x marcs	ϵ_x marcs	R_y marcs	ϵ_y marcs
1979.28	7	-18	8	-46	8
1980.24	8	+36	5	+34	4
1981.23	3	-54	4	+44	2
1982.30	6	-26	9	-28	6
1983.25	7	+20	7	-2	10
1984.29	4	+44	8	+36	7
1985.30	5	-46	7	-2	6
1986.24	3	-22	10	-34	4
1987.25	7	+32	6	-14	4
1988.30	8	+16	8	+56	2
1989.27	5	-28	6	0	12
1990.30	9	- 2	5	-24	3

The mean errors of the mean yearly place are 0."0069 in X and 0."0057 in Y for "Fantasy" as the mean number of the plates is about 6 and reaches 0."0020 in the some cases for the system I .The corresponding errors for "Ascorecord " equal 0."0076 and 0."0086.

Our investigation of the dependence of the accuracy for the different systems of reference stars has shown that some decrease the accuracy begins when the number of reference stars is less than 6 ones. The mean errors of one normal place for all of set 1979 – 1995 for "Fantasy" with the system II is 0."0082 for two coordinates, the corresponding values for "Ascorecord" were equal 0."0115.

Also the first astrometric observations of 51 Peg (Gliese 882)

$$[\ R.A. = 22^h55.^m0 \quad Decl. = +20^o30' \ \ (1950.0)$$
$$m_v = 5.^m5\ ;\ \text{sp G4Y}; \quad \pi = 0.''073\ ]\ \text{(Gliese, 1969)}$$

were made with 26-inch refractor in 1995-1996. One is known that this star has a planet orbiting it with an orbital period about 4. 2 days with a mass of about 0.5 Jupiter mass (Mayor *et al.*, 1995), but the precise radial velocities are indicated a presence of second component having a mass greater than 10 Jupiter masses and period greater than 1 year, (Marcy, 1995) (private communication). After this communications we have decided to include this star in the Pulkovo programme of stars with unseen components because we may hope to verify its second component by means of precise astrometric observations for some years.

Now this material consisted of 12 plates with 7 exposures and 13 reference stars . This star has a lower Decl. and its altitude reaches only 51° over the horizon during the observations near the meridian at Pulkovo, when the corresponding altitude for Gliese 623 is about 78°.

A neutral filter with Δm equaled 3 magnitudes has been used for the weakening the brightness of the main object . The results of the measurements by means of "Fantasy" have errors of one exposures about 2.2 microns and the internal error of one plate is estimated as 0.75 micron or 0."015.

We intend to continue the observations of this star and hope that this accuracy will allow one to detect any perturbations with an amplitude about 0."01 – 0."03 on the basis of observations for some years. Taking into account a value of parallax of this star we may hope to detect the influence of the satellites with a small mass on its motion.

3. Conclusions

The automatic mashine "Fantasy" has shown an increase of the internal accuracy of measurements of a single star in 1.3–1.4 times and the external accuracy which is characterized by the mean errors of the yearly normal plate in 1.2 times .

Also this machine has confirmed the orbital motion of Gliese 623A and gives it with enough high precision. This test gives us the hope for a treatment of all our series of the stars with dark companions and also parallactic sets with great productivity and with most high precision which may give the ground photographic observations.

4. Acknowledgements

We thank G.Marcy for his information about 51 Peg, also we thank all observers of 26-inch refractor for taking part in observarions, A.A.Kisselev for useful discussion and I.I.Kanaev for an attention to this work and support with technical equipement.

References

Gliese W., (1969) Catalogue of Nearby Stars. *Veröff. of Astr.Rech.Inst.Heidelberg* **Vol 22,** *p.p. 3–116*

Kisselev A.A., (1982) A Determination of Trigonometric Parallaxes of Stars from Observations in Hour Angles", *Izv. GAO* n_o 199, *pp. 3–11*

Marcy G.W. and Moore D.,(1989) The Extremely Low Mass Companion to Gliese 623*Astroph.Journ.*n_o 341,*pp. 961–967*

Marcy G.W,(1995) (private communication)
Mayor F. and Quelos D.(1995), *Nature,n_o378.pp. 355–359*
Shakht N.A.,1995, The Observations of Gliese 623 and Some Other Objects with Suspected Unseen Components. *Proc.Coll.IAU n_o* 166, *,p. 359*
Shakht N.A., A Study of the Motion of the Star Gliese 623 on the Basis of Observations at Pulkovo, *Astr. and Aph. Trans.* (in the Press)

UNE NOUVELLE MÉTHODE DE DÉTERMINATION DES ÉLÉMENTS DE POSITION D'UNE ÉTOILE DOUBLE VISUELLE À PARTIR D'UNE IMAGE ACQUISE AVEC UN DISPOSITIF À TRANSFERT DE CHARGE.

EDGAR J. SOULIÉ
CEA-DSM-DRECAM-SCM et URA 331 du CNRS,
CEA/Saclay
F-91191 Gif-sur-Yvette Cedex, France.

AND

GUY MORLET
Société Astronomique de France
3, rue Beethoven 75016 Paris, France.

Abstract. Thanks to a CCD camera, the light map of a visual double star is acquired as an array of numbers. A new empirical function is proposed to model the light intensity from a star as a function of the distance of a point in the field to the center of the star image, namely the product of a Cauchy-Lorentz function and of a Gauss-Laplace function. Using the least squares criterion and the algorithm of Levenberg-Marquardt for the minimization of a sum of squares, the coordinates of the components and their difference of magnitude are determined. Software in the FORTRAN and C programming languages are developed.

Résumé. Grâce à une caméra avec dispositif à transfert de charge (DTC), la carte des éclairements d'une étoile double visuelle est acquise sous forme d'un tableau de nombres. Une nouvelle fonction empirique est proposée pour modéliser l'éclairement dû à une étoile en fonction de la distance d'un point du champ au centre de l'image stellaire, à savoir le produit d'une fonction de Cauchy-Lorentz et d'une fonction de Gauss-Laplace. En utilisant le critère des moindres carrés et l'algorithme de Levenberg-Marquardt pour la minimisation d'une somme de carrés, on détermine les coordonnées des composantes et leur différence de magnitude. Des logiciels en FORTRAN et en langage C sont en cours de développement.

Keywords: visual double stars, CCD, Levenberg-Marquardt algorithm, minimisation, position elements, difference of magnitude

1. Introduction

Deux sortes de méthodes permettent la mesure des étoiles doubles visuelles:
— les méthodes micrométriques (à fils, à double image, à grille de diffraction, etc.),

J. A. Docobo et al. (eds.), Visual Double Stars: Formation, Dynamics and Evolutionary Tracks, 107–111.

— les méthodes impersonnelles: photographie, caméra avec dispositif à transfert de charge (en anglais CCD pour "charge coupled device", en français DTC).
La caméra DTC présente les avantages suivants:
— elle autorise une durée d'acquisition courte, le plus souvent comprise entre un dixième de seconde et quelques secondes,
— elle permet la détermination précise des éléments de position et de la différence de magnitude.

Par rapport aux méthodes micrométriques traditionnelles, elle présente cependant l'inconvénient d'une moindre résolution qui restreint actuellement la possibilité de mesurer des couples serrés.

Par rapport à la photographie, la caméra DTC présente aussi plusieurs avantages:
— alors que la photographie demande à être développée et fixée, l'image produite par la caméra DTC est immédiatement observable sur un écran d'ordinateur; l'observateur peut faire défiler à l'écran les images successives d'un même couple, vérifier le cadrage et la mise au point et finalement enregistrer seulement les images que la turbulence atmosphérique n'a pas trop dégradées,
— le domaine des éclairements pouvant être enregistrés est plus étendu (plus grande dynamique),
— elle dispense de la mesure du cliché avec un densitomètre, car elle fournit une image numérisée se prêtant à des traitements par ordinateur.

2. Acquisition des images

Le matériel nécessaire se compose:
— d'un télescope ou d'une lunette dont la longueur focale est au moins de l'ordre d'une dizaine de mètres, éventuellement obtenue grâce à un amplificateur de Barlow,
— d'un dispositif à miroir basculant, permettant le pointage, le cadrage et la mise au point à l'oeil,
— d'une caméra DTC, dont les caractéristiques les plus importantes sont des photosites (en anglais pixels) carrés et de petite taille (de l'ordre de la dizaine de microns) et le refroidissement (par effet Peltier ou par circulation d'un fluide réfrigérant),
— d'un logiciel d'acquisition de données, spécifique du modèle de caméra,
— d'un ordinateur.

Les images utilisées pour la mise au point de la méthode et des logiciels de réduction décrits au paragraphe 3 ont été acquises en septembre 1995 à l'observatoire du Club d'Astronomie d'Ajaccio avec le matériel suivant:
— un télescope Schmidt–Cassegrain de 35 cm d'ouverture (Celestron 14) équipé d'un amplificateur de Barlow portant la longueur focale à 12,63 m,
— une caméra DTC de type Hi–Sis 22, conçue par Christian Buil; on trouvera des indications détaillées sur les caméras DTC dans son livre "CCD Astronomy" (Buil 1989).

3. Méthode

3.1. MÉTHODE D'ORIGINE

Vers 1980, H. Jenkner a écrit un logiciel appelé SCAN; ce logiciel fut amélioré ensuite par A. Schermann et surtout par D. Sinachopoulos (Sinachopoulos, 1995). Ce logiciel avait pour objet de réduire les images numérisées d'étoiles doubles. La méthode utilisée consistait à projeter l'image successivement sur deux axes perpendiculaires. Chacune des deux projections fournissait un profil des éclairements observés. Ce profil était comparé au profil des éclairements calculés à l'aide d'une fonction empirique proposée par Franz, dépendant de trois paramètres ajustables. Une somme de ces fonctions, pondérées par deux autres paramètres ajustables, représentait la variation d'éclairement d'un point du profil en fonction de sa position. Le logiciel ajustait les paramètres selon le critère des moindres carrés, appliqué à la différence entre les profils observé et calculé. En substituant deux profils à une image, ce logiciel évitait la manipulation de grands tableaux de nombres, mais au prix d'une perte d'information importante; lorsque la direction de la droite joignant les deux composantes est voisine de celle d'un des deux axes de projection, un composante est masquée par l'autre.

3.2. NOTRE MÉTHODE

L'évolution des techniques et matériels informatiques a rendu possible l'utilisation de tableaux de nombres de grande taille, et par conséquent la représentation d'une image dans sa totalité. Nous avons donc choisi de comparer l'ensemble des valeurs échantillonées par la caméra DTC à l'ensemble des valeurs calculées à l'aide du modèle décrit ci-après.

Chaque composante est représentée par une surface de révolution engendrée par une courbe choisie empiriquement pour tenir compte de l'agitation atmosphérique, dont l'effet est généralement prépondérant sur celui de la diffraction. Plutôt que la fonction de Franz, qui n'est pas partout dérivable, nous avons utilisé une fonction ayant la forme du produit d'une fonction de Cauchy-Lorentz et d'une fonction de Gauss- Laplace:

$$f(p,q,x) = \exp(-\ln 2.(x/q)^2)/(1+(x/p)^2)$$

Cette fonction ne dépend que des deux paramètres ajustables p et q, et représente mieux qu'une fonction de Franz le profil d'une étoile réelle. L'éclairement au point de coordonnées cartésiennes X,Y est calculé comme la somme des éclairements dûs aux deux composantes et de l'éclairement du fond du ciel:

$$E^{\text{calc}}(X,Y) = EA \cdot f(p,q,dA) + EB \cdot f(p,q,dB) + R$$

où EA = éclairement au photocentre de la première composante
EB = éclairement au photocentre de la seconde composante
dA = distance du point courant de coordonnées X,Y au photocentre de la première composante
dB = distance du point courant au photocentre de la seconde composante
R = éclairement du fond du ciel

Si les erreurs sur les valeurs échantillonnées des éclairements acquis par la caméra DTC suivent la loi de distribution de Gauss, les valeurs les plus probables des paramètres ajustables correspondent au minimum de la somme des carrés des écarts entre les éclairements observés et calculés. Bien que cette hypothèse ne soit généralement pas vérifiée, le critère des moindres carrés a été retenu, comme dans la méthode d'origine, en raison de sa simplicité.

La fonction que l'on minimise dépend de neuf paramètres ajustables et a pour expression:

$$F = \sum_i \sum_j (E^{\text{calc}}(X_i, Y_j) - E^{\text{obs}}{}_{i,j})^2$$

Cette fonction dépend de façon quadratique des trois paramètres EA, EB et R. En raison de cette dépendance particulière, on détermine les valeurs de ces trois paramètres qui minimisent la fonction ci-dessus en résolvant un système de trois équations à trois inconnues. Après détermination de ces trois valeurs, la somme des carrés des écarts dépend de six paramètres, à savoir les quatre coordonnées cartésiennes des photocentres et les paramètres p et q de la fonction empirique. Toutes les méthodes de minimisation d'une fonction non linéaire sont itératives. Dans le cas où la fonction à minimiser a la forme d'une somme de carrés de fonctions non linéaires deux fois dérivables, l'algorithme de Levenberg–Marquardt (voir, par exemple, Press et al. 1992) permet d'obtenir efficacement le minimum. Pour passer d'une itération à la suivante, Levenberg puis Marquardt ont proposé un compromis dynamique entre l'algorithme du gradient de Cauchy et l'algorithme de Newton. L'un de nous (Soulié 1986) avait proposé d'appliquer leur algorithme à la détermination des orbites des étoiles doubles visuelles, en faisant un changement de variable sur un des paramètres ajustables, qui n'est pas nécessaire dans le cas présent.

4. Les aspects pratiques

Quel que soit le langage de programmation choisi, et quel que soit le problème traité, un logiciel de minimisation d'une fonction comporte trois parties. La première partie, appelée *exécutant,* pilote l'ensemble des calculs. Cette partie appelle une fois et une seule la seconde partie appelée *optimiseur,* qui met en oeuvre l'algorithme de minimisation choisi, en l'occurence l'algorithme de Levenberg–Marquardt. A chacune des itérations successives, l'optimiseur appelle la troisième partie, appelée *simulateur,* qui calcule la fonction que l'on cherche à minimiser, en l'occurence, la somme des carrés des écarts entre les éclairements observés et calculés.

L'un de nos logiciels a été développé en FORTRAN 77 (E. Soulié) et l'autre en langage C (G. Morlet et P. Bacchus). Dans les deux logiciels, les dérivées partielles premières de la fonction "éclairement calculé" sont approximées par des différences finies. Les images acquises par la caméra HiSis-22 sont au format PIC et peuvent être converties au format FITS. Etant donné que la norme du langage FORTRAN 77(voir référence) ne comporte pas la possibilité de lire un fichier dans l'un ou l'autre de ces formats, la réduction d'une image à l'aide du logiciel en FORTRAN 77 est précédée de la conversion de cette image en un fichier de nombres représentés par des caractères ASCII, réalisée grâce à un logiciel auxiliaire en langage BASIC. Un autre logiciel auxiliaire écrit en BASIC permet d'afficher sur l'écran d'ordinateur une image acquise avec un DTC (ou "image observée"), une image calculée ou une "image-différence". Ce logiciel de visualisation d'image permet une première estimation des coordonnées des composantes de l'étoile double. Des valeurs initiales doivent également être données aux paramètres p et q. Pour déterminer des valeurs initiales appropriées de p et q, il est nécessaire de simuler l'image et de comparer visuellement les images observée et calculée avant de procéder à l'ajustement de paramètres. Le logiciel d'ajustement écrit en langage C est autonome et effectue l'ensemble des fonctions assurées par le logiciel en FORTRAN 77 et les deux logiciels en BASIC. Il fonctionne sur micro-ordinateur "compatible PC" (système d'exploitation MS-DOS). Le logiciel d'ajustement écrit en FORTRAN 77 fonctionne d'une part sur micro-ordinateur "compatible PC" et d'autre part sur station de travail SUN (sous le système d'exploitation SOLARIS). Etant écrit conformément à la norme précitée, le logiciel en FORTRAN 77 devrait en principe fonctionner sur tout ordinateur équipé d'un compilateur du FORTRAN 77 pouvant traiter des tableaux de grande taille. Il est avantageux de "fenêtrer" les images pour réduire à la fois la place nécessaire en mémoire et le temps de calcul.

5. Premiers résultats

Les logiciels d'ajustement mentionnés au paragraphe 4 ont d'abord été appliqués à quelques couples relativement écartés, par exemple ADS 2578 et Psc 65. En cinq itérations, la somme des carrés des écarts a sensiblement diminué et les paramètres ajustables se sont stabilisés. Au cours des itérations suivantes, la somme des carrés des écarts diminue très faiblement, les paramètres ne changeant plus de façon significative. Dans chaque cas, les images observée et calculée se ressemblent. Cependant, en raison de l'agitation atmosphérique, les images observées des composantes ne sont pas toujours de révolution - contrairement aux images calculées. L'application au couple STF 3062 a donné un angle de position de 326 degrés, une séparation de 1,5 seconde de degré, une différence de magnitude de 0,8. Ces valeurs sont proches de celles données dans les catalogues. L'obtention d'images utilisables dans le cas d'un couple aussi serré exige que l'agitation atmosphérique soit faible.

6. Conclusion et perspectives

Lorsque la longueur focale est de l'ordre de 12 mètres, les logiciels de détermination des éléments de position écrits en C et en FORTRAN 77 fournissent de bons résultats jusqu'à une séparation des composantes comprise entre 1,5 et 2 secondes de degré. Lorsqu'une dizaine d'images d'un même couple ont été acquises, il devient possible de calculer la médiane, la moyenne et l'écart quadratique moyen de l'angle de position, de la séparation et de la différence de magnitude, et ainsi d'évaluer les incertitudes correspondantes. En principe, pendant un temps d'observation donné, grâce aux caméras DTC, des données plus nombreuses que par la méthode traditionnelle de l'observation à l'oeil avec un micromètre à fils pourront être acquises. Aucune comparaison n'a encore été faite entre la différence de magnitude déterminée par ajustement de paramètres et sa valeur mesurée à l'aide d'un micromètre à double image utilisé comme photomètre différentiel (Durand 1987).

Remerciements.- Les auteurs remercient M. Pierre Bacchus pour de fructueuses discussions, MM. Michel Grenon et Maurice Salaman qui ont permis le contact avec M. Edouard Oblak, et ce dernier qui avait conseillé l'utilisation du logiciel SCAN. Ils remercient également Florence et Pascal Mauroy du Club d'Astronomie d'Ajaccio pour leur collaboration. Enfin, les auteurs remercient M. Paul Couteau qui a aimablement présenté leur communication à Saint-Jacques de Compostelle.

Références

Buil C., CCD astronomy, Willmann-Bell, Richmond, Virginie, Etats-Unis, traduction de l'édition originale publiée en français en 1989.

Durand P. Les étoiles doubles in Les techniques de l'astronomie d'amateur, coordonnateur Patrick Martinez, Société d'Astronomie Populaire, Toulouse, 1987. Tome 2, chapitre XII, §4.9 Le photomètre à double image, p. 695

Press W. H., Teukolsky S. A., Vetterling W. T., Flannery B. P., Numerical recipes in FORTRAN. The art of scientific computing, Cambridge University Press, New-York, seconde édition, 1992. §15.5, pp.675-683.

Sinachopoulos D. 1995, communication privée.

Soulié E. 1986, Astron. & Astrophys., 164, 408

Le langage de programmation FORTRAN, norme NF Z 65-110, AFNOR, Courbevoie, juin 1983; cette norme est équivalente à la norme ISO 1539:1978.

SECTION II

FORMATION AND EVOLUTION

STATISTICS OF YOUNG VISUAL BINARY STARS: IMPLICATIONS FOR (BINARY) STAR FORMATION

HANS ZINNECKER
Astrophysikalisches Institut Potsdam
An der Sternwarte 16
D-14482 Potsdam
Germany

AND

WOLFGANG BRANDNER
Astronomisches Institut der Universität Würzburg
Am Hubland
D-97074 Würzburg
Germany

1. Young binary stars: why care?

It is well known that most main sequence stars are members of binary and multiple systems (e.g. Abt 1983, Herczeg 1984). Recent survey work in nearby star forming regions (such as Taurus and Ophiuchus) has demonstrated that this is also the case for young low-mass pre-main-sequence stars (Zinnecker, Brandner, Reipurth 1992; Reipurth and Zinnecker 1993, Leinert et al. 1993, Ghez, Neugebauer, Matthews 1993, Simon et al. 1995). Therefore, it is fair to say that binary star formation is the rule in star formation, while single star formation (cf. Shu, Adams, and Lizano 1987) seems to be the exception. Statistical studies of pre-main-sequence binary stars are privileged over those of Main Sequence stars in that the former generally carry a closer memory of star formation and of binary formation mechanisms in particular. Also the memory of environmental influences on binary star formation (differences among young binaries born in T vs. OB associations or young star clusters) is not yet lost. We can also turn the question around and ask if there is something about Main Sequence binaries that we can use to learn about binary star formation without resorting to young pre-main-sequence objects. Indeed, the distribution of mass ratios might be such a property, provided mass ratios are frozen in since stellar birth. This is because on the Main Sequence there is a well-calibrated luminosity mass relation, while this relation is time-dependent for contracting pre-main-sequence objects; furthermore, accretion luminosity can contaminate the stellar luminosity in the case of actively accreting T Tauri stars, messing up any mass-luminosity relation.

Fig. 1 shows an example (Sz 24) of a T Tauri star binary system resolved by direct CCD imaging at the 3.5m ESO New Technology Telescope at La Silla. The component separation is 0.″69. This is just wide enough to secure optical spectra of the individual components enabling us to place the two components separately on the H-R diagram. A number of other similar binary objects could also be placed on the H-R diagram (Fig.2).

J. A. Docobo et al. (eds.), Visual Double Stars: Formation, Dynamics and Evolutionary Tracks, 115–125.

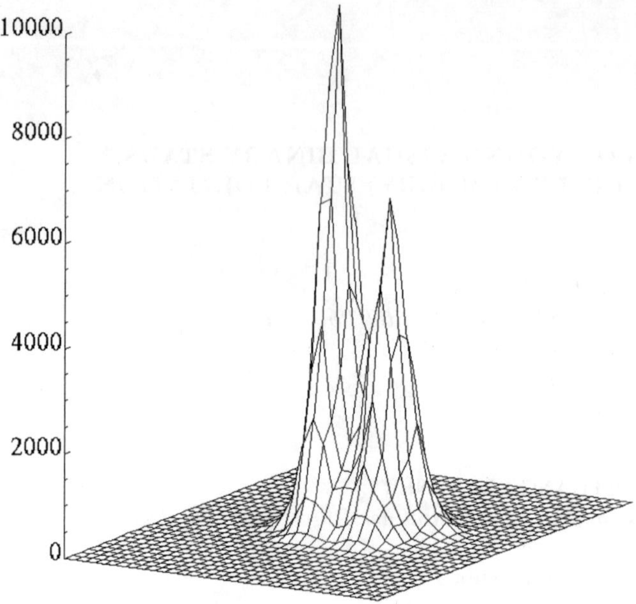

Figure 1. The young visual binary star Sz 24, also known as VW Cha, resolved into its components (projected separation $0''.69$).

One can easily see that if young binaries are not recognized as such, the masses and ages of these systems would be substantially misjudged (cf. Simon, Ghez, and Leinert 1993). This then shows that unresolved young binaries can cause errors in the determination of the Initial Stellar Mass Function and also in the reconstruction of the star formation history.

2. Observational methods

Figure 3 shows the frequency distribution as a function of semimajor axis for solar-type main-sequence binaries (adapted from Duquennoy & Mayor 1991). If we assume the same distribution to be valid for T Tauri stars at a distance of 150 pc, we can estimate which percentage of binaries remains unresolved. Using conventional observing techniques (i.e. direct imaging) under typical seeing conditions (FWHM $1''$) we detect only 30% of all binaries.

In order to achieve a higher spatial resolution more sophisticated observing techniques such as speckle imaging or lunar occultation need to be employed. In the following we give a brief summary of the different techniques used to search for PMS binaries.

$1''$ resolution: **direct imaging** in the optical and near infrared offers seeing limited resolution.

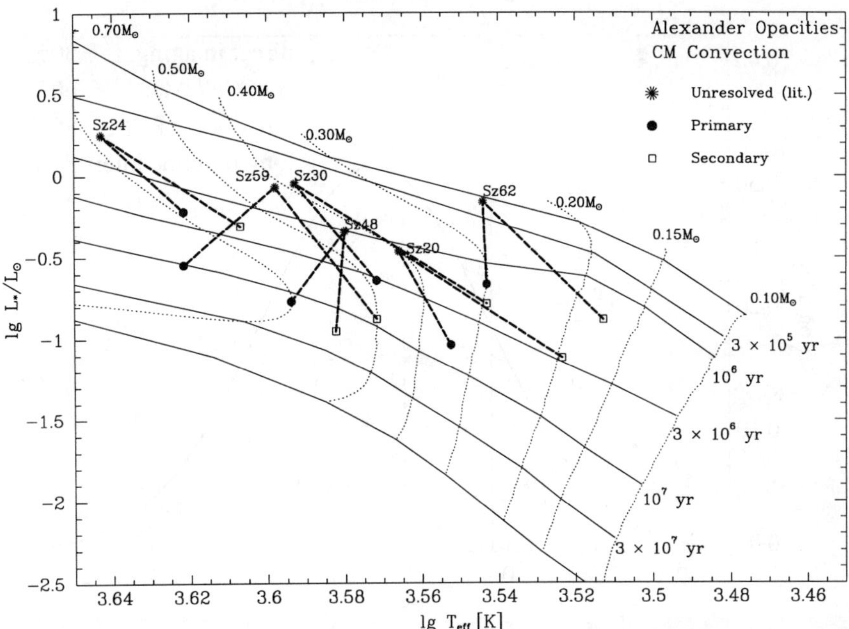

Figure 2. A sample of young binaries in the H-R diagram before and after resolving them into their individual components. It is seen that unresolved binaries lead to an underestimate of the age of a T Tauri star population (Brandner & Zinnecker 1997).

0.″1 resolution: **speckle imaging & adaptive optics** both compensate for the wavefront distortions induced by earth's atmosphere and produce diffraction limited images (e.g. $\lambda/D \approx 0.''1$ for $\lambda = 2.2\mu$m and D = 4m telescopes).

0.″01 resolution: **lunar occultation & HST-Fine Guidance Sensors** (cf. contribution by Otto Franz) provide even higher resolution independent of the seeing, but dependent on the diameter of the telescope and system throughput (effectively the total light collecting power).

0.″001 resolution: **long baseline interferometry** is currently only applicable to bright sources. The new interferometers now under construction (Keck, CHARA, VLTI, LBT) will be able to resolve PMS binaries with orbital periods of a year or even less (a separation of 0.″007 corresponds to 1 AU at a distance of 150 pc).

3. Results

- The **binary frequency** among pre-main-sequence stars is at least as high as among main-sequence stars. Individual low-mass star forming regions (i.e. Taurus-Auriga) show a clear (factor of 2) overabundance of pre-main-sequence binaries with separations $\geq$ 15 AU in comparison to the main-sequence (Leinert et al. 1993, Ghez et al. 1993, 1997). Other regions (i.e. Ophiuchus, Trapezium cluster) appear to have a similar binary frequency as on the main sequence (Simon et al. 1995, Prosser et al. 1994). For binaries with separations $\geq$ 150 AU the overabundance is mild or non-

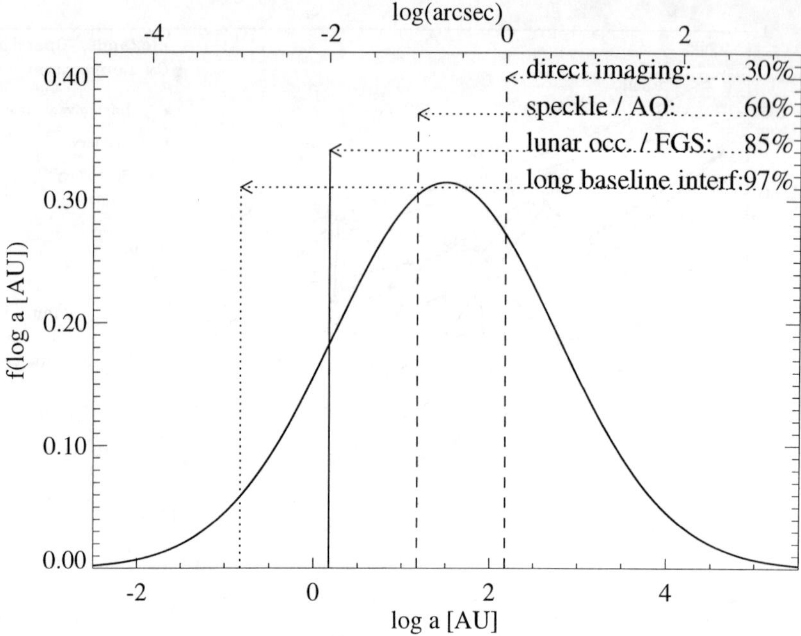

Figure 3. Distribution of semimajor axes for nearby solar-type main-sequence binaries according to Duquennoy & Mayor (1991). The resolution limits of various detection techniques are indicated for an assumed distance of 150 pc (i.e. the distance to the nearest star forming regions).

existent within the statistical uncertainties (Reipurth and Zinnecker 1993, Mathieu 1994). However, there are variations in the binary frequency within the boundaries of certain associations, such as the Sco-Cen OB association; here localized areas exist where the PMS binary frequency is enhanced or suppressed with respect to the main sequence (Brandner et al. 1996).

- The histogram of infrared (1μm) **brightness ratios** I_2/I_1 (Fig. 4a) of the components of wide visual PMS binaries ($\geq$ 150 AU) exhibits a small gradual increase towards smaller ratios, the median being around 0.3 (Reipurth and Zinnecker 1993, Brandner et al. 1996). The corresponding histogram of PMS speckle binaries (now measured at 2μm) looks flat for ratios I_2/I_1 in the range 0.4–1.0 (Leinert et al. 1993). The distribution seems to rise slightly (Fig. 4b) for smaller ratios, but this may be a fluctuation which goes away when a coarser binning is used. However, there is a selection effect against the detection of the smallest brightness ratios at the smallest separations (near the diffraction limit). Therefore it is possible that the frequency distribution of brightness ratios for PMS stars indeed rises towards the smallest ratios (cf. Ghez et al. 1993, 1997).
- The **surface density of companions** for the 40 young Taurus speckle binaries rises with the 2.4 power towards smaller projected separations (Leinert et al. 1993, see also Larson 1995 who finds a slope of 2.15 when combining the data sets of Leinert et al. 1993 and Ghez et al. 1993). Thus the corresponding **semi-major axis distribution** of nearby PMS binaries roughly follows a 1/a distribution in the range of separations

0.″1 – 12″, i.e. 15 AU – 1800 AU. Similarly, the systematic CCD survey among visual T Tauri stars in southern star forming regions produced a sample of 42 visual binaries and 1 triple system, whose distribution of separations also follows closely a 1/a law between 150 AU – 1800 AU (Reipurth & Zinnecker 1993). This is essentially Öpik's law, originally found and confirmed for main sequence binaries (see Allen, Poveda & Herrera, these Proceedings).

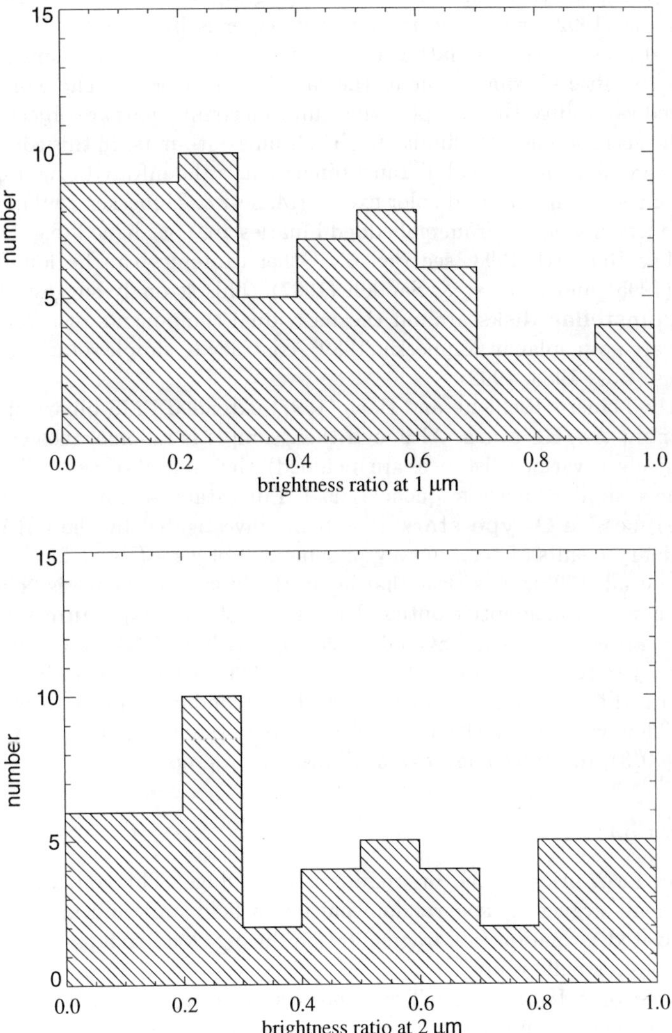

Figure 4. Top: Brightness ratio I_2/I_1 of the components of visual pre-main-sequence binaries at 1 μm (Reipurth & Zinnecker 1993). Bottom: Brightness ratio I_2/I_1 of the components of speckle pre-main-sequence binaries at 2 μm (Leinert et al. 1993).

- The components of almost all close visual pre-main-sequence binaries appear to be coeval to within 1 Myr or less (Brandner & Zinnecker 1997, see Fig. 2). Previous work on this topic (Hartigan et al. 1994) which concluded that only 2/3 of the chosen PMS binary sample had **coeval components** used much wider visual binaries some of which may be optical pairs.
- There is a subclass of young binaries in which some optically visible T Tauri stars have an **infrared companion** (i.e. a companion which is invisible or just barely visible in the optical). Examples include the T Tau system itself, Haro 6-10, UY Aur, and a few others such as VV CrA (Fig. 5). For an early review see Zinnecker & Wilking (1992), while the most recent paper is by Koresko, Herbst, and Leinert (1997). In particular, the latter authors noted that UY Aur changed its state from visible to infrared companion in the last 50 years or so. Therefore, based on the observed variability, they propose that infrared companions are objects which undergo episodic accretion events, similar to FU Orionis outbursts. In this context, it may also be relevant that in classical T Tauri binaries it is the infrared bright component that carries most of the infrared color excess (Moneti & Zinnecker 1991).
- Double jets emanating from embedded binaries are not aligned (e.g. HH 111/HH 121, Gredel & Reipurth 1993; see Fig. 6). Other examples can be found in Hodapp & Ladd (1995) and in Knee & Sandell (1997). This strongly suggests that the planes of **circumstellar disks** around the individual components of a pre-main-sequence binary are not coplanar (Zinnecker 1989), contrary to simple rotational fragmentation schemes (Boss 1988, 1992).
- Recently, Leinert, Richichi, and Haas (1997) extended PMS binary statistics to some 30 **Herbig Ae/Be stars**. Although their sample is inhomogeneous and incomplete (i.e. objects at various distances are included), their initial survey indicates that Ae/Be stars have similar binary frequencies as T Tauri stars. As for the binary frequency of (young) massive **O-type stars**, it is being investigated by the CHARA group and preliminary results of their survey of some 200 objects ($V<8$) have been announced (Mason et al. 1996). It is clear that again the binary frequency is "high".
- Latest news: 2 μm adaptive optics observations of the **Trapezium stars** in the Orion Nebula carried out at the ESO 3.6m telescope with ADONIS have shown Θ^1A to be a $0\farcs2$ binary (a result first found by Petr et al. 1996 with speckle techniques). Moreover, Θ^1B turned out to be a $0\farcs7$ binary, with the secondary being a (very red) $0\farcs2$ binary itself. Thus Θ^1B is a spatially resolved triple system. Finally, Θ^2A is a sub-arcsec binary ($0\farcs3$), too (McCaughrean & Zinnecker, in prep.).

4. Implications

1. Binaries form prior to the pre-main-sequence stage. This and the coevality of the components argues against random pairing of field stars and in favour of some sort of correlated formation (i.e. fragmentation during cloud collapse or independent condensation in small groups).
2. The excess of PMS binaries in T associations like Taurus over field star Main Sequence binary frequencies in the same range of semi-major axes may indicate that quiescent T associations are not the dominant birth place for the field star population. Perhaps PMS stars in dense clusters (like those in Orion) or PMS stars in OB associations (like Sco-Cen), which exhibit less of an excess, are more representative (binary) star formation sites (cf. Mathieu 1996). At any rate, the fact that binary frequencies can

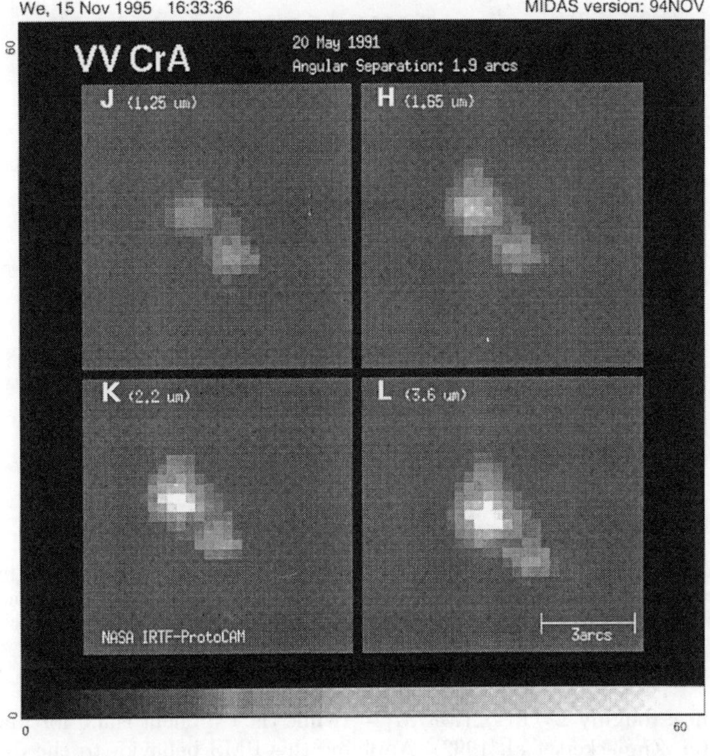

Frame	: mosaic2
Identifier	: image built from polygon
ITT-table	: ramp.itt
LUT-table	: heat
Coordinates	: 0, 0 : 60, 60
Pixels	: 1, 1 : 512, 512
Cut values	: -60000, 200000
User	: stanke

ESO-MIDAS(94NOV); We, 15 Nov 1995 16:33:36

Figure 5. Infrared array images of the VV CrA young binary system (from Stanke & Zinnecker 1997). The infrared companion becomes increasingly brighter at longer infrared wavelengths. At optical wavelengths (not shown here) the infrared companion is barely visible.

differ in different star formation environments may allow us to trace the main origin of the field star population via some kind of "inverse binary population synthesis".

3. The distribution of brightness ratios of PMS binary components reflects the distribution of mass ratios only very crudely, better so at 1 μm than at 2 μm (at 1 μm only the stellar photosphere contributes, while at 2 μm the hot dust in the cirumstellar disk also causes non-stellar emission). Furthermore, the luminosity–mass relation for low-mass coeval PMS stellar pairs is shallower than that for the same pair on the Main

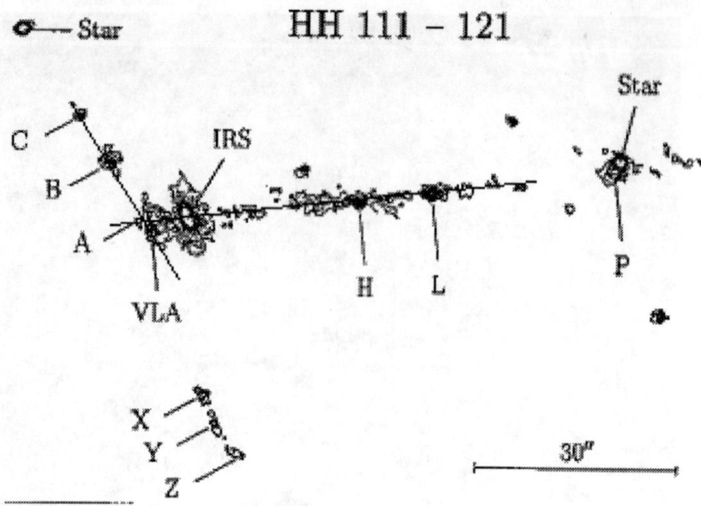

Figure 6. Jets and Herbig-Haro objects arising from the individual components of pre-main-sequence binaries appear to be misaligned. The $2\mu m$ image above (taken from Gredel & Reipurth 1993) shows two series of knots of shock excited molecular hydrogen emission at an angle of $\approx 120°$, emanating from a deeply embedded putative binary system in Orion (IRAS 05491 +0247). The misalignment indicates that circumstellar disks in binary systems may not always be coplanar.

Sequence: roughly $L_2/L_1 \propto (M_2/M_1)^{1.6}$ while the exponent is 3.5 for Main Sequence pairs (cf. Zinnecker et al. 1992). Applying this PMS behavior to the distribution of brightness ratios shown in Fig. 4 yields a histogram of mass ratios where the small mass ratios get depleted, resulting in an approximately flat distribution; also the median brightness ratio of 0.3 become a median mass ratio of 0.5. Then in the range of separations considered all mass ratios are equally probable, arguing against random pairing from a reasonable IMF (slightly so for a Miller-Scalo IMF, strongly so for a Salpeter IMF). However, because of all the PMS uncertainties involved, we recommend to infer the distribution of binary mass ratios from observations of detached Main Sequence binaries, perhaps using Hipparcos data.

4. The distribution of semi-major axes of visual PMS binaries, taken at face value, is scale-free (1/a distribution) down to the resolution limit (15 AU). Alternatively the observed distribution is also consistent with a Gaussian in log a, with a peak around 30 AU (cf. Duquennoy & Mayor 1991). It is important to discriminate by further observations which of the two shapes is the correct one, as each shape requires a different type of explanation. Several theories for the origin of semi-major axis or period distribution have been suggested depending on the adopted shape. A broad Gaussian in log a can be explained by models combining magnetic braking and ambipolar diffusion in protostellar clouds (Mouschovias 1977) or by models of a multi-step hierarchical fragmentation (Bodenheimer 1978, Zinnecker 1984). On the other hand, the scale-free 1/a distribution – valid over a wide range (two orders of magnitude) – may point to orbital decay induced by gravitational drag (dynamical friction) and related angular momentum transport effects (Larson 1996, Silk 1978).
5. It must be noted that the 1/a distribution diverges at small separations which would

seem to suggest that at small separations ($\leq$ 10 AU) a second binary formation process must take over, perhaps related to the scale of circumstellar disks (disk fragmentation, see e.g. Bonnell 1994). At first glance, this second process should produce a different distribution of mass ratios, most likely biased towards rather small mass ratios (because the mass in the disk is small). However, continuing accretion of the higher angular momentum envelope material preferentially onto the "secondary" may tend to equalize the mass ratio, consistent with the existence of PMS spectroscopic binaries with normal mass ratios (see Mathieu's 1996 summary of the census of PMS spectroscopic binaries). The formation of the very close binary systems (a $\leq$ 1 AU) may be yet another story (Bonnell & Bate 1994).

6. The origin of the infrared companions is still a matter of debate. Bonnell & Bastien (1992) and Bate & Bonnell (1997) relate the infrared companion to accretion events onto the more massive binary component, while Artymowicz & Lubow (1994) and Koresko, Herbst, and Leinert (1997) favor the less massive of the two components to be the accretor and hence the infrared companion. While Koresko et al. may well be on the right track with their episodic accretion theory for infrared companions (see above), differential mass-dependent evolutionary effects as well as differential disk inclination effects between the components may also play a role (Zinnecker & Wilking 1992).
7. The proposition of two different binary formation mechanisms is not a consequence of PMS binary statistics but goes back to studies of Main Sequence binaries. It was originally put forward by Abt & Levy (1976) based on their result that wide binaries had a mass ratio distribution different from close binaries, a result later challenged by Halbwachs (1987) and Duquennoy & Mayor (1991). Thus the situation of whether one or two binary formation processes are needed remains unclear; neither Main Sequence nor PMS binary statistics have answered this important question beyond any doubt.
8. High resolution observational studies dedicated to the binary statistics of young massive stars should be pursued more intensively, in an effort to investigate the mass range of their companions. For example, OB stars may have low-mass T Tauri companions which can be detected with the help of adaptive optics in the infrared (e.g., Zinnecker 1996).

5. Future prospects

Up to now, mass estimates for PMS stars are primarily based on the comparison with theoretical PMS evolutionary tracks in an H-R diagram. Different sets of theoretical PMS evolutionary tracks based on different input physics (e.g. D'Antona & Mazzitelli 1994, Forestini 1994, Palla & Stahler 1993), however, yield ambiguous results. In addition, star-disk interactions (accretion luminosity) and uncertain intrinsic colours for young stars makes it often difficult to place these stars in an H-R diagram at all.

With a typical distance of 150 pc to the most nearby star forming regions and a diffraction limit of $0\farcs1$ at 2.2 μm at a 4m class telescope, only pre-main-sequence binaries with separations $\geq$ 15 AU have been resolved so far. Continuous observations over a period of 50 yr or even longer are necessary in order to determine the orbital parameters of these binaries. The new generation of 8m to 10m class telescopes and the large optical interferometers now under construction have the power to resolve PMS binaries in nearby star forming regions with semimajor axis of 1 AU and less. This will enable us to determine orbital parameters, mass ratios, and individual masses of pre-main-sequence stars within a timespan of a year. Thus, these observations will allow for a mass calibration of (theoretical) PMS tracks.

In addition, these very high angular resolution observations will allow us to decide whether the semi-major axis distribution is unimodal or bimodal. From this we will ultimately learn whether there is a single binary formation mechanism which can explain the full range of component separations or two mechanisms – one for wide binaries (filament fragmentation?) and one for close binaries (disk fragmentation?).

References

Abt H.A. (1983) *ARA&A* 21, 343
Abt H.A. & Levy S.G. (1976) *ApJS* 30, 273
Artymowicz P., Lubow S.H. (1996) *ApJ* 467, L77
Bate M.R., Bonnell I.A. (1997) *MNRAS* in press
Bodenheimer P. 1978, *ApJ* 224, 488
Bonnell I.A., Bastien, P. (1992) *ApJ* 401, L31
Bonnell I.A., Bate, M.R. (1994) *MNRAS* 271, 999
Boss A.P. (1988) *Comments in Ap.* 12, 169
Boss A.P. (1992) in *Close Binary Stars*, eds. J. Sahade, G. McClusky, Y. Kondo (Kluwer)
Brandner W., Alcalá J.M., Kunkel M., Moneti A., Zinnecker H. (1996) *A&A* 307, 121
Brandner W., Zinnecker H. (1997) *A&A* in press
D'Antona F., Mazzitelli I. (1994) *ApJS* 90, 467
Duquennoy A., Mayor M. (1991) *A&A* 248, 485
Forestini M. (1994) *A&A* 285, 473
Ghez A.M., Neugebauer G., Matthews K. (1993) *AJ* 106, 2005
Ghez A.M., McCarthy, D.W., Patience, J.L., Beck, T.L. (1997) *ApJ* in press
Gredel R., Reipurth B. (1993) *ApJ* 407, L29
Hartigan P., Strom K.M., Strom S.E. (1994) *ApJ* 427, 961
Herczeg T. (1984) *Ap&SS* 99, 29
Halbwachs J.L. (1987) *A&A* 183, 234
Hodapp K.-W., Ladd E.F. (1995) *ApJ* 453, 715
Knee L.B.G., Sandell G. (1997) *A&A* in press
Koresko, C.D., Herbst, T.M., Leinert, C. (1997) *A&A* in press
Larson R.B. (1995) *MNRAS* 272, 213
Larson R.B. (1996) in *Structure and Evolution of Stellar Systems*, ed. V.V. Orlov, St. Petersburg Univ. Press, in press
Leinert Ch., Zinnecker H., Weitzel N. et al. (1993) *A&A* 278, 129
Leinert Ch., Richichi A., Haas (1997) *A&A* in press
Mason, B.D. et al. (1996) in *The origins, evolution, and destinies of binary stars in clusters*, eds. E.F. Milone & J.-C. Mermilliod, ASP Conf. Ser. 90, p. 40
Mathieu R.D. (1994) *ARA&A* 32, 465
Mathieu R.D. (1996) in *The origins, evolution, and destinies of binary stars in clusters*, eds. E.F. Milone & J.-C. Mermilliod, ASP Conf. Ser. 90, p. 231
Moneti, A., Zinnecker, H. (1991) *A&A* 242, 428
Mouschovias T. (1977) *ApJ* 211, 147
Palla F., Stahler S.W. (1993) *ApJ* 418, 414
Petr M., Beckwith S.V.W., Coude du Foresto V., McCaughrean M.J., Richichi A. (1996), AG Abstract Ser. 12, 175
Reipurth B., Zinnecker H. (1993) *A&A* 278, 81
Shu F.H., Adams F.C., Lizano S. (1987) *ARA&A* 25, 23
Silk J.I. (1978) *Protostars & Planets I*, p. 172
Simon M., Ghez A.M., Leinert Ch. (1993) *ApJ* 408, L33
Simon M., Ghez A.M., Leinert Ch. et al. (1995) *ApJ* 443, 625
Stanke Th., Zinnecker H. (1997) *A&A* sub.
Zinnecker H. (1984) *MNRAS* 210, 43
Zinnecker H. (1989) in *Low-mass Star Formation and Pre-Main Sequence Evolution*, Proc. ESO Workshop, ed. B. Reipurth, p. 447
Zinnecker H., Brandner W., Reipurth B. (1992) in *Complemantary Approaches to Double and Multiple Star Research*, eds. H.A. McAlister & W.I. Hartkopf, p. 50

Zinnecker H., Wilking B. (1992), in *Binaries as Tracers of Stellar Formation*, Cambridge Univ. Press, eds. A. Duquennoy & M. Mayor, p. 269

Zinnecker H. (1996) in *The origins, evolution, and destinies of binary stars in clusters*, eds. E.F. Milone & J.-C. Mermilliod, ASP Conf. Ser. 90, p. 57

THE MAXIMUM AGES OF TRAPEZIUM SYSTEMS

HELMUT A. ABT
Kitt Peak National Observatory, National Optical Astronomy Observatories, Box 26732, Tucson, AZ 85726-6732, USA

AND

CHRISTOPHER J. CORBALLY, S. J.
Vatican Observatory Research Group, University of Arizona, Tucson, AZ 85721, USA

Abstract. Allen et al. (1977) compiled a catalog of 968 possible Trapezium systems by scanning the IDS for systems of three or more stars whose relative separations within the system did not exceed a factor of 3. We explored whether a sample of 265 of those systems were actually physical systems. We used CCD pictures in U, B, and V magnitudes to obtain evidence for common proper motions and consistent color-magnitude and color-color diagrams. We also obtained a spectral classification for at least one star in each system to get the space reddening and distance.

We found only 14 possible Trapezium systems; the remainder are 126 purely optical systems, 109 physical pairs, 14 hierarchical systems, and two small clusters. Nine of the Trapezium systems have ages less than 25 million yr, so that probably is the maximum age of such unstable systems. The remaining five systems are probably hierarchical systems appearing to be Traepzium systems in projection. The reason why so few of the proposed Trapezium systems are physical systems is because the faint IDS magnitudes are both systematically and accidentally very incorrect, causing the attempt by Allen et al. to eliminate faint background stars to fail in most cases. But for the Allen et al. systems with O5-B2 primaries, 45% were found to be true Trapezium systems.

1. Introduction

Trapezium systems consist of three or more stars with roughly equal separations while hierarchical systems are ones with very unequal separations. Typically a hierarchical system consists of a pair with a distant third star, or a pair with a distant pair, or more complex configurations. Because multiple-star systems tend to be spherical (Worley 1967), rather than coplanar, obviously there will be some hierarchical systems that appear in projection against the plane of the sky to be Trapezium systems. Ambartsumian (1954) called those pseudo-Trapezium systems and estimated that 9% of the hierarchical systems will appear to be Trapezium systems.

J. A. Docobo et al. (eds.), Visual Double Stars: Formation, Dynamics and Evolutionary Tracks, 127–131.

Trapezium systems are dynamically unstable; the orbits of the stars are not closed and they will evolve into hierarchical systems or disperse. Allen and Poveda (1974, 1975) have estimated that their maximum ages should be a few million years.

Early studies of Trapezium systems were made by Ambartsumian (1954) and Sharpless (1954). They compiled catalogs of such systems and found the best examples to occur in nebulosities, such as in the Orion Nebula, whose central cluster lends its name to the term "Trapezium systems."

In order to find a large sample of Trapezium systems independent of age and association with nebulosities, and to study their characteristics, Allen, Tapia, and Parrao (1977) scanned the IDS catalog (Jeffers, van den Bos, and Greeby 1963) in punched-card form for such systems. Realizing that the background of faint stars can badly contaminate multiple-star systems with optical companions, they devised a conservative "1% filter." That is, they required that for a given star in the area of a multiple-star system, the chances of finding a field star of that magnitude within that area must be less than 1%. With that filtering assumption they scanned the 53,836 systems in the IDS and isolated 968 in which the ratio of maximum to minimum separations was less than 3. That list has not been published because of the unexpected characteristics of those systems and one test of their reality.

Allen et al. (1977) discussed the characteristics of the 968 possible Trapezium systems and found most of them to have primaries of intermediate types, namely A to K. Such systems are not young unless the primaries are supergiants and it is unlikely that there are so many undiscovered supergiants.

Abt (1986) studied spectroscopically a sample of those proposed Trapezium systems and found that two-thirds of them were badly contaminated by optical companions. In other words, the "1% filter" was not working to eliminate them. He attributed that to the poor IDS magnitudes that are based on the Bonner Durchmusterung magnitudes and are often too bright. He realized that the solution to this delemma lay in obtaining better magnitudes.

2. Method and results

The present authors devised a project to obtain CCD pictures in the Johnson-Morgan U, B, V system for a sample of the 968 proposed systems. Those observations were obtained with a Kitt Peak 0.9-meter telescope during 10 clear nights spread out over one year. The CCD chip had the limitation that all stars must be fainter than V = 7 because brighter stars are saturated during the minimum exposure time of 1 sec. Also pairs closer than the seeing limit of 1-2" or more would not give good photometric measures. Finally, systems farther south than a declination of -20° would not give good photometry. We ended up observing 265 systems. The preliminary results from 64 of them was given during the Korean meeting in 1990 (Abt and Corbally 1993).

We found that the all-sky photometry based on Landoldt (1973) standards and using the IRAF reduction program gave results of reasonable accuracy. For stars of V = 7-14 mag the mean errors are ±0.04 mag in V (see Fig. 1), ±0.03 mag in B-V, and ±0.04 mag. in U-B.

However for stars fainter than V = 14 mag the results deteriorate rapidly and most of those data were not used.

The accuracy of the astrometry was surprisingly good as judged by data on 69 pairs observed twice. For stars brighter than V = 10 the errors in position angle (see Fig. 2)

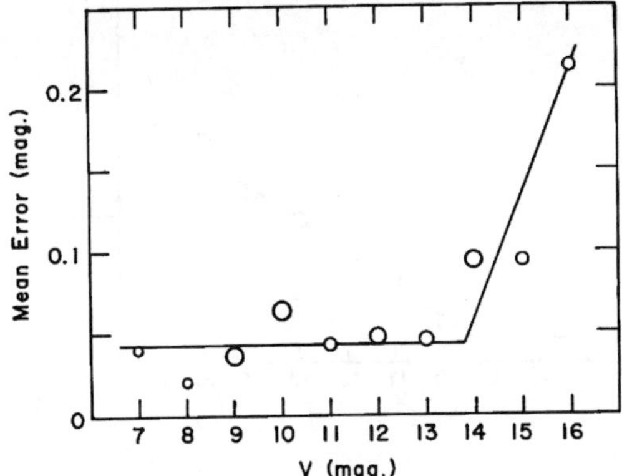

Figure 1. The measured standard deviations in V magnitudes based upon multiple measures of 89 stars. The sizes of the circles indicate the relative numbers of observations per point.

averaged $\pm 0.015°$ and were less than $\pm 0.1°$ to V = 14. The errors in separation (see Fig. 3) were ± 0.01" for stars brighter than V = 11 and remained <0.05" to V = 14. These measures were compared with the visual ones usually obtained roughly 100 years ago to search for common proper motions.

For each system we observed not only the three or more stars listed in the Allen et al. catalog but also all stars within the 2.5'x4.1' field and brighter than about V = 14 mag. The average is 6.2 stars per system. Although half of those do not have published astrometry in the past, ours can be used by future double-star observers. In addition we obtained with the Kitt Peak 2.1-m and the Steward Observatory 2.3-m the spectrum for classification of at least one star in each system to determine the space reddening toward the system and whether the primary is a main-sequence star or more luminous.

For a star to be considered a physical member of a system it had to pass up to three tests: (1) for stars with older astrometry the position angle should have remained constant during roughly 100 yr to within 2° and the separation should have been constant to less than 0.7"; (2) the star's position in a V vs B-V diagram should fit that of the other members within ± 1 mag in V; (3) the star's position in a U-B vs B-V diagram should fit within ± 0.15 mag. or 4 times the estimated deviation. Illustrations of these tests were given by Abt and Corbally (1993). We found that in 126 of the 265 systems the companions of the primaries are all optical. In 109 systems the primary had one physical companion only, the remaining being optical. Two systems are small clusters with 8 and 12 members. Fourteen systems are hierarchical in the sense that they consist only of pairs and single stars that have ratios of separations greater than 3. Finally there are 14 systems that are Trapezium or Trapezium + hierachical systems.

For the 14 Trapezium or Trapezium + hierarchical systems we noted the primary types. If the primaries are off the main sequence, we assumed a 1.5-mag. bolometric brightening from the main-sequence turnoff.

For nine of the 14 Trapezium systems the primary types or main-sequence turnoff types are O5-B2. Those represent 45% of observed systems with such primary types, so a large fraction of the O5-B2 systems are indeed Trapezium systems. The remaining five Trapezium systems have B6-F8 primaries and are probably projected hierarcical systems. How

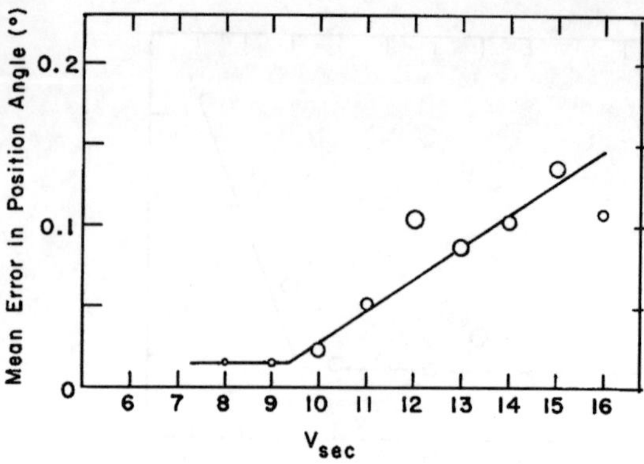

Figure 2. The measured standard deviations in position angles from multiple measures of 69 pairs.

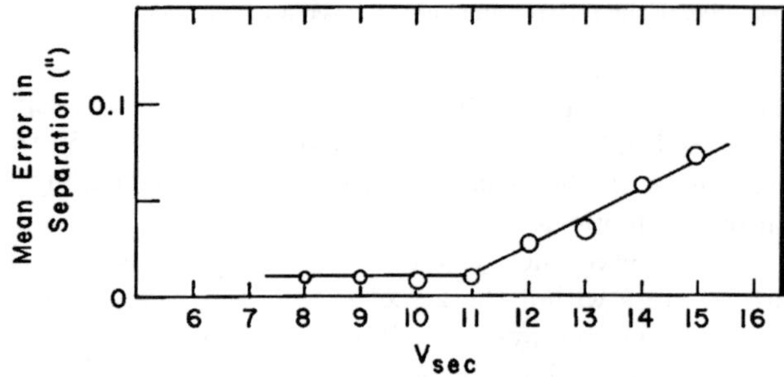

Figure 3. Fig. 3-The measured standard deviations in separations from multiple measures of 69 pairs.

does that compare with Ambartsumian's prediction of 9% pseudo-Trapezium systems? We do not know because we do not know how many hundreds of hierarchical systems Allen et al. bypassed when they compiled a catalog of possible Trapezium systems. Even a conservative estimate of the known hierarchical systems will easily account for five pseudo-Trapezium systems.

What are the maximum ages of O5-B2 stars? The isochrones by Bertelli et al. (1994), using a log T_{eff} = 4.29 for a B2 star, give an age of 25 million yr. That is a maximum age because not all of the B2 stars are about to leave the main sequence, and the O stars are younger. Therefore the maximum age of Trapezium systems is 25 million yr.

Why is such a large fraction, namely 89%, of the proposed Trapezium systems not even physical multiples? The answer, again, is in the poor magnitudes of the IDS catalog. For 300 stars we compared the stated visual magnitudes with our CCD V magnitudes. The results are shown in Figure 4. We see that at V = 7 the IDS too faint by 0.4 ±0.4 mag (s.e.). From V = 8-10 they are systematically roughly correct but with a similar scatter. But by V = 12-14 they are too bright by 1.0 ±1.0 mag. This means that 84% of the stars stated by the IDS to be V = 12.0 mag are actually fainter than that. This systematic error plus the large scatter means that the Allen et al. 1% filter let through far too many faint background stars.

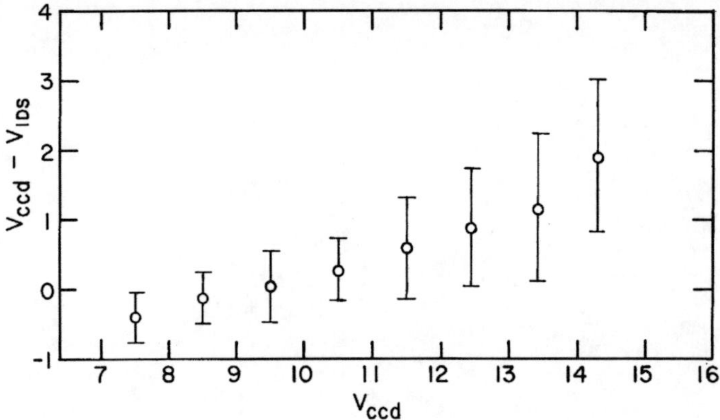

Figure 4. The mean systematic errors (circles) and their measured standard deviations between the CCD and IDS visual magnitudes for 300 stars.

One can wonder about the seemingly small number of Trapezium systems, but we can note that among those with O5-B2 primaries (or similar main-sequence equivalents) the success rate for finding them was 45%; we should not find any true Trapezium systems for late-type primaries.

References

Abt, H. A. 1986, *ApJ*, **304,** 688.

Abt, H. A., and Corbally, J., S. J. 1963, in New Frontiers in Binary Star Research, ed. K.-C. Leung and I.-S. Nha (ASP Conf. Series 38), **72**.

Allen, C., and Poveda, A. 1974, in IAU Symp. 62: The Stability of the Solar System and Small Stellar Systems, ed. Y. Kozai (Dordrecht: D. Reidel Pub. Co.), p. 239 — 1975, PASP, 87, 499.

Allen, C., Tapia, M., and Parrao, L. 1977, *Rev. Mex. Aston. Ap.*, **3,** 119.

Ambartsumian, V. A. 1954, Contr. Byurakan Obs., **15,** 3.

Bertelli, G., Bressan, A., Chiosi, C., Fagotto, F., and Hasi, E. 1994, *A&AS*, **106,** 275.

Jeffers, H. M., van den Bos, W. H., and Greeby, F. M. 1963, Pub. Lick Obs., **21,** 1.

Landoldt, A. U. 1973, *AJ*, **78,** 959.

Sharpless, S. 1954, *ApJ*, **119,** 334.

Worley, C. E. 1967, in On the Evolution of Double Stars, ed. J. Dommanget, p. 221, Comm. Obs. Roy. Belgique, Ser. B, No. **17**.

THE DISTRIBUTION OF SEPARATIONS OF WIDE BINARIES

CHRISTINE ALLEN, A. POVEDA AND M.A. HERRERA
Instituto de Astronomía
Universidad Nacional Autónoma de México

Abstract. Several recent observational determinations of the distribution of separations of wide binaries are critically reviewed and compared with the distributions inferred from observational work on pre-main sequence stars and from theoretical ideas on binary formation. It is concluded that it is not very fruitful to depict the whole distribution function of separations by one law, and that it is preferable to represent separately the two main mechanisms of binary formation, namely first-collapse fragmentation (for $s > 25$ AU) and disk fragmentation (for $s < 25$ AU). It is also found that the best representation for the group of wide binaries ($s > 25$ AU) is given by Öpik's law $\rho(s) = ks^{-2}$, but that this distribution has a cutoff that is age-dependent. This cutoff is interpreted in terms of the dynamical evolution of wide binaries.

1. Introduction

The observed distribution of separations of binaries is a result both of conditions prevailing during the star formation phase, and of effects of subsequent dynamical evolution. Fortunately, it has now become possible to begin to disentangle both effects, because recent observational work on pre-main sequence binaries and multiples allows us to have a glimpse at stars that have formed very recently, and that presumably have not evolved significantly from their initial state. Of top importance among the properties of such stars is the primeval distribution of separations. We shall therefore begin by reviewing the observational properties of pre-main sequence stars, and comparing them with those of main-sequence stars.

Companions to pre-main sequence stars have been detected in large numbers. This indicates that binary and multiple star formation is pervasive, and that binaries are formed as early as during the protostellar collapse phase, by fragmentation of the protostellar cloud. Indeed, recent surveys of the frequency of companions to pre-main sequence stars (henceforth PMS) indicate that this frequency (about 50% or larger) is at least equal to, if not greater than, the observed binary frequency for main sequence stars. This implies that the great majority or even the totality of binary stars originate prior to the pre-main sequence phase, i.e., during protostellar collapse. Thus, the classical mechanisms of orbital capture and rotational fission postulated to occur during or after the phase of contraction to the main sequence have now become largely superfluous. In view of the traditional difficulties that these mechanisms encountered this is a welcome development.

J. A. Docobo et al. (eds.), Visual Double Stars: Formation, Dynamics and Evolutionary Tracks, 133–143.

We study first the main observational results on PMS binaries, as well as modern theoretical mechanisms of binary star formation. With these results as guidelines, we then compare some PMS binary properties with those of main sequence binaries.

Section 2 presents a summary of the main observational results on PMS stars. Section 3 is devoted to a brief description of recent theoretical work on binary and multiple star formation. Within the framework set by both observational and theoretical work, some properties of main sequence stars are presented and discussed in Section 4. The main results and conclusions are summarized in Section 5.

2. Observational data on pre-main sequence binaries

Studies of pre-main sequence binaries began fairly recently -about a decade ago. The reason for this late start is that the required high angular resolution techniques, namely infrared speckle and imaging devices, as well as facilities for lunar occultation work became widely available only quite recently. Recall that $0.5''$ corresponds to about 50 AU at the distance of the nearest star-forming regions.

PMS binaries have been detected in many nearby star-forming regions, most extensively in Taurus-Auriga (Simon et al. 1992, Simon 1992, Ghez et al. 1993, Leinert et al. 1993, Richichi et al. 1994), in Ophiuchus-Scorpius (Ghez et al. 1993) and in a number of other southern star forming regions (Reipurth and Zinnecker 1993, Brandner et al. 1996). Speckle as well as lunar occultation techniques have yielded at least 66 new systems in the Taurus-Auriga region, and similar numbers in other regions. The frequency of binaries is at least 51% (and may be as high as 60%) in the separation range between about 2 and 1800 AU (ibid.). Reipurth and Zinnecker, for instance, find from a survey of 11 southern star forming regions a frequency of 16% in the range of separations they observe, which they extrapolate to a total binary frequency of 80%. A similar fraction, 16.4% was found by Brandner et al. (1996) for a sample of X-ray and H-alpha selected T-Tauri stars. Prosser et al. (1994), in contrast, find for the Trapezium cluster a binary frequency quite comparable to the main sequence value. The fraction of multiples is also high, ranging from about 13 to 35%. It is interesting to note that these multiples are generally hierarchical.

The distribution of separations for these PMS binaries (Figures 1, 2) appears to be quite flat for the longer periods ($P > 100$ y), when plotted as a function of the logarithm of the periods or the observed separations (Reipurth and Zinnecker 1993, Leinert et al. 1993, Ghez et al. 1993, Padgett 1995, Prosser et al. 1995, Mathieu 1996). A distribution of the form $g(\log s) = \mathit{constant}$, as found for these larger separations, is equivalent to $f(s) = ks^{-2}$ (where s is the surface density of companions), a distribution first found by Öpik (1924) to represent well the wide main sequence binaries. Many authors have compared the distribution of separations they find for PMS binaries to the one established by Duquennoy and Mayor (1991) for main sequence stars, which has approximately a Gaussian form. However, most of the surveys of PMS binaries become seriously incomplete for the short periods, whereas for the longer periods they tend to be more complete. In this region the flat shape of the frequency plot as a function of the logarithm of the observed separations is quite striking. This is also the separation region for which the Duquennoy and Mayor's distribution becomes indistinguishable from Öpik's distribution (see discussion in Section 4, below, and also Poveda et al., this Volume).

PMS short period binaries have also been searched for by spectroscopic work. A number of radial velocity surveys have been carried out (see, e.g., Mundt et al. 1983, Mathieu et al. 1989, Mathieu 1992, Mathieu 1994). They are found in fractions comparable to those of main sequence solar-type stars. However, much work remains to be done to reliably establish the distribution of separations of such close PMS pairs.

Within regions of recent or ongoing star formation, PMS binaries are found at all ages, even at the limit of star formation, i.e., at the stellar birthline. Embedded binaries and protobinary candidates have also been observed (Mathieu 1994). Many studies of such regions have shown that the principal properties of the main sequence binary population have already been established at the PMS phase. Some examples are:

1. Periods extending from days to about 10^6 years (corresponding from a few to more than 10^4 AU) are found.
2. The period distribution for the whole range of periods observed can be represented approximately by a bell-shaped curve, with a median of a about a few hundred years. However, the period distribution for the wide pairs is flat in $\log p$.
3. The eccentricities of the shortest-period binaries tend to be zero, whereas for longer periods a broad range of eccentricities is found, with maximum eccentricity increasing with increasing period.
4. A large range of secondary masses is found among the long-period binaries, with a distribution that resembles the initial mass function.
5. Multiple systems are abundant. Many are of hierarchical type. The orbital planes of multiple systems wider than about 30 AU seem to be randomly oriented (Bonnell, this Volume).

In addition to these results, which indicate that many important properties of main sequence binaries are established very early in their lifetimes, it is also found that circumstellar and circumbinary material is common among PMS binaries; in many cases, accretion upon the stellar surfaces has been detected. A few PMS show infrared companions.

3. Theoretical studies on binary and multiple star formation

Larson's (1995) studies on star formation in groups indicate that a dense molecular cloud undergoes two phases of collapse: a self-similar isothermal collapse followed by a subsequent, thermally supported contraction of each individual Jeans mass fragment. The changeover occurs at separations of about 8000 AU. These two phases would correspond to the regimes of formation of clusters and binary or multiple stars, respectively (Figure 3). This separation is very roughly the maximum separation observed for PMS or bound main sequence binary stars, and is also the length at which the observed surface density of companions plotted versus their separation shows a break in slope (Ghez et al. 1993, Leinert et al. 1993, Gomez et al. 1993, Simon 1992). If this break is confirmed by further research, it would set a natural limit for the maximum separation of a binary at birth, which would be about 10 000 AU. If wider binaries are observed, they must then result from mass loss or from dynamical evolution. The distribution of separations for the binary regime studied by Larson is of the form $\rho(s) = ks^{-2}$ (where $\rho(s)$ is the surface density of companions), and that found for the cluster regime is $\rho(s) = ks^{-0.62}$. Thus, both distributions are clearly different, and the binary distribution agrees with that found by Öpik for main sequence stars, as well as with the flat, $g(\log s) = constant$ distribution found by other workers for PMS binaries.

The most realistic models of star formation start from a prolate spheroidal (2:1 axial ratio) molecular cloud with a Gaussian density profile (Boss 1993). Uniform and power-law density laws have also been modelled, but uniform profiles are clearly unrealistic, and power-law profiles do not in general favor fragmentation into multiple stars. Uniform, initially elongated clouds produce chains of condensations. Calculations with 3-dimensional hydrodynamic codes of the collapse of Gaussian profile clouds do produce fragments corresponding in separation to wide binaries (a few hundred AU). Multiples, both of hierarchical and of trapezium type, can also form at this stage. This is referred to as the first dynamical collapse, and it leads to the formation of the first protostellar core, or cores. Again, it is interesting to note that extremely wide binaries seem not to be produced. A second collapse then takes place, after the dissociation of molecular hydrogen reduces the gas pressure support of the first protostellar core. The hydrodynamical evolution of a variety of first-core models has been followed. The results are either rotationally flattened disks, or rings. The disks usually contain central cores, and mass accretion onto them produces trailing spiral arms, which can then condense to form companions; rings tend to fragment directly into several protostellar objects.

In summary: the most sophisticated 3-dimensional hydrodynamical codes have followed the first collapse of a cloud, up until the first stellar cores are established. Frequently, cloud fragmentation takes place. This process gives rise to a variety of wide binaries and multiples, both of hierarchical and of trapezium type. Characteristic separations at this stage range from a few hundred AU to a few thousand AU.

A second collapse of the individual fragments then ensues. This second collapse leads to the formation of rings or disks around the stellar core. These disks interact with the central core and accrete more material, which processes result in the formation of spiral arms that develop condensations to form close companions. Further evolution of the rings leads directly to fragmentation. The scales involved in this second collapse give companions with separations as small as a few solar radii.

Much work remains to be done, but it is tempting to conclude both from the observational properties of PMS binary stars discussed in Section 2, and from the theoretical studies on cloud collapse, that the widest PMS binaries observed may be either the low-limit end of the process that gives rise to clusters, or else a result of mass loss of the components or of dynamical interactions of the PMS binary with its dense environment. Within this scheme, less extreme PMS binaries with separations of a few tens to a few thousands AU (henceforth referred to as wide binaries) would form as a result of the first collapse of a dense protostellar cloud, whereas the shorter period pairs (the close binaries) would originate during the second collapse, as a result of disk or ring fragmentation.

4. Observational data on main-sequence stars

Work on this topic has a long and distiguished history. Already back in 1924, Öpik published a comprehensive study on binary star statistics, based on a preliminary, unpublished version of the Aitken Catalogue of Visual Binary Stars. He established that the distribution of separations of such binaries can be well represented by the law $\rho(s) = ks^{-2}$, where $\rho(s)$ is the observed surface density of companions. Obviously, the frequency distribution for the number of companions will follow a law of the form $f(s) = ks^{-1}$. Later, in 1937, Ambarsumian showed that the distributions that result from statistical equilibrium are $\rho(s) = ks^{-3/2}$ for the observed separations and $f(e) = ke$ for the eccentricities. Therefore, the observed distribution of separations for visual binaries does not correspond to the one predicted for dynamical equilibrium, and hence it is unlikely to be the result of dynamical interactions either in clusters or with field stars. Ambartsumian re-established the validity of Öpik's law, based on more complete material available in the final Aitken Catalogue. He also showed that if the distribution of the spatial separations of components about their primaries follows a law $\rho(s) = ks^{-n}$, where s is the separation in AU and n is any number, then the distribution of observed (i.e., projected) separations will follow a law of the form $\rho(s) = ks^{-(n-1)}$, where s is now the separation in arcseconds; furthermore, the summing up of such distributions for different volume elements along the line of sight and for different directions in the sky will also lead to a law of the form $\rho(s) = ks^{-(n-1)}$ (Ambarsumian 1937).

These results, which we reproduce here because they are not widely known, are extremely useful when comparing theoretical distributions with observed ones, because they allow us to infer the true space distribution of companions from the observed separations, as long as they can be represented by power laws. They will be further discussed in a forthcoming paper (Poveda et al., in preparation).

Different laws for the distribution of separations have been proposed. Kuiper, back in 1942, found that taking together close and wide pairs the separation distribution could be fitted by a bell-shaped curve with a median separation of about 30 AU. Of course, Kuiper had in mind one single mechanism of formation for all binaries, from the closest spectroscopic pairs to the widest common proper motion pairs. Due to the enormous range of angular momentum represented by the wide range of separations of these systems, a single formation mechanism has never been successful (Bodenheimer 1995). Nonetheless, a bell-shaped curve in the logarithm of separations has been used repeatedly as a convenient representation, most recently by Duquennoy and Mayor (1991) in their extensive compilation of nearby solar-type binaries. This is fine if all one seeks is a mathematically simple representation of the observed data. However, if one wants to draw connections between the observed distributions and possible scenarios of formation or dynamical evolution, it is more fruitful to fit separately the regimes corresponding to diferent cosmogonical or dynamical scenarios, i.e., the close and the wide binaries. Our studies have followed this approach.

Our interest in this topic goes back to at least 1982, when we published a study on the statistics of the IDS catalogue (Poveda, Allen and Parrao 1982). The distribution of observed separations we found for the visual binaries (after removing pairs likely to be optical) was $\rho(s) = ks^{-2.04}$, in close agreement with that of Öpik (Figure 4), and the distribution found for the close companions of hierarchical triples was $\rho(s) = ks^{-2.07}$ (Poveda 1987). A sample of wide binaries from the LDS catalogue (Luyten 1969) gave as best fit $\rho(s) = ks^{-2.05}$ (Figure 5). All these studies deal with the part of the distribution of separations larger than about 25 AU, i.e., the wide binaries. According to the scheme envisioned above, these binaries would correspond to the PMS binaries formed by cloud fragmentation during the first collapse.

The regime corresponding to fragmentation during first collapse encompasses separations from a few tens to a few thousands AU. A convenient lower limit seems to be given by separations of about 25 AU, which is the separation found originally by Abt and Levy (1976) to delimit different secondary mass distributions. Secondary masses of binaries wider than about 25 AU follow Salpeter's law (or their luminosities the van Rhijn luminosity function), whereas closer pairs tend to favor more nearly equal masses. Support for the hypothesis that the individual components of wide systems are independently generated by the luminosity function was found, among others, by Close et al. (1990) and Duquennoy and Mayor (1991). However, the matter of the distribution of secondary masses of the close pairs has become quite controversial in recent years, with the controversy being focused on the close separations. So, for example, confirmation of Abt and Levy's result has been given by Mazeh and Goldberg (1992), ant Mazeh et al. (1992), whereas discrepant results have been found most notably by Duquennoy and Major (1991).

The difficulties of deriving a secondary mass distribution for limited samples of single-lined spectroscopic binaries have been stressed by Heacox (1995), who arrives at the pessimistic conclusion that an increase of an order of magnitude in the available number of stars is necessary before reliable results can be given. Nevertheless, Heacox (ibid.) states that his analysis is "probably compatible with the mass ratio distribution reported by Mazeh et al. (1992)". So, although the matter clearly merits further study, the evidence indicates that the secondary mass distribution of the close pairs does not correspond to that of Salpeter, whereas that of the wide pairs does, and that the dividing line occurs at separations of about 25 AU, or periods of about 100 years. Thus, the differences found for the secondary mass distributions of close and wide binaries are compatible with the scheme of binary star formation in which the two types of binaries originate by distinct physical processes during the collapse of protostellar clouds, namely first-collapse fragmentation and disk fragmentation. This view lends support to our approach of separately studying the properties of wide and close binaries, particularly their separations.

There have been several recent studies of binary stars in the solar neighborhood, seeking to obtain inventories complete within certain specified limits. Thus, for example, Close, Richer and Crabtree (1990) have examined a sample of 39 nearby, probably physical, wide binaries, which they state is complete within well defined boundaries ($s > 64$ AU, parallax > 0.040 arcsec, $M_v > 9.0$, and north of -12 degrees declination). For this sample, they have investigated the distribution of separations. They find that the separations of their binaries are well fitted by a power law of the form $\rho(s) = ks^{-2.4}$. If they introduce a cut-off for separations greater than 20 000 AU the exponent of their best fit is 2.3.

Duquennoy and Mayor (1991) completed a study of nearby solar-type stars and found that the period distribution could be well fitted by a Gaussian curve with a median separation of about 30 AU, very similar to the one found by Kuiper long ago. However, if following our approach we restrict ourselves to the wide binaries, specifically, those with separations between 25 AU and 8000 AU, we find that this portion of the distribution is very well represented by a law of the form $\rho(s) = ks^{-1.99}$. In fact, the fit is remarkably good; the Kolmogorov-Smirnov estimator of goodness of fit is $Q = 0.9998$ (Figure 6).

Our group has recently compiled a catalogue of nearby, wide binaries and multiples, which contains an estimation of the ages of the systems listed (Poveda et al. 1994). For this group of binaries, again, the distribution of separations we obtain is $\rho(s) = ks^{-2.04}$. In this study we also find that young binaries follow Öpik's distribution to separations significantly larger than do old binaries. This effect had been found before (Poveda 1988, Poveda et al. 1993; see also Poveda et al., this Volume), and interpreted as a result of the dynamical evolution that the wide binary experiences through encounters with massive molecular clouds, clusters, spiral arms, etc., which will change the shape of the distribution especially at the wide separations, and which can entirely dissociate a binary. All of the results quoted above indicate that the wide part ($s > 25$ AU) of the distribution of separations of both PMS and main sequence binaries is very well represented by Öpik's law. The upper limit for the validity of Öpik's law decreases with increasing age of the binary population from about 8000 AU for the youngest systems to about 2400 AU for the oldest.

5. Conclusions

We briefly summarize now the main results that have emerged from studies of both PMS binaries and their main-sequence counterparts.

1. Observations of many regions of recent or ongoing star formation show that binary stars are present in fractions at least as great as the main-sequence fraction. It has become clear that many main-sequence binary properties are established shortly after binary star formation.
2. The distribution of separations of wide binaries is established during the phase of first dynamical collapse. This phase roughly corresponds to the process earlier envisioned as independent condensations within a protostellar cloud.
3. The distribution of separations of close binaries originates during the second dynamical collapse, as a result of the fragmentation of disks or rings. Therefore, it can be expected to follow a quite different law from the one describing the wide binaries. The mass distribution of secondaries is also different for both types of systems. The relative orientation of orbital planes for multiple systems tends to be random for the wide pairs, and coplanar for the close ones.
4. The distribution of separations of both PMS and main sequence wide binaries studied by a variety of authors is well represented by a power-law, with an exponent very close to -2, corresponding to Öpik's law. The lower limit for this fit is about 25 AU, the upper limit decreases with increasing age of the binary population

References

1. Abt, H.A. and Levy, S. 1976, *Astroph. J. Suppl.*, **30**, 273.
2. Ambarsumian, V. 1937, *Astron. Zh.*, **14**, 207
3. Bodenheimer, P. 1995, *Ann. Rev. Astron. Astroph.*, **33**, 199
4. Boss, A. P., 1993, *Astroph. J.*, **410**, 157.
5. Brandner, W., Alcala, J., Kunkel, M. Moneti, A. and Zinnecker, H. 1996, *Astron. Astroph.*, **307**, 121
6. Close, Richer and Crabtree, D.R. 1990, *Astron. J.*, **100**, 1968
7. Duquennoy, A. and Mayor, M. 1991, *Astron. Astroph.*, **248**, 485
8. Ghez A.M., Neugebauer, G., and Matthews, K. 1993, *Astron. J.*, **106**, 2005
9. Gómez, M., Hartmann. L., Kenyon, S.J. and Hewett, R. 1993 *Astron. J.* **105** 1927
10. Heacox, W.D. 1995, *Astron. J.*, **109**, 2670
11. Kuiper G.P. 1935, Pub. **Ast. Soc. Pacif**, **47**, 149
12. Larson , R.B. 1995, *M.N.R.A.S.*, **272**, 213
13. Leinert, C. et al 1993, **Astron. Astroph 278**, 129
14. Luyten, W.J. 1941-1963, *Publ. Astr. Obs. U. Minn.*, Vol. III, Nrs. 1 to 12
15. Mathieu, R.D. 1992, in *Complementary Approaches to Double and Multiple Star Research*, H.A. McAlister and W.I. Hartkopf, Eds., p. 30
16. Mathieu, R.D. Walter, F.M., and Myers, P.C. 1989, *Astron. J.*, **98**, 987
17. Mathieu, R.D. 1994, *Ann. Rev. Astron. Astroph.*, **32**, 465
18. Mazeh, T. and Goldberg, D. 1992, *Ap. J.*, **394**, 592
19. Mazeh, T. Goldberg, D. Duquennoy, A. and Mayor, M. 1992, *Ap. J.*, **401**, 265
20. Mundt, R. et al. 1983, *Astroph. J.* **269**, 229
21. Öpik, E. J. 1924, *Tartu Observatory Publications*, **25**
22. Padgett, D.L., Strom, S.E., Edwards, S., Dougados, C., Hartigan, P., Ghez, A. and Strom, K. 1996, in *Disks and Outflows around Young Stars*, eds. J. Staude and S.V.W. Beckwith, Springer Verlag, in press.
23. Poveda, 1988, *Astroph. Space Sci.*, **28**, 43
24. Poveda, A. Allen and Parrao, L. 1982, *Astroph. J.*, **258**, 589
25. Poveda, A. Herrera, M.A., Allen, C., Cordero, G., and Lavalley, C. 1994, *Rev. Mexicana Astron. Astrof.*, **28**, 43
26. Prosser et al. 1994, *Astroph. J.*, **421**, 517
27. Richichi, A., Leinert, C., Jameson, R. and Zinnecker, H. 1994, *Astron. Astroph.*, **287**, 145

28. Reipurth, B. and Zinnecker, H. 1993, *Astron. Astroph.*, **278**, 81
29. Simon, M. Chen, W. P., Howell, R.R., and Slovik, D. 1992, *Astroph. J.*, **384**, 212

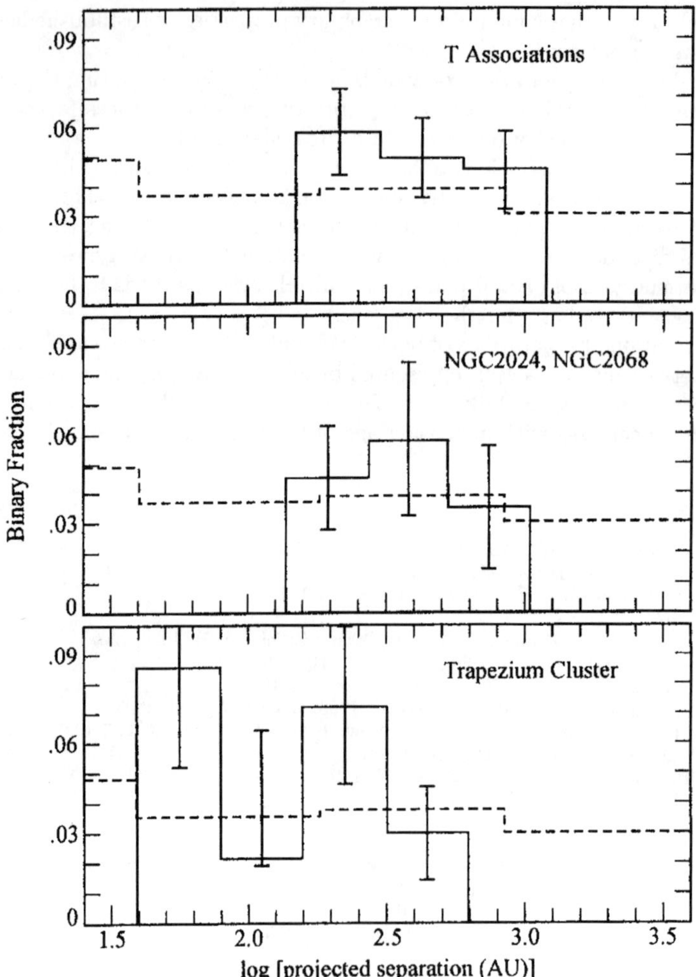

Figure 1. Distribution of observed (projected) separations for pre-main sequence stars in several star forming regions. The top panel shows results by Reipurth and Zinnecker (1993). The middle panel is due to Padgett (1995). The bottom panel shows results obtained by Padgett (1995) by reworking the data of Prosser et al. (1994). Figure adapted from Mathieu (1994).

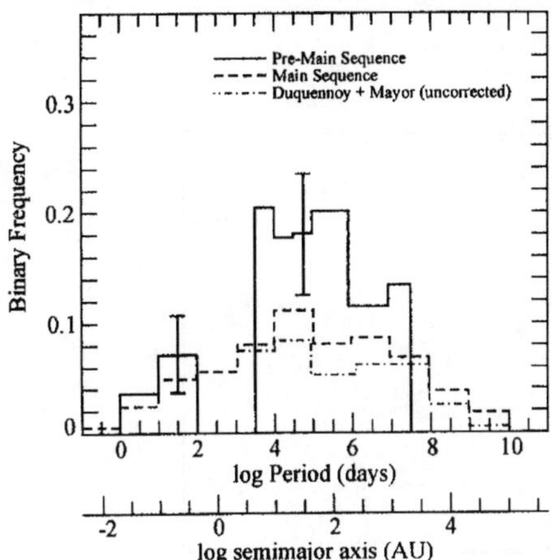

Figure 2. Frequencies among pre-main sequence binaries, as compared with main sequence binaries. The dotted line represents the uncorrected data of Duquennoy and Mayor (1991). Figure adapted from Mathieu (1994).

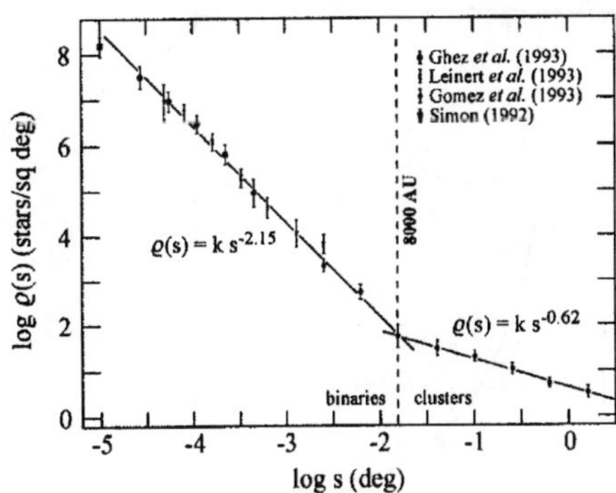

Figure 3. Surface densities of companions for several star forming regions. The break in slope occurs at separations of about 8000 AU. The region corresponding to the wide binaries ($s < 8000$ AU) is well represented by Öpik's law. (Larson, 1995).

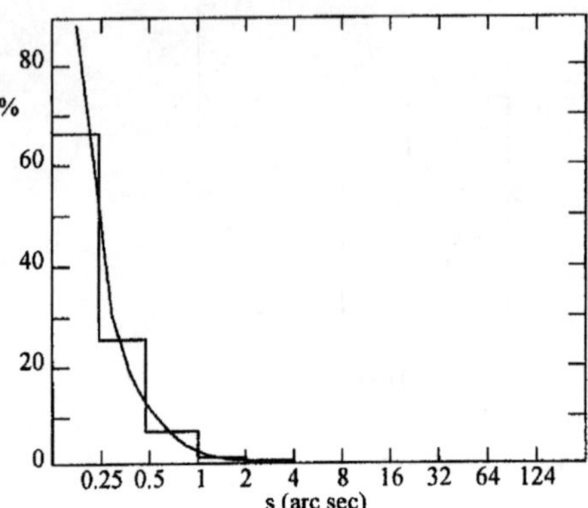

Figure 4. Distribution of observed separations for binaries from the IDS, after removing most optical systems. The histogram corresponds to the surface densities of secondaries. The line is the best fit, $\rho(s) = 2.9s^{-2.04}$, for $s < 32''$. (Poveda 1988).

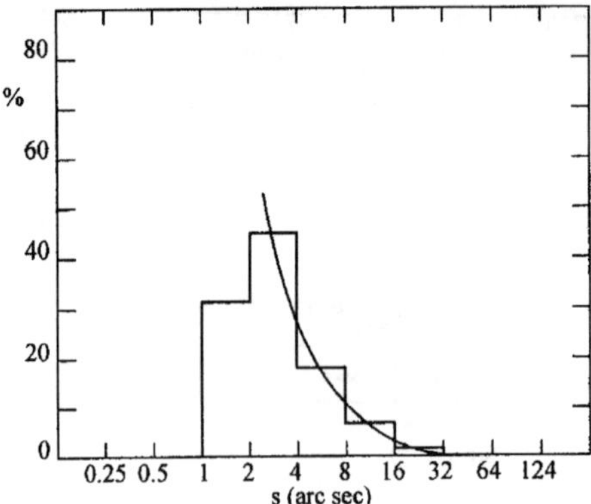

Figure 5. Distribution of separations for a sample of binaries from the LDS catalogue. The ordinates are normalized surface densities. The full line is the best fit $\rho(s) = 544.8s^{-2.05}$, for $s < 128''$. (Poveda 1988).

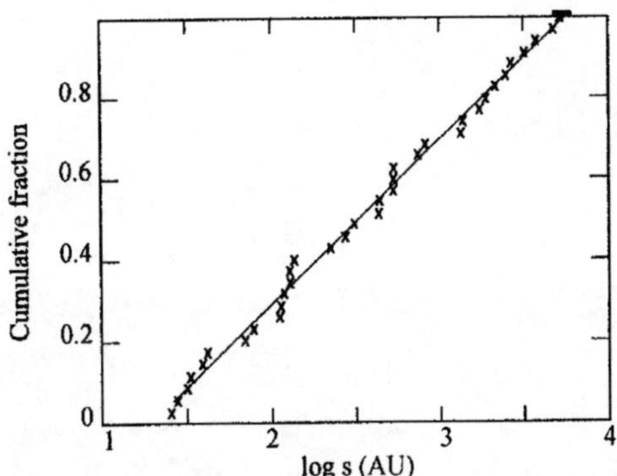

Figure 6. Power law fit to uncorrected data from Duquennoy and Mayor (1991) between separations 25 AU and 8000 AU. The exponent is -1.99, and the KS estimator of goodness-of-fit is $Q = 0.9998$.

THE EFFECT OF ACCRETION ON YOUNG HIERARCHICAL TRIPLE SYSTEMS

K.W. SMITH
Queen Mary & Westfield College,
Department of Physics,
Mile End Road,
London E1 4NS,
UK.

I.A. BONNELL
Institute of Astronomy,
Madingley Road,
Cambridge, CB3 0HA,
UK.

AND

M.R. BATE
Max-Planck-Institut für Astronomie,
Königstuhl 17,
D-69117 Heidelberg,
Germany.

Abstract. We consider the possible effects of accretion on the stability of young triple systems. The dynamics of the accretion depend critically on the specific angular momentum of the infalling gas relative to that of the triple system and on the component masses. We consider a selection of scenarios, with differing specific angular momentum accretion onto systems of various mass and separation ratios. A simple analytical argument is used for cases where the angular momentum of the accreting gas is too low to affect the angular momenta of the triple orbits. For higher angular momentum accretion, we employ the results of a previous numerical study of the effect of accretion onto a binary system. We show that accretion can help stabilise unstable triple systems, or destabilise stable ones. We discuss our results in the context of two multiple star formation mechanisms.

1. Introduction

Studies of multiplicity in stellar systems show that most main sequence stars are in binary systems (e.g. Duquennoy & Mayor 1991) with an even higher proportion among pre-main

J. A. Docobo et al. (eds.), Visual Double Stars: Formation, Dynamics and Evolutionary Tracks, 145–151.

sequence stars such that each star has, on average, at least one companion (Ghez 1995). This implies that a large fraction of the binary systems are members of triple or higher order systems (e.g. ≈ 35 %, Ghez et al. 1993). The frequency of triples on the main sequence is uncertain as most radial-velocity surveys are not designed to search for higher order systems than binaries. However, it has been suggested that at least one third of spectroscopic binaries have visible tertiaries (Herczeg 1988). The higher multiplicity among pre-main sequence stars could then be a result of undiscovered main-sequence triples.

Triple systems are most probably formed (as are binary stars) through a fragmentation process, either through a prompt fragmentation scenario (Pringle 1989), fragmentation during collapse (e.g. Bonnell et. al. 1991; Boss 1992), disc fragmentation after collapse (Bonnell & Bate 1994; Burkert & Bodenheimer 1996) or a combination of the above (e.g. Bonnell & Bastien 1992; Bate, Bonnell & Price 1995). The resultant triples are typically not in a very hierarchical configuration, casting doubt on the survival of the system. Furthermore, the three protostars comprise a small fraction of the total mass of the system, with the majority of the mass in the form of a gaseous envelope, a remnant of the initial cloud core that collapsed. The infall and accretion of this gaseous envelope has significant effects on the system, modifying the component masses and their separations.

2. Stability considerations

Hierarchical triple systems can be considered as comprising a binary system (masses M_1 and M_2) and a 'stand-alone' component (mass M_3) in orbit about each other. The most important feature of a stable triple system is that the binary component is sufficiently hard that the closest approach of the single component cannot disrupt it. We use a stability criterion given by Harrington (1977), which states that a triple with binary separation, $d_{\rm binary}$, and binary-single separation, $d_{\rm triple}$, is stable if

$$\frac{d_{\rm triple}}{d_{\rm binary}} > K\left\{1 + A\ln\left[\frac{2}{3}\left(1+\frac{M_3}{M_1+M_2}\right)\right]\right\}, \tag{1}$$

with $K = 3.5$ and $A = 0.7$ for a co-revolving system.

A binary's separation is a function of the component masses, M_1 and M_2, and the total orbital angular momentum, L, in the system,

$$d = \frac{L^2}{G}\frac{(M_1+M_2)}{(M_1M_2)^2}. \tag{2}$$

We apply this equation, separately, to both the binary separation and the binary-single separation to determine the evolution of the separation ratio for the system. This is used to assess the stability or instability of the system using equation 1.

3. Angular momentum of accreting material

There are three different critical values of the specific angular momentum in a triple system. The first, $J_{\rm binorb}$, is the specific angular momentum of the binary system around the triple's centre of mass. The second, $J_{\rm binary}$, is the binary's specific angular momentum around its barycentre and the third, $J_{\rm single}$, is the specific angular momentum of the single component around the triple's centre of mass. We consider first the case of zero angular momentum infall using a simple analytical argument. We then apply the results of Bate & Bonnell (1996) to examine the effects of higher angular momentum accretion.

Figures 1 and 2 show the effect of different amounts of accreted material for a range of initial separation ratios and for different values of $J_{\rm infall}$. Figure 1 is for a system with a massive binary and less massive single star. Figure 2 is for a system in which the mass of the binary and single components are approximately equal, and the masses of the binary components are unequal. The specific angular momentum of the accreting material is measured in units of $J_{\rm circ} = \sqrt{GM_{\rm triple}d_{\rm triple}}$.

3.1. ZERO ANGULAR MOMENTUM INFALL.

We consider the situation where the accreting material has insignificant specific angular momentum compared to any of the triple orbital angular momenta.

This type of situation might be expected to occur if the triple system forms from a post-fragmentation event, such as capture (three-body or star-disc capture) in a small cluster (e.g. McDonald & Clarke 1995). In this scenario, where the triple forms in a somewhat clustered environment, accretion comes predominantly from the gaseous envelope in which the whole cluster is embedded. The angular momentum of this gas need not be correlated to that of the triple, because the system's motion in the cluster is independent of the pre-fragmented velocity distribution. Thus, this matter has low specific angular momentum relative to the triple system (e.g. Bonnell et. al. 1996).

3.1.1. *Case 1; Massive binary, lower mass companion*

Here we consider the case of a low mass single component in orbit about a much more massive binary system. The binary component will lie near the barycentre of the triple system. Zero angular momentum material will accrete preferentially onto the binary component. we assume initially that $M_1 = M_2$. Applying equation 2 we find that the separation of the binary evolves as

$$\tilde{d}_{\rm binary} = \frac{d_{\rm binary}}{(1+\alpha)^3}. \qquad (3)$$

The binary-single separation is affected less as the single accretes no mass. Thus

$$\tilde{d}_{\rm triple} = \frac{d_{\rm triple}}{(1+\alpha)}, \qquad (4)$$

and the separation ratio therefore evolves as

$$\frac{\tilde{d}_{\rm triple}}{\tilde{d}_{\rm binary}} = (1+\alpha)^2 \frac{d_{\rm triple}}{d_{\rm binary}}. \qquad (5)$$

Thus, as the binary accretes mass but no angular momentum, the separation ratio increases and the system becomes more stable.

The argument given above hinges around the fact that, when we apply equation 2 to the the binary component, both components' masses increase significantly, whereas when we consider the binary-single orbit, only one of the masses increases. If the mass ratio M_2/M_1 (hereafter $q_{\rm binary}$) changes this need no longer be true. In particular, if the binary mass ratio falls below the single/binary mass ratio ($M_{\rm single}/M_{\rm binary}$, hereafter $q_{\rm triple}$), then the stabilising effect may disappear or even act to destabilize the system if the distant single star accretes more mass than the least massive of the binary components.

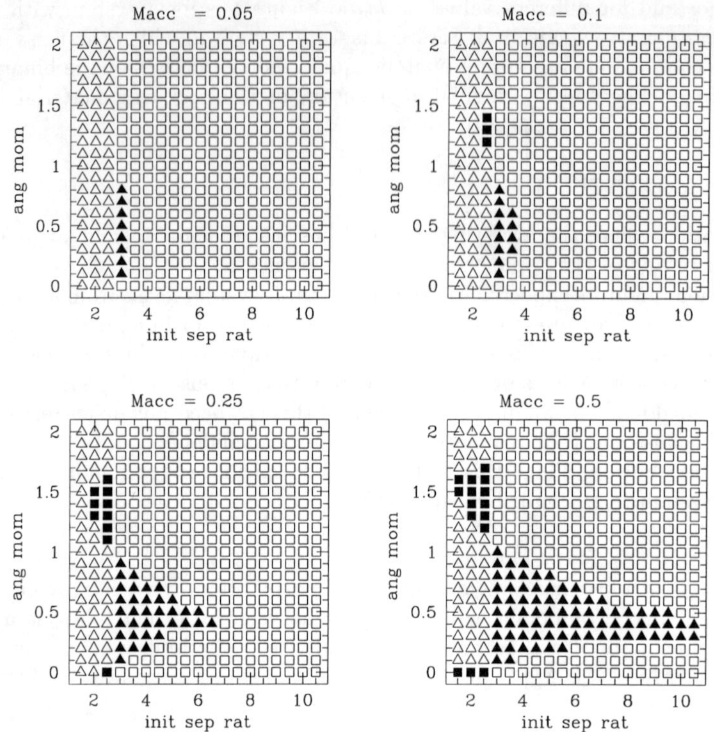

Figure 1. The effect of accretion on the stability of triple systems as a function of the initial separation ratio and the relative specific angular momentum of the accreted material for systems where $M_{\text{single}} = 0.2M_{\text{binary}}$. Open squares denote initially stable systems that remain stable. Open triangles denote initially unstable systems that remain unstable. Filled squares denote initially unstable systems that are stabilised through accretion and filled triangles denote initially stable systems that are destabilised through accretion. The amount of mass accreted, in terms of the total system mass, is given at the top of each panel. Angular momentum is in units of the initial J_{circ}.

3.1.2. *Case 2; Massive single star with orbiting lower mass binary*

In this case, the massive single star will sit close to the triple barycentre, and the zero angular momentum material will accrete preferentially onto it. In many ways, this is the opposite of the case 1 scenario; the single-barycentre separation shrinks, but the binary separation does not. The system then becomes more unstable.

3.1.3. *Case 3; Binary and single components of equal masses*

Since the binary and single components are equidistant from the barycentre in this case, they should each accrete an equal amount of material. If $q_{\text{binary}} = 1$, there should be no effect on stability. If $q_{\text{binary}} < 1$, the system will tend towards instability. (See figure 2).

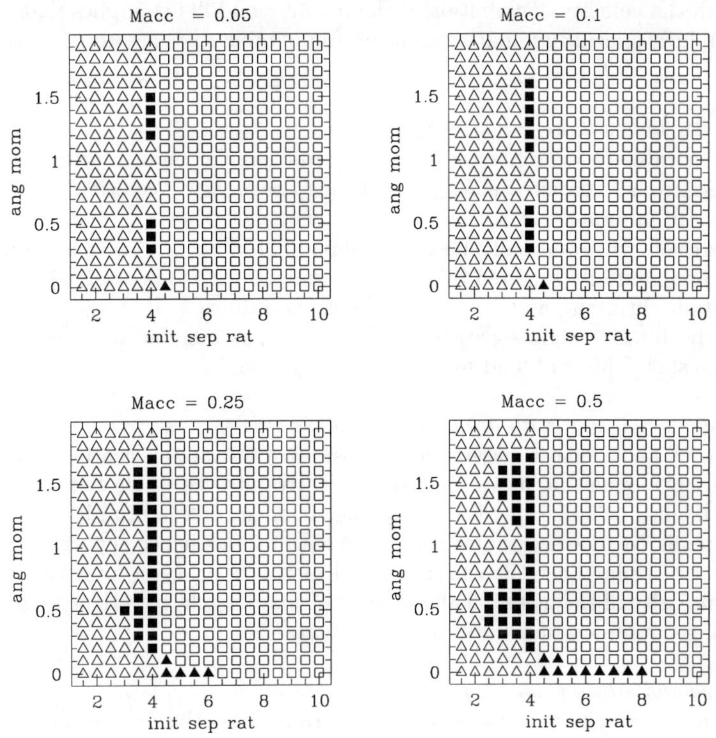

Figure 2. As for Figure 1 except for with $M_2 = 0.2M_1$ and $M_{\text{single}} = M_{\text{binary}}$.

3.2. NON-ZERO ANGULAR MOMENTUM INFALL

If the infalling material has significant specific angular momentum relative to any of the angular momenta of the triple (J_{binorb}, J_{binary} or J_{single}), we can no longer use the arguments outlined above to model the situation. There are two reasons for this. Firstly, the proportion of material accreted onto each component will now be a function of relative specific angular momentum of the accreting material. Secondly, the accretion will alter the appropriate value of L in equation 2, affecting the separations.

To model the effects of accretion in such a complex scenario, we make use of the results of Bate & Bonnell (1996), who carried out a numerical study on the effects of accretion on young binaries. We assume that the separation ratio of our triple systems is sufficiently large that we can regard them as two binary systems. We can then consult the findings of Bate & Bonnell to discover what proportion of the accreting material falls onto each component, the single star and the binary. We then consider the binary component alone, and determine, again from Bate & Bonnell, how much of the binary's share is accreted by each of the components.

Non-zero angular momentum accretion might occur if the triple forms from an isolated fragmentation process. The orbital motions of the resultant system are then likely to be highly correlated with the velocity distribution of the parent cloud. This implies that the infalling material will have specific angular momentum comparable to or greater than that of the triple system ($J_{\rm infall} \gtrsim J_{triple}$).

3.2.1. *Case 1; Massive binary, lower mass companion*

If $J_{\rm infall}$ becomes comparable to the binary's intrinsic specific angular momentum, $J_{\rm binary}$, the accretion will begin to act to increase $J_{\rm binary}$, widening the binary. Accretion onto the distant single star is still minimal, and so the system is destabilized.

Higher $J_{\rm infall}$ accretion will eventually become comparable to $J_{\rm single}$. In this regime, the material will be preferentially accreted by the single component. This can lead to the binary-single separation shrinking and the system being destabilized. If $J_{\rm infall}$ becomes larger than $J_{\rm single}$, the binary-single separation will increase as angular momentum is accreted by the single star. This will tend to stabilise the system.

3.2.2. *Case 2; Massive single star with orbiting lower mass binary*

The single star is now by far the most massive star and sits at the system's centre of mass. Accretion of low angular momentum matter is predominantly onto the single star, pulling the binary closer to it until it is disrupted. High angular momentum material is accreted by the binary, but tends to increase the angular momentum, and hence separation, of it. This destabilizes the triple. There is a narrow band of values of $J_{\rm infall}$ for which the binary accretes the majority of the mass but the individual components accrete no angular momentum, hardening the binary and stabilising the system.

3.2.3. *Case 3; Binary and single components of equal masses*

In this case, the binary is no closer to the system's centre of mass than the single component. Accretion onto the binary and onto the single is equal (since they have equal masses) independent of the specific angular momentum of the infalling gas. With high specific angular momentum infall, the binary accretes angular momentum as well as mass and its separation grows comparably to that of the binary-single. The effect of accretion on stabilising the system is thus reduced. For higher values of $J_{\rm infall}$, the infalling gas is not able to accrete onto the binary and instead forms a circumbinary disc. This limits the growth of the binary's separation while the binary-single separation continues to grow and thus the system moves towards stability.

4. Conclusions

We have considered the effects of accretion onto young triple systems. We have shown that, for zero specific angular momentum accretion, systems with $q_{\rm triple} < q_{\rm binary}$ evolve into more stable states, whereas systems with $q_{\rm triple} > q_{\rm binary}$ evolve into less stable states. Zero angular momentum accretion can occur if the system forms in a clustered environment where the angular momentum of the accreting gas is not linked to the angular momentum of the multiple system. Our findings have clear implications for the expected mass ratios of surviving triples forming in such circumstances.

The accretion of higher angular momentum material, which is expected if the system forms according to some fragmentation model, can act to stabilise or destabilize systems, depending on the exact parameters involved. One clear finding is that accretion onto a system with a very massive single component and a much less massive binary almost always acts to destabilize the system. Such configurations are therefore unlikely to survive significant accretion in their youth.

Applying these findings in reverse, statistical information on the mass ratio and separation ratio distributions of main sequence triples could be used to infer the likely formation environments for multiple systems.

References

Bate M. R., Bonnell I. A., 1997, MNRAS, 285, 33

Bate M. R., Bonnell I. A., Price N. M., 1995, MNRAS, 277, 362

Bonnell I., Bastien P., 1992, ApJ, 401, 654

Bonnell I. A., Bate M. R., 1994, MNRAS, 269, L45

Bonnell I. A., Bate M. R., Clarke C. J., Pringle J. E., 1997, MNRAS, 285, 201

Bonnell I. A., Martel H., Bastien P., Arcoragi J.-P., Benz W., 1991, ApJ, 377, 553

Boss A. P., 1992, in *Close Binaries*, eds J. Sahade, G. McCluskey, Y. Kondo, Kluwer, Dordrecht

Burkert A., Bodenheimer P., 1996, MNRAS 280 1190

Duquennoy A., Mayor M. 1991, A&A, 248, 485

Ghez A. M., 1995, in *Processes in Binary Stars*, eds R. A. M. J. Wijers, M. B. Davies and C. A. Tout, Kluwer, Dordrecht, p. 1

Ghez A. M., Neugebauer G., Matthews K., 1993, AJ, 106, 2005

Harrington R.S., 1977, AJ, 82, 753

Herczeg, 1988, Ap&SS, 142, 89

McDonald J. M., Clarke C. J., 1995, MNRAS, 275, 671

Pringle, J.E. 1989, MNRAS, 239, 361

THE EFFECTS OF ACCRETION DURING BINARY STAR FORMATION

MATTHEW R. BATE
Max-Planck-Institut für Astronomie
Königstuhl 17, D-69117 Heidelberg, Germany

AND

IAN A. BONNELL
Institute of Astronomy
Madingley Road, Cambridge CB3 0HA, United Kingdom

Abstract.

We consider the effects of accretion during binary star formation. When a protobinary system forms within a collapsing molecular cloud core, its final state is determined by the accretion of, or interaction with, the remaining gas as it falls on to the system. The binary's mass ratio and orbit and the formation of circumstellar and circumbinary discs all depend on the dynamics of this accretion process.

We study the effects of accretion on a binary's mass ratio and separation. We also investigate the formation of circumstellar and circumbinary discs and find that, under some circumstances, a large circumstellar disc forms around the primary while the secondary has only a small circumstellar disc, or indeed no disc at all. The observational implications are discussed.

1. Introduction

Surveys of both main-sequence (Duquennoy & Mayor, 1991; Mayor et al., 1992; Fischer & Marcy, 1992) and pre-main-sequence (Simon et al., 1995) stellar systems show that the star formation process predominantly forms binary stellar systems. The favoured mechanism for producing the majority of these systems is the fragmentation of a molecular cloud core during its gravitational collapse (e.g. Boss & Bodenheimer 1979; Boss 1986; Bonnell et

J. A. Docobo et al. (eds.), Visual Double Stars: Formation, Dynamics and Evolutionary Tracks, 153–164.

al. 1991, 1992; Bonnell 1994; Bonnell & Bate 1994b). However, the binary protostellar systems that are formed in such fragmentation calculations typically contain only a small fraction of the total mass of the original cloud. Furthermore, the size of this fraction decreases with the system's separation (Boss 1986; Bonnell & Bate 1994b). Thus, the final state of such a binary system depends on the accretion of, and interaction with, the remaining cloud material as it falls on the system. As the system grows in mass, its mass ratio and orbit change. The infalling material may also form circumstellar and/or circumbinary discs.

We investigate the effects that accretion has on a protobinary system. We do not attempt to follow the evolution of a protobinary system from its formation, via the fragmentation of a progenitor cloud, until all the mass from the cloud has been accreted. The number of possible initial cloud configurations and binary parameters, and the computational effort required to follow such calculations prohibits this. Rather, we assume that a protobinary is formed via a fragmentation process and seek to determine, given the accretion of a small amount of gas with known characteristics, how the system changes. We determine the effects of accretion on a protobinary's mass ratio and separation as functions of the initial mass ratio of the binary and the specific angular momentum of the infalling gas. We also study circumstellar and circumbinary disc formation in the accreting protobinary system.

2. Calculations

The calculations presented here were performed using a three-dimensional smoothed particle hydrodynamics (SPH) code based on a version originally developed by Benz (Benz, 1990; Benz et al., 1990). The code uses a tree to find the nearest neighbours of particles and has individual time steps for each particle (Bate, Bonnell & Price, 1995). In addition to the standard SPH code, sink particles are used to model the protostars. Sink particles were developed (Bate et al., 1995) to allow calculations with large density contrasts (e.g. fragmentation calculations) to be followed in a realistic amount of computational time. A high-density, gravitationally-bound fragment, modelled by many particles with small time steps, is replaced by a single, massive particle which has their combined mass and momentum. This sink particle accretes any SPH gas particles if they fall within a certain radius, the accretion radius $r_{\rm acc}$, of the sink particle, and if they are bound to the sink particle. In this way, the high-density gas within bound fragments is ignored and it does not contribute to the computations. Boundary conditions between the gas and the sink particles are included when necessary.

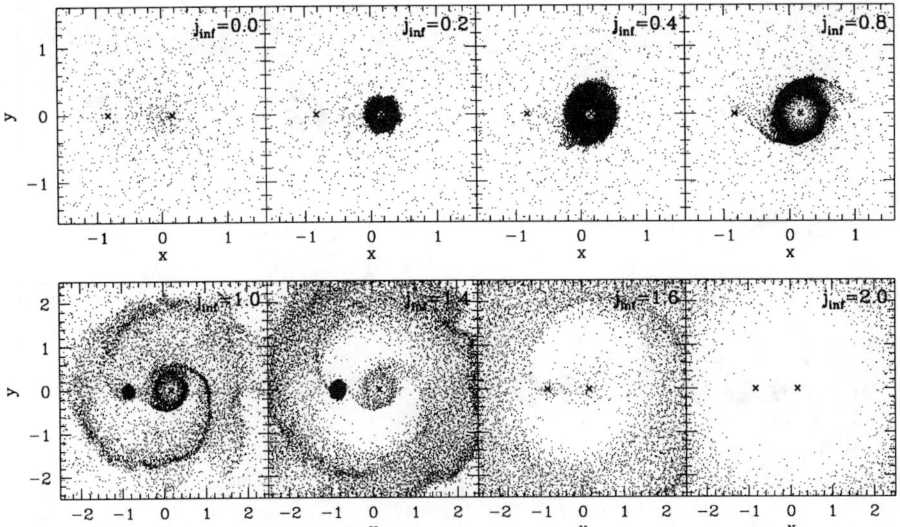

Figure 1. The gas distributions around a protobinary with mass ratio $q = 0.2$ as a function of the specific angular momentum of the infalling gas $j_{\rm inf}$. The positions of the protostars are marked with crosses, with the primary on the right. Distance is in units of the binary's separation.

In the calculations, gas is injected around a binary protostellar system, falls on to the system, and is accreted by the binary's components or forms discs around them. The binary system is modelled by sink particles. The masses of the primary and secondary are M_1 and M_2, respectively. The binary has mass ratio $q = M_2/M_1$, total mass $M_{\rm b}$, and separation a. Only binaries with initially circular orbits are considered. Natural units are used with the gravitational constant $G = 1$ and, initially, $M_{\rm b} = 1$ and $a = D = 1$. For this choice of units, a particle that is in a circular orbit around a mass $M_{\rm b}$ at a radius $r = D$ has a specific angular momentum j of unity (i.e. $j_{\rm circ} = \sqrt{GM_{\rm b}D} = 1$). Also, note that since $M_{\rm b} = 1$ and $a = 1$ initially, then $\dot{M}_{\rm b}$ and $\dot{a}$ are the *fractional* changes in $M_{\rm b}$ and a with time. Thus, for example, $\dot{q}/\dot{M}_{\rm b} = \dot{q}/(\dot{M}_{\rm b}/M_{\rm b})$ and $\dot{a}/\dot{M}_{\rm b} = (\dot{a}/a)/(\dot{M}_{\rm b}/M_{\rm b})$. The sink particles have accretion radii of $r_{\rm acc} = 0.05D$; circumstellar discs with radius less than this are not resolved.

The axis of rotation of the infalling gas is the same as that of the binary and the gas either falls in radially or rotates in the same sense as the binary. These assumptions are justified as it is envisaged that the infalling gas is from the cloud out of which the protobinary formed. Throughout each calculation, the gas is injected far from the binary with the kinetic energy it would have if it had fallen from infinity, and with a fixed specific angular momentum $j_{\rm inf}$ about the binary's centre of mass. The latter allows the effects of accretion to be studied as a function of the specific angular

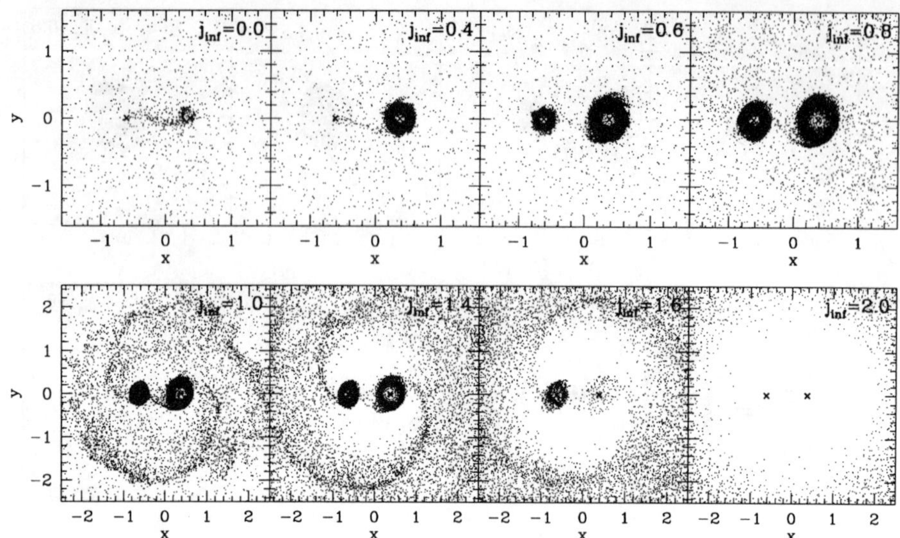

Figure 2. The same as in Figure 1, except that the protobinary has a mass ratio of $q = 0.6$. Note also, that the values of $j_{\rm inf}$ differ slightly.

momentum of the infalling gas. The rate of infall of gas on to the binary is $\dot{M}_{\rm inf}$. The gas has an isothermal equation of state and is cold (i.e. the sound speed is much less than the orbital velocity of the stars). Finally, the gas is non-self-gravitating (i.e. the total gas mass that is injected is very much less than $M_{\rm b}$). Further details are given by Bate & Bonnell (1997).

3. Accretion Flows

Calculations were performed of binaries with mass ratios $0.1 \leq q \leq 1.0$ accreting gas with specific angular momentum $0 \leq j_{\rm inf} \leq 2$. Figures 1 and 2 show the distributions of gas around binaries with mass ratios of $q = 0.2$ and $q = 0.6$ once steady-state accretion on to the binary has been established. The distributions are given as functions of $j_{\rm inf}$.

A general progression is observed with increasing $j_{\rm inf}$ (e.g. Fig 2). With most mass ratios, under infall with zero angular momentum, no discs are resolved around the protostars. When the angular momentum of the cloud is increased, a circumprimary disc is formed, but there is still no circumsecondary disc. Note that even if no resolved disc is formed around a protostar, it may still accrete significantly via a Bondi-Hoyle-type accretion stream directly on to the star or on to a small, unresolved disc. For still higher angular momenta, both circumprimary and circumsecondary discs are formed. The value of $j_{\rm inf}$ above which a circumsecondary disc forms depends on the binary's mass ratio (c.f. Figures 1 and 2). With angular

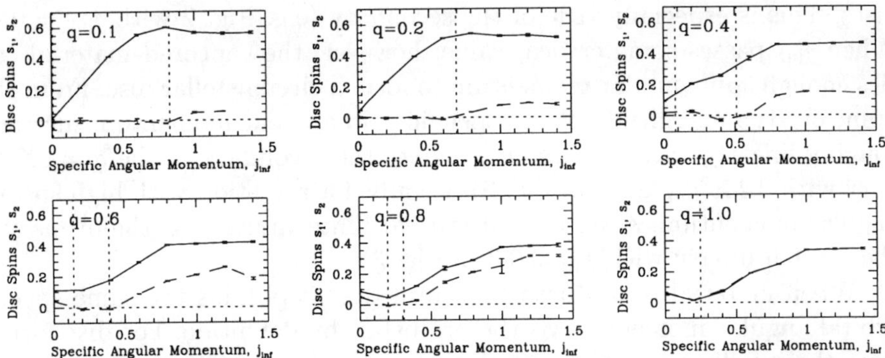

Figure 3. The specific spin angular momentum of the material captured by the primary s_1 (solid line) and secondary s_2 (dashed line) as functions of the binary's mass ratio q and $j_{\rm inf}$. Estimated error bars are given. The vertical lines give the specific orbital angular momenta of the primary j_1 (left) and secondary j_2 (right).

momenta of $j_{\rm inf} \approx 1$, two circumstellar discs are formed, and the formation of a circumbinary disc begins. Spiral density waves are produced in the circumbinary material via gravitational torques from the binary. At still higher $j_{\rm inf}$, a region of exclusion appears near the binary where little gas is present. The infalling gas has too much angular momentum to fall directly into this region. However, accretion does proceed on to the circumstellar discs from the circumbinary material via streams of gas which lose angular momentum due to gravitational torques from the binary. The amount of material accreted by the circumstellar discs decreases with increasing $j_{\rm inf}$. Finally, for material with the highest specific angular momentum, all the gas goes into a circumbinary disc. The binary is unable to perturb this disc strongly enough for gas to be accreted on to the circumstellar discs.

4. Disc Formation

4.1. CIRCUMSTELLAR DISCS

The capture of gas by a protostar can occur either via a Bondi-Hoyle-type accretion flow, or the formation of a circumstellar disc. Also, larger-radii circumstellar discs are formed from infall with higher specific angular momentum. The presence and size of the circumstellar discs can be quantified by determining the mean specific spin angular momentum of the material that is captured by the protostars (s_1 and s_2 for the primary and secondary, respectively). The spins, s_1 and s_2, are given in Fig. 3 as functions of the mass ratio and $j_{\rm inf}$. Three distinct regimes are apparent. With low $j_{\rm inf}$, the protostars capture gas that essentially has *no* spin. This gas is accreted via Bondi-Hoyle-type accretion streams rather than forming circumstellar discs. This is especially true of the secondary (e.g. Fig. 2 with $j_{\rm inf} = 0.4$). When $j_{\rm inf}$ passes some critical value, however, the captured material carries enough spin angular momentum to form a circumstellar disc. From this point on, the magnitude of the specific spin angular momentum increases roughly linearly with increasing $j_{\rm inf}$ (e.g. the secondary of the $q = 0.6$ binary with $0.4 \lesssim j_{\rm inf} \lesssim 1.0$ in Fig. 3). Finally, for accretion with high specific angular momentum, s_1 and s_2 saturate at finite values (e.g. the primary of the $q = 0.6$ binary with $j_{\rm inf} \gtrsim 0.8$ in Fig. 3).

When material is captured by a protostar it obtains the same specific orbital angular momentum as the protostar, by definition. For disc formation, the infalling material must have $j_{\rm inf}$ greater than the specific orbital angular momentum of the protostar that captures it so that its excess angular momentum provides spin angular momentum about the protostar (c.f. Fig. 3). Thus, the criterion for disc formation is that

$$j_{\rm inf} \gtrsim j_* \quad \text{where } j_* = \begin{cases} j_1 & \text{for the primary} \\ j_2 & \text{for the secondary} \end{cases} \tag{1}$$

and j_1 and j_2 are the specific orbital angular momenta of the primary and secondary, respectively. If $j_{\rm inf} < j_*$ no disc is formed as the gas must obtain the additional angular momentum it needs from the protostar it is captured by. This angular momentum is gained as it falls on to the protostar in a Bondi-Hoyle-type accretion stream.

When $j_{\rm inf} \gtrsim j_*$, more of the specific angular momentum of the infalling gas is converted to spin angular momentum as $j_{\rm inf}$ increases. This results in the roughly linear increase of s_1 and s_2 with $j_{\rm inf}$. Also, because $j_1 < j_2$ (with the exception of $q = 1$) then $s_1 > s_2$. Note that equation (1) is only approximate as the material gains a little angular momentum from the binary as it falls in. The primary forms a resolved disc under infall with zero angular momentum in binaries with mass ratios of $0.3 \lesssim q \lesssim 0.6$ because of this gain of angular momentum. Finally, in all cases, the specific spin angular momentum of the primary s_1 and the secondary s_2 saturate at finite values for high values of $j_{\rm inf}$. The spins saturate because the sizes of the discs are limited roughly to the sizes of the protostars' Roche lobes by truncation due to resonances (Artymowicz & Lubow, 1994).

4.2. CIRCUMBINARY DISC

For infall with high specific angular momentum, not all of the gas is captured by the individual components of the binary; some of the gas forms a circumbinary disc. The fractions of the infalling material that are captured by the primary, secondary and/or a circumbinary disc are given in Fig. 4 as functions of the binary's mass ratio and the cloud's specific angular

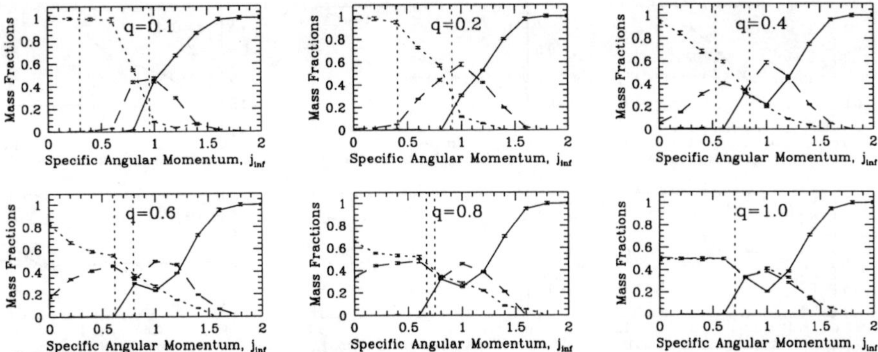

Figure 4. The fractions of the infalling gas that are captured by the primary $\dot{M}_1/\dot{M}_{\rm inf}$ (dotted line), the secondary $\dot{M}_2/\dot{M}_{\rm inf}$ (dashed line), and the circumbinary disc $\dot{M}_{\rm cb}/\dot{M}_{\rm inf}$ (solid line). The fractions are given as functions of the specific angular momentum of the cloud $j_{\rm inf}$ for various mass ratios q. Estimated error bars are given. The vertical lines give the specific angular momentum required for the gas to form a circular orbit at the radius of the primary (left) or secondary (right).

momentum. Circumbinary disc formation begins with lower $j_{\rm inf}$ infall for higher mass ratios because the secondary is closer to the centre of mass of the binary and, therefore, the infalling material does not require as much specific angular momentum to form a disc outside its orbit. The criterion for circumbinary disc formation is that $j_{\rm inf}$ be approximately equal to that required to form a circular orbit at the radius of the secondary. In fact, circumbinary disc formation begins when $j_{\rm inf}$ is slightly below this value as some of the gas gains angular momentum due to gravitational torques from the binary. As $j_{\rm inf}$ increases, the fraction of the infalling material reaching the circumstellar discs generally decreases, with all the gas going into a circumbinary disc for $j_{\rm inf} \gtrsim 1.8$.

5. Effect of Accretion on the Mass Ratio

Due to the different rates at which the primary and secondary capture infalling gas (Fig. 4), the mass ratio of the binary may be altered. The rates of change of mass ratio per unit mass captured by the binary's components $\dot{q}/\dot{M}_{\rm b}$ are given in Fig. 5. Note that we assume all the material in a protostellar disc is eventually accreted by the protostar it orbits. This occurs on a longer (viscous) time scale than that which is considered here.

Qualitatively, for infall with low specific angular momentum, the primary accretes most of the material and the mass ratio is decreased, while, when the cloud's angular momentum is increased, the secondary accretes more gas relative to the primary and the mass ratio increases. The variation in the relative accretion rates with $j_{\rm inf}$ is simple to understand. For

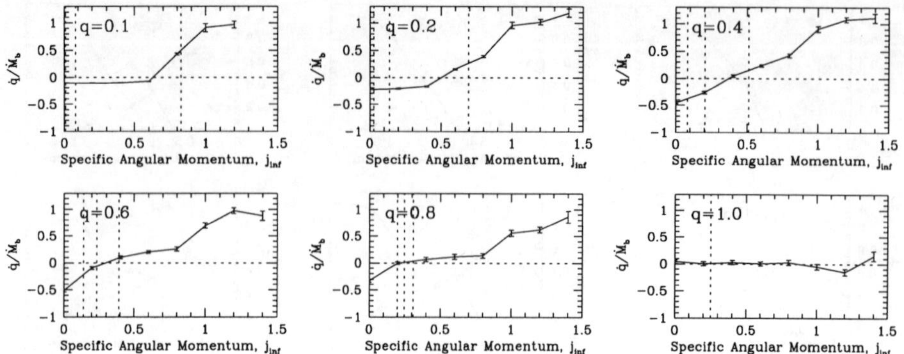

Figure 5. The rates of change of mass ratio per unit mass captured by the binary's components $\dot{q}/\dot{M}_b$, for binaries of mass ratio q accreting gas with specific angular momentum j_{inf}. Estimated error bars are given. The vertical lines give the specific orbital angular momenta of the primary j_1 (left), binary j_b (middle), and secondary j_2 (right).

infall with low angular momentum, the gas essentially falls into the centre of mass of the binary and, hence, is mainly accreted by the primary (which is closest to the centre of mass in a system with a low mass ratio). For infall with higher angular momentum, the gas can only fall in to its periastron distance from the centre of mass of the system (in the absence of gravitational torques). This makes accretion by the secondary easier, as the gas does not need to gain as much angular momentum to be captured. When a distinct circumbinary disc is formed, the secondary captures more gas than the primary, even though it has a lower mass, because it is closer to the radius to which the gas falls in and, thus, is able to perturb the gas more strongly.

The rate of change of mass ratio is found to change sign when j_{inf} is slightly less than the specific orbital angular momentum of the secondary j_2 (Fig. 5). Thus, a binary of mass ratio $q \gtrsim 0.7$ always increases its mass ratio if the specific angular momentum of the infalling material is greater than the specific angular momentum of the binary j_b. For lower mass ratios, j_{inf} must be increasingly larger than j_b for the mass ratio to increase.

6. Effect of Accretion on the Separation

As well as the role accretion plays in the formation of discs and its effect on the masses of the binary's components, accretion alters the separation of the binary. Two separate effects are involved. First, there is the accretion itself which changes the separation due to the addition of mass and angular momentum and the changing of the mass ratio. Second, if a circumbinary disc is formed, the binary may transfer orbital angular momentum to the material in the disc via gravitational torques (Artymowicz et al., 1991). This loss

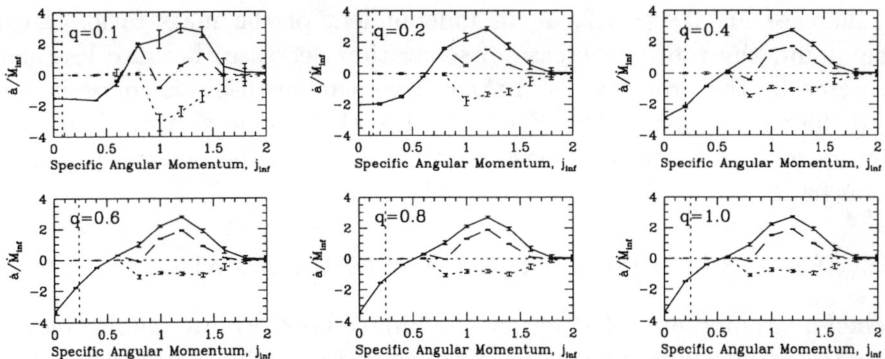

Figure 6. The rates of change of separation per unit mass of infall. The separation is affected both by accretion $\dot{a}_{\rm acc}/\dot{M}_{\rm inf}$ (solid line), and by the loss of orbital angular momentum to the circumbinary disc $\dot{a}_{\rm grav}/\dot{M}_{\rm inf}$ (dotted line). The combination of these two effects $\dot{a}/\dot{M}_{\rm inf}$ is also given (dashed line). Estimated error bars are given. The vertical line gives the specific orbital angular momentum of the binary $j_{\rm b}$.

of orbital angular momentum reduces the separation of the binary. Thus, the rate of change of separation of the binary is given by $\dot{a} = \dot{a}_{\rm acc} + \dot{a}_{\rm grav}$, where $\dot{a}_{\rm acc}$ and $\dot{a}_{\rm grav}$ are the rates of change of separation due to accretion and due to gravitational torques from the circumbinary disc respectively. Note that $\dot{a}_{\rm grav}$ depends linearly on the viscosity in the calculations (Lubow & Artymowicz, 1996) while $\dot{a}_{\rm acc}$ is independent of the viscosity. Thus, it is important to separate these two effects. The viscosity of SPH is known to be large (we estimate the effective viscosity parameter $\alpha \sim 0.1$), and thus the values of $\dot{a}_{\rm grav}$ presented here are likely to be higher than is physical.

Fig. 6 gives the rates of change of separation per unit mass of infall, as functions of the mass ratio and $j_{\rm inf}$. The total effect of gaseous accretion on the separation $\dot{a}/\dot{M}_{\rm inf}$ is given, along with the separate effects due to accretion only $\dot{a}_{\rm acc}/\dot{M}_{\rm inf}$, and due to the transfer of orbital angular momentum to the circumbinary disc $\dot{a}_{\rm grav}/\dot{M}_{\rm inf}$.

6.1. EFFECTS DUE TO ACCRETION

First we consider the rate of change of separation due to accretion only $\dot{a}_{\rm acc}/\dot{M}_{\rm inf}$ (Fig. 6, solid lines). Qualitatively, the separation decreases for infall with low specific angular momentum, and increases with high $j_{\rm inf}$. Quantitatively, the functions are relatively *independent* of the mass ratio. The separation decreases for $j_{\rm inf} \lesssim 0.6$. Thus, for all mass ratios, there is a region where the separation decreases even though $j_{\rm inf} > j_{\rm b}$. For a $q = 0.1$ binary the specific angular momentum of the infalling material has to be more than 7 times greater than that of the binary for the separation to increase! When $j_{\rm inf} \gtrsim 0.6$ the separation increases, with a peak rate of increase at $j_{\rm inf} \approx 1.2$, again independent of the mass ratio. Beyond this point, the rate of increase of separation decreases because less gas is accreted by the protostars and their circumstellar discs and more is going into the circumbinary disc. For $j_{\rm inf} \gtrsim 1.8$, there is no accretion on to the circumstellar discs for any mass ratio (Fig. 4) and the binary's separation is unchanged.

6.2. EFFECTS DUE TO CIRCUMBINARY DISC FORMATION

When a circumbinary disc forms, the binary loses orbital angular momentum to the disc via gravitational torques and the orbit shrinks ($\dot{a}_{\rm grav}/\dot{M}_{\rm inf}$ in Fig. 6). The maximum transfer rate generally occurs near the lowest value of $j_{\rm inf}$ for which a circumbinary disc forms ($j_{\rm inf} \approx 0.8 - 1.0$ depending on q). As $j_{\rm inf}$ increases, the transfer rate generally decreases again because the circumbinary disc is not as close to the binary and the gravitational torques rapidly decrease in strength with increasing radius.

6.3. COMBINED EFFECTS ON THE BINARY'S SEPARATION

When the effects of accretion and interactions with the circumbinary material are combined, the rate of change of separation for $j_{\rm inf} \lesssim 0.7 - 0.9$ (depending on q) is determined purely from accretion. Above this, the gravitational interaction with the circumbinary material gives a lower rate of increase of separation than that purely from accretion. This can even cause the separation to decrease again in the range $0.8 \lesssim j_{\rm inf} \lesssim 1.0$! However, as mentioned above, the rate of decrease of separation due to the circumbinary disc depends linearly on the viscosity which is likely to be higher than is physical in these calculations.

7. Observational Implications

Two obvious observational implications arise from the results presented here. First, if a protobinary forms within a collapsing molecular cloud core and gains a large proportion of its final mass via accretion from the rest of the cloud material, then medium to long-period binaries are more likely to have low mass ratios, while close binaries are more likely to have equal mass ratios. This is because the specific angular momentum of the infalling gas, relative to the binary, is likely to be higher for close binaries than wide binaries. In addition, close binaries are likely to accrete a greater fraction of their final mass since the initial mass of binaries formed via fragmentation decreases with separation (see Section 1). Observations show that low mass ratios are favoured for medium to long-period binaries (Duquennoy & Mayor, 1991) and there is weak evidence that equal mass ratios may be favoured for close binaries (Mazeh et al., 1992), although this needs to be confirmed. Note, however, that if a massive circumbinary disc is formed around the binary it may fragment (Bonnell & Bate, 1994a) which would complicate this picture.

The second implication is for the presence of circumstellar discs in pre-main-sequence systems. If a protobinary grows to its final mass via the accretion of gas with low specific angular momentum, the primary may have a large circumstellar disc while the secondary is essentially naked. This offers an explanation for the existence of infrared companions to optically visible T-Tauri stars (Zinnecker & Wilking, 1992). The optically visible star in this scenario would be the secondary, whereas the embedded object would be the primary viewed through its edge-on circumstellar disc. Note that, if there is still infall on to the system, the secondary may even show significant accretion luminosity due to accretion from its own very small circumstellar disc or even directly from a Bondi-Hoyle-type accretion stream. The infrared companion systems formed by this method are expected to be fairly wide binaries ($\gtrsim$100 AU separation) and there should be no significant circumbinary disc due to the relatively low specific angular momentum of the accreted material. Alternately, for binaries where a significant circumbinary disc is formed, both stars are expected to form circumstellar discs. However, the masses of these discs may differ considerably and, if the circumstellar discs evolve quickly, the secondary may still be accreting material with high angular momentum from the circumbinary disc (Artymowicz & Lubow, 1996) long after the circumprimary disc has been dispersed. In this case the secondary may appear as the infrared companion.

8. Conclusions

We have studied the effects of non-self-gravitating, gaseous accretion on circular protobinary systems. We find the general behaviour of a binary under accretion is described by a few simple relationships.

First, a circumstellar disc forms around a component of the binary only if the specific angular momentum of the infalling gas $j_{\rm inf}$ is greater than the specific orbital angular momentum of that component about the centre of mass of the binary. If $j_{\rm inf}$ is lower, the protostar accretes via a Bondi-Hoyle-type accretion stream instead. In many cases this results in the primary having a large circumstellar disc while the the secondary has none. If two circumstellar discs are formed, the primary always has a larger radius disc than the secondary and the radii of the circumstellar discs increase with $j_{\rm inf}$. The circumstellar discs have maximum sizes of approximately the Roche-lobe radius of the protostars.

Second, circumbinary disc formation begins when $j_{\rm inf}$ approaches the specific angular momentum required for the gas to form a circular orbit at the distance of the secondary from the centre of mass of the system.

Third, the mass ratio decreases when gas with low specific angular momentum is accreted, but starts to increase when $j_{\rm inf}$ approaches the specific orbital angular momentum of the secondary j_2.

Finally, the separation decreases for $j_{\rm inf} \lesssim 0.6$ and generally increases above this value, independent of the mass ratio. This means that, for all mass ratios, there is a region where the separation decreases even though the infalling material has significantly more specific angular momentum than the binary itself ($j_{\rm inf} > j_{\rm b}$). When a circumbinary disc is present, the binary loses orbital angular momentum to it via gravitational torques and this decreases the rate of change of separation over that given purely by the accretion of gas. This effect generally is not strong enough to decrease the separation for $j_{\rm inf} \gtrsim 0.6$, but this conclusion depends on the viscosity of the gas.

References

Artymowicz P., Clarke C.J., Lubow S.H., Pringle J.E. 1991, ApJ, 370, L35

Artymowicz P., Lubow S.H., 1994, ApJ, 421, 651

Artymowicz P., Lubow S.H., 1996, ApJ, 467, L77

Bate M.R., Bonnell I.A., 1997, MNRAS, in press

Bate M.R., Bonnell I.A., Price N.M., 1995, MNRAS, 277, 362

Benz W., 1990, in Buchler J.R., ed., The Numerical Modeling of Nonlinear Stellar Pulsations: Problems and Prospects. Kluwer, Dordrecht, p. 269

Benz W., Bowers R.L., Cameron A.G.W., Press W., 1990, ApJ, 348, 647

Bonnell I.A., 1994, MNRAS, 269, 837

Bonnell I.A., Bate M.R., 1994a, MNRAS, 269, L45

Bonnell I.A., Bate M.R., 1994b, MNRAS, 271, 999

Bonnell I., Arcoragi J.-P., Martel H., Bastien P., 1992, ApJ, 400, 579

Bonnell I., Martel H., Bastien P., Arcoragi J.-P., Benz W., 1991, ApJ, 377, 553

Boss A.P., 1986, ApJS, 62, 519

Boss A.P., Bodenheimer P., 1979, ApJ, 234, 289

Duquennoy A., Mayor M., 1991, A&A, 248, 485

Fischer D.A., Marcy G.W., 1992, ApJ, 396, 178

Lubow S.H., Artymowicz P., 1996, in Wijers R.A.M.J., et al. ed., Evolutionary Processes in Binary Stars. Kluwer Academic Publishers, Dordrecht, p. 53

Mayor M., Duquennoy A., Halbwachs J.-L., Mermilliod J.-C., 1992, in McAlister H.A., Hartkopf W.I., ed., Complementary Approaches to Double and Multiple Star Research (IAU Colloquium 135). ASP, San Francisco, p. 73

Mazeh T., Goldberg D., Duquennoy A., Mayor M., 1992, ApJ, 401, 265

Simon M., Ghez A.M., Leinert, Ch., Cassar L., Chen W.P., Howell R.R., Jameson R.F., Matthews K., Neugebauer G., Richichi A., 1995, ApJ, 443, 625

Zinnecker H., Wilking B.A., 1992, in Duquennoy A., Mayor M., ed., Binaries as Tracers of Stellar Formation. Cambridge University Press, Cambridge, p. 526

HIERARCHICAL SYSTEMS IN OPEN CLUSTERS

R. DE LA FUENTE MARCOS
Universidad Complutense de Madrid, E-28040, Madrid, Spain
fiast05@emducms1.sis.ucm.es

AND

S. J. AARSETH, L. G. KISELEVA AND P. P. EGGLETON
Institute of Astronomy,
Madingley Rd., Cambridge CB3 0HA, UK

Abstract. In this paper we study the formation, evolution and disruption of hierarchical systems in open clusters. With this purpose, N-body simulations of star clusters containing an initial population of binaries have been carried out using Aarseth's NBODY4 and NBODY5 codes.

Stable triples may form from strong interactions of two binaries in which the widest pair is disrupted. The most frequent type of hierarchical systems found in the cluster models are triples in which the outer star is single, but in some cases the outer body is also a binary, giving a hierarchical quadruple. The formation of hierarchical systems of even higher multiplicity is also possible. Many triple systems are non-coplanar and the presence of even a very distant and small outer companion may affect the orbital parameters of the inner binary, including a possible mechanism of significant shrinkage if the binary experiences a weak tidal dissipation.

The main features of these systems are analyzed in order to derive general properties which can be checked by observations. The inner binaries have periods in the range $1 - 10^3$ days, although rich clusters may have even smaller periods following common envelope evolution. For triple systems, the outer body usually has a mass less than 1/3 of the binary, but is sometimes a collapsed object with even smaller mass. The formation of exotic objects, such as blue stragglers and white dwarf binaries, inside hierarchical triple systems is particularly interesting. An efficient mechanism for generating such objects is the previous formation of a hierarchical system in which the inner binary may develop a very short period during a common envelope phase, which finally results in a stellar collision.

J. A. Docobo et al. (eds.), Visual Double Stars: Formation, Dynamics and Evolutionary Tracks, 165–178.

1. Introduction

In the past few years significant evidence for a large number of binary and multiple stellar systems in galactic clusters has been obtained from observational surveys (Mathieu et al. 1990; Mermilliod et al. 1992; Ghez et al. 1993; Mason et al. 1993; Mayer et al. 1994; Mermilliod et al. 1994; Simon et al. 1995), both for zero age main sequence and pre-main-sequence stars. For multiple systems the frequency estimated for star forming regions is up to 35 % and for the field is up to 20 %. However, the fraction of these systems detected in open clusters is smaller, although the improvement in observational techniques has increased in the past few years. Currently, the majority of multiple systems discovered in open clusters are triples and quadruples. These systems are usually highly hierarchical. Triple (or even higher multiplicity) systems are found in the Pleiades (Mermilliod et al. 1992), the Hyades (Griffin & Gunn 1981, Griffin et al. 1985, Mason et al. 1993), Praesepe (Mermilliod et al. 1994), M67 (Mathieu et al. 1990), and NGC 1502 (Mayer et al. 1994).

The majority of binary systems observed in open clusters are thought to be primordial, but there is no preferred formation mechanism for multiple systems (dynamical or primordial) at present. Most of the multiple systems studied are hierarchical because of the intrinsic stability of these systems. The origin of observed multiple systems has not been clear since the beginning of the study of these systems. Duquennoy (1988) analyzed a sample of 17 systems (14 triples and 3 quadruples) in the solar neighbourhood. He obtained a linear correlation between the logarithm of the inner and outer binary period. This was interpreted as an indication of preferential primordial origin for these systems. From a theoretical point of view, Boss (1991) has suggested that the formation of hierarchical systems occurs during the collapse of protostellar cores. Recently, Mermilliod et al. (1994) have found significant period ratios ($X = P_{out}/P_{in} \simeq 250$) in clusters which suggest a dynamical origin. Although the question of the origin of these hierarchical systems is far from being answered, we assume here that all the hierarchical systems formed in open clusters have a purely dynamical origin.

The formation and dynamical evolution of hierarchical systems in open clusters can be studied within the context of N-body simulations because complex interactions between stars can readily be followed in detail by numerical methods. This approach has recently been adopted (Aarseth 1996a, Kiseleva et al. 1996, Eggleton & Kiseleva 1996, Kiseleva 1996). In this paper the results of almost a hundred cluster models are analyzed with the purpose of studying the formation, evolution and final destinies of hierarchical systems in clusters. These models have been obtained using direct N-body integration by the standard workstation code NBODY5 (Aarseth 1985, 1994) and the more recent NBODY4 (Aarseth 1996b).

NBODY5 consists of a fourth-order predictor corrector scheme with individual time steps. In order to account for stellar evolution and mass loss (stellar winds and supernova events), we use the fast fitting functions of Eggleton, Fitchett and Tout (1989) and Tout (1990) for population I stars.

NBODY4 is based on the so-called Hermite scheme (Makino 1991) which forms the basis of a new generation of special-purpose computers (Makino, Kokubo & Taiji 1993). Mass loss by stellar winds is now treated according to a modified Reimers (1975) expression (Tout 1990). Chaotic tidal motion (Mardling 1995), tidal circularization (Mardling & Aarseth 1996), exchange of mass (Roche overflow) in binaries and magnetic braking are also included.

The models with NBODY5 have been studied on a DEC 2100 4/275 AXP system. All the calculations performed with NBODY4 were made at Cambridge on the HARP-2 computer.

2. Stability of hierarchical systems

The main aim of this paper is to study the formation and evolution of hierarchical systems in open clusters. Considered as isolated from field stars, such systems can survive over a long time. Small non-hierarchical systems, except in a few special cases, are always unstable in the long term (Marchal 1990). Even for hierarchical triple systems, stability is not an easy question. There are a number of criteria to identify triple systems as stable or unstable, obtained analytically or numerically (see e.g. Kiseleva 1996). For historical reasons, the models described here employ two criteria which are in fact fairly similar for most mass ratios involving stars. Thus the older NBODY5 code employs the Harrington (1975) criterion in the form

$$F^{min} = A\left(1 + B\,log\left(\frac{1 + m_3/(m_1+m_2)}{3/2}\right)\right), \qquad (1)$$

with $A = (2.65+e)(1+m_3/(m_1+m_2))^{1/3}$ modified by the inner eccentricity, e, according to Bailyn (1984) and $B = 0.7$. Here F^{min} denotes the critical ratio of the outer periastron distance of the mass m_3 to the inner apastron distance of $m_1 + m_2$.

In the NBODY4 code, we adopt the stability criterion of Eggleton and Kiseleva (1995, hereafter EK). In either case, we use their definition of stability: that a hierarchical triple system is stable if it persists continuously in the same configuration (which excludes exchange and disintegration).

The stability criterion of EK used for the NBODY4 models is based on the critical period ratio for stability, $X_0^{min} = P_{out}/P_{in}$. It was derived from a numerical study which examined a wide range of parameters: (a) eccentricities of both inner and outer orbits, e_{in} and e_{out}; (b) relative inclinations of inner and outer orbits, from prograde to retrograde; (c) initial relative phase; (d) both mass ratios, $q_{in} = m_1/m_2 \geq 1$ and $q_{out} = (m_1 + m_2)/m_3$. This criterion can be written in two forms. Let Y_0^{min} again be the critical ratio F^{min}. For computational purposes Y_0^{min} is a more relevant parameter, since it depends *only* (to a certain level of approximation) on the two mass ratios. The period ratio X_0^{min} also depends on the two eccentricities and is more useful for observed triples, since their periods and eccentricities, rather than semi-major axes, are normally determined by observation. Note that the inclination and phase are not included explicitly in the criterion. The two forms of the stability criterion are given by:

$$(X_0^{min})^{2/3} = (\frac{q_{out}}{1 + q_{out}})^{1/3} \frac{1 + e_{in}}{1 - e_{out}} Y_0^{min}, \tag{2}$$

$$Y_0^{min} \approx 1 + \frac{3.7}{q_{out}^{1/3}} - \frac{2.2}{1 + q_{out}^{1/3}} + \frac{1.4}{q_{in}^{1/3}} \frac{q_{out}^{1/3} - 1}{q_{out}^{1/3} + 1}. \tag{3}$$

This criterion discriminates between those triples that are likely to last a long time in the absence of strong external perturbations and those which may break up rapidly. Comparison of the two criteria used shows reasonable agreement, except for the less probable case of a massive outer component.

There are two main conditions for identifying the 'birth' of a hierarchical subsystem in our simulations:

- The inner binary must be hard and the outer component must also form a hard binary with the inner binary.
- The criterion (3) for hierarchical stability must be satisfied.

If both the above conditions are satisfied simultaneously the subsystem is considered as a hierarchical triple (or higher multiplicity) system and a number of parameters of this system are recorded. The subsequent treatment consists of combining the inner binary into one body to permit a two-body treatment (recursively if double hierarchy).

A hierarchical system is terminated when one of the following situations occurs:

- The stability criterion (3) is violated because, for example, of significant changes of e_{out} due to small secular external perturbations.
- The external perturbation exceeds a critical value.
- Effects of stellar evolution become important or the predicted pericentre distance becomes too small.

In the last case, however, the same hierarchical system usually appears again, with slightly different parameters (e.g. during gentle mass loss) and we take this situation into account in the statistical analysis of results.

In the present calculations it is possible to produce triple systems, quadruple systems and double hierarchies. Triple systems consist of an inner binary and an outer single star, quadruple systems also have a binary as an outer body and in double hierarchies we have a nested system (binary + outer body + another outer body) with up to six stars. Hence, we can study even hierarchical sextuple systems. Recently Mardling and Aarseth (1996) have developed a new stability criterion based on a chaos description (Mardling 1995) which contains the inner and outer eccentricities explicitly as well as the mass ratio $m_3/(m_1+m_2)$ for coplanar orbits. Except for large e_{out}, it gives values that are somewhat smaller than EK.

3. Cluster models

In order to obtain realistic results it is important to choose a general mass function (hereafter IMF). In all the models the IMF used is described by:

$$f(m) = \frac{0.3\ \mathrm{X}}{(1 - \mathrm{X})^{0.55}}, \tag{4}$$

where X is a random variable in $[0,1]$. This IMF is a fit to Scalo's (1986) results. For binaries we introduce a correlation $(m_1/m_2)' = (m_1/m_2)^{0.4}$, subject to the sum being constant, which yields mass ratios closer to unity than for random samples.

The initial NBODY5 models have a mass range of 0.1-15.0 $m_\odot$ for the single stars. Spherical symmetry and constant density are assumed, with the ratio of total kinetic and potential energy fixed at 0.25. All these models have random and isotropic initial velocities. Stars outside twice the classical tidal radius are assumed to escape and are removed from the calculations.

The models with NBODY4 have a mass range of 0.2-10.0 $m_\odot$ for the single stars. Initial coordinates and velocities are generated from an equilibrium King model in an external galactic field (Heggie and Ramamani 1992). In each code the cluster is assumed to be in a circular orbit in the solar neighbourhood, with a linearized tidal force added to the equations of motion (Aarseth 1985, 1994).

The probability of formation of hierarchical systems in a cluster with only single stars is very small; hence in order to produce a relatively large number of such systems, a significant initial binary population is needed (de la Fuente Marcos 1996). NBODY4 models can have an arbitrary number of binaries; for example models with 5100 stars have 100 primordial binaries. However, for NBODY5 models ($N = 100, 500$ and 1000) two binary fractions, 10 % and 50 %, have been studied. Apart from the binary fraction some other parameters must be specified for the initial binary population; in particular, the semi-major axis and eccentricity must be chosen. For an initial population of hard binaries the semi-major axis is about $-GM^2/4EN$, where G is the gravitational constant, M is the total mass, E is the energy.

The semi-major axis is taken from a uniform distribution: $a_b = a_b^0\ 10^{-q}$, where a_b^0 is a parameter whose value is $1/N$ in units of the virial radius, and q is equal to $X \log R$. X is a random number uniformly distributed in the interval $[0,1]$ and R is a parameter. The latter gives the spread in semi-major axes and thus the spread in energies and periods. For example, in the $N = 5100$ models with NBODY4 the binary semi-major axis is in the range 0.1 – 100 AU. For NBODY5 models, all the runs have been repeated with $R = 5, 10, 50, 100$, with maximum values of the semi-major axes in the range 230 – 1000 AU. Moreover, all the runs have been computed twice: one time with the chain regularization excluded and another time with this option included. Chain regularization (Mikkola & Aarseth 1990, 1993) is a numerical treatment of close encounters in compact subsystems in which the external perturbation due to nearby stars is taken into account.

4. Results

In this section we discuss some representative results. We are mainly concerned with results which can be checked directly with observational ones.

4.1. OVERALL EVOLUTION

Although there are three sets of different simulations, we find some common features. As expected, the majority of the systems formed are triple, but some NBODY4 models show almost the same tendency to form quadruples. Hierarchies do not form at a preferred cluster evolution stage. Usually in clusters with primordial binaries the first hierarchies appear at about 0.025 - 0.05 T_{cl}, where T_{cl} is the total life-time of the cluster, and in the cluster remnant (when N is a few tens) there are sometimes long-lived hierarchies. The distribution of inclinations for hierarchies is fairly symmetrical, with a pronounced peak centred on 90° (Kiseleva 1996). In large clusters at a late stage of the evolution, many short-lived systems form via repetitive triple-binary and triple-triple exchanges. Sometimes a hierarchical system leaves the cluster, escaping before the disruption of the system takes place. Poor clusters ($N = 100$) with wide binaries do not show any tendency to form hierarchical systems since we require the outer orbit to be a hard binary. Sometimes the outer body is a collapsed object (white dwarf). The largest number of hierarchies observed at the same time is five.

First we present some results for NBODY5 models. In the triple systems which form, the inner binary is usually not primordial. The quadruple systems found in poor clusters are more eccentric in models with no chain treatment. These systems form typically at the late stages of cluster evolution in rich clusters and inside the cluster core. For systems which include the chain treatment the quadruple systems formed in rich clusters are more eccentric. They also form preferentially inside the core. For triple systems the outer star usually has a mass less than one-third of the binary mass and sometimes 1/15, when the outer body is a collapsed object. The typical lifetime for the systems formed is a few million years, or 10^3 outer periods. The inner binaries of the hierarchies have periods from a few days to 10^3 days.

NBODY4 models also show preferential formation of triples but now quadruples are relatively abundant (up to 40 % of all hierarchical systems in $N = 5100$ models). Quadruples tend to form at late stages of the cluster evolution. On the other hand, the corresponding percentage of systems with primordial inner binaries is significantly greater than for the NBODY5 models (about 70 % vs 40 %) because of the shorter periods. Recent models include a new technical feature of so-called double hierarchies. In such configurations an outer body (single or binary) is added to an existing stable triple or quadruple, provided the stability criterion is satisfied. These systems tend to occur during very late stages when the core has expanded significantly; however, they tend to be short lived in the models studied so far with termination due to significant external perturbation.

The formation of hierarchical systems may have interesting observational consequences (Aarseth 1992). In such models, after the formation of a hierarchical triple system the pericentre of the outer body is usually close enough to the inner binary to perturb its eccentricity. This systematic perturbation often promotes a stellar collision of the inner binary. This process has been observed in several NBODY5 models with $N = 1000$ and 50 % primordial binaries, and is quite common in NBODY4 models which contain binaries with periods of days.

4.2. THE PERIOD – ECCENTRICITY PLANE

One of the most interesting diagrams for displaying the results of binary observations is the plot of the eccentricity versus the logarithm of the period because it provides more insight into binary astrophysics than other distributions of orbital elements and is expected to be equally useful in the study of hierarchies. We have plotted the eccentricity of the outer body (single or binary) against the logarithm of its period (in days). The results depend on the type of model and also on the membership and initial orbital elements of the binary population. As remarked above, our models can be classified into three different groups but several features are common to all:

- A lower cut-off period below which no orbits are observed.
- An upper cut-off period above which no orbits are observed.
- A zone of forbidden eccentricities for systems in which the inner binary is not primordial.

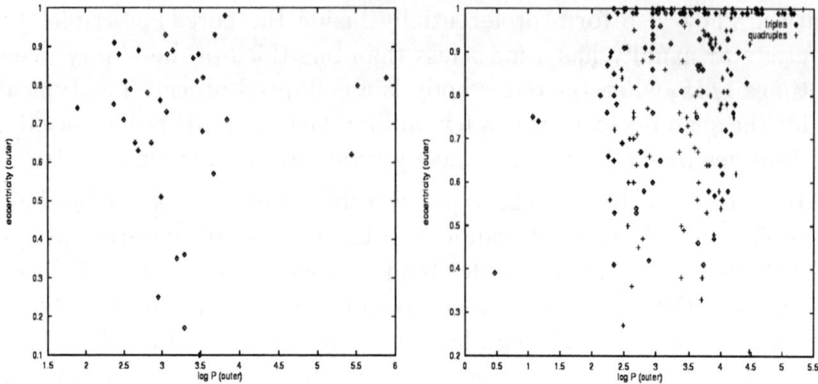

Figure 1. Eccentricity – period diagram of the outer binary in days. Left panel is for an NBODY5 model, the right one is for an NBODY4 model. High eccentricity systems are short-lived in all the models.

The short period cut-off. Its value is dependent on the initial values of the initial primordial binary parameters. For the NBODY4 models common envelope evolution is included so a few systems are observed below the cut-off period; most of these suffer a physical collision which sometimes produces a blue straggler. For models without Roche overflow and common envelope evolution wide inner binaries implies longer outer periods; i.e. models with very hard binaries form hierarchical systems which permit shorter periods for the outer body. Moreover, the cut-off is larger if the inner binary is not primordial. In fact, even for the NBODY4 models, we observe a clear cut-off in the period ratio with no exceptions, due to common envelope evolution or collisions. Another interesting feature is the dependence on the nature of the outer body (single or binary). The cut-off is larger for systems in which the outer body is a single star.

The long period cut-off. This is mainly due to the stability limit against perturbations from the neighbouring stars. Models with wider binaries show higher values of the upper period cut-off. The width of the period distribution depends strongly on the initial distribution of the semi-major axes of the binaries; small binaries generate sharp distributions of periods for the outer body. The high cut-off also depends on the nature of the inner binary. Systems with an inner primordial binary show slightly higher periods and the same is found for systems in which the outer body is also a binary.

The forbidden region. A dependence of the eccentricity distribution on the character of the inner binary is observed. When the inner binary is not primordial no systems are observed with eccentricities smaller than 0.20. Compared with real systems, this can be considered as a limit below which all the systems observed should contain a primordial binary. This must reflect the fact that non-primordial binaries, although usually formed by exchanges or disruptions of primordial systems, are wider and less energetic than primordial ones. The existence of real systems below this limit with a non-primordial inner binary may be explained as an effect of tidal interaction. There is also a correlation between the membership of the cluster and the lower eccentricity observed. Finally, systems formed in poor clusters are more eccentric on average.

4.3. THE (LOG P_B, LOG P_O) PLANE

In order to compare our results with those from the observational literature, we consider the plane log P_O vs log P_B, where P_O is the outer orbital period and P_B the inner period. We compare mainly with a sample of triples (fig. 2) in the solar neighbourhood by Duquennoy (1988). He finds a linear correlation for a sample of 13 triples with components of solar type. The slope of the straight line is 0.68 with a correlation coefficient of 0.91. From the relation between the periods he concludes that triple systems are rarely formed by capture, but rather by fragmentation processes, because capture would produce random combinations of periods. He suggests dynamical evolution as another possible cause in order to reproduce the observed distribution of periods. However, triples in our models have a purely dynamical origin and we find a similar behaviour.

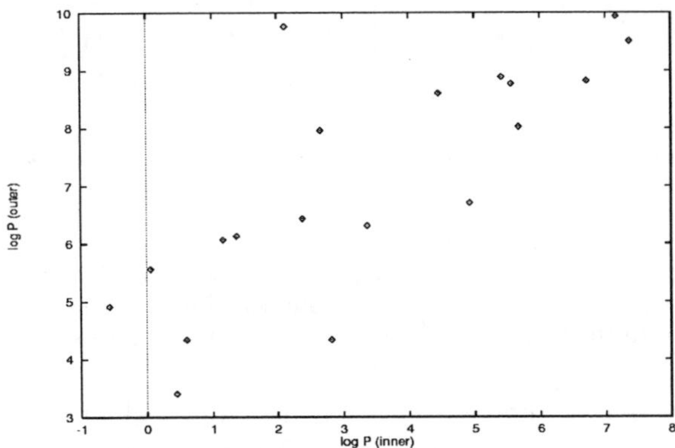

Figure 2. Period diagram for the observational sample of multiple systems listed in Duquennoy (1988). Note the larger outer periods in comparison with fig. 3 because most of them are estimated parameters in the original paper.

For an NBODY5 model with 1500 stars and 500 primordial binaries (fig. 3) we obtain a slope of 0.68 ± 0.09. For an NBODY4 model with 5100 stars and 100 primordial binaries we obtain a slope of 0.53 ± 0.03. Both results include systems in which the outer body is either a single or a binary. However, the number of quadruple systems in the first case is about 12 % (4:34) and about 45 % in the second. We analyze the two subsamples (triples and quadruples) in the latter model. For triple systems we obtain a slope of 0.63574 ± 0.00008 and for quadruple systems we find 0.4219 ± 0.0001. The results suggest that there are two kinds of correlations depending on the nature of the outer body.

Figure 3 shows the plot for an NBODY4 model; the upper contour is clearly linear and is connected with the stability criterion. Our results seem to be compatible with those from the observations in spite of our theoretical sample of hierarchical systems containing a wide range of masses and different stages of evolution. The lack of systems with large outer periods but small inner period is of theoretical and observational interest. This is probably connected with the formation mechanism in which other stars act as perturbers during wide encounters. Thus it has been noted from the calculations that hierarchical formation mainly takes place when two or more binaries suffer a close encounter. It would be desirable to compare our results with those from systems observed in real open clusters. Unfortunately, there are only three such systems with orbital elements for both the inner binary and the outer body. Two are in the Hyades, vB 75 and μ Orionis; the latter is a quadruple. The other system is a quadruple in NGC 1502, SZ Cam.

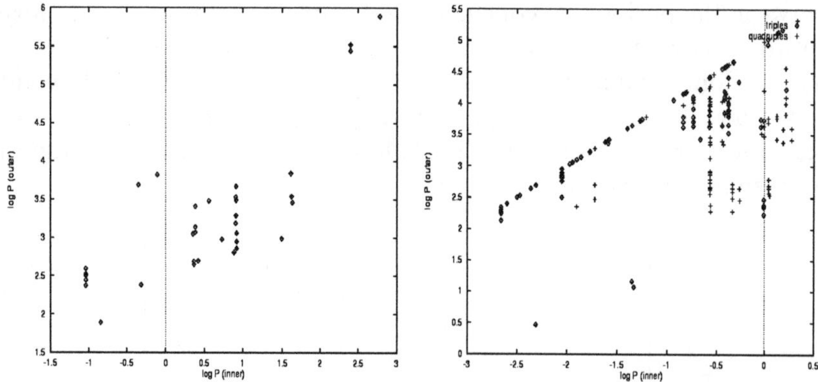

Figure 3. The left figure shows the plane log P_O vs log P_B for a model studied with NBODY5. The right figure is for a model performed with NBODY4. The linear upper contour is due to the criterion for stability. Note that inner binaries in NBODY4 have smaller periods due to the astrophysical phenomena included (see the text).

4.4. THE (LOG P_O, M_B/M_O) PLANE

Although this plane (fig. 4) also contains much astrophysical information, it is not easy to obtain the relevant data from observations. Here M_B is the mass of the inner binary and M_O is the mass of the outer body. For the NBODY4 models the majority of the systems formed fall in the mass ratio range 1–2.2. There are a few systems with higher ratios and the upper cut-off is about 14. The number of systems with mass ratios below 1 is very small. As for the other correlations there is a clear distinction between quadruple systems and triples. The range of mass ratios for quadruples is 0.4–2.2, which is expected because in most cases both binaries are primordial. For triple systems there are several systems with ratios greater than 2.2 but none below 0.8. As regard the nature of the inner binary, systems with no primordial binaries have a range 0.8–3.0 with only one system above (14). For systems with a primordial inner binary the range is wider.

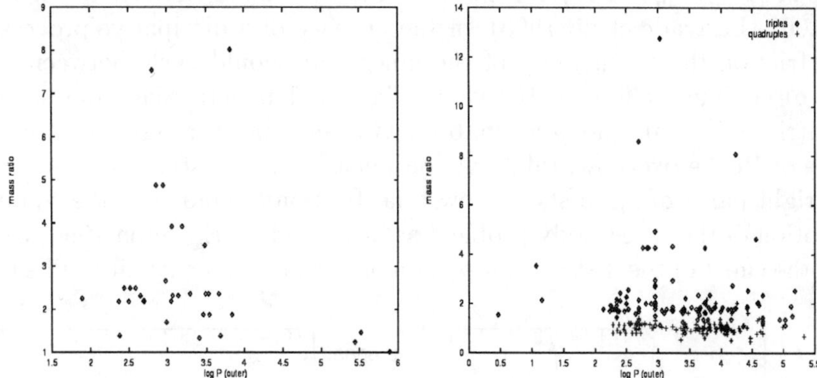

Figure 4. Mass ratio versus logarithm of the period of the outer body for an NBODY5 model (left) and an NBODY4 model. Note the sharp mass ratio for quadruple systems in the latter.

5. Tidal dissipation in triple systems

Neither primordial binaries, nor those that are formed in clusters later (for example by exchange), have exactly circular orbits to start with. However, tidal friction will circularise binary orbits on a rather short time-scale compared with the nuclear time-scale, provided that at least one star of the binary has a radius comparable to the separation between binary components and the dynamical influence of other stars on the binary orbit is negligible. The situation can be very different for binaries within hierarchical triple systems, especially those where the inner and outer orbits are nearly perpendicular (as our numerical simulations shows, not a rare situation). It can be shown analytically and numerically (Marchal 1990, Heggie 1996, Kiseleva 1996) that for non-coplanar triple systems there is a quasi-periodic change of the inner eccentricity (on a time-scale $\sim P_{out}^2/P_{in}$) during which it reaches a maximum value $e_{\rm in}^{\rm max}$. This value only depends on the inclination i between the two orbital planes; other parameters affect only the time-scale. If $i \approx 90^o$, $e_{\rm in}^{\rm max} \approx 1$ and the two stars may collide or suffer a strong tidal interaction.

This effect cannot be neglected in numerical studies of triple stars in clusters. The combined influence of tidal friction and of the third component on the binary orbit may produce interesting and even dramatic results, such as for example a severe shrinking of the orbit.

In order to investigate the interaction between tidal friction and the gravitational dynamics of point masses in a hierarchical triple system we consider two well-known isolated triples β Per (Algol) and λ Tau, which have well-defined orbital parameters. The influence of the distant third body induces eccentricity in the orbit of the close pair. Even if the third body is distant, as in β Per ($P_{\rm out}/P_{\rm in} \approx 237$), its effect need not be small. Because the observed $i \approx 100°$ (Lestrade et al. 1993), in the absence of a dissipative process like tidal friction the eccentricity of the inner pair should cycle between 0 and ~ 1, on a time-scale of $\sim 10^3$ years (fig 5, left panel). Since the observed eccentricity is ~ 0, and presumably has been small throughout its current phase of Roche overflow, tidal friction must act successfully in this system. The right panel of fig. 5 shows how tidal friction (included in the equations of motion of the three-body problem as described by Eggleton 1996) can reduce the effect of the distant companion on the inner eccentricity. The recipe

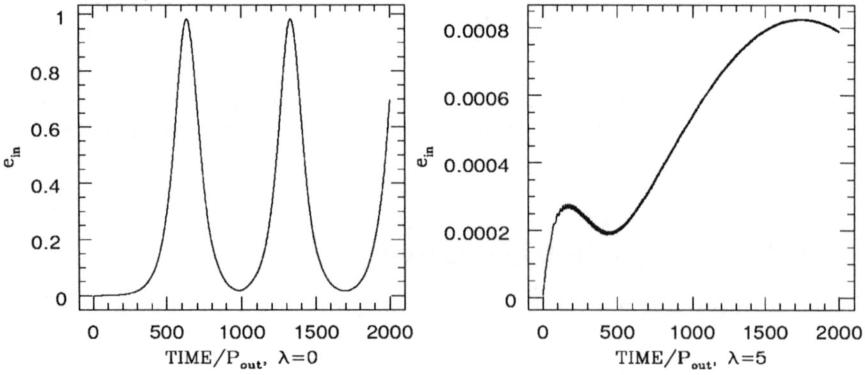

Figure 5. The effect of tidal friction on long-term modulations of the inner eccentricity in β Per triple systems. The parameter λ defines the strength of tidal friction and is described in Eggleton (1996).

for tidal friction dissipates orbital energy but conserves angular momentum; it decreases the semi-major axis and orbital period of the inner binary, and hence increases the period ratio. In many cases this effect can be rather insignificant. For β Per (and other similar systems with high inclination) we find a rather narrow range of tidal friction values which is not strong enough to prevent significant quasi-periodic variations of the inner eccentricity, and yet is strong enough to decrease sharply the binary semi-major axis every time the eccentricity reaches its local maximum. Figure 6 shows this effect.

This is a possible mechanism for the production of close binaries and/or other exotic objects, particularly in clusters.

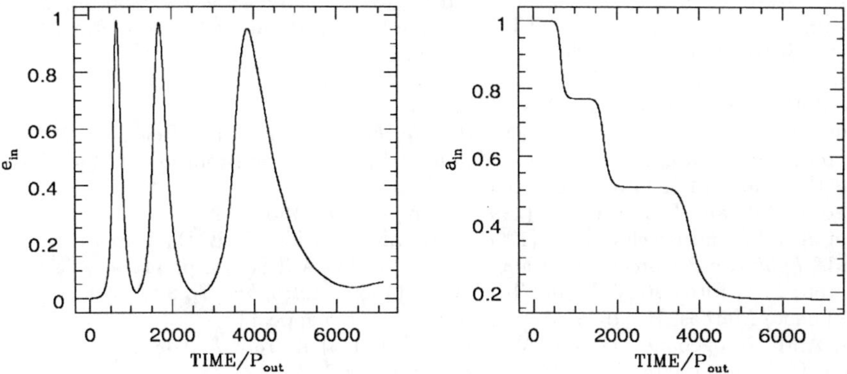

Figure 6. The possible shrinking of the binary orbit in a triple system like β Per under the influence of rather weak tidal friction.

6. Conclusions

This work has allowed us to discuss various aspects of hierarchical systems in open clusters. We have found that it is, indeed, possible to reproduce some observational properties (such as the linear correlation of periods) of hierarchical systems as well as to predict some characteristics of these systems for observations. Although the results described in this paper are encouraging, it is still not clear how the fraction of primordial binaries influences the formation rate of hierarchical systems and how the tidal effects, which were only discussed briefly here, affect the stability of systems with short periods. These questions will be left for future developments.

7. Acknowledgements

R.F.M. thanks the Department of Astrophysics of Universidad Complutense de Madrid for providing excellent computing facilities. L.G.K. thanks the NATO for a Collaborative Research Grant. Our thanks are due to J.-C. Mermilliod who provided most of the observational data both personally and across his database (BDA) at Lausanne University.

References

Aarseth, S.J. (1985) in J.U. Brackbill and B.I. Cohen ed., *Multiple Time Scales*, Academic Press, New York, p. 377

Aarseth, S.J. (1992) in A. Duquennoy and M. Mayor, ed., *Binaries as Tracers of Stellar Formation*, (CUP), p. 6

Aarseth, S.J. (1994) in G. Contopoulos, N.K. Spyrou, and L. Vlahos, ed., *Galactic Dynamics and N-Body Simulations*, (Springer-Verlag), p. 277

Aarseth, S.J. (1996a) in E.F. Milone and J.-C. Mermilliod, ed., *The Origins, Evolution, and Destinies of Binary Stars in Clusters*, ASP Conference Series v. 90, p. 423

Aarseth, S.J. (1996b) in P. Hut and J. Makino, ed., *Dynamical Evolution of Star Clusters*, IAU Symp. 174 (Kluwer, Dordrecht), p. 161

Bailyn, C.D. (1984) (unpublished)

Boss, A.P. (1991) *Nature*, **351**, 298

Duquennoy, A. (1988) in M. Valtonen, ed., *The Few Body Problem*, Reidel, p. 257

Eggleton, P.P. (1996),in P. Hut and J. Makino, ed., *Dynamical Evolution of Star Clusters*, IAU Symp. 174 (Kluwer, Dordrecht), p. 213

Eggleton, P.P., and Kiseleva L.G. (1995) *Astrophys. J.*, **455**, 640

Eggleton, P.P., and Kiseleva L.G. (1996) in R.A.M.J. Wijers, M.B. Davis and C.A. Tout, ed., *Evolutionary processes in binary stars*, NATO ASI Series, p. 345

Eggleton, P.P., Fitchett, M.J., and Tout, C.A. (1989) *Astrophys. J.*, **347**, 998

de la Fuente Marcos, R. (1996) *Astron. Astrophys.* (to appear)

Ghez, A.M., Neugebauer, G., and Matthews, K. (1993) *Astron. J.*, **106**, 2066

Griffin, R.F., and Gunn, J.E. (1981) *Observatory*, **106**, 35

Griffin, R.F., Gunn, J.E., Zimmerman, B.A., and Griffin, R.E.M. (1985) *Astron. J.*, **90**, 609

Harrington, R.S. (1977) *Astron. J.*, **82**, 753

Heggie, D.C. (1996), in P. Hut and J. Makino, ed., *Dynamical Evolution of Star Clusters*, IAU Symp. 174 (Kluwer, Dordrecht), p. xxx

Heggie, D.C., and Ramamani, N. (1992) *Mont. Not. Roy. Ast. Soc.*, **272**, 317

Kiseleva, L.G. (1996) in P. Hut and J. Makino, ed., *Dynamical Evolution of Star Clusters*, IAU Symp. 174 (Kluwer, Dordrecht), p. 233

Kiseleva, L.G., Aarseth, S.J., Eggleton, P.P., and de la Fuente Marcos, R. (1996) in E.F. Milone and J.-C. Mermilliod, ed., *The Origins, Evolution, and Destinies of Binary Stars in Clusters*, ASP Conference Series v. 90, p. 433

Lestrade, J.-F., Phillips, R. B., Hodges, M. W. & Preston, R. A. (1993) ApJ, 410, 808

Makino, J. (1991) *Astrophys. J.*, **369**, 200

Makino, J., Kokubo, E., and Taiji, M. (1993) *Public. Astron. Soc. Pac.*, **45**, 349

Marchal, C. (1990) *The Three Body Problem*, Amsterdam: Elsevier

Mardling, R.A. (1995) *Astrophys. J*, **450**, 722

Mardling, R.A., and Aarseth, S.J. (1996), (in preparation)

Mason, B.D., McAlister, H.A., Hartkopf, W.I., and Bagnuolo, W.G., Jr. (1993) *Astron. J.*, **105**, 220

Mathieu, R.D., Latham, D.W., and Griffin, R.F. (1990) *Astron. J.*, **100**, 1859

Mayer, P., Lorenz, R., Chochol, D., and Irsmambetova, T.R. (1994) *Astron. Astrophys.*, **288**, L13

Mermilliod, J.-C., Rosvick, J.M., Duquennoy, A., and Mayor, M. (1992) *Astron. Astrophys.*, **265**, 513

Mermilliod, J.-C., Duquennoy, A., and Mayor, M. (1994) *Astron. Astrophys.*, **283**, 515

Mikkola, S., and Aarseth, S.J. (1990) *Celest. Mech. Dyn. Astron.*, **47**, 375

Mikkola, S., and Aarseth, S.J. (1993) *Celest. Mech. Dyn. Astron.*, **57**, 439

Reimers, D. (1975) *Mem. Roy. Soc. Liège, 6e Ser*, **8**, 369

Scalo, M.J. (1986) *Fundam. Cosmic Phys.*, **11**, 1

Simon, M., Ghez, A.M., Leinert Ch. et al. (1995) *Astrophys. J.*, **412**, L33

Tout, C.A. (1990) Ph.D. thesis, University of Cambridge

ABSOLUTE DIMENSIONS IN VISUAL BINARIES WITH DOUBLE-LINED ECLIPSING COMPONENTS

A. GIMÉNEZ
LAEFF-INTA, Villafranca del Castillo
Apartado 50727, 28080 Madrid, Spain

1. Introduction

Many young visual binaries are in fact multiple systems in which one or both of the components are close binaries themselves. These close systems may show eclipses as well as radial velocity variations, leading to an accurate measurement of the absolute dimensions of their component stars: masses, radii, effective temperatures, etc. This information is, no doubt, of high importance to understand the evolutionary status, chemical composition, formation and age of the eclipsing stars, but also of the whole stellar system.

The purpose of this talk is to call the attention of those interested in visual double stars to the importance of eclipsing binaries for their astrophysical research. I have been working in the field of close binary systems for almost 20 years now and my interest in this type of interacting stars was based on the possibility of investigating in some detail the theory of stellar structure and evolution, one of the biggest assets of modern astrophysics. The reason to choose close binaries instead of wide separated double stars was certainly related to the detection of eclipses and radial velocity variations within a reasonably short span of time. A detailed analysis of these photometric and spectroscopic variations lead to the determination of accurate values for the stellar radii and masses, which are the basis for the test of stellar evolutionary models.

Therefore, the best candidates for the determination of good absolute dimensions are well-detached double-lined eclipsing binaries with orbital periods between 1 day and 2 weeks (for practical monitoring reasons). These systems also happen to be generally quite young, with main sequence components, since more evolved binaries are bound to have strong interactions

J. A. Docobo et al. (eds.), Visual Double Stars: Formation, Dynamics and Evolutionary Tracks, 179–186.

due to the little space available between the two stars, thus changing their normal evolution as individual stars.

For those interested in visual binaries (with longer periods and larger separations), their ages, evolution, and chemical composition, the presence of an eclipsing system as at least one of their components is of the highest importance because fundamental data can be obtained independent of calibrations. But also for those studying close binaries, the presence of a visual companion, not contaminated by interactions, may provide important reference data.

2. The case of Algol and related binaries

Algol itself was the first eclipsing binary to be discovered by Goodricke some 200 years ago. It is also one of the best studied stars in the sky and it happens to be a multiple system where the close eclipsing pair, with an orbital period of 2.86 days, rotates around the center of mass formed with another star of 1.5 solar masses in a eccentric orbit and a period of 1.86 years.

The absolute dimensions of the binary system are well determined from the observed radial velocity and photometric light curves. The system is the prototype of the Algol stars, semidetached binaries where the originally more massive component filled its Roche lobe loosing mass and transferring part of it to its companion, now the more massive but less evolved star.

Close binaries with relatively short orbital periods along their evolution, easily become interacting systems where the more massive star fills its Roche lobe and starts a phase of mass and angular momentum transfer towards the less massive companion and outside of the system, to the interstellar medium. In these binaries, the behaviour of the component stars after some time does not show indeed the structure and history of an individual star of the same mass and effective temperature.

The comparison of positions of the individual components in the log g - log T plane is the basis for the determination of the age of the binary system if both components are, of course, close to the same isochrone. But in systems with mass transfer episodes this comparison with theoretical isochrones is not possible and a third, not contaminated, point is very important to estimate a realistic value for its age. This point is easily provided by additional individual stars if the binary is a member of an open cluster (Giménez, 1996) through photometric indices. But, still better, in visual binaries, if the third component remains a "normal" star, a similar procedure can be applied and also the mass can be measured. Algol-type binaries are of special interest for the estimation of evolutionary ages since this is the best test of different evolutionary scenarios based on theoretical computer simulations of mass and angular momentum loss/transfer leading to the actual dimensions observed (De Greve, 1993; De Loore and De Greve, 1992). In the case of Algol itself, a detailed comparison was done by Sarna (1993).

In addition, studies of the chemical composition in Algol-type binaries (see e.g., Tomkin et al., 1993) have increased interest of age estimations in order to allow the interpretation on evolutionary terms of the observed abundance ratios, C in particular, in each of the member stars. The semidetached system V505 Sgr (Tomkin, 1992) has been detected spectroscopically, using Reticon observations of the NaD lines in the secondary, to be a triple system.

3. The case of V772 Her and other active binaries

Even in systems, like chromospherically active binaries, where no mass transfer has occurred the behaviour of the component stars may differ significantly from that of individual cases with the same mass and temperature simply because of tidal effects. Induced fast rotation, due to tidal synchronization, changes the magnetic activity of the components and may also affect other physical characteristics. As an example, stars with the same spectral type and luminosity class of the Sun are found in some late-type close binaries showing distinct features of enhanced activity.

In all these close binaries, the existence of a visual companion, physically linked to the close system, provides important additional information for the astrophysical interpretation of their structure and evolution. For example, using the different indicators of chromospheric activity available, it can be seen that the visual components of binary systems with chromospherically active stars, not close binaries themselves, do not show the same levels of activity even with the same masses, ages, and formation scenario.

Moreover, it has been recently shown that the depletion of Lithium in stellar atmospheres is somehow inhibited in close binaries which are also chromopherically active. The comparison of the behaviour of the components in the close binary with that of a distant visual companion, with no tidal interaction, also appears to show that short periods are the reason for such an anomaly. Theoretical calculations of Li depletion in individual stars is now used for the estimation of stellar ages, and this can be checked in visual binaries with eclipsing components. The estimated age for the individual star can be compared with that derived from isochrones for the close binary. Unfortunately this kind of work has not yet been performed in a systematic way.

V772 Her is a member of the multiple system ADS 11060, also named as member A, which shows indications of stellar activity and very shallow eclipses (Reglero et al., 1991). It has a visual companion, member B, and an orbital period of 0.88 days. Member C is another close binary, not eclipsing, showing high levels of chromospheric activity and later spectral types. It is found from the comparison of absolute magnitudes and distances that only A, B and C are linked physically, the rest (D, E and F) being only optical coincidences in a crowded field. The age of the system, estimated from the absolute dimensions of the close components, is approximately 2 billion years while their abundance of lithium is closer to what is predicted for ages of the order of 50-100 times lower.

4. The case of IU Aurigae and other detached systems

We have seen some examples in which the study of the physics taking place in a close binary is supported by the presence of a companion star in visual double system. But the study of this type of multiple stars is also improved by the fact that it may contain a double-lined eclipsing binary. An example is the early type eclipsing binary IU Aur recently studied in light of a new treatment of the reflection effect in its light curve (Drechsel et al., 1994) and the radiation pressure on the configuration of the Roche surfaces. A third component was detected both in the light curve and the radial velocity curve and its found to be gravitatinally linked to the close binary as deduced from the observation of precessional motion of the orbital plane. The long term inclination change with time was investigated by an analysis of the variation of eclipse minimum depth. The nodal rotation period of 335 years and the inclination of the third body are consistent with a mass of 17-18 solar masses. Because of the flux of the third light it is deduced that the third body should itself be also a binary system.

In the case of CW Cep, also with a visual companion, the light curve analysis by Clausen and Giménez (1991) was based on four-colour photometry light curves obtained excluding the visual star 20" apart. Nevertheless, a fainter and redder than the main component stars should be present though it could not be detected in a CCD inspection of the diaphragm field (28"). The effect on the derived orbital inclination as a function of wavelength was the hint for discovery.

Andersen et al. (1990) have studied the eclipsing components of the triple system V1031 Ori which shows an orbital period of 3.4 days. Masses and radii were obtained with accuracies better than 1% and the stars are found to be in a very short-lived evolutionary phase, past the central hydrogen exhaustion at an age of around 0.5 billion years if models without convective overshooting are used. Considering a moderate amount of core overshooting though, the components are still in the main sequence with an age of 0.7 billion years. Speckle observations show significant orbital motion of the visual companion separated by only 0.16 arcsec.

In these binaries, stellar evolutionary tracks can be easily checked, even some tests of the input physics of the models can be attempted. But from the point of view of the visual system, it is important that good estimations of the stellar ages and the initial chemical composition of the member stars can be obtained.

5. Calibration of colour indices versus physical parameters

In multiple systems in which a component is an eclipsing binary, we can determine the separate colour indices of the close binary components and the visual individual companion. Thus, we have known values of the colour indices and the masses and radii of the close stars. The same calibration may be applied to the separated visual companion. If a large enough number of candidates is available, a consistent system of equations can be constructed to determine a good calibration of the photometric system providing dereddening, distances and bolometric corrections. Calibrations permit to determine from photometric indices most of the physical parameters of astrophysical relevance: mass, radius and log g if an appropriate correction for second order effects is taken into account.

The determination of the mass-luminosity relation in low mass stars is one of the major contributions to stellar astrophysics provided by visual binaries. Popper in fact made some years ago a detailed evaluation of available data to publish his calibration tables (Popper, 1980). But the comparison of the secondaries of some eclipsing binaries (namely, V477 Cyg, TX Her, CM Lac, RR Lyn, EE Peg and IQ Per), late type stars but without photometric complications (something relatively difficult to find), shows according to Lacy et al. (1987) an anomaly in the sense that if the primaries are assumed to lie on the calibration curve, then the secondaries are less bright than expected. Otherwise, if the secondaries are assumed to lie on the calibration curve, then the primaries are brighter than expected. It does not seem possible at this time to find an unambiguous interpretation of this anomaly.

Distances to double-lined eclipsing binaries provided by the European satellite Hipparcos will contribute to our knowledge of the relations between colour indices and absolute dimensions, but their determination in binaries members of visual systems will allow the estimation of distances and thus orbital periods of many of these separated systems.

Moreover, the observation of eclipses may help the study of a multiple system from the point of view of its composition, formation, and hierarchy, though in some cases, like DN UMa (Popper, 1986), the determination of the eclipsing component may have proven to be difficult.

6. Eccentric binaries in multiple systems

Eccentric close binaries are very important when the absolute dimensions of the component stars can be accurately determined. Then, an additional physical parameter of the component stars can be measured, namely, the average internal structure constant derived from the observed rate of apsidal motion. Of course, the analysis of the apsidal motion of a given system for a long time may also provide evidences of previously unknown third components that may even become visually detected: e.g. U Oph (Kamper, 1985), RU Mon (Khaliullina et al., 1985), or AO Vel (Clausen et al., 1995). This is similar to the case of third bodies detected by the behaviour of the times of minima used for the calculation of ephemerides. The displacement of eclipses may also provide accurate determinations of the orbital period and eccentricity of the "visual" system.

In addition, we have preliminarly found though a detailed statistical study is under way, that the number of eccentric close binaries which have a close visual companion or a third body is much higher than in other types of binaries. This may of course be a bias linked to age since most eccentric eclipsing binaries should be young binaries in order to avoid tidal effects and, of course, keep a well-detached configuration. Examples of highly eccentric eclipsing binaries with short orbital periods are V1647 Sgr (3.28 days, e = 0.413) or NO Pup (1.26 days, e = 0.123).

Another way to see the same effect was mentioned by Mazeh (1990). It is well known that there is a cut off period in all stellar groups of binary stars. Systems with periods below the cut-off value should have circular orbits and those above may have any orbital eccentricity. This is clearly shown in a general sample of binary stars but it is more evident in well defined subgroups with similar composition, like star clusters or the galactic halo.

Nevertheless, it is often found that some binaries that should have been circularized long time ago according to their orbital period and age, still present significant eccentricities. This spureous eccentricities are attributed by Mazeh to a third star, wich can modulate the binary eccentricity with a relatively long timescale. In fact, Mazeh and Shaham (1979) estimated this modulation to be approximately periodic and of the order of several thousand years if the period of the third body is of the order of years, i.e. relatively close. For visual components, with periods of the order of hundred years, the modulation has periods of the order of tidal circularization time scales, thus not intruducing any effect.

7. Some statistics

The first catalogue of variable stars in visual double systems was compiled by Plaut (1934, 1940) and is made of 137 cases. A second catalogue by Baize

(1962) included 160 binaries with a variable companion. Later, Proust et al. (1981) compiled a catalogue with 300 stars, while the catalogue by Baize and Petit (1989) included a subgroup of that by Worley and Heintz (1983), plus some later additions, of visual binary stars with calculated orbits. A total of 171 of these systems show one or both of their components to be photometrically variable.

A study by Herczeg (1988) indicated that as a many as about a third of the spectroscopic binaries might have a distant visual companion. Therefore, the frequency of triple, close and wide systems together, might be as high as 50% of the short period binaries.

About 80 eclipsing binaries are known to be components of multiple-star systems according to Chambliss (1992). This is less than 2% of all known eclipsing binary systems, but the actual total number must certainly be much higher than this as the majority of the stars under review were in fairly bright, well-studied systems. Clearly, eclipses can be detected at large distances (even outside of our Galaxy) while visual components can only be found at short distances from the observer.

Triple systems are by far the most common of these eclipsing binaries in visual double stars, but at least one system in five is a quadrupole or even higher-order multiple system. In fact, both V772 Her and DN UMa mentioned above are members of quintuple systems, while YY Gem is a rare example of a sextuple star system. The most common type of eclipsing binary to be found in multiple systems is that composed of two early-type stars (B5 or earlier). W UMa are also well represented though at a lower level while RS CVn and cataclysmic variables are quite rare.

8. Interferometric observations of close binaries

Interferometric measurements are, of course, the future for the determination of many more orbits with high spatial resolution. Speckle interferometry is providing measures of binary systems with separations down to 30 milliarcsec while the spacecraft Hipparcos has permitted the study of visual double stars with a resolution of the order of 0.002 arcsec in the optical range. A larger scale interferometric instrumentation in space will improve the available results significantly. A new perspective for high precision interferometry from space is now open by the new missions of the European Space Agency, ESA, within the Horizon 2000 planning of its Science Programme. This envisages to achieve positional accuracies two orders of magnitude better than Hipparcos in the optical range, or in the infrared using a set of cooled telescopes.

Interferometric methods are in fact closing the gap between "visual" and spectroscopic or eclipsing binaries and the distinction will be completely eliminated with the long-baseline space projects. An example of this kind of studies in radio wavelengths is the VLBI astrometry of the multiple system Algol carried out by Lestrade et al. (1993).

For the future, it is therefore very important that new light curves of eclipsing binaries with enough accuracy are secured, in particular for members of visual systems. This requires a systematic approach of the observers to perform photometric measurements of visual binaries in a long term basis.

References

Andersen, J., Clausen, J.V. and Nordström, B. (1990), *Astron. Astrophys.*, **278**, 365
Baize, P. (1962), *J. Observateurs*, **45**, 117
Baize, P. and Petit, M. (1989), *Astron. Astrophys. Suppl. Ser.*, **77**, 497
Chambliss, C.R. (1992), *P.A.S.P*, **104**, 663
Clausen, J.V. and Giménez, A. (1991), *Astron. Astrophys.*, **241**, 98
Clausen, J.V., Giménez, A. and van Houten, C.J. (1995), *Astron. Astrophys. Suppl. Ser.*, **109**, 425
De Greve, J.P. (1993), *Astron. Astrophys. Suppl. Ser.*, **97**, 527
De Loore, C. and De Greve, J.P. (1992), *Astron. Astrophys. Suppl. Ser.*, **94**, 453
Drechsel, H., Haas, S., Lorenz, R. and Mayer, P. (1994), *Astron. Astrophys.*, **284**, 853
Giménez, A. (1996), in *The Origins, Evolution, and Destinies of Binaries in Clusters*, A.S.P. Conf. Ser., **90**, 109
Herczeg, T. (1988), *IAU Colloq.*, **97**, 89
Kamper, B.C. (1985), *I.B.V.S.*, **2800**
Khaliullina, A.I., Khaliullin, Kh.F. and Martynov, D.Ia. (1985), *Mon. Not. R.A.S.*, **216**, 909
Lacy, C.D., Frueh, M.L. and Turner, A.E. (1987), *Astron. J.*, **94**, 1035
Lestrade, J.F., Phillips, R.B., Hodge, M.W. and Preston, R.A. (1993), *Astrophys. J.*, **410**, 808
Mazeh, T. (1990), *Astron. J.*, **99**, 675
Mazeh, T. and Shaham, J. (1979), *Astron. Astrophys.*, **77**, 145
Plaut, L. (1934), *Bull. Astron. Inst. Netherlands*, **227**, 181
Plaut, L. (1940), *Bull. Astron. Inst. Netherlands*, **234**, 49
Popper, D.M. (1980), *Ann. Rev. Astron. Astrophys.*, **18**, 115
Popper, D.M. (1986), *P.A.S.P.*, **98**, 1312
Proust, D., Ochsenbein, F. and Pettersen, B.R. (1981), *Astron. Astrophys. Suppl. Ser.*, **44**, 179
Reglero, V., Fernández, M.J., De Castro, E., Giménez, A. and Fabregat, J. (1991), *Astron. Astrophys. Suppl. Ser.*, **88**, 545
Sarna, M.J. (1993), *Mon. Not. R.A.S.*, **262**, 534
Tomkin, J. (1992),*Astrophys. J.*, **387**, 631
Tomkin, J., Lambert, D.L. and Lemke, M. (1993), *Mon. Not. R.A.S.*, **265**, 581
Worley, C.E. and Heintz, W.D. (1983), *US Naval Obs. Publ.*, **24**

CALIBRATION OF THE MASS-LUMINOSITY RELATION, II: DATA FROM THE GENERAL CATALOGUE OF TRIGONOMETRIC STELLAR PARALLAXES, 1995

W. F. VAN ALTENA, J. T. LEE, E. D. HOFFLEIT, T. M. GIRARD

AND

E. P. HORCH

Yale Astronomy Department

P. O. Box 208101, New Haven, CT 06520

Abstract. The recent completion of the Fourth Edition of the General Catalogue of Trigonometric Stellar Parallaxes (YPC) has enabled us to re-calibrate the Mass-Luminosity Relation (MLR) for main sequence stars. Since the YPC has been carefully evaluated and corrected for systematic errors in both the absolute parallaxes and their error estimates, we believe that the derived MLR will be significantly better than previous calibrations, especially for the intrinsically and apparently faint stars of the lower main sequence where ground-based trigonometric parallaxes are the primary source of distances. For the upper main sequence, a combination of eclipsing and spectroscopic binaries and the new Hipparcos parallaxes will still provide a superior MLR.

Key words: Binaries; parallaxes; masses; mass-luminosity relation.

1. Introduction

In Paper I, (van Altena, *et al.* 1992) data from the Preliminary Version of the YPC91 was used to calibrate the MLR. In addition, the Lutz-Kelker (1973) corrections for the Luminosity were extended to the Mass axis of the MLR. In this paper, we will utilize the parallaxes in the recently completed Fourth Edition of the General Catalogue of Trigonometric Stellar Parallaxes by van Altena, Lee and Hoffleit (1995), hereafter referred to as the YPC, and the extended Lutz-Kelker corrections to derive an improved MLR.

2. Trigonometric parallaxes and orbital data

The characteristics, errors and system of the new edition of the YPC have been detailed in two papers by van Altena, Lee and Hoffleit (1993, 1994) and in the Introduction of the YPC. While the data included in the YPC are very heterogeneous, every effort was made to put the ensemble on a uniform system with weights that represent the external accuracy of the different observations. Due to our lack of knowledge of the details involved

J. A. Docobo et al. (eds.), Visual Double Stars: Formation, Dynamics and Evolutionary Tracks, 187–190.

in the determination of the parallax for each star, the system and observatory weights are meant only to be overall averages and may misrepresent the case for an individual star. On the other hand, since we are interested here in the calibration of the MLR using a relatively large number of stars, this global approach should be adequate. As noted in our publications concerning the YPC, any user interested in an individual star should refer to the original parallax publications cited in the YPC to evaluate the reliability of the various determinations of that parallax.

In this discussion, we draw from approximately 500 binaries with orbits listed in the YPC, of which 150 have mass fractions determined. While most of these data are listed in the YPC, we have taken additional astrometric data and magnitude differences of the binary star components from the excellent paper by Henry and McCarthy (1993) and other papers in the literature. In order to avoid the uncertainties in the Bolometric Corrections, we will limit this analysis to a discussion of the absolute visual magnitude axis.

3. Discussion

It has been known for many years that the direct use of trigonometric parallaxes to calibrate the luminosities of stars can result in values biased towards lower luminosities, when the sample is chosen from stars with measured parallaxes larger than some minimum parallax. The origin of this error is in the rapidly increasing number of stars encountered per unit parallax as one moves to larger and larger distances. When a true parallax distribution is convolved with an observational error, the result is that many more stars with small true parallaxes are scattered towards larger observed parallaxes than conversely. As a result, an uncorrected calibration uses parallaxes that are on the average too large, which results in a luminosity or absolute magnitude that is too dim. Lutz and Kelker (1973) formalized the problem and analytically calculated the corrections to be applied to the derived absolute magnitudes for a uniform distribution of stars in space. In Paper I, van Altena, Girard and Lee (1992) extended the Lutz-Kelker (1973) corrections for luminosities to masses. Hanson (1979) had already generalized the uniform space distribution restriction to one involving various selection effects, by noting that for a statistical sample of stars with a well- defined velocity dispersion, the parallax is directly proportional to the proper motion. For a typical dispersion of 35 km/sec, the proper motions will be about eight times larger than the corresponding parallaxes thus making it possible to accurately infer the distribution of parallaxes from the distribution of proper motions.

A uniform distribution of stars in space results in a parallax, or proper motion, distribution that is proportional to the -4.0 power of the parallax or proper motion. The proper motion distribution of our binary stars is found to be proportional to the -2.5 power, indicating that the sample is, as expected, seriously incomplete due to selection effects in their discovery and observation. As a result, the Lutz-Kelker corrections are about a factor of two smaller for a given ratio of the parallax error to the parallax (σ/π) so that we can use smaller parallaxes or accept larger parallax errors and practically speaking, include more stars in our sample. The use of more stars is important so that we can have a representative sample of stars in the solar neighborhood to calibrate the MLR and not have to rely only on those stars that happen to lie very close to the sun at the present time and whose metallicities may not be characteristic of the general solar neighborhood.

4. Conclusions

If we apply the corrections to both axes as outlined in Paper I, but only for the stars with mass fractions, then we obtain the absolute magnitude versus log mass (MLR) diagram shown in Figure 1 The data plotted are for binary star systems, A-components as squares and B-components as circles, with $\pi \geq +0.025''$, $(\sigma/\pi) \leq 0.25$; the data in both axes have been corrected for the Lutz-Kelker effect. Binary components that have been classified as giants have been eliminated from the diagram but the white dwarf components are indicated. Also plotted are the one-sigma error bars, including the uncertainty in the corrections, the MLR mean relation derived by Henry and McCarthy (1993), and the 0.075 solar mass limit for nuclear burning. The Henry-McCarthy MLR is defined by the spectroscopic and eclipsing binaries for Mv $< +5$, but for stars fainter than +5, it is the trigonometric parallaxes that define the relation. It is too early to say at this stage of our analysis if a significant modification of the MLR is required for Mv $> +5$, but there are certainly hints in Figure 1 that such changes may be necessary. The new calibration will be discussed in a forthcoming paper.

5. The Future

It is clear that the parallaxes obtained with the Hipparcos Astrometric Satellite will make dramatic improvements to the MLR, given the average parallax accuracy of 1 mas. However, due to the magnitude limit of Hipparcos, its impact on the lower main sequence will be minimal and that is emphasized by the fact that of the 8800 stars in the YPC, approximately 2200 are not in the Hipparcos Catalogue and most of those are the intrinsically faint stars of the lower main-sequence. For the foreseeable future, the calibration of the low-mass end of the MLR will be dominated astrometric data and magnitude differences of the binary star components from ground-based parallaxes of faint stars. In addition, simulations by Meyer (1996) at Yale show that the fundamental limitations to the accuracy with which the MLR is known will be set by our knowledge of the semi-major axes of the orbits and the magnitude differences of the components once the more accurate parallaxes are used. Meyer is developing techniques which should make it possible to accurately determine the magnitude differences, but there will be a continuing need for more and better observations of the separations and position angles of binary stars. Increased efforts should be made to expand the ground-based interferometric observations of binaries so that the MLR is no longer limited by the lack of ground-based observations.

6. Acknowledgements

This research was supported in part by grants from the National Science Foundation to the Yale Southern Observatory.

References

Hanson, R. B. (1979) *MNRAS*, **186**, 875
Henry, T. J., & McCarthy, Jr., D. W. (1993) *AJ*, **106**, 773
Lutz, T. E., & Kelker, D. H. (1973)*PASP*, **85**, 573
Lutz, T. E. (1979) *MNRAS*, **189**, 273
Meyer, R. D. (1996) Yale Univ. Ph.D. thesis in progress.
van Altena, W. F., Girard, T. M., & Lee, J. T. (1992) in *Complementary Approaches to Double and Multiple Star Research*, **IAU Coll. 135**, Eds. H. A. McAlister & W. I. Hartkopf, *ASP Conf. Series*, **32**, 276

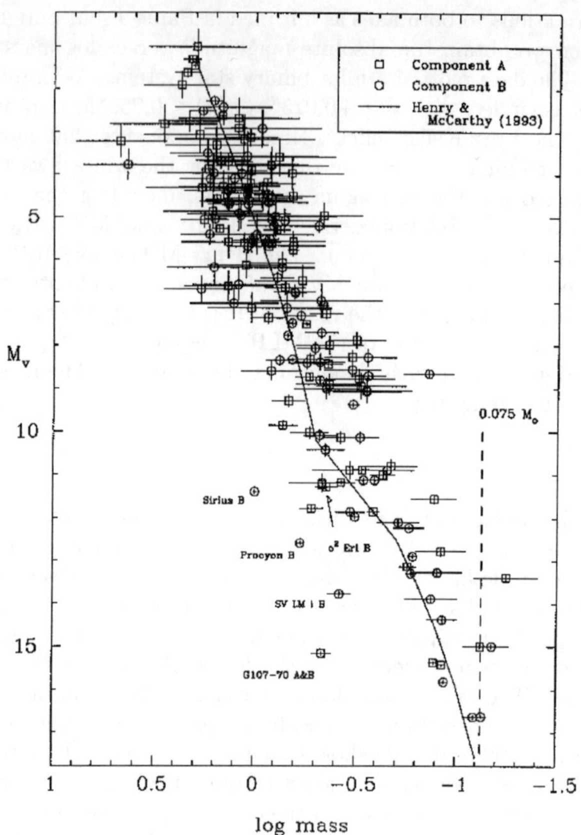

Figure 1. The absolute visual magnitude versus log mass in solar units diagram. The A-components are plotted as squares and the B-components as circles, both with one-sigma error bars. The Lutz-Kelker (1973) corrections have been applied to the absolute visual magnitudes, while the extended Lutz-Kelker corrections for the masses are from van Altena, *et al.* (1992). The solid line is the mass-luminosity relation from Henry and McCarthy (1993) and the dashed line is the mass-limit for hydrogen burning stars. Binary components with a spectral classification of giant have been eliminated, but some of the white dwarf components are indicated.

van Altena, W. F., Lee, J. T., & Hoffleit, E. D. (1993) in *Workshop on Databases for Galactic Structure*, Eds. A. G. D. Philip, B. Hauck, and A. R. Upgren, (L. Davis Press: Schenectady) 65

van Altena, W. F., Lee, J. T., & Hoffleit, E. D. (1994) in *Galactic and Solar System Optical Astrometry*, eds. L. V. Morrison and G. F. Gilmore, Cambridge University Press, 50

van Altena, W. F., Lee, J. T., & Hoffleit, E. D. (1995) *The General Catalogue of Trigonometric Parallaxes*, Fourth Edition, Yale University Observatory, New Haven.

EVOLUTIONARY EFFECTS IN THE SEPARATIONS OF WIDE BINARIES

A. POVEDA, CHRISTINE ALLEN AND M.A. HERRERA
Instituto de Astronomía, UNAM
México DF, 04510. México

Abstract. Utilizing the authors' recent catalogue of wide double and multiple stars for the solar vicinity (Poveda et al. 1994) we investigate the distribution of separations of their components. We find that the distribution of separations, $f(s)$, is well represented by Öpik's relation $f(s) \propto s^{-1}$ for separations smaller than 2,400 AU. We find an age-dependent critical separation S_c which divides those binaries that have been affected by energy exchanges in encounters with perturbing objects from those whose separations are still unchanged from their primeval distribution. For the group of youngest systems from our catalogue $S_c = 7,800$ AU, and for the oldest, $S_c = 2,400$ AU. The velocity dispersions for the first group give a mean age of 2.5×10^9 years, while those for the oldest systems yield 7×10^9 years.

1. Introduction

The present-day distribution of orbital and physical parameters of wide double and multiple stars (henceforth WDMS) is the result both of star formation processes and of the subsequent dynamical evolution of the system. Furthermore, the degree of duplicity and multiplicity in a stellar population of a given age is also the result of the interplay of cosmogonical processes and dynamical interactions of the components of double and multiple systems.

In the dynamical evolution of a WDMS at least two phases can be perceived: an "internal one", where the mutual dynamical interactions of the components of a multiple system transform the eccentricities, the major semiaxes and the multiplicities of the systems, and an external one, where encounters of a WDMS with massive galactic objects cause the gradual modification of the original orbital parameters of the system.

In the present paper we investigate the distribution function of projected separations $f(s)$ for the sample of WDMS in the solar vicinity contained in the authors' recent catalogue of wide double and multiple stars (Poveda et al. 1994). The systems of the catalogue constitute the most comprehensive list of WDMS in the solar vicinity; in addition, each system is classified as being either "young" or "old"; this age classification allows one to investigate the time evolution of the distribution of major semiaxes.

When analysing the distribution $f(a)$ of semimajor axis, we face the limitation that in most cases no orbit is known for a given binary; this is particularly so for WDMS. We can overcome this limitation by noting that there is a well established relation between

J. A. Docobo et al. (eds.), Visual Double Stars: Formation, Dynamics and Evolutionary Tracks, 191–198.

the expected value of the observed separation s of a sample of binaries and the expected value of the semimajor axis for the same sample, i.e.,

$$E(\log a) - E(\log s) = 0.146$$

The above formula was derived by Couteau (1960) on theoretical grounds. He tested its observational validity using a catalogue of 410 orbits, for which the constant turned out to be 0.150. It is not too different from the empirical relation derived earlier by Kuiper (1942) and widely used since, where the constant has a value of 0.11.

In the present paper, we will work with the projected separations s (in astronomical units) among the components of WDMS, with the understanding that the distributions of separations s are equivalent to the appropiate distribution of semimajor axis a.

When investigating the dynamical evolution of the distribution of separations $f(s)$ it would be desirable to first establish the primeval distribution, in order to determine how this original distribution has been distorted with the passage of time. To find the original distribution of separation we need to ascertain whether there are several distinct distributions corresponding to different modes of WDMS formation. The work of Abt and Levy (1976) established with reasonable certainty that there are indeed at least two modes of double star formation. Binaries with periods greater than about 100 years show a mass spectrum for their secondaries corresponding to Salpeter2s distribution of stellar masses, which indicates that they were formed from the pairing of independent condensations, each component extracted from the Salpeter distribution. On the other hand, binaries with periods less than 100 years have their mass ratios $q = m_2/m_1$ distributed with a maximum around $q = 1$, which suggests that they were formed by a process of disk fragmentation.

In the present investigation we will study the distribution $f(s)$ for large separations, corresponding to WDMS formation by independent condensations. Thus we shall restrict ourselves to binaries with periods larger than about 100 years, or major semiaxes larger than 25 AU. Our recent catalogue was constructed precisely with this mode of WDMS formation in mind.

In the past, there have been two conflicting representations of $f(s)$, namely Öpik's (1924) classical distribution $f(a) = ka^{-1}$, and Kuiper's (1942) Gaussian distribution in log(a), namely $\phi(\log a) = k \exp(-\log a)^2$. An investigation of the filtered IDS (Poveda et al., 1982), showed that the distribution of separations followed Öpik's relation. The same conclusion was arrived at by Poveda (1988) with a sample of common proper motion binaries from Luyten's LDS Catalogue. In recent times, the ambiguity as to which of the distributions is closer to the observed data has persisted. In fact, Close et al. (1990) found for a sample of binaries taken from Woolley's (1970) catalogue of nearby stars that $f(s) = ks^{-1.3}$, resembling Öpik's distribution within the uncertainties of the data, while Duquennoy & Mayor (1991) found from a sample of nearby binaries with primaries from Gliese's catalogue (1991), that $f(\log p)$, or equivalently, $f(\log a)$ had a Gaussian distribution, as proposed by Kuiper.

The material to be studied in the present investigation comprises many more wide binaries than either Close et al. or Duquennoy & Mayor were able to study. We shall show clear evidence indicating that the primeval distribution follows Öpik's law up to a maximum S_c beyond which the distribution becomes modified by the effects of perturbations during encounters with massive objects. We shall further show that S_c is different in the two age groups, specifically, that S_c (youngest systems) $>$ S_c (oldest systems), as might be expected.

It can be shown by a Kolmogorov-Smirnov (KS) test that the distribution $f(s) = ks^{-1}$ is consistent with both Close's and Duquennoy & Mayor's data (within the uncertainties inherent to their small samples), as long as one considers separations corresponding to WDMS and smaller than S_c.

In Section 2 we examine the distribution of separations for the whole sample of our catalogued WDMS and find that for separations smaller than $S_c = 2,400$ AU, Öpik's distribution is satisfied. In Section 3 we investigate the distribution of separations for the youngest systems and for the oldest ones and find that there is an age dependent critical separation S_c beyond which the distribution departs from Öpik's; for the youngest systems this critical separation is about three times larger than for the oldest systems. In Section 4 we investigate the velocity dispersion of the youngest systems and compare it with that of the oldest. From this comparison we deduce that the mean age of the oldest systems is about three times that of the youngest ones. Conclusions are presented in Section 5.

2. The primeval distribution of separations

The process of binary (and multiple) star formation seems to occur very early in the pre-main sequence phase of stellar evolution (Mathieu, 1994); with the passage of time, the primeval distribution of semiaxes major and eccentricities is modified by dynamical interaction within the system (in case of a multiple star) and by perturbations by nearby stars within the star forming region; later on, the WDMS disperse into the general galactic field where they are subject to perturbations for a much longer period of time.

We find convenient to define the primeval distribution of elements (separations) as the one that results from the convolution of the "internal" perturbations on the distribution established at birth, i.e., the distribution that is delivered to the general field. The reason for this definition is that the binaries we are about to study have been subjected to perturbers in the field for a much longer time than in the parental cluster.

It has been shown by Wasserman (1988) that an isolated binary subject to perturbations by field stars, molecular clouds, etc. is disrupted, after a given time, with a probability that increases with the semimajor axis. From Wasserman's work, it can be seen that binaries with separations smaller than 2,000 AU have a survival probability of nearly one after 10^{10} years. On the other hand, binaries with major semiaxes of 5,000 AU, have a survival probability of 1/2 after 10^{10} years, while those systems with a $\approx$ 12,000 AU, have a survival probability of only 0.1 after 10^{10} years.

From this result we may feel confident that if we choose to study the distribution of separations of field binaries with separations smaller than 2,000 AU, we will have a sample which has not been perturbed appreciably by encounters in the field and therefore will exhibit the primeval distribution of separations.

Our catalogue of wide binaries and multiple systems in the solar vecinity (Poveda et al. 1994), is the most appropiate source of systems to study both the primeval distribution of separations, and its behaviour with the passage of time, because it includes also an age classification for the vast majority of the systems. When investigating the distribution of separations of WDMS a convenient representation consists in placing all the primaries in one point and the secondaries at their corresponding separations, the polar angle being uniformly distributed. The surface density of secondaries $\rho(s)$ can be represented in general by the function $\rho(s) = As^{-\alpha}$, where the exponent α is the critical parameter to be determined; Öpik's distribution $f(s) \propto s^{-1}$ corresponds to $\alpha = 2$. The number $N(s)$ of binaries with separations smaller than s is given by

$$\log N(s) = \text{const} + (2 - \alpha) \log s \qquad \alpha \neq 2 \tag{1}$$

and

$$N(s) = \text{const} \log s, \qquad \alpha = 2. \tag{2}$$

For every separation s (a triple corresponds to 2 separations) we fitted the best linear relation $\log N(s) = const + (2 - \alpha)$ log s and found the largest separation for which the fit was acceptable by the Kolmogorov-Smirnov (KS) test (see below). We found that for separations smaller than 4500 AU the best fit of equation (1) corresponds to $\alpha = 2.08$ and the KS test gives the probability $Q = 0.91$ that such a sample of separations ($s \leq 4,500$ AU) has been extracted from a distribution $\rho(s) = As^{-2}$. Moreover a fit of equation (1) to the set of separations $s < 8,000$ AU gives $\alpha = 2.106$ and the KS estimator drops to $Q = 0.26$. We conclude from the above that the distribution $\rho(s) = As^{-2}$ is a good representation for the parent population from which the sample of binaries ($s \leq 4,500$ AU) has been extracted. Figure 1 shows the best fit of equation (2) for $s \leq 4,500$ AU; an eye inspection of this figure is consistent with the result of the KS test mentioned above, i.e., that the primeval distribution of separations for a mixture of binaries of all ages, as represented in the solar neighborhood, follows Öpik's relation $\rho(s) \propto s^{-2}$, $f(s) \propto s^{-1}$.

3. The evolution of $f(s)$ with time

In order to investigate the effect of perturbations on binaries (by field stars, molecular clouds and other massive objects), we proceed to compare the distribution of separations $f(s)$ for two extreme age subgroups of our catalogue, i.e., for the *youngest* systems (YS) and for the *oldest* systems (OS).
(i) A system is called *youngest* (YS) if one or more of its components satisfies at least one of the following properties:

1. It is a main sequence star of type earlier than, or equal to, F5V.
2. It has Hα emission
3. It is a flare star
4. Its rotational velocity $V \sin i$ is greater than 6 km s^{-1}
5. It is a member of a moving group or supercluster younger than the Sun
6. Its age has been determined from its chromospheric activity, lithium abundance, etc., and it is younger than the Sun.
7. The intensity of its H and K emission has been determined to be larger than +2 (Wilson and Woolley, 1970)

(ii) A system is called *oldest* (OS) according to the following criteria:

1. It is marked with an O in the catalogue.
2. It is marked as Y? or O? in the catalogue, but has a W-velocity larger than 25 km s^{-1}.
3. It is marked as an o in the catalogue, but has a W-velocity larger than 25 km s^{-1}

The above criteria for age classification are similar to the ones explained in our catalogue (see Poveda et al. 1994 for details). Here we have modified some of the characteristics to increase the age constrast between young and old systems.

As before, we have ordered the youngest systems by increasing separations and the cumulative fraction $N(s)$ as a function of $\log s$, is shown in Figure 2 for a total of 47 pairs; the best fit of an Öpik's distribution for the first 42 separations, i.e., $s \leq 7,862$ AU, is shown. The agreement with Öpik's distribution (right up to 7,862 AU) is excellent as indicated by the KS estimator $Q = 0.99$.

A similar analysis was performed with the oldest systems; in Figure 3 the cumulative distribution $N(s)$ as a function of $\log s$ is fitted with Öpik's distribution. Again the fit is excellent for the first 78 separations, corresponding to $s \leq 2,409$ AU. The KS estimator is $Q = 0.997$. If we extend the fit to the next pair, with separation $s = 2,964$ AU, the KS estimator drops to $Q = 0.77$ which indicates that for the oldest systems Öpik's distribution extends to a critical separation S_c of about 2,500 AU.

The comparison of the above results (see Figs. 2 and 3), i.e. $S_c(YS) = 7,862$ AU and $S_c(OS) = 2,409$ AU, clearly shows how the disruptive perturbations on old binaries have depleted the population of the oldest systems with $s > S_c(OS) = 2,409$ AU, while for the youngest systems the effect of perturbations begins to be noticed at the more weakly bound pairs with $s > S_c(YS) = 7,862$ AU.

The difference by more than a factor of 3 between the critical separation S_c for the oldest and the youngest systems is so large that we can safely conclude that we are witnessing here, in a very clear way, how the disruption of binaries proceeds with time and how Öpik's primeval distribution is eroded.

4. The mean age of the youngest and the oldest systems

In order to be more confident in the above interpretation, it is desirable to estimate and compare the ages of the two groups of binaries, which we do by determining the velocity dispersion of each group; we estimate the velocity dispersion by the method of percentiles. The velocity dispersions σ_u, σ_v, σ_w and σ_T, determined by the average of the 56%, 68% and 80% percentiles for the 36 youngest systems for which we have the full kinematic information are $\langle\sigma_u\rangle = 17.3$ km s^{-1}, $\langle\sigma_v\rangle = 12.9$ km s^{-1}, $\langle\sigma_w\rangle = 9.3$ km s^{-1}, $\langle\sigma_T\rangle = 23.5$ km s^{-1}.

Similarly for the 23 oldest systems, the velocity dispersions are $\langle\sigma_u\rangle = 43.0$ km s^{-1}, $\langle\sigma_v\rangle = 27.5$ km s^{-1}, $\langle\sigma_w\rangle = 15.6$ km s^{-1} and $\langle\sigma_T\rangle = 53.4$ km s^{-1}. To avoid biasing the results, were considered only systems whose assignment to the OS group did not depend on their kinematics.

We can compare the above velocity dispersion with that of the F4V and F5V stars in Gliese Catalogue. For these stars we obtain $\sigma_T = 28$km s^{-1}, which is somewhat larger than the velocity dispersions for the youngest systems. Meynet et al. (1990) have determined the main sequence lifetime for the F4 stars to be 5 Gy which, under the assumption of constant birth rate, gives a mean age of 2.5 Gy; we conclude that the youngest systems have a mean age smaller than 2.5×10^9 years.

On the other hand, we can compare the velocity dispersion of the oldest systems with that of the main sequence G stars from Gliese Catalogue (taking only parallax qualities a and b) which turns out to be $\sigma_T = 47$ km s^{-1}. The mean age of these stars, again under the hypothesis of constant rate of star formation, is one half the age of the galactic disk (assumed to be 14×10^9 years). We conclude therefore that the mean age of the oldest systems is larger than about 7×10^9 years.

From the above discussion, we see clearly that the ages of the oldest pairs are about 3 times larger than the youngest systems.

5. Conclusions

From the study of the distribution of separations of wide double and multiple systems from our catalogue we may conclude the following:

1. The primeval distribution of separations (semiaxes major) follows Öpik's distribution $f(s) \propto s^{-1}$.
2. The youngest systems follow Öpik's distribution up to separations $S_c(YS) \approx 8,000$ AU.
3. The oldest systems follow Öpik's distribution up to $S_c(OS) \approx 2,400$ AU.
4. The difference of a factor of 3 between $S_c(YS)$ and $S_c(OS)$ can be interpreted as due to the process of disruption of weakly bound systems by encounters with massive objects in the galactic field.
5. It is interesting to point out that the critical separation $S_c(OS)$ at which the distribution of separations departs from Öpik's coincides with the separation which , according to Wasserman's computation, a binary will survive against perturbations throughout 10^{10} years.
6. The mean age of the youngest systems is smaller than 2.5×10^9 years.
7. The mean age of oldest systems is larger than 7×10^9 years.

An extended version of this paper will be published elsewhere.

References

1. Abt, H.A. & Levy, S. 1976, *Ap. J. Suppl.*, **30**, 273
2. Close, L.M., Richer, H.B. and Crabtree, D.R. 1990, *Astron. J.*, **100**, 1968
3. Couteau, P. 1960, *J. des Observateurs*, **43**, No. 3
4. Duquennoy, A. & Mayor, M. 1991, *Astron. & Astroph.*, **248**, 485.
5. Gliese, W. and Jahreiss, H. 1991, in *Selected Astronomical Catalogs*, eds. L.E. Brotzman & S.E. Gessner, Vol. I, NSSC, NASA, GSFC (GJ91).
6. Kuiper, G.P., 1942, *Ap. J.*, **95**, 201
7. Mathieu, R.D. 1994, *ARA & A*, **32**, 465
8. Meynet, G., Mermilliod, J.C., and Maeder, A. 1990, in *Astrophysical Ages and Dating Methods.* Eds. E. Vangloni: Flam, M. Cassé, J. Audouze and J. Tran Thanh Van. (Ed. Frontiéres), p. 91
9. Öpik, E. 1924, *Publ. Tartu Univ. Obs.*, **25**, No. 6
10. Poveda, A., Allen, C. & Parrao, L., 1982, *Ap. J.*, **258**, 589.
11. Poveda, A. 1988, *Ap. & Sp. Sci.*, **142**, 67
12. Poveda, A., Herrera, M.A., Allen, C., Cordero, G. and Lavalley, C. 1994, *Rev. Mex. Astron. Astrof.*, **28**, 43-89
13. Wasserman, I. 1988, *Astroph. & Sp. Sc.*, **142**, 267
14. Wilson, O. & Woolley, R. 1970, *MNRAS*, **148**, 463
15. Woolley, R., Epp, E.A., Penston, M.J., & Pocock, S.B. 1970, *Royal Obs. Annals*, 5

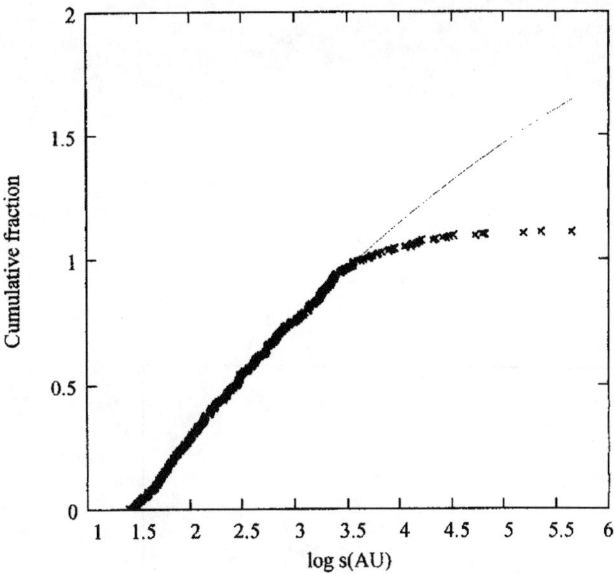

Figure 1. Power law fit $\rho(s) = AS^{-\alpha}$ to all separations in our catalogue smaller or equal than 4,500 AU (329 separations). The exponent is $\alpha = 2,083$ and the KS estimator of the fit is $Q = 0.91$.

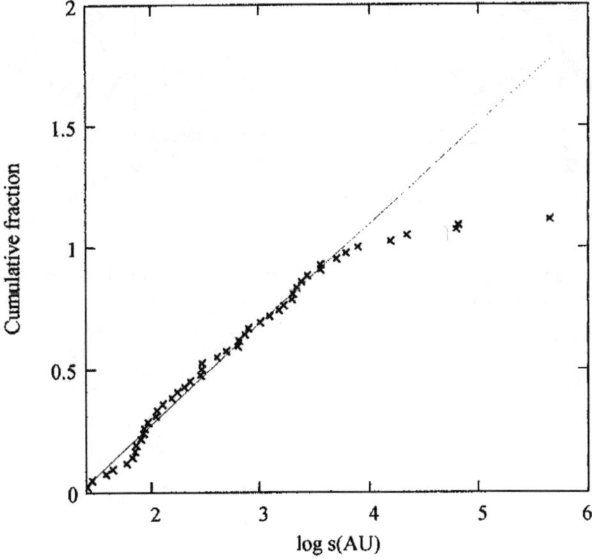

Figure 2. Fit of an Öpik's distribution to the first 42 separations of the sample of the youngest systems. The KS estimator is $Q = 0.994$ and the largest fitted separation is 7,862 AU.

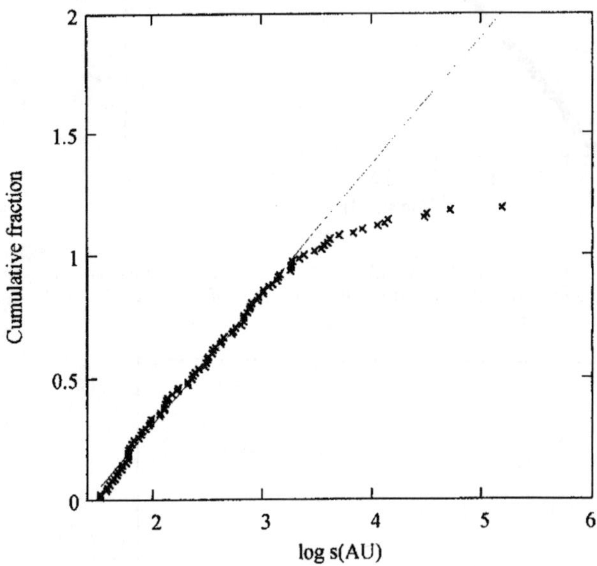

Figure 3. Fit of an Öpik's distribution to the first 78 separations of the sample of the oldest systems. The KS estimator is $Q = 0.997$ and the largest fitted separation ios $2,409$ AU.

WIDE BINARIES IN THE HYADES CLUSTER

M.A. HERRERA, A. POVEDA, A. NIGOCHE AND A. SEGURA
Instituto de Astronomía, UNAM
México DF, 04510. México

Abstract. With the purpose of studying the time evolution of the distribution of separations of binary systems, a long term effort is being made to obtain homogeneous samples of binaries of different ages. In this work, the results of a search for binaries in the Hyades cluster are presented. The search was done by means of the nearest neighbor statistic, complemented with numerical simulations. The data base was Luyten's list of 929 probable Hyades. A total of 53 probable binaries was detected; 22 of them are the 22 binaries in Luyten's list that were previously known, and the other 31 are new probable identifications.

1. Introduction

The time evolution of the distribution of separations of selected samples of binary stars should throw light on the nature of their interactions with surrounding matter and on the nature of the objects with which they interact. Since the effect of these interactions is expected to be more evident in the less bounded systems, a long term effort is being made to derive the distribution of separations of samples of wide binaries of different ages. A first step in this direction was the elaboration of a quasicomplete catalogue of wide binary and multiple systems in the solar vicinity (Poveda et al., 1994). A second step, now in process, is the derivation of the distributions of separations of other samples with different ages. This work presents preliminary results of a search for wide binaries in the Hyades cluster based on Luyten's sample of 929 probable members of the cluster (Luyten, 1981).

2. Theory

The identification of probable binaries was made following the classical approach of comparing the observed angular separations of the stars in the sample with those expected for a two-dimensional random distribution of points. In particular, we based our analysis in the 2-D nearest neighbor statistic. If we assume that the projected positions of the stars on the sky follow a Poisson distribution, then it is easy to show that, for any star, the probability of its nearest neighbor lying at an angular distance between s and $s + ds$ is $p(s)ds = 2\pi s \exp(-s^2/2S^2)ds$, where S, the average angular separation between two nearest neighbors, is related to the mean surface density of stars, n, by $S = 1/(2n^{1/2})$. For each star, the probability of finding its nearest neighbor closer than a given distance s

J. A. Docobo et al. (eds.), Visual Double Stars: Formation, Dynamics and Evolutionary Tracks, 199–203.

is $P(s) = (2/\pi)\exp(-s^2/2S^2)$, and the expected total number of nearest neighbors closer than s, if the sample consists of N stars (randomly ditributed on a plane), is $NP(s)$. The basic idea of the method is that when this number is much less than the number of nearest neighbors closer than s observed in the actual sample, then we may assume that most of these pairs are real (physical) binaries.

3. Numerical simulations

In practice, the application of this theoretical distribution to a given sample yields only an order of magnitude idea of the number of actual binaries in the sample because it does not take into account possible edge effects and also because real clusters do not show the constant density profile assumed in the theory. Therefore, we decided to test the range of applicability of the theoretical formulation in a number of randomly built artificial clusters with boundaries and with a given non-constant density profile. To this end, we first found an analytical fit to the density profile of the stars in our sample in the form of the well known formula proposed by King for the light intensity of globular clusters (King, 1962). The best fit was obtained for a core radius $r_c = 4.915$ pc and a tidal radius $r_t = 16.352$ pc (here, and through he rest of this work, we assume a distance of 45 pc to the Hyades). The observed densities and the analytical fit are shown in Fig. 1, where it is readily seen that the fit is rather good (in fact, it is better than 2% for all points except 5).

To simulate the real cluster, we then constructed an artificial cluster of 900 "stars" -500 "singles" and 200 "binaries"- by the following procedure. First, we placed the 500 "singles" randomly in an area equal to that occupied by our sample, using a Monte Carlo technique to assure that they followed statistically the previously derived analytical density distribution (King's formula). Then, we added the 200 "primaries", also randomly distributed and also following the real cluster's density profile. Finally, we added to each primary its corresponding secondary at a randomly selected position angle, uniformly distributed between 0 and 2π, and with a separation taken from a set of 200 random separations built in such a way that they followed an Öpik distribution $f(s) \sim s^{-1}$ (Öpik, 1924). This procedure was used to build three different "clusters" of 900 "stars" and with the density profile of the real sample. One of them is shown on Fig. 2. It may be compared with the real cluster (Fig. 3).

In each of the three artificial clusters we then found, for each star, the distance to its nearest neighbor. Measuring these distances in units of the average separation $S(r)$ *at the distance of the star from the center of the cluster*, we then counted the number of nearest neighbors with separations less or equal than 1/10, 2/10, etc. Finally, we compared these numbers with those given by the theory, always taking into account the dependence of the average separation S on the position of the star. The results of this comparison showed that it is convenient to define a "probable binary" as a pair of stars which are mutual nearest neighbors, i.e., pairs A-B where A is the nearest neighbor of B and B is the nearest neighbor of A. If this definition is adopted, we find that: 1. 100% of the original binaries are never recovered and there are always spurious binaries. 2. The best compromise between number of recovered binaries and proportion of recovered original binaries vs. spurious binaries occurs for separations less or equal to three tenths of the average separation between stars. In this case, an average of 74% of the original binaries is recovered and the inevitable number of spurious binaries amounts on the average to 7% of the original binaries. 3. All the original binaries farther than half the tidal radius from the center of the cluster are recovered. 4. 99.5% of the original binaries with separations

$< 25,000$ AU are recovered (remember that this number assumes a distance of 45 pc to the Hyades).

4. Results and conclusions

The statistical nature of the present work made it desirable to work with the largest possible number of stars. Therefore, from the numerous lists of probable members of the Hyades cluster (van Bueren, 1952, van Altena, 1966, etc.), we chose as our working sample the one with the largest number of entries, i.e., the list of 929 probable Hyades published by Luyten in 1981 (Luyten, 1981). In this list, 624 entries are Luyten's own identifications on the Palomar plates, 184 come from van Bueren's classical work and the rest were taken from several other sources (all the relevant references may be found in Luyten's work). The sample covers an area of approximately $30° \times 30°$ on the sky. Figure 3 shows all the stars in an $\alpha - \delta$ plot.

The application to the sample of the "probable binarity" criterion discussed in the previous section (mutual nearest neighbors closest than three tenths of the average separation) yielded the following results: 1. There are a total of 301 pairs which are mutual nearest neighbors, but only 53 of them have separations less or equal than three tenths of the average separation corresponding to their distance to the center of the cluster and are, therefore, probable real systems. 2. As far as we know, 22 of these 53 probable real binaries have not been mentioned before as probable binaries. 3. The remaining 31 pairs, previously recognized in the literature as probable systems, seem to be *all* the previously known systems in Luyten's list. In other words, the statistical method recovered 100% of the previously known binaries in the sample. This is important because it may be interpreted as meaning that the 22 new candidates have a good chance of also being real systems. The list of the 53 probable binaries found will be published in a forthcoming more detailed paper. Finally, we state some preliminary results that follow from assuming that these 53 probable binaries are, indeed, members of the cluster (and assuming, as always, a distance of 45 pc to the cluster): 1. The minimum separation of the binaries in the Hyades is 67.5 AU. 2. 50% of the binaries in the Hyades have separations less or equal than 3,871 AU. 3. 80% of the binaries in the Hyades have separations less or equal than 9,455 AU. 4. The separations of the binaries in the Hyades accumulate in the range 3,200 - 4,525 AU.

References

1. King, I., 1962, *AJ*, **67**, 471.
2. Luyten, W.J., 1981, *A catalogue of 929 possible candidates for Hyades membership.* University of Minnesota, Minneapolis.
3. Öpik, E., 1924, Publ. Tartu Univ. Obs., 25, No. 6.
4. Poveda, A., Herrera, M.A., Allen, C., Cordero, G. y Lavalley, C., 1994, *Rev. Mex. Astron. Astrof.*, **28**, 43.
5. van Altena, W.F., 1966, *AJ*, **71**, 482.
6. van Bueren, H., 1952, *Bull. of the Astronomical Institute of the Netherlands*, **11**, 385.

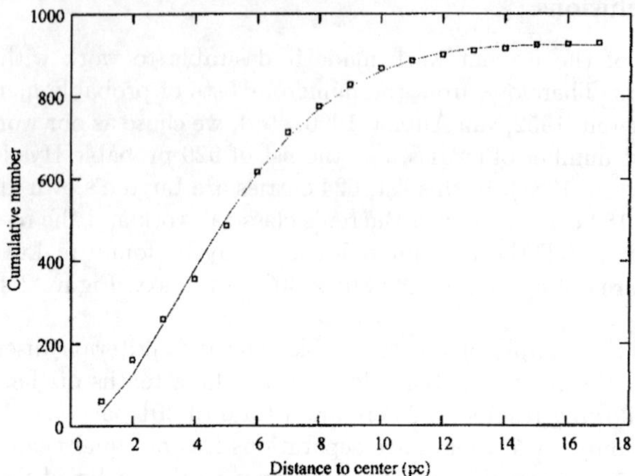

Figure 1. Observed accumulated fraction of stars in Luyten's list (squares) and analytical fit to them by means of King's formula (solid line) as a function of the distance to the center of the cluster

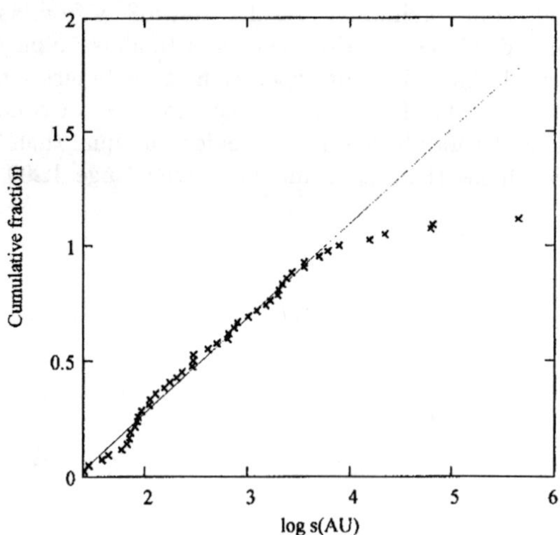

Figure 2. Artificial cluster of 900 "stars" in a Right Ascension-Declination plane. The points are distributed according to the analytical density profile fitted to the actual cluster

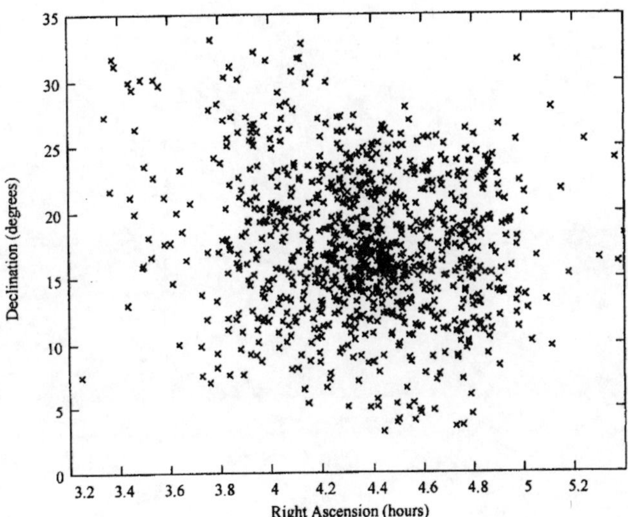

Figure 3. Luyten's 929 probable Hyades in a Right Ascension-Declination plane.

ON THE NATURE OF THE SECONDARIES OF THE AM STARS

J.M. CARQUILLAT, N. GINESTET
Observatoire Midi-Pyrenées-Toulouse, France

AND

C. JASCHEK
Observatoire de Strasbourg-France

Abstract. The paper presents the preliminary results of a study of infrared spectra of Am stars in the 8400-8800 region. This is part of a large program of classification of the cool components of composite spectra. We analize the infrared spectra of 33 Am stars and show that 22 can be explained as composites of an Am star plus a normal dwarf, whereas the remaining 11 are formed by an Am plus a metal deficient star. We discuss some of the difficulties found and speculate on the consequences of our findings.

We have started at the Observatoire de Haute Provence an extended observational program for the determination of the spectral types of the components of composite spectra.

The first part of this program consisted in the obtention of the infrared spectra of 180 composites in the 8400 - .8800 Å wavelength region, on CCD material. To guide our classifications, we had observed previously a set of MK standards in the same region (Ginestet et al,1994) and we have also collected the standard stars in an Atlas. The first part of this Atlas (O to GO) has already been published (Andrillat et al 1995), whereas the second part (F5 to M) has been accepted for publication. (Carquillat et al 1996). The classification of the cool components of the composites has also been accepted for publication.

We are now starting a second program for observing the ultraviolet-blue part of the spectrum in order to determine the spectra of the hot components.

Within this program of composite spectra we observed a number of Am stars, because in practice the distinction between certain composites and Am stars is not at all obvious and because Am stars must be composites since they are all binaries.

The object of the present paper is to report on our results regarding the 33 Am stars included in our program.

The standard technique we have used consists in supposing known the spectral types of the primary, from the blue spectra we had taken. The usual three types for the Ca ll, the hydrogen lines and the metallic lines were assignated from a comparison with the set of standards given in our Atlas of photographic spectra (Ginestet et al, 1992).

Using these blue spectral types we know what the equivalent of the Ca ll, the Paschen lines and the metallic lines should be in the near infrared. If these values are substracted from the observed equivalent widths, what remains belongs to the spectrum of the

J. A. Docobo et al. (eds.), Visual Double Stars: Formation, Dynamics and Evolutionary Tracks, 205–208.

secondary. However the remainder depends also on the magnitude difference (delta V) between both components. The way we proceeded was by iterations, starting with one assumed value of delta V and changing until the value of delta V and the values of the different residual equivalent widths are simultaneously satisfied. A check on the whole procedure is imposed by the fact that the magnitude difference found cannot have an arbitrary value, but must satisfy the values of absolute magnitude for stars of class V or Am stars.

Although the description of the procedure sounds complicated, the whole calculation takes usually less than five minutes per star. In figures 1 and 2 are illustrated the spectra of some composite-spectrum stars and of some Am stars in the infrared. Both figures are from the forthcoming infrared Atlas.

Results

The results of the procedure outlined, when applied to our 33 stars are the following.

Two thirds of the stars of the material (22 stars) have main sequence companions, with magnitude differences between the components varying between 1 and 2.5 magnitudes, with an average of 1.75 . It should be added that the limit of 2 . 5 is the practical limit for detection of companions.

The remaining third of the group, eleven objects, have an Am companion, and in most of the cases the metallic line type is earlier than the hydrogen type, implying that the star is deficient in metals. The magnitude difference between both components varies between 0 and 1.2, with an average of 0.5.

The latter results concerning the magnitude differences are due to the fact that if both companions are Am stars, their absolute magnitude should be similar and the magnitude difference should be small. In the case of dwarf companions, the difference in absolute magnitude is larger simply because Am stars lie above the main sequence.

A look at the radial velocities shows that of the 33 stars of our sample, only 18 have an orbit- 6 with Am and 12 with dwarf secondaries. Apparently most dwarf secondaries- except two- have periods between 1 and 10 days, wheras the six with Am secondaries are much more spread out over period (from 0.9 to 1054 days) Since the number of objects is rather small, we prefer not to elaborate on the point.

Our most important result is that the Am companions seem to have systematically weak metals, a fact which is new. That there do exist some Am stars in which the spectral type of metals is earlier than that of hydrogen has been found by Abt and Morrell(1995) - in fact about 5Am stars do have this characteristics. However in our case the majority of the cooler components has this characteristics, the metallic types being several tenths earlier than those of hydrogen. Since also the Ca-line spectral type is earlier than the hydrogen type, we have thus an object in which all lines are weaker than those of the hydrogen, which is the definition of a metal weak star.

Such a fact could be related to interactions between both components in some recent past. It presents a difficulty for the diffusion type explanation of Am stars (Alecian 1996), since in this theory the abundance of metals is a function of time, which in the case of coeval binaries would produce similar anomalies in both components.

Let us comment briefly upon the main difficulties encountered. The crucial point of our procedure is the accuracy of the spectral types of the (primary) Am star. Now a comparison of our classifications with those of Abt and Morrell (1995) shows that all

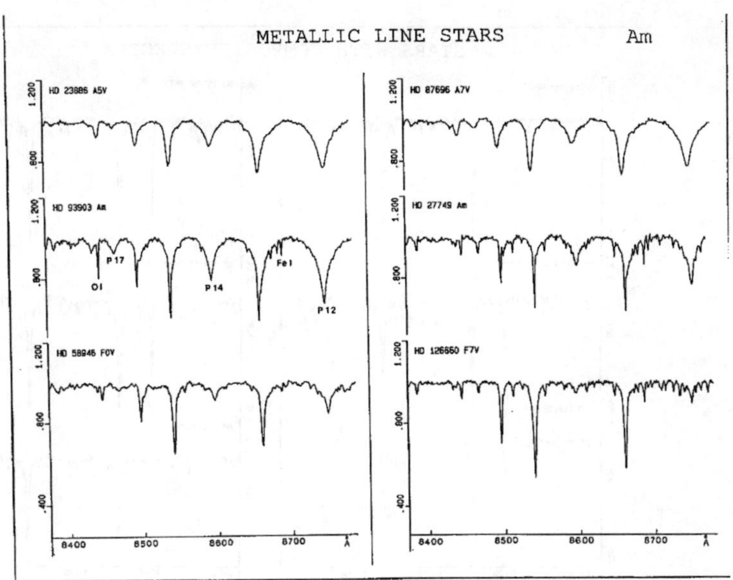

Figure 1. Metallic line stars

three spectral types are rather imprecise. It would be an easy way out if one could reject the classification of one of the groups (Abt-Morrell or ours) as being of lesser quality. However the problem is simply part of the more general one that classifiers do not agree on the classification, and not even on the belonging of a given object to the group of Am. This makes that the area occupied by Am stars can be considered as an"problem area". After turning around the problem for several years we reached the conclusion that the only way out is to measure equivalent widths of the relevant lines in the blue region of the Am stars and in a number of MK standards, so as to provide a good calibration of equivalent width as a function of spectral type, using CCD material. The program is now under way and will lead us in a reasonable time to be able to decide on the basis of objective criteria what the classifications of Am stars are and which objects should be put into this group.

A second problem which also came up is that the magnitude differences calculated from the IR spectra sometimes do not agree with what one expects. What one expects is given by the absolute magnitude difference between an Am star and its main sequence or Am star companion. This leads us naturally to the question if there does exist something like a well defined sequence of Am stars - located for instance 0.7 magnitudes above the main sequence. It could well be that Am stars fill an area, rather than to lie on one single sequence. This problem will probably be solved in a few months when the HIPPARCOS parallaxes become available.

In summary, we have used the infrared region to determine the spectral type of the cool components components, which are elusive in the blue spectral region. Our most important conclusion is that a large part of the Am companions present a metal deficiency.

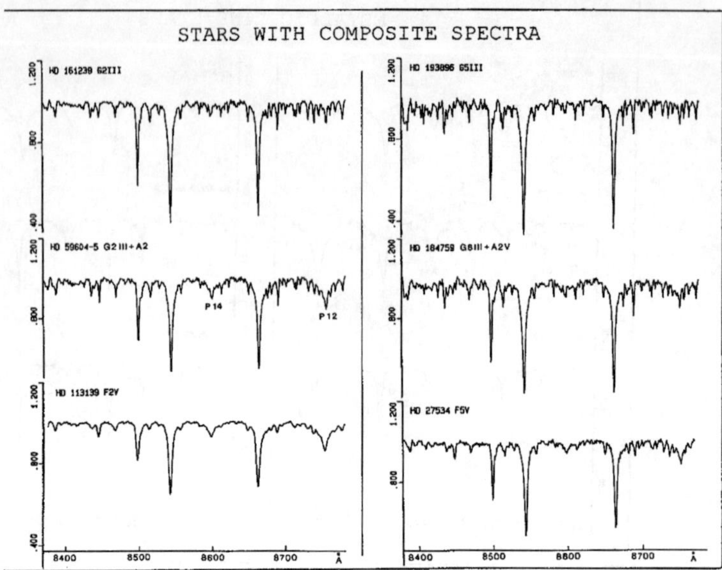

Figure 2. Stars with composite spectra

References

Abt H. and Morrell N. (1995) *ApJ S.* **99,** 135

Andrillat Y, C.Jaschek and M.Jaschek (1995) *AA S.* **112,** 475

Alecian G. (1996) *AA* **310,** 872

CarquillatJ.M., C. Jaschek, M.Jaschek and N.Ginestet (1996) *AA S.* (accepted)

Ginestet N,Carquillat J.M., Jaschek M., Jaschek C., Pédoussaut A. and J.Rochette (1992) *Atlas of stellar spectra* Observatoire Midi-Pyrenées- Observatoire Strasbourg

Ginestet N., Carquillat J.M., Jaschek M. and Jaschek C. (1994) *AA S.* **108,** 359

HOT SUBDWARFS IN BINARIES: EVOLUTION AND PHYSICAL PARAMETERS

A. ULLA
Instituto de Astrofísica de Canarias, 38200 La Laguna, Tenerife, Spain and LAEFF-INTA, 50727, 28080 Madrid, Spain

AND

P. THEJLL
Nordita, Blegdamsvej 17, DK-2100 Copenhagen Ø, Denmark

1. Introduction

Hot subdwarf B (sdB), O (sdO) and sdOB stars (see Greenstein&Sargent (1974) for pioneered work in this field) are subluminous evolved stars, on their way to becoming white dwarfs. They have long been known as stars burning helium in their cores, covered by a very thin hydrogen layer with negligible luminosity. In the log(g)-Teff plane, sdB and sdOB stars are narrowly distributed along the so-called 'Extended Horizontal Branch'.

The sdBs are probably HB stars that have undergone near-total loss of the hydrogen-envelope after the red giant branch evolution and these stars will evolve into white dwarfs without ascending the AGB. About 50% of the sdB are known to be binary (Allard et al. 1994). The sdOs may be a mixture of object types: HB stars with even smaller envelopes - or Helium Main Sequence objects of various masses created when the envelope was stripped away by close binary interaction - or merged white dwarfs that coalesce because of decay of the orbits due to gravitational wave radiation - or post-AGB objects that cooled slowly compared to the time it took the planetary nebulae to disperse. Scarce previous binary studies were available for the sdO population (Thejll et al. 1995).

In spite of their future fate, the origin of hot sds is an unsolved problem. Theory offers two explanations for the extreme mass loss required: i) extreme mass loss in a canonical single star evolution scenario; or ii) mass exchange in close binaries. By combining new IR colours and UV-to-optical broad spectral distributions obtained from a massive search through astronomical archives for a large sample of hot subdwarfs, the binary scenario has been tested and many any new binary companions have been discovered, along with confirmation of suspected candidates.

1.1. POTENTIAL USEFULNESS OF HOT SUBDWARFS

The Hot Subdwarfs can be used for the further understanding of:

– Mass-loss and evolution near the tip of the red giant branch
– Inputs to the white dwarf cooling sequence

J. A. Docobo et al. (eds.), Visual Double Stars: Formation, Dynamics and Evolutionary Tracks, 209–212.

* Do the He-rich WDs come directly from the He-rich sdOs?
- Structure of Galaxy
- UV-flux from galaxies and clusters
- Close binary evolution with common envelope phases
- The understanding of stellar pulsations

2. The project

We have gathered and analysed JHK and optical broad-band photometry and UV data for 89 hot sdB and sdOs in a search for IR excesses and binary companions. The IR photometry was carried out with the 1.5m *Carlos Sánchez* Telescope (CST) of the Observatorio del Teide (Tenerife), in February, June and October, 1994. The total number of hot subdwarfs here stud ied basically corresponds to the total possible number of objects observable, as extracted from Kilkenny et al. (1988), given the limiting magnitude of the instrument ($\sim$ 13.5 mag in K) employed at the CST. This total sample is, to our knowledge, the most complete JHK hot subdwarf catalogue available to date. Most of our targets were observed in the IR for the first time.

The investigation is complemented with comprehensive bibliographic and other archival data searches, including SIMBAD and the Low Dispersion Unified Archive (ULDA) for IUE data.

On the grand average as many as 40% display larger JHK fluxes than those expected for a single hot blue object. The wide energy distributions of our targets, from UV to the K band, have been analyzed by means of fitting Kurucz (1993) atmospheric models for the hot and cool components. Temperature estimates, when available, have been obtained from the literature literature or our own values otherwise. Our method (see Thejll et al. (1995), Ulla&Thejll (1996) for details) derives upper limits for the log$g$ of the hot subdwarf on the basis of assuming that the cooler companion is a ZAMS star, as well as the spectral type of the latter.

Table 1 shows the range of cool companion spectral classes that correspond to the extracted excess colours, as estimated from comparison to the Bessel&Brett (1988) tables of $J-H$ and $J-K$ colors. Only systems for which the excesses were significant at the 2σ level in each of the J, H and K bands have been included. Confirmation of previously suspected binaries as well as new discoveries were found.

UBVRI CCD images were also obtained on two telescopes and with three different cameras in the hope of finding statistically significant figures about the existing percentage of detached binaries among the hot subdwarfs. This data is still under analysis. In a separate collaboration (Heber as PI), an HST snapshot proposal has been awarded for 1996 aimed at resolving binaries with separations as short as 0.1 arcsec only. Depending upon the percentage of detached systems found, constraints on the evolutionary models predicting mergers or close interacting binaries for the sds population (see, eg, Iben&Tutukov, 1986a,b), will be imposed. This part of the project is yet to be tackled.

Acknowledgements

A.Ulla is at present a postdoctoral fellow at the Instituto de Astrofísica de Canarias (IAC; Tenerife, Spain), within the ISO team there and acknowledges the Spanish "Becas de Reincorporación" grant programme for financial support. P.Thejll acknowledges financial support from Nordita (Copenhagen, Denmark). The IR observations here presented were obtained with the CST operated on the island of Tenerife by the IAC in the Spanish Observatorio del Teide. The staff at the CST are thanked for their skilled assistance during the observations. This work has made use of information in the SIMBAD databank, operated at CDS (Strasbourg, France) and of the RDAF facility at the GSFC at Greenbelt Maryland and of the ULDA at the LAEFF in Madrid (Spain).

References

Allard,F., Wesemael,F., Fontaine,G., Bergeron,P., Lamontagne,R., 1994 , AJ, 107, 1565
Bessel,M., Brett, J.M., 1988, PASP 100, 1134
Greenstein, J.L., Sargent, A.I., 1974, ApJSS 28, 157
Iben, I.Jr., Tutukov, A.V., 1986a, ApJ 311, 742
Iben, I.Jr., Tutukov, A.V., 1986b, ApJ 311, 753
Kilkenny, D, Heber, U., Drilling, J., 1988, SAAO Circ. No. 12
Kurucz, R.L., 1993, Kurucz CD-ROM No. 13.
Thejll, P.A., Ulla, A., and MacDonald J. 1995, A&A 303, 773-784
Ulla, A., Thejll, P.A., 1996, in preparation

TABLE 1. The excess fluxes in significant cases (coloumn 1). Coloumn 2 contains the excess J flux and its error, in units of $10^{-24} erg/cm^2/s/Hz$, in coloumns 3 and 4 are given the $J-H$ and $J-K$ colours, respectively, of the excess flux. The Bessel&Brett (1988) dwarf colour types are given in coloumn 5. When an overlap in type existed only the overlapping types are given while the two type-ranges are shown in all cases where an overlap was not found. Notes: 'v.e.' and 'v.l.' indicate ranges that are outside the ranges of the Bessel&Brett table II - either 'very early' (i.e. before B8 type) or 'very late' (i.e. after M5). →X indicates a range including all previous spectral type up to the X type. X→ indicates all classes later than type X.

Name (1)	J±error (2)	$J-H$ (3)	$J-K$ (4)	color type (5)
BD-3 5357	20.09± 0.000	0.54± 0.00	0.69± 0.00	K3 K4
BD-7 5977	4.13± 0.079	0.50± 0.02	0.61± 0.02	K1-K2
BD-11 162	0.58± 0.022	0.25± 0.05	0.31± 0.05	F4-G0
BD-13 842	0.44± 0.010	-0.22± 0.08	0.44± 0.09	v.e. F8-K0
BD+10 2357	6.08± 0.182	0.09± 0.04	0.09± 0.04	A4-A8
BD+25 4655	0.13± 0.021	-0.25± 0.19	-0.03± 0.26	v.e.
BD+29 3070	1.92± 0.019	0.21± 0.02	0.23± 0.03	F3-F4
BD+33 2642	0.12± 0.028	0.28± 0.23	0.29± 0.32	→K2
BD+34 1543	2.07± 0.093	0.32± 0.06	0.36± 0.06	F6-G5
BD+37 1977	0.12± 0.044	0.05± 0.46	-0.09± 0.49	→G4
BD+48 1777	0.11± 0.030	-0.42± 0.38	-0.78± 0.42	v.e.
Feige 34	0.09± 0.016	0.79± 0.21	0.79± 0.21	K3→
Feige 80	0.39± 0.056	0.30± 0.16	0.19± 0.17	F0-F9
GD 274	0.40± 0.005	0.43± 0.02	0.53± 0.03	G8-G9
GD 299	0.13± 0.039	0.20± 0.33	0.20± 0.32	→G9
HD 4539	0.25± 0.029	-0.14± 0.15	-0.08± 0.14	→A0
HD 45166	1.22± 0.039	0.08± 0.05	0.23± 0.06	F0
HD 113001	3.06± 0.065	0.11± 0.03	0.06± 0.03	A5
HD 128220	17.38± 0.340	0.32± 0.03	0.41± 0.02	G3-G5
HD 149382	0.72± 0.080	0.42± 0.13	0.57± 0.13	G6-K3
HD 185510	55.64± 0.000	0.65± 0.01	0.79± 0.01	K6-K7
HD 216135	0.24± 0.035	-0.51± 0.25	-0.60± 0.42	v.e.
HDE283048	3.40± 0.201	0.22± 0.08	0.29± 0.08	F2-F9
KP2109+440	1.55± 0.073	1.16± 0.05	1.47± 0.05	v.l.
LSI+63 198	19.79± 0.366	0.66± 0.02	0.80± 0.03	K6-K8
LSIV+10 9	0.48± 0.016	0.22± 0.07	0.32± 0.06	F4-F6
LSV+22 38	7.64± 0.000	0.37± 0.01	0.57± 0.01	G5-G6 K1
MRK509C	0.07± 0.025	0.23± 0.39	1.00± 0.40	K2-K5
PB 8555	0.44± 0.018	0.26± 0.05	0.41± 0.06	F8-G1
PG 0011+221	0.08± 0.015	-0.15± 0.24	-0.49± 0. 38	v.e.
PG 0110+262	0.15± 0.027	0.09± 0.25	0.24± 0. 30	→G4
PG 0116+242	0.68± 0.014	0.34± 0.03	0.45± 0. 06	G4-G5
PG 0232+095	0.56± 0.073	0.34± 0.15	0.50± 0. 17	F6-K1
PG 0314+103	0.15± 0.018	0.62± 0.13	0.89± 0. 15	K5→
PG 2110+127	0.19± 0.011	0.21± 0.08	0.17± 0. 10	F0-F4
PG 2118+126	0.12± 0.014	0.04± 0.16	0.08± 0. 23	→F3
PG 2148+095	0.18± 0.022	-0.18± 0.23	0.18± 0. 42	→A4
PG 2151+100	0.17± 0.011	0.70± 0.08	1.10± 0. 09	v.l.
PG 2219+094	0.11± 0.018	0.02± 0.19	0.04± 0. 28	→F4
PHL 1079	0.13± 0.025	0.39± 0.20	0.61± 0.21	G4-K4
SB 7	0.08± 0.037	0.58± 0.43	0.47± 0.47	F1-M4
TON 139	0.20± 0.033	0.23± 0.19	0.23± 0.24	A3-G7

SECTION III

DYNAMICS

MUTUAL PERTURBATIONS OF THE PLANETARY COMPANIONS OF PULSAR B1257+12

S.FERRAZ-MELLO
Instituto Astronômico e Geofísico,
Universidade de São Paulo,
Caixa Postal 9638, São Paulo, SP, Brasil.
E-mail sylvio@iagusp.usp.br

Abstract. According to Wolszczan, the variations in the observed pulse arrival times of the pulsar B1257+12 are consistent with the motion of the pulsar around the barycenter of a planetary system formed by two Earth-size and one small Moon-size planets. The two largest planets have orbital periods nearly commensurable as 2:3, and the observed pulse arrival times reproduce the expected variation of the periods of the Keplerian orbits of the two largest planets due to their mutual perturbations.

The indirect detection of planetary companions of a star is founded on the analysis of the kinematical behaviour of the central star, as observed from the Earth. As the observed parameters are always indicators of the motion of the central star relative to the Earth, the informations on the motion of the central star are mixed with those of the motion of the Earth and many of the past "discoveries" of planetary companions vanished when the motion of the Earth was properly taken into account into the reduction of the observational data. Some elementary calculations may help to assess the difficulties of the task. Consider a planetary system like the Sun-Earth system placed far away of us and edge on. The light received on Earth comes from the motion of the star around the system barycenter. The radial velocity of the star will have, in this example, a yearly variation of 18 cm/sec – which is some hundreds of times less than the typical resolution achieved in modern differential single-line Doppler spectroscopy. If the central star is a pulsar, the time of arrival of the pulses will also show a yearly variation, the pulses received when the star is receding will be delayed and will accumulate, in the half year between the closest and the farthest positions, a delay of ~0.003 sec, which is a measurable effect and large enough to allow the detection of such an earthly planetary companion.

However, the observations are not done from an inertial observer and we may enumerate a ten of factors which will also introduce variations in the time of arrival of the pulses, at least as large as those due to the motion of the source, starting with the biggest one, the orbital motion of the Earth which, for a pulsar located near the ecliptic, may reach an amplitude of 1000 sec. Compare this value to the 0.003 sec which one wants to observe and the immediate conclusion is that we need to know the motion of the Earth around the barycenter of the solar system with a relative precision much better than 10^{-6}. Other effects to be considered to reach such a precision are the rotation of the Earth (since the

J. A. Docobo et al. (eds.), Visual Double Stars: Formation, Dynamics and Evolutionary Tracks, 215–220.

TABLE 1. The putative companions of pulsar PSR B1257+12

Orbital parameters and masses	*A*	*B*	*C*
Semi-major axis(*) (light ms)	0.0035(6)	1.3106(6)	1.4121(6)
Eccentricity	0.0	0.0182(9)	0.0264(9)
Epoch of periastron (JD)	2448754.3(7)	2448770.3(6)	2448784.4(6)
Orbital period (s)	2189645(4000)	5748713(90)	8486447(180)
Longitude of periastron (deg)	0.0	249(3)	106(2)
Planet mass ($M_\oplus$)	$0.015\xi_A$	$3.4\xi_B$	$2.8\xi_C$
Distance from the pulsar (AU)	0.19	0.36	0.47
Orbital period (days)	25.34	66.54	98.22

(*) refer to the motion of the star around the system barycenter.
Figures in parentheses are 3σ statistical uncertainties on the last digit quoted.

observer is turning with it) as well as the fact that the used clock is also moving with the Earth and therefore does not give the same time as a fictitious clock placed in the barycenter of the solar system (for the list of effects to be taken into account to reach a μsec precision see Chandler, 1996). Even in the case of detection by spectroscopic means, whose accuracy is lesser, one must take into account all perturbations of the motion of the Earth with a relative magnitude 10^{-4}, which includes the perturbations due to the orbital motion of the major planets and the motion of the Earth around the barycenter of the Earth-Moon system.

The number and magnitude of the "local" effects affecting the measured time of arrival of the pulse or, even, the radial velocity of the central star was responsible for many of the planetary companions whose discovery was announced in the past, just to be withdrawn a few months later, after the improvement of the reduction of the measurements. It is worthwhile mentioning that the recent announcements of some Jupiter-size planetary companions of stars discovered by spectroscopic means, were founded on observations reduced with more recent codes, which largely benefited from the errors of the past, and are likely to be confirmed as true companions even if some of them may be rather brown-dwarf stars than planets.

Among pulsars, the most spectacular announcement are the planets orbiting around PSR B1257+12, a pulsar about 400 pc away from the Sun (Wolszczan and Frail, 1992). The orbital parameters and masses of planets A, B, and C derived from 3.5 years of observations with the 305-meter Arecibo radiotelescope are given in Table I. In this table, the masses of the planets were estimated adopting 1.4 $M_\odot$ for the mass of the central neutron star. Thus, instead of the usual factors $1/\sin i_k$, due to the unknown inclination i_k of the plane of the orbital motion of the $k^{\rm th}$ planet to the tangent plane to the celestial sphere at the star, we have introduced the factors

$$\xi_k = \frac{1}{\sin i_k}\sqrt[3]{\frac{1.4M_\odot}{M_{\rm psr}}}.$$

Figure 1 shows the residuals of pulse arrival times during 2.5 years of observations with the 305-m Arecibo telescope, properly corrected of the motion of the Earth and the proper motion and rotational parameters of the pulsar. The residuals fit well (within 3

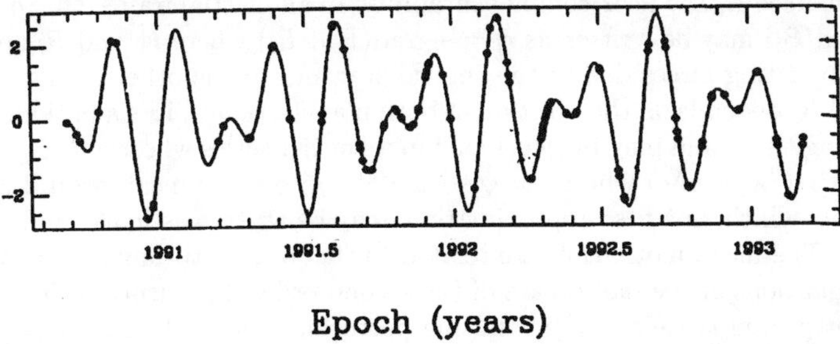

Figure 1. Residuals (in milliseconds) of the pulse arrival times from PSR B1257+12 corrected of the motion of the Earth and the rotational parameters of the pulsar (figure taken from Wolszczan, 1994a)

μsec r.m.s.) to a model which consists of the pulsar reaction to two orbiting planets (the third-planet hypothesis came later). The consequences of the perturbation are below the resolution of the graph.

The existence of planets around neutron stars is an unexpected fact and it immediately raised some controversies (see Wolszczan, 1994a). The fact that the residuals may be fitted to the sum of two near sinusoids corresponding to the displacements of the neutron stars due to the Keplerian orbits of two planetary bodies is a strong clue in favour of the two-planet hypothesis. They gave to astronomers enough confidence to believe in the actual existence of this planetary system, but the fitting to the sum of two near sinusoids was not, by itself, enough to rule out other possibilities.

However, the putative planets B and C have the conspicuous feature of having near commensurable periods. As it was soon pointed out by Rasio *et al.*(1992) and Malhotra *et al.*(1992), the closeness to the 2:3 period commensurability should produce perturbations large enough to be observed from Earth, thus allowing the very existence of the planets to be confirmed in a few years of continued observations. This possibility is of particular importance since these perturbations may be very accurately predicted and their observation, or not, is a crucial test to the planetary nature of the observed residuals. Its detection would yield a unique proof that the residuals result from the orbital dynamics of two planet-sized bodies.

Motions whose periods are close to the ratio of two small integers are very common in the solar system and, for this circumstance, they are well known. The most interesting cases in Celestial Mechanics are those of deep resonance, in which the two motions are locked one to another (see Sessin and Ferraz-Mello, 1984), like Pluto and Neptune. Would the pulsar planets form a resonant system of this kind, it would gain in importance. But the ratio of the periods of the pulsar planets is 1.4762, that is, it is close to 2/3 but not too close (in the Neptune-Pluto pair, for instance, the mean periods are in the 2/3 ratio with a difference less than 10^{-4}). The motions are just near-commensurable. The dynamics of the system is not essentially different of that of a fully non-resonant system, but the near-commensurability may enhance the effects of the mutual gravitational attraction and give rise to important perturbations (as the great inequalities in the mean longitudes of Jupiter and Saturn due to the 2/5 near-commensurability of their periods).

An elementary reasoning is enough to anticipate how the effects of the mutual perturbations is enhanced by the near-commensurability. In an inertial frame, the motion of each planet (P_1 and P_2) may be written as $\ddot{\vec{r}}_i = -\mathrm{grad}(F_0 + R_i)$ where F_0 and R_i are the potentials of the attracting forces due to the central star and to the other planet, respectively. The potential R_i depends on the position of both planets, hence, it is a periodic function of their mean longitudes and may be developed in a Fourier series whose generic term may be written $a_{jk} \exp \iota(j\lambda_1 + k\lambda_2)$ where $\iota = \sqrt{-1}$ and λ_1, λ_2 are the mean orbital longitudes of the two planets which, in a first approximation, may be written as linear functions of time: $\lambda_i = n_i + \mathrm{cte}$. The mean motions n_i are related to the periods through $n_i = P_i/2\pi$. As the differential equations of the motion are of the second order, the terms of the Fourier series have to be integrated twice and the resulting perturbations will be inversely proportional to $(jn_1 + kn_2)^2$. If the motions of the two planets are near-commensurable, there will be some pair of integers j, k for which $(jn_1 + kn_2)^2$ is a very small number. For instance, in the case of the pulsar planets B and C: $(2n_{\mathrm{B}} - 3n_{\mathrm{C}})^2 = 9.24 \times 10^{-6}\mathrm{day}^{-1}$. Accordingly, the velocities (and periods), which are obtained with just one integration, will be inversely proportional only to $2n_{\mathrm{B}} - 3n_{\mathrm{C}}$ (not to its square).

An important point in near-commensurable motions is the existence of some invariance geometrical laws imposing that, in coplanar motions, each coefficient a_{jk} in the Fourier series is formed by homogeneous monomials in the eccentricities of the 2 orbits whose degree of homogeneity is, at least, equal to $|j + k|$. Thus, in a first-order near-commensurability, since $|j+k| = 1$, the coefficients of the critical terms will include monomials which are proportional to each of the two eccentricities; this property makes these near-commensurabilities much more important than others where $|j+k|$ is a larger number (for instance, in the pair Jupiter-Saturn, the critical terms have argument $2n_{\mathrm{Jup}} - 5n_{\mathrm{Sat}}$ and thus the corresponding coefficient will include products of the eccentricities of Jupiter and Saturn always in powers whose sum is, at least, equal to 3).

The main observable effect of the near-commensurability of the periods of B and C is the variation of the periods of the two planets. The predicted mutually perturbed periods of B and C, derived from numerical integrations for the epoch of the Arecibo observations, are shown in Figure 2 (from Wolszczan, 1994). The main harmonic variation of the periods is associated with the angle $2\lambda_{\mathrm{B}} - 3\lambda_{\mathrm{C}}$ and has a 5.66 years period. The main harmonic of the period of B was maximum around 1990.6, at the time of the first observations, and minimum around 1993.4. The behaviour of the period of C is inverted, being minimum when that of B is maximum, and maximum when the other is minimum. The relative variation of the periods (2.5×10^{-4} and 3.3×10^{-4}, respectively) is larger than the 10^{-5} relative precision of the determination of the periods. In fact these variations shall be multiplied by ξ_{C} and ξ_{B} respectively, since the perturbations of each of the planets is proportional to the mass of the other, and that the given values of these masses are affected by the unknown inclination of their planetary orbits to the line of sight.

Complete theories of these effects were given by Malhotra (1993) and Peale (1993). Figure 3, taken from Malhotra's paper, is a summary of the expected observable effects. It shows the difference in the pulse arrival times obtained for the mutually perturbed three-body solution and those obtained from a model with independent fixed Keplerian orbits. The results shown are for edge-on orbits, but similar graphs were also given for inclined non-coplanar orbits. As figure 3 shows, it is impossible to detect the mutual perturbation of the two bodies in 2 years of observations. However, the phase residual begins to accumulate and reaches an amplitude of 20 μsec after $\sim$3.5 years. Similar graphs were computed by Peale. However, while Malhotra's graphs consider as fixed reference

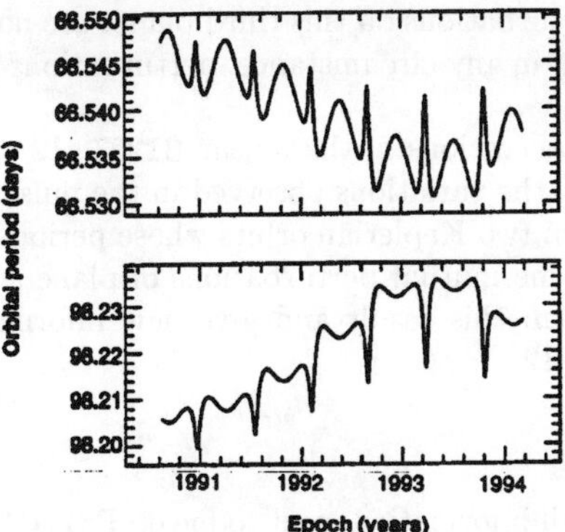

Figure 2. Predicted periods of the mutually perturbed motion of B (top) and C (bottom), for the epoch of the Arecibo observations (figure taken from Wolszczan, 1994b)

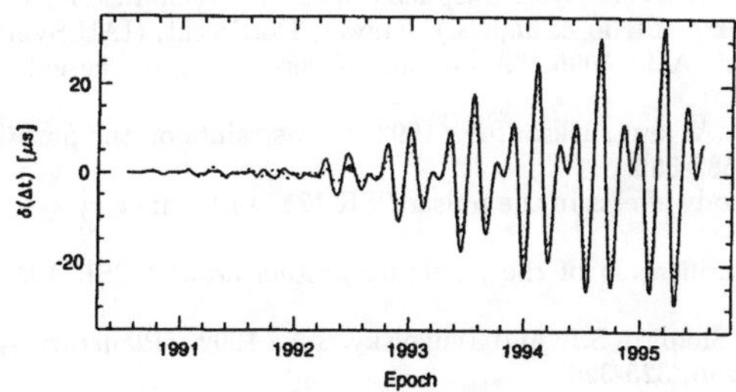

Figure 3. Difference in the pulse arrival times obtained for the mutually perturbed three-body solution and those obtained from a model with independent fixed Keplerian orbits (figure taken from Malhotra, 1993)

Keplerian orbits those of the beginning of the time interval, Peale's consider as reference the fixed Keplerian orbits which best fit the ensemble of observations in a given period. In this case the differences rarely exceed 10 μsec.

Peale's codes were used by Wolszczan to study 3.5 years of observations made with the 305-meter Arecibo radiotelescope. The best-fit orbits thus obtained are those given in Table I. Wolszczan (1994b) detected the mutual perturbations and, assuming coplanar orbits, showed that the χ^2 curve of the residuals of the comparison of the observations to numerical simulations of the three-body solutions shows a well defined minimum at

$$\xi = 1.025 \pm 0.033$$

Analyzing the remaining residuals, Wolszczan also showed that they also fit a near-sinusoid which could correspond to a third planet, planet A, with a moonlike mass. More recently, the residuals of the available Arecibo observations were frequency analyzed by Konacki and Maciejewski (1996). Although the amplitude of the signal caused by the planet A is only about 0.2 percent of the total signal, its frequency was easily detected. However, the three-body perturbations on this third planet are not enhanced by any small divisor and we do not expect, in any circumstance, perturbations larger than 1 μsec in the pulse arrival times.

The available Arecibo observations of the pulsar B1257+12 span about 4.5 years; according to Wolszczan (1994), the variations observed in the pulse arrival times are consistent with those expected from two Keplerian orbits whose periods are varying accordingly with what is expected from the mutual perturbations of planets B and C. We hope that new observations may confirm this result and give new informations on the system of planets around PSR B1257+12.

Acknowledgements

I thank Dr.A.Wolszczan for all informations sent to me on PSR B1257+12 and for his comments on an early version of this note and Drs. A.Elipe and J.Docobo for their hospitality in Zaragoza and Santiago de Compostela.

References

Chandler, J.F.: 1996, "Pulsar and Solar-System Ephemerides", In *Dynamics, Ephemerides and Astrometry of the Solar System* (S.Ferraz-Mello *et al.*,eds.), Kluwer, Dordrecht, (IAU Symposium 172), 105-112.

Konacki, M. and Maciejewski, A.J.: 1996, "A method of verification of the existence of planets around pulsars", *Icarus*, in press.

Malhotra, R., Black, D., Eck, A. and Jackson, A.: 1992, "Constraints on the putative companions to PSR 1257+12", *Nature*, **355**, 583-585.

Malhotra, R.: 1993, "Three-body effects in the pulsar PSR 1257+12 planetary system", *Astrophys. J.* **407**, 266-275.

Peale, S.J.: 1993, "On the verification of the planetary system around PSR 1257+12", *Astron. J.* **105**, 1562-1570.

Rasio, F.A., Nicholson, P.D., Shapiro, S.L. and Teukolsky, S.A.: 1992, "Planetary system in PSR 1257+12: A crucial test", *Nature*, **355**, 325-326.

Sessin, W. and Ferraz-Mello, S.: 1984, "Motion of two planets with periods commensurable in the ratio 2:1", *Celest. Mech.*, **32**, 307-332.

Wolszczan, A. and Frail, D.A.: 1992, "A planetary system around the millisecond pulsar PSR 1257+12", *Nature*, **355** 145-147.

Wolszczan, A.: 1994a, "Towards planets around neutron stars", *Astrophys. Sp. Sci.* **212**, 67-75.

Wolszczan, A.: 1994b, "Confirmation of Earth-mass planets orbiting the millisecond pulsar PSR B1257+12", Science, **264**, 538-542.

Wolszczan, A.: 1996, "Probing Planetary Dynamics with a Pulsar Clock", *Trans. Intern. Astron. Union*, **22 B**, 133-134.

SEARCHES FOR PLANETS AROUND NEUTRON STARS

A. WOLSZCZAN
The Pennsylvania State University, Department of Astronomy & Astrophysics, 525 Davey Laboratory, University Park, PA 16802, USA

Abstract. We review the methodology of searches for planet–mass bodies around neutron stars observable as radio pulsars and discuss the current status of these searches. PSR B1257+12, the 6.2–millisecond pulsar, remains the only neutron star accompanied by confirmed planets. It is possible that there is a fourth distant planet in the 1257+12 system. The best of the other candidates for pulsar planets under consideration is a distant, possibly Jovian–mass companion to PSR B1620-26, a 11-millisecond pulsar in the globular cluster M4.

1. Introduction

A detection of planetary companions to stars other than the Sun has been one of the most challenging tasks of modern observational astrophysics. Searches for extrasolar planets directly address fundamental problems related to the origin of the Solar System and they are instrumental in the process of understanding the relation of Earth and terrestrial life to the rest of the Universe.

The first planets beyond the Solar System have been detected around a billion year old neutron star, a 6.2–millisecond radio pulsar, PSR B1257+12 (Wolszczan & Frail 1992; Wolszczan 1994). The three companions to PSR B1257+12 remain the only known system of terrestrial mass planets orbiting a star other than the Sun. More recent results include growing evidence for a Jupiter–mass object orbiting a binary pulsar, PSR B1620-26, in the globular cluster M4 (Arzoumanian *et al.* 1996) and a possibility for an Earth–mass body around PSR B0329+54 (Shabanova 1995).

Radio pulsars, especially those of the millisecond period variety, which have been "spun–up" or "recycled" by transfer of matter and angular momentum from their binary stellar companions, are extremely stable rotators (see Phinney & Kulkarni (1994) for a recent review). The intrinsic rotational stability of these objects and the corresponding steady repetition rate of the observed pulses of radio emission (Fig. 1) makes them the most precise "cosmic clocks" known, with performance rivaling that of the best terrestrial time standards. The shortest period, 1.57–ms pulsar clock, PSR B1937+21, exceeds a fractional frequency stability of 10^{-14} on a time scale of several years (Kaspi, Taylor & Ryba 1994).

In this paper, we present the pulse timing as a high–precision method of detection of planetary mass bodies around neutron stars and summarize the results of searches for

J. A. Docobo et al. (eds.), Visual Double Stars: Formation, Dynamics and Evolutionary Tracks, 221–231.

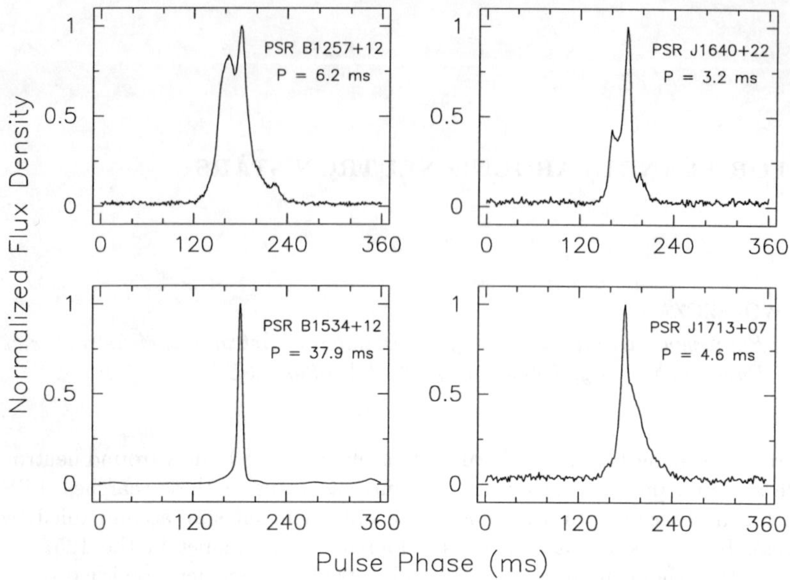

Figure 1. Integrated pulse profiles of four millisecond pulsars observed with the Arecibo 305–m radiotelescope at 430 MHz.

pulsar planets using this technique.

2. Pulse Timing Analysis

Pulse timing is a precise method of monitoring the behavior of pulsar clocks by means of the time of arrival (TOA) measurements of pulsar pulses. Since the pulsar's reflex motion induced by orbiting planets translates itself into variations in the arrival times of pulsar pulses, a capability of the pulse timing analysis to detect these variations provides a powerful method of the indirect planetary detection.

The topocentric TOA measurements are made at a telescope which takes part in Earth's rotation and in its motion within the solar system. To correct for the effects of these motions, the TOAs are reduced to the solar system barycenter. In addition, the pulsar's rotational characteristics have to be taken into account and a dispersive delay of the signal as it propagates through the interstellar plasma must be corrected for. Furthermore, errors in the pulsar position and proper motion have to be taken into account. Finally, if the pulsar is a member of a binary system, or it has planetary companions, their orbital characteristics have to be incorporated in the timing model.

Practical translation of the topocentric TOA's to the arrival times at the barycenter is accomplished by a widely used timing analysis program TEMPO (Taylor & Weisberg 1989), which uses a solar system ephemeris to determine dynamical parameters of the telescope at the time of observation. Ephemeris data are extracted from either the Center for Astrophysics PEP740R or the Jet Propulsion Laboratory DE200 ephemeris. All other necessary corrections are also made in the process of calculating the barycentric TOAs.

All the timing model parameters are determined as corrections to their initial values in the process of a linearized least–squares fit of the timing model to the arrival time data. A goodness of this fit is assessed from a χ^2 statistic given by:

$$\chi^2 = \sum_{i=1}^{N} \left[\frac{\phi_i - N_i}{\sigma_i/P}\right]^2, \tag{1}$$

where N_i is the closest integer number of pulse periods, P, corresponding to each computed phase, ϕ_i, and σ_i is the uncertainty of the arrival time.

Timing analysis of planet pulsars must accomodate the fact that orbital motion involves additional delays in the pulse arrival times. In the case of a single non–relativistic Keplerian orbit, five parameters characterizing the binary motion have to be taken into account. These are the orbital period, P_b, the eccentricity, e, the longitude of periastron, ω, the semi–major axis, $a_1 \sin(i)$, where i is the orbital inclination, and the time of periastron passage, T_0. For multiple orbits, additional sets of these parameters have to be included in a timing model. If gravitational interaction between planetary orbits is significant enough to be detectable, numerically generated perturbed orbits must be taken into account, and the timing analysis becomes considerably more complicated (Wolszczan 1994, and references therein).

3. Detecting Neutron Star Planets

Precise measurements of the pulse arrival time variations generated by the reflex motion of a pulsar due to orbiting planets provide a powerful method of the indirect planet detection. This method is closely related to the single–line Doppler spectroscopy and astrometry which are widely used in planetary searches around ordinary stars. For a circular orbit and the pulsar mass, $M_{psr} = 1.35 M_\odot$, a relationship between the planetary mass, m_2, the planet's orbital period, P_b, and the semi–amplitude, Δt, of the corresponding TOA variations is:

$$m_2 sin\ i = 21.3 M_\oplus \left(\frac{\Delta t}{1\ ms}\right)\left(\frac{P_b}{1\ day}\right)^{-2/3}, \tag{2}$$

where $M_\oplus$ is the Earth mass, i is the orbital inclination and $m_2 << M_{psr}$.

A practical sensitivity of the pulse timing method to the presence of planets around a pulsar is demonstrated in Fig. 2 which shows timing residuals from simulated observations of the Solar System planets with the Sun replaced by a $1.4 M_\odot$ neutron star. Clearly, the Jovian planets generate large residuals, the terrestrial planets are detectable with a $\sim$ 1 ms precision provided by "normal", slowly rotating pulsars ($P \sim 1$ s) and they are extremely easy to detect with millisecond pulsars due to their microsecond timing accuracy. Moreover, as demonstrated in Fig. 2c, with a sub–microsecond timing precision attainable with the fastest and strongest millisecond pulsars, the largest asteroids become barely detectable as well!

It is illuminating to compare a planet detection power of the pulsar timing with the current capabilities of optical methods. A radial velocity precision required to detect a Moon–like body in an inner planet–sized orbit around a solar–mass star would be $\sim$1 mm s^{-1}. This corresponds to a timing residual amplitude of a few microseconds (Eq. (2)) which would be easy to measure with a millisecond pulsar (e.g. Wolszczan 1994). The most advanced Doppler searches for planets around normal stars have recently achieved a $\leq$ 5m s^{-1} accuracy (Butler *et al.* 1996) which is sufficient to detect Jupiters and "super–Jupiters". Further technical improvements may make it possible to lower this limit to $\sim$1

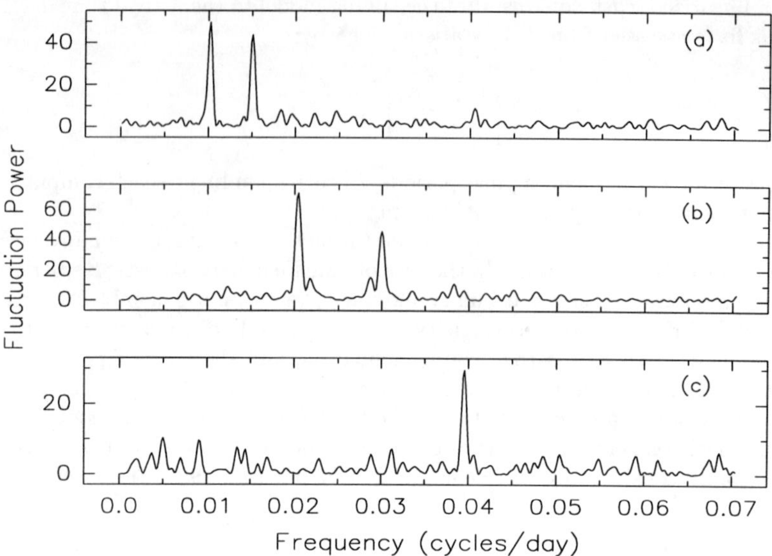

Figure 2. Simulated timing residuals from observations of the Solar System planets around a $1.4M_{\odot}$ pulsar, after subtraction of the best–fit model including the initial phase, the pulsar rotation period, P, and its slowdown rate, $\dot{P}$. *(a)* All planets present, the residuals are dominated by Jupiter. *(b)* Outer planets fitted out, the residuals for a slow pulsar (filled circles, timing accuracy 0.5 ms) show the presence of the Earth and Venus. The solid line represents the same detection with a slow pulsar replaced with a millisecond pulsar (0.1 μs timing precision). *(c)* All planets removed, the residuals from a millisecond pulsar (defined as above) reveal the presence of Ceres, the largest asteroid.

m s^{-1} and gain access to Saturn–mass bodies. Undoubtedly, in a foreseeable future, the pulse timing method will remain unique in its ability to detect low–mass planetary objects outside the Solar System and to study their dynamics.

Practical methods of detection of the TOA variations caused by orbiting planets include direct fits of Keplerian and real orbits (e.g. Thorsett & Phillips 1992; Wolszczan 1994; Lazio & Cordes 1995) and model–independent frequency domain approaches based on Fourier transform techniques (Konacki & Maciejewski 1996; Bell *et al.* 1997). In fact, it appears that it is best to search for periodicities in TOAs (or residuals) by examining Lomb–Scargle periodograms of the data and then refine the search by fitting orbits in time domain using the initial orbital parameters derived from frequency domain analysis. The Lomb–Scargle algorithm (Lomb 1976; Scargle 1982) provides an efficient way to compute spectra of unevenly spaced data.

This combination of methods has been successfully applied by Wolszczan and Frail (1992) and Wolszczan (1994) to detect planets around PSR B1257+12. Further refinements include a promising implementation of the frequency domain analysis in which contributions from any periodic TOA variations are successively subtracted from the data to reveal lower level fluctuations (Konacki & Maciejewski 1996; Fig. 3).

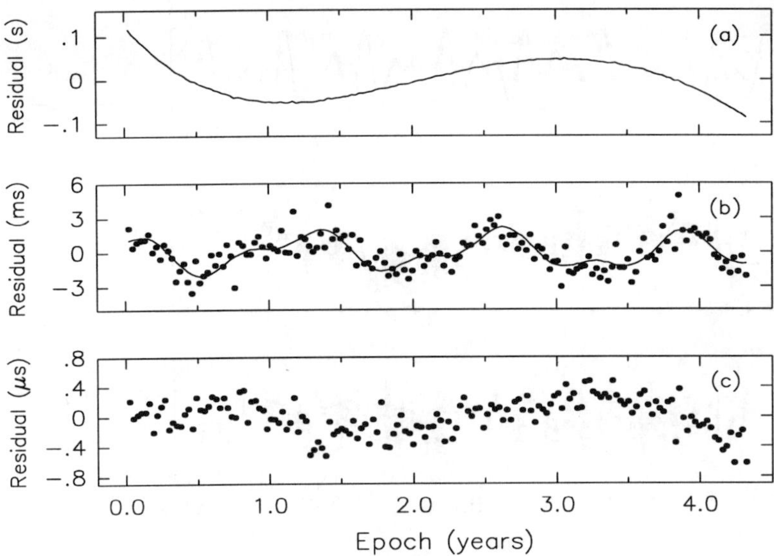

Figure 3. An example of the frequency domain analysis of pulse arrival times from the planet pulsar, PSR B1257+12, observed with the Arecibo telescope at 430 MHz. *(a)* A periodogram of the original TOA set. The two spectral peaks correspond to the orbits of planets B and C. *(b)* A periodogram of TOA variations with the fundamental frequencies of planets B and C removed. The peaks are first harmonics due to non–zero eccentricities of the orbits. *(c)* The fundamental freqencies and their first harmonics removed. A periodicity due to planet A becomes clearly visible (courtesy of M. Konacki).

4. Planets Around PSR B1257+12

A 6.2–millisecond pulsar, PSR B1257+12, was discovered in 1990 during a pulsar search conducted with the 305–m Arecibo radiotelescope (Wolszczan 1991). The analysis of the follow–up timing observations of this pulsar has led to a detection of the first extrasolar planetary system (Wolszczan & Frail 1992) later confirmed by a detection of planetary perturbations between planets B and C (Wolszczan 1994). The system consists of three planet–mass bodies, A, B, and C, with orbital characteristics listed in Table 1. Timing residuals due to planets B and C are shown in Fig. 4a (planet A is not discernible on this scale).

Residuals from the least–squares fit of a model including the spin parameters and astrometric parameters of the pulsar, three orbits, and planetary perturbations are shown in Fig. 4b. Clearly, the three–planet timing model for PSR B1257+12 requires a fit for a second–order derivative of the pulsar spin period, $\ddot{P}$, to correctly predict the observed pulse arrival times (Fig. 4c). A very intriguing possibility is that the observed $\ddot{P}$ is due to a dynamical influence of a distant, long–period fourth planet in the pulsar system (Wolszczan 1996).

As discussed by Joshi and Rasio (1996), one can establish analytical relationships between the measured pulsar spin frequency derivatives and orbital elements of an outer planet, and use them to constrain its orbit in a straightforward manner. For PSR B1257+12, numerical values of the spin frequency and its first three derivatives are: f = 160.8 Hz,

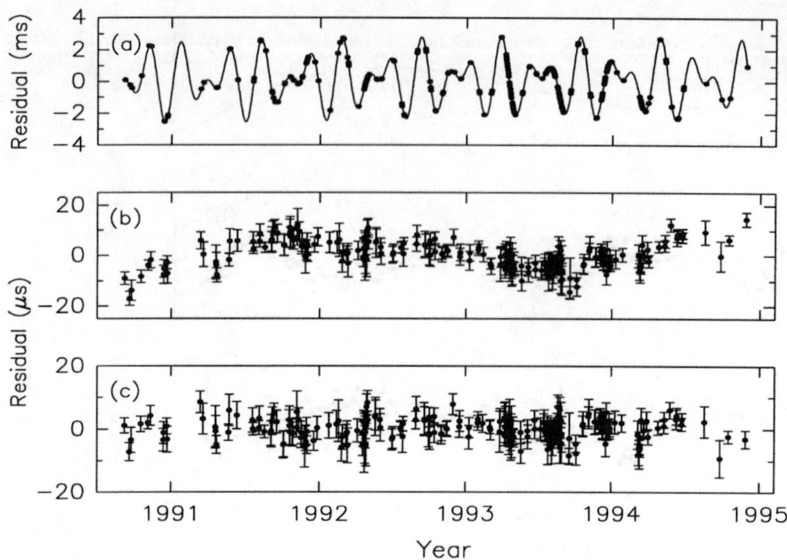

Figure 4. Timing residuals for PSR B1257+12. at 430 MHz. *(a)* A fit for the spin parameters and astrometric parameters only. *(b)* A fit including standard pulsar parameters, three planets and perturbations between planets B and C. *(c)* A fit with the second period derivative taken into account.

$\dot{f} = -8.6 \times 10^{-16}$, $\ddot{f} = (-1.25 \pm 0.05) \times 10^{-25}$, and $\dddot{f} = (1.1 \pm 0.3) \times 10^{-33}$, respectively.

With the pulsar mass of 1.35 $M_{\odot}$, and the assumption that the observed $\dot{f}$ is dominated by the effect of orbital acceleration, these values of spin parameters give a planet in a ~170 year orbit, with the orbital radius of ~35 A.U. and the planetary mass of ~95 $M_{\oplus}$, which would be a Saturn–mass object at a Pluto–like distance from the pulsar. Obviously, smaller acceleration contributions to $\dot{f}$ will lead to correspondingly different planetary masses and orbital elements. For example, a hypothetical fourth planet could have a low, Mars–like mass and orbit the pulsar at ~9 AU (Joshi and Rasio 1996).

TABLE 1 Parameters of the PSR B1257+12 planetary system

Keplerian orbital parameters			
	A	*B*	*C*
Semi–major axis (light ms)	0.0035(6)	1.3106(6)	1.4121(6)
Eccentricity	0.0	0.0182(9)	0.0264(9)
Epoch of periastron (JD)	2448754.3(7)	2448770.3(6)	2448784.4(6)
Orbital period (s)	2189645(4000)	5748713(90)	8486447(180)
Longitude of periastron (deg)	0.0	249(3)	106(2)
Parameters of the planetary system			
Planet mass ($M_{\oplus}$)	0.015/sin i_1	3.4/sin i_2	2.8/sin i_3
Distance from the pulsar (AU)	0.19	0.36	0.47
Orbital period (days)	25.34	66.54	98.22

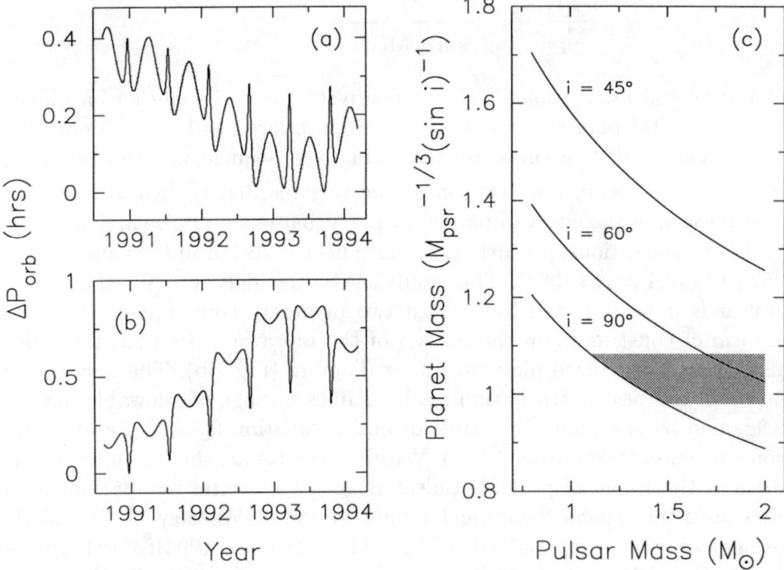

Figure 5. *(a)* Predicted perturbations of orbital periods derived from numerical integration of equations of motion of planets B and C. *(b)* Constraints on the pulsar mass, the masses of planets B and C and the common orbital inclination. The grey–shaded area contains the most likely combinations of these parameters.

Another type of microsecond–level variability in the timing residuals of PSR B1257+12 is related to gravitational perturbations between planets B and C. The question of detectability of this effect has been raised soon after the announcement of the PSR B1257+12 system (Rasio *et al.* 1992; Malhotra *et al.* 1992). It has been pointed out that an approximately 3:2 ratio of the orbital periods of the two larger planets creates a near–resonance condition which leads to accurately predictable and possibly measurable periodic perturbations of the two orbits.

Detailed analyses of the dynamics of the PSR B1257+12 planetary system and their effects on pulsar timing have been carried out by Malhotra (1993), Rasio *et al.* (1993) and Peale (1993). The observable consequences of a mutual gravitational interaction between the two planets include near–resonant, periodic variations of the elements of their orbits and the superimposed short–term, nonresonant fluctuations (Fig. 5a). If the planetary masses are not too big, the relevant period is simply given in terms of the mean angular velocities of the planets, n_1 and n_2, as $2\pi/(2n_1 - 3n_2) \; = \; 5.56$ years and the predicted maximum amplitude of the corresponding timing residuals (after fitting out the two non–interacting orbits) is a function of the ratios of planetary masses and the pulsar mass and the total time span of observations (Peale 1993). Using the timing model of Table 1, and with the planetary masses, $m_{1,2}$ expressed in terms of the Earth mass and M_{psr} in units of 1.4 $M_\odot$, these mass ratios are given by:

$$\frac{m_1}{M_{psr}} = \frac{3.4}{\sin i_1 \, M_{psr}^{1/3}} \tag{3}$$

$$\frac{m_2}{M_{psr}} = \frac{2.8}{\sin i_2 \, M_{psr}^{1/3}}, \tag{4}$$

for planets B, C and orbital inclinations i_1, i_2, respectively. Since the effect of a mutual inclination of the two orbital planes would have to be quite large to become detectable (Malhotra 1993), it is reasonable to assume coplanar orbits. Consequently, with $i_1 = i_2 = i$, the perturbation amplitude becomes a function of one variable, $(\sin i)^{-1} M_{psr}^{-1/3}$, which is the only parameter governing the effect of planetary perturbations on pulse arrival times.

After three years of observations, perturbations have been detected in the timing residuals of PSR B1257+12 (Wolszczan 1994). This result has been widely recognized as a final proof that the pulsar is indeed orbited by at least two planetary companions. It can be presented in the form of constraints on the masses of the pulsar and its planetary companions as a pulsar mass/normalized planetary mass diagram (Fig. 5b). The uncertainty of the determination of the best perturbation model defines a range of allowable masses. It is further constrained by the highest possible orbital inclination ($i = 90°$) and by the maximum neutron star mass (taken to be $2M_\odot$). Within these limits, the minimum pulsar mass is $\sim$1.2 $M_\odot$ and the range of possible pulsar masses for orbital inclinations close to $i = 90°$ includes both the typical theoretical value ($\sim 1.3M_\odot$) (Woosley 1987) and the observed average neutron star mass ($1.35 \pm 0.27 M_\odot$) (Thorsett *et al.* 1993). Furthermore, the masses of planets B and C must be similar to their respective "canonical", Earth–like values of $3.4M_\oplus$ and $2.8M_\oplus$ (Table 1 and Eqs. 3, 4) and the orbital inclinations are unlikely to be less than 60° for any reasonable choice of a neutron star mass.

5. Planets Around Other Pulsars

The discovery of planets around PSR B1257+12 has stimulated further searches for planetary companions to other neutron stars observable as radio pulsars. In addition to pulse timing measurements which allow a detection of isolated orbiting bodies, there have been several attempts to make direct observations of possible debris disks around a nearby pulsars (Zuckerman 1993; Phillips & Chandler 1993).

It is important to emphasize that there are a number of phenomena related to physics of the neutron star interiors and magnetospheres, the interstellar propagation, and the Solar System dynamics, which can affect pulse arrival times and are capable of mimicking planetary signatures in the timing residuals. In particular, a "timing noise" due to the seismology of a neutron star and a precession–induced wobble of its rotation axis may lead to quasiperiodic TOA variations (see Cordes (1993) for an extensive review).

At the time of this writing, the best candidate for another planet pulsar is PSR B1620-26, the 11–ms pulsar in a 191–day neutron star–white dwarf binary system located in the globular cluster M4 (Arzoumanian *et al.* 1996, and references therein). The timing residuals for this pulsars (Fig. 6a) show a deterministic, non–linear behavior that is most naturally accounted for by the presence of another orbiting mass around the inner binary. Indeed, analyses of the systematically growing set of TOAs for this pulsar, which have culminated in a reliable determination of its first four spin period derivatives, indicate that a third body in a long–period, eccentric orbit around the inner binary is present in this system. Arzoumanian *et al.* (1996) and Joshi and Rasio (1996) make a compelling case for a substellar (10^{-3}–$10^{-2} M_\odot$) companion, but they also demonstrate that a stellar mass object cannot be entirely ruled out. A future measurement of the fifth–order period derivative will provide the final solution to this exciting problem.

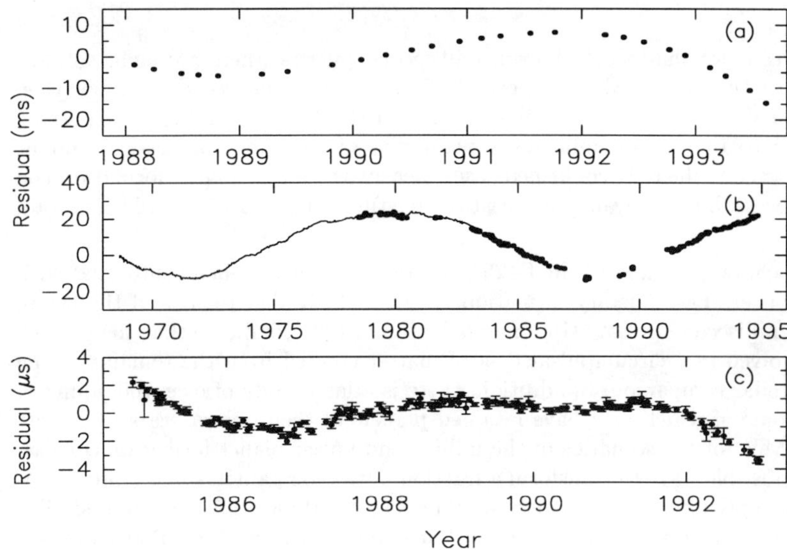

Figure 6. Long–term variations of pulse timing residuals for three pulsars. *(a)* PSR B1620-26 (courtesy of D. Backer), *(b)* combined JPL data (solid line) and Pushchino data (filled circles) for PSR B0329+54 (courtesy of V. Shabanova), and *(c)* PSR B1937+21 (Princeton public–access data base).

A case for a terrestrial–mass planet around a relatively young, "slow" pulsar, PSR B0329+54, has been recently made by Shabanova (1995). Curiously, the first suggestion that this object can have a planetary companion has appeared as early as 1979. It was based on the detection of a ~3–year periodicity in the timing residuals of the pulsar (Demiański & Prószynski 1979) later confirmed by Bailes, Lyne and Shemar (1993). Shabanova's analysis suggests that the 3–year periodicity of residuals practically disintegrates after 3–4 cycles, and that the data collected until 1995 are best explained in terms of a $\sim 2M_\oplus$ planet in a 17–year, eccentric orbit around the pulsar (Fig. 6b). Of course, in view of a distinct possibility that other effects, such as those mentioned above, can contribute to the observed TOA behavior of PSR B0329+54, further long–term observations are necessary to either confirm or dismiss this intriguing case.

Another millisecond pulsar which has been considered as a possible candidate for planetary companions is the 1.57–ms pulsar PSR B1937+21. This apparently isolated object, the fastest–rotating neutron star detected so far, has been shown by Kaspi, Taylor and Ryba (1994) to exhibit microsecond–level, long–term TOA fluctuations (Fig. 6c). In an unpublished analysis, Fukushima (1995) has demonstrated that it is possible to remove these fluctuations by including a second–order period derivative and orbital parameters of a Ceres–mass asteroid in the timing model of this pulsar. As in the cases discussed above, a verification of this model will require many more TOA measurements and a much longer data span.

6. Discussion

At present, among more than thirty known millisecond pulsars, there are eight solitary objects that have either managed to dispose of their binary stellar companions, or they have been created without the aid of binary evolution (Bailes *et al.* 1997). If a missing stellar companion to the pulsar indicates a possibility of "leftover" planets around it, PSR B1257+12 remains the only confirmed case. Clearly, the pulsar planet formation is a low–efficiency process, but the available statistics are still insufficient to reliably constrain it.

The observed characteristics of PSR B1257+12 and its planets, when confronted with standard ideas concerning planetary formation (Levy 1993, Ruden 1993) and the origin and evolution of millisecond pulsars (Phinney & Kulkarni 1994), indicate that the planets have probably evolved in a circumpulsar disk of matter created from the remains of the pulsar's binary stellar companion. In addition, there is a large body of over 700 younger, "slow" pulsars, some of which may have retained planets of their parent stars (Thorsett & Dewey 1993). Most of the scenarios of the millisecond pulsar planet formation concern themselves with possible ways to transform a fraction of the companion's mass into a protoplanetary disk, implying that the planets would subsequently form in a manner similar to that envisioned for the origin of planetary systems around normal stars (Podsiadlowski 1993; Phinney and Hansen 1993). One such scenario envisions that pulsar planets could be created as a by–product of a white dwarf merger which may produce a rapidly rotating neutron star and a suitable debris disk (Livio, Pringle & Saffer (1992). Another interesting alternative, described by Phinney and Hansen (1993), postulates a stellar companion disruption triggered by a close encounter with the pulsar formed in an asymmetric supernova explosion. This mechanism could produce a high–velocity single neutron star with a planetary system created out of the remnants of the former binary companion. In this context, it is interesting to note that PSR B1257+12, with its $\sim$300 km s^{-1} proper motion is the fastest–moving millisecond pulsar (Bailes *et al.* 1997). It is conceivable that the exceptionally high velocity of PSR B1257+12 and the fact that it has planets are closely related.

The above discussion indicates an interesting possibility that planetary systems around solar–type stars and pulsars may differ in their physical and chemical characteristics, but the fundamental features of their dynamics should be similar. The existence of pulsar planets, and a series of spectacular discoveries of giant planetary–mass objects around solar–type stars (e.g. Mayor & Queloz 1995; Marcy & Butler 1996) provide compelling evidence that extrasolar planetary systems exist in a variety of forms that would be impossible to predict on the basis of our present knowledge of the Solar System.

This research was supported by NASA under grant NAGW–3405 and by the NSF under grant AST–9317757.

References

Arzoumanian, A., Joshi, K., Rasio, F. A & Thorsett, S. E. (1996) Proc. IAU Symp. No. 160 *Pulsars: Problems and Progress*, S. Johnston *et al.* (eds.), ASP Conf. Ser.**105**, 525

Bailes, M., Lyne, A. G. & Shemar, S. L. (1993) in *Planets around Pulsars*, ed. J. A. Phillips, S. E. Thorsett & S. R. Kulkarni, ASP Conf. Ser.**36**, 19

Bailes, M. *et al.* (1997) *Astrophys. J.* , in press

Bell, J. F. *et al.* (1997) *Mon. Not. R. astr. Soc.* , in press

Butler, R. P., Marcy, G. W., Williams, E., McCarthy, Ch. & Dosanjh, P. (1996) PASP, **108**, 500

Cordes, J. M. (1993) in *Planets around Pulsars*, ed. J. A. Phillips, S. E. Thorsett & S. R. Kulkarni, ASP Conf. Ser.**36**, 43
Demiański, M. & Prószyński, M. (1979) *Nature* **282**, 383
Joshi, K. J. & Rasio, F. A. (1996) preprint
Kaspi, V. M., Taylor, J. H. & Ryba, M. (1994) *Astrophys. J.* **428**, 713
Konacki, M. & Maciejewski, A. J. (1996) *Icarus*, **122**, 347
Lazio, T. J. W. & Cordes, J. M. (1995) *Mercury*, March/April 23
Levy, E. H. (1993) in *Planets around Pulsars*, ed. J. A. Phillips, S. E. Thorsett, S. R. Kulkarni, ASP Conf. Ser. **36**, 181
Livio, M., Pringle, J. E. & Saffer, R. A. (1992) *Mon. Not. R. astr. Soc.* **257**, 15P
Lomb, N. R. (1976) *Ap. Space Sci.*, **39**, 447
Malhotra, R., Black, D., Eck, A. & Jackson, A. (1992) *Nature* **356**, 583
Malhotra, R. (1993) *Astrophys. J.* **407**, 266
Marcy, G. W. & Butler, R. P. (1996) *Astrophys. J. (Letters)* **464**, L147
Mayor, M. & Queloz, D. (1995) *Nature* **378**, 355
Peale, S. J. (1993) *Astron. J.* **105**, 1562
Phillips, J. A. & Chandler, C. J. (1993) *Astrophys. J.* **420**, L83
Phinney, E. S. & Hansen, B. M. S. (1993) in *Planets around Pulsars*, ed. J. A. Phillips, S. E. Thorsett & S. R. Kulkarni, ASP Conf. Ser.**36**, 371
Phinney, E. S. & Kulkarni, S. R. (1994) *Ann. Rev. Ast. Ap.*, **32**, 591
Podsiadlowski, P. (1993) in *Planets around Pulsars*, ed. J. A. Phillips, S. E. Thorsett & S. R. Kulkarni, ASP Conf. Ser.**36**, 149
Rasio, F. A., Nicholson, P. D., Shapiro, S. L. & Teukolsky, S. A. (1992) *Nature* **355**, 325
Rasio, F. A., Nicholson, P. D., Shapiro, S. L. & Teukolsky, S. A. (1993) in *Planets around Pulsars*, eds J. A. Phillips, S. E. Thorsett and S. R. Kulkarni, ASP Conf. Ser.**36**, 107
Ruden, S. P. (1993) in *Planets around Pulsars*, ed. J. A. Phillips, S. E. Thorsett, S. R. Kulkarni, ASP Conf. Ser. **36**, 197
Scargle, J. D. (1982) *Astrophys. J.* **263**, 835
Shabanova, T. W. (1995) *Astrophys. J.* **453**, 779
Taylor, J. H. & Weisberg, J. M. (1989) *Astrophys. J.* **345**, 434
Thorsett, S. E., Arzoumanian, Z., McKinnon, M. M. & Taylor, J. H. (1993) *Astrophys. J. (Letters)* **405**, L29
Thorsett, S. E. & Dewey, R. J. (1993) *Astrophys. J. (Letters)* **419**, L65
Thorsett, S. E. & Phillips, J. A. (1992) *Astrophys. J.* **387**, L69
Wolszczan, A. (1991) *Nature* **350**, 688
Wolszczan, A. & Frail, D. A. (1992) *Nature* **355**, 145
Wolszczan, A. (1994) *Science* **264**, 538
Wolszczan, A. (1996) Proc. IAU Symp. No. 160 *Pulsars: Problems and Progress*, S. Johnston *et al.* (eds.), ASP Conf. Ser.,**105**, 91
Woosley, S. E. (1987) in *The Origin and Evolution of Neutron Stars, IAU Symp. 125*, (eds Helfand D J and Huang J H Reidel, Dordrecht), p. 255
Zuckerman, B. (1993) in *Planets around Pulsars*, ed. J. A. Phillips, S. E. Thorsett & S. R. Kulkarni, ASP Conf. Ser.**36**, 303

STABLE PLANETARY ORBITS IN BINARY SYSTEMS

D. BENEST
C.N.R.S. U.R.A. 1362 Cassini
O.C.A. Observatoire de Nice, B.P. 4229
F-06304 NICE Cedex 4 (FRANCE)

1. Introduction

Recent discoveries, although indirect, of more than a dozen extrasolar planets allow us to expect the existence of planets around many stars other than the Sun. On an other hand, double (and multiple) star systems are established to be more abundant than single stars (such as the Sun), at least in the Solar Neighborhood.

I wish to present here some thoughts about the possibility of existence of planets in binary star systems.

This possibility is related to two main astronomical questions : first, can planet form in a double star early environment ? for example, do formation of a binary leave enough matter for accretion of planets ? this is a cosmogonical problem, which I will not deal here with (nevertheless, about the formation process of double stars, see e.g. Benest and Fulconis, 1982); second, assuming that planets have formed, do the gravitational laws warrant them to circulate safely on stable trajectories for billions of years ? this is a dynamical problem, which I will develop in the following.

This possibility is also a fascinating topic in bioastronomy (i.e. the study of the attempts of evaluation of the probabilities for extra-terrestrial life). More than half of the stars are indeed double or even multiple, at least in the solar neighborhood but most astronomers think that this sample is probably a fairly good representative. Let us then imagine, for instance, a multicolored double sunset. Moreover, life needs, as far as we know – and limiting ourselves to organic (= carbon-based) biochemistry –, a habitable planet, i.e. a planet where liquid water exists permanently; therefore, arises a last question : among the stable planetary orbits (if exist), are there some habitable ones, i.e. which stays inside the so-called habitable zone as defined by Hart in 1979.

J. A. Docobo et al. (eds.), Visual Double Stars: Formation, Dynamics and Evolutionary Tracks, 233–240.

2. The dynamics

Dynamically speaking, what kind of orbits – in binaries – can we expect for our planets ?

Let us take the example of the Solar System, where we have a star (the Sun) and a majestuous planet (Jupiter); let us consider these two bodies as a binary (this would be less unrealistic if the jovian mass had increased up to 0.05 solar mass, like a brown dwarf) : we can then say that the heliocentric orbits of the "inner" planets (Mercury to Mars, with the asteroids) and the jovicentric orbits of the galilean satellites (Io to Callisto, with all the other satellites and rings of Jupiter) are of the same kind, which we may call the "S-type", i.e. regular revolutions around one of the two components of the binary; there exist also "outer" planets (Saturn to Pluto, and beyond) which revolve around the binary as a whole, and we may call them of "P-type".

Note that two groups of asteroids (the Trojans) form with the Sun and Jupiter two equilateral triangles which rotate like a solid together with the binary, usually called of "L-type" (they are on or near the triangular stable equilibrium Lagrangian points L_4 and L_5); comets show often trajectories which alternate heliocentric motions and "captures" by Jupiter, motion called "interplay" and generally unstable. Other kinds of orbits have been considered by the dynamicians, less usual and/or purely academic; an example is the so-called Sitnikov problem, where the motion of a little body is limited along a perpendicular to the orbital plane of the binary and passing by the barycenter of the binary (here, generally, the two stars have equal masses); for that matter, the first paper which refers explicitely (to my knowledge) to planets in binary (Pavanini, 1907) is about the Sitnikov problem.

Since then, the litterature dedicated to these possible planetary orbits has grown enormously. However, most of the analytical studies have been obtained in the case where the binary's orbit is circular; this is a strong limitation, as a typical binary's orbit is plainly eccentric. Taking into account this eccentricity is indeed very difficult in pure analysis; therefore, numerical (or at least semi-numerical) methods are to be applied, but are extremely time-consuming when made by hand. Preliminary studies were relatively pessimistic about the existence of stable planetary orbits in double and multiple star systems.

Fortunately, the development of the computers in the early 60's allowed this latter studies, so more and more numerical results were obtained (see e.g. Dole, 1961; Hénon, 1965). Finally, modern computations have shown that many such stable planetary orbits do exist (but possible chaotic behaviour), either around each component of the binary (S-type) or around the binary itself (P-type). Here, only the S-type and the P-type will be considered.

For example, Harrington (1968, 1972) showed in particular that 3-D orbits become unstable only when the inclination approaches 90°. He concluded (Harrington, 1978) that S-type orbits can be stable up to some critical maximum dimensions, and that P-type orbits can be stable down to some critical minimum dimensions; however, these limits were still to be computed more precisely. More recently, Dvorak et al. (1989) undertook a task on this question, particularly for the P-type where the most impressive result is that stable planetary orbits exist around the binary down to 3 or 4 times the separation of the two stars (Dvorak, 1984, 1986); note that this lead to the conclusion (from the point of view of the last question in the Introduction : "are there habitable orbits ?") that P-type orbits may be habitable only around close binaries, otherwise they lie far out the habitable zone (an extensive bibliography on the subject can be found in Paprotny et al., 1980, 1982, 1984; see also Benest, 1988, 1993).

Now, let us talk about the S-type planetary orbits.

3. The model

Stricty speaking, the dynamical model to use is the general three-body problem, where three point masses evolve under the newtonian laws of motion; the parameters are the values of the three masses, and the initial conditions are the positions and velocities of the three bodies at $t = 0$ (i.e. 18 quantities).

But, as planetary masses are very much lighter than stellar masses (at least, the mean values), we may approximate – and simplify – the problem by considering that the mass of the planet is negligible with respect to the masses of the two stars : we are then in the so-called classical Restricted Three-Body Problem, which is the simplest realistic non-integrable (in the sense of Poincaré) problem. As the third body (the planet) does not perturb the motion of the two stars, the orbit of the binary is purely keplerian; the Restricted Three-Body Problem is then the study of the motion of the planet in the gravitational field of the binary; usually, equations are settled in a rotating-pulsating frame where the two stars lie quiet on the X-axis respectively at $X = 0$ – for the sun of the planet – and $X = -1$ – for the other star – (see e.g. Szebehely, 1967).

One parameter is the eccentricity e of the binary; as we consider only S-type orbits, we may choose the reduced mass μ of its "sun" as the second parameter. The initial conditions are the position and the velocity of the planet at $t = 0$ (i.e. 6 quantities). Moreover, we may limit ourselves the motion of the planet in the binary's orbital plane : we are then in the so-called "plane case" and the number of initial conditions decreases to 4. Finally, it can be shown (see e.g. Benest, 1978) that we do not reduce the generality of the problem when the planet starts from the half-axis $X > 0$ and perpendicularly to it : the initial conditions are then $X_o > 0$ and $\dot{Y}_o = V_0$; therefore, we may represent in the half-plane of initial condition $(X_o > 0, V_0)$ the set of points corresponding to stable orbits, so that it is easy to visualize.

Our definition of stability is : a planetary orbit is said stable when the planet revolves regularly around its sun during a given time T_M without any collision with a star and without escape. T_M has been taken as 100 revolutions of the binary. The equations are integrated numerically using a classical 4th-order Runge-Kutta scheme with variable step size.

In the so-called "circular case" (i.e. $e = 0$), the exploration has been made for every value of mu, i.e. from 0 to 1. The main result was (Benest, 1978) that stable planetary orbits exist always up to distance of their sun of the order of the binary's separation.

Let us turn now to the more realistic elliptic case ($e > 0$).

4. Planetary orbits in the elliptic case

Up to now, three nearby binaries has been explored : α Centauri (Benest, 1988b), Sirius (Benest, 1989) and η Coronae Borealis (Benest, 1996). For each binary, stable planetary orbits have been searched around respectively the lighter and the heavier components (B and A) : $\mu = 0.45$ and 0.55 for α Centauri and η Coronae Borealis, $\mu = 1/3$ and 2/3 for Sirius. The eccentricities were fairly high for the first two ($e = 0.52$ and 0.592 for respectively α Centauri and Sirius), although the orbital eccentricity of η Coronae Borealis is less important (0.28).

Thses numerical studies confirm globally the main result of the circular case, i.e. the existence of stable planetary orbits at very large distance of their sun : for α Centauri and η Coronae Borealis, the largest stable planetary orbits were found at, respectively around the B and the A component, 2/3 and 3/4 of the periastron separation of the binary, although some of these large orbits seem to show a chaotic behaviour (see also Lohinger et al., 1993); for Sirius, these largest orbits were found at 1/2 (around B) and 4/5 (around A). At least for these 3 examples, the influence of μ seems to be more important that the one of the eccentricity, but many more cases have to be explored before setting any certain conclusions (nevertheless, recently, Holman and Wiegert, have studied the α Centauri system during much longer numerical integration – 10 millions years –, and their result confirm globally what we have obtained for an 8000 years integration).

Moreover, for α Centauri and η Coronae Borealis, among these stable planetary orbits, a set was found to be "nearly circular", that is the orbital eccentricity of the planet around its sun stays under a value (0.05) which is not much higher than the heliocentric orbital eccentricity of the Earth (0.016). These nearly circular orbits exist up to 1/4 of the periastron separation of the binary, and therefore some of them lie in the habitable zone for each component of these two binaries. A planetary system, analogous to the system of the inner planets in our Solar System (from Mercury to asteroids) has even been able to be proposed around each component of the α Centauri system (Benest, 1991; see also Croswell, 1991).

As example, Figure 1 shows the stability zone for each component of η Coronae Borealis, with the set of nearly circular orbits. Coordinates are dimensionless : unit abscissa is the periastron separation of the binary, unit ordinate is such that the period of revolution of the binary is 2π. The (X_o, V_o) plane is explored for $0 < X_o \leq 1$ and $-3 \leq V_o \leq 3$ with a grid $\Delta X_o = 0.0125$ and $\Delta V_o = 0.05$; establishing with a fairly good precision the limit between stable and unstable regions (but avoid too long computation time) requires then the computation of almost 10 000 orbits for each graph. Stable orbits are materialized by a square filled at least partially (a blank square inside the stability zone indicates an isolated unstable orbit, i.e. a "lake" of instability); the darkest squares correspond to nearly circular orbits.

References

Benest D. (1978) *Orbites de satellite dans le problème restreint*, Thesis, Nice.

Benest D. (1988) Stable planetary orbits around one component in nearby binary stars, *Cel. Mech.*, **43**, 47-53.

Benest D. (1988b) Planetary orbits in the elliptic restricted problem I. The α Centauri system, *Astron. Astrophys.*, **206**, 143-146.

Benest D. (1989) Planetary orbits in the elliptic restricted problem II. The Sirius system, *Astron. Astrophys.*, **223**, 361-364.

Benest D. (1991) Habitable planetary orbits around α Centauri and other binaries, in *Bioastronomy, the search for extraterrestrial life - the exploration broadens*, eds. J. Heidmann and M.J. Klein, *Lecture Notes in Physics*, **390**, 44-47.

Benest D. (1993) Stable planetary orbits around one component in nearby binary stars II, *Cel. Mech.*, **56**, 45-50.

Benest D. (1996) Planetary orbits in the elliptic restricted problem III. The η Coronae Borealis system, *Astron. Astrophys.*, **314**, 983-988.

Benest D. and Fulconis M. (1982) L'évaporation des systèmes triples à l'origine de binaires peu serrées, *Mitt. Astr. Ges.*, **57**, 265-273.

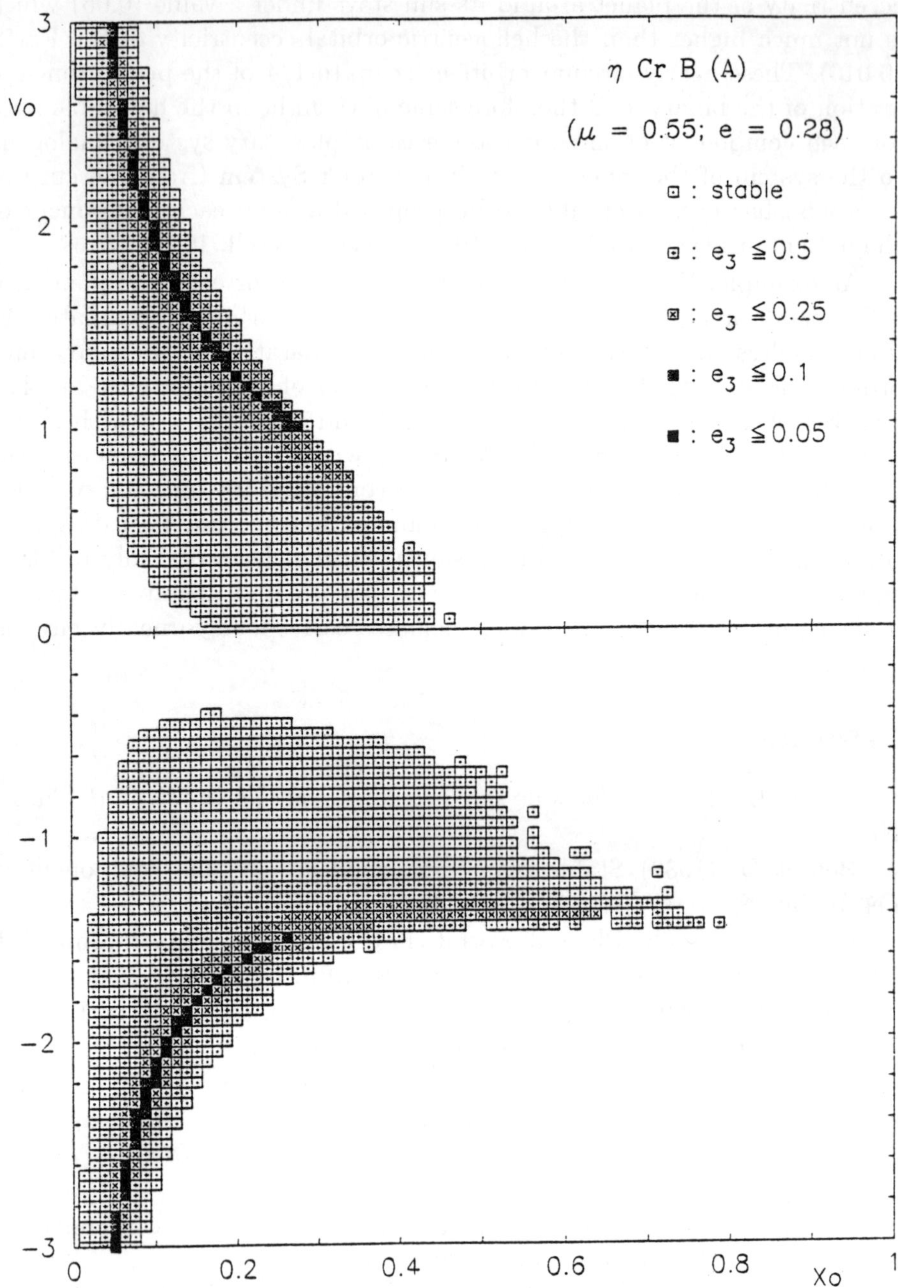

Figure 1. **Stable planetary orbits around η Coronae Borealis A.**

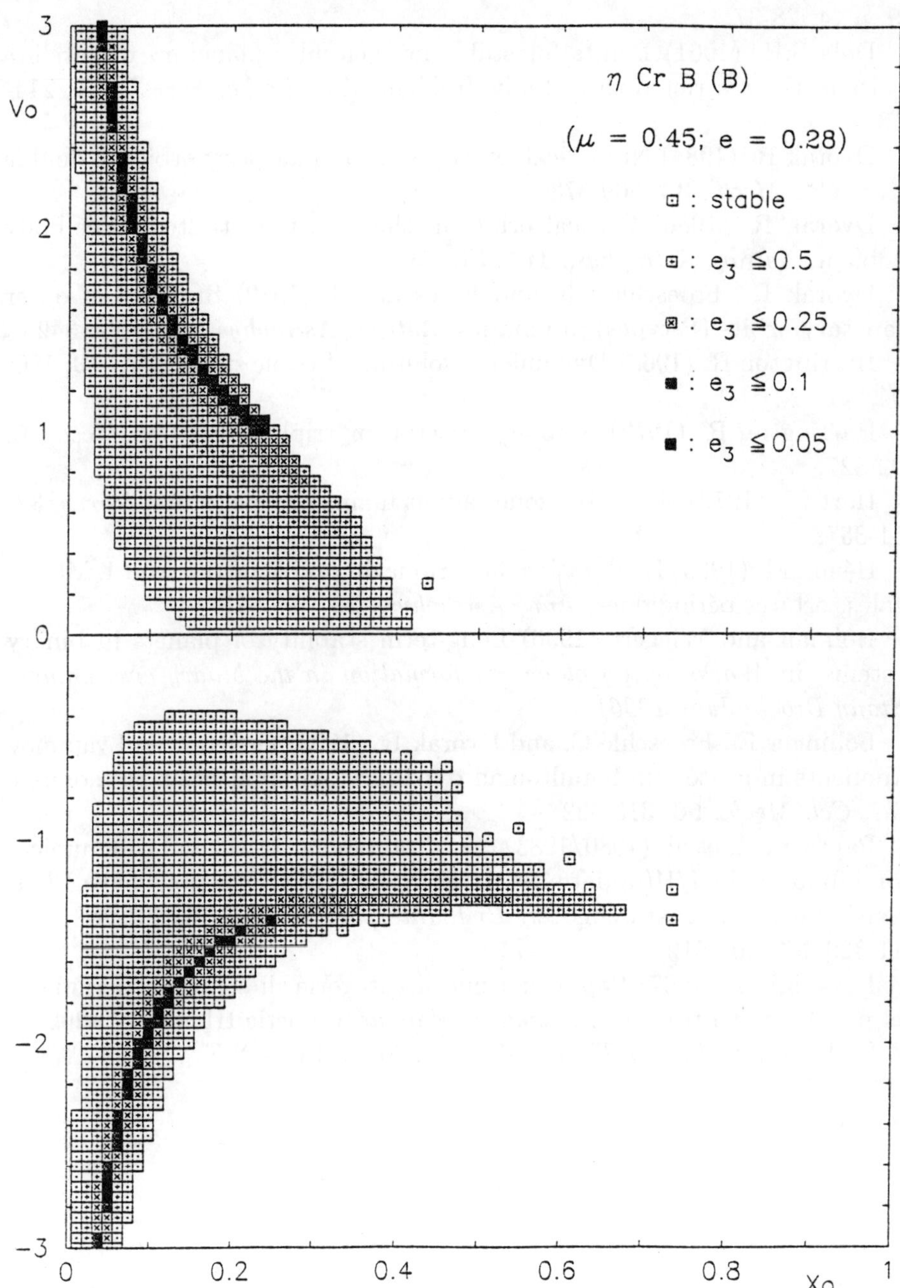

Figure 2. Stable planetary orbits around η Coronae Borealis B.

Croswell K. (1991) Does α Centauri have intelligent life ?, *Astronomy*, **19**, n° 4, 28-37.

Dole S.H. (1961) Limits for stable near-circular planetary or satellite orbits in the restricted three-body problem, *Am. Rocket Soc. J.*, **31**, 214-219.

Dvorak R. (1984) Numerical experiments on planetary orbits in double stars, *Cel. Mech.*, **34**, 369-378.

Dvorak R. (1986) Critical orbits in the elliptic restricted three-body problem, *Astron. Astrophys.*, **167**, 379-386.

Dvorak R., Froeschlé Ch. and Froeschlé C. (1989) Stability of outer planetary orbits (P-types) in binaries, *Astron. Astrophys.*, **226**, 335-342.

Harrington R. (1968) Dynamical evolution of triple stars, *A.J.*, **73**, 190-194.

Harrington R. (1972) Stability criteria for triple stars, *Cel. Mech.*, **6**, 322-327.

Hart M. (1979) Habitable zones about main sequence stars, *Icarus*, **37**, 351-357.

Hénon M. (1965) Exploration numérique du problème restreint I. Masses égales, orbites périodiques, *Ann. Astrophys.*, **28**, 499-511.

Holman and Wiegert (1996) Long-term stability of planets in binary systems, in *Workshop on planetary formation in the binary environment (Stony Brook, June 1996)*

Lohinger E., Froeschlé C. and Dvorak R. (1993) Generalized Lyapunov exponents indicators in hamiltonian dynamics: an application to a double star, *Cel. Mech.*, **56**, 315-322.

Paprotny Z. et al. (1980/1983/1984) Interstellar travel and communication bibliography I/II(update 1982)/III(update 1984). (section 05.04: Planets of multiple star systems), *J. Brit. Interplanet. Soc.*, **33**, 201-248/ **36**, 311-329/**37**, 502-512.

Pavanini G. (1907) Sopra una nuova categoria di soluzioni periodiche nel problema dei tre corpi, *Annali di Matematica* Serie III, **13**, 179-202.

Szebehely V. (1967) *Theory of orbits*, Acad. Press N.Y.

WIDE BINARIES FROM FEW-BODY INTERACTIONS

MAURI VALTONEN
Tuorla Observatory, University of Turku
21500 Piikkiö, Finland

Abstract. The orbital properties of wide binaries are modified by stellar encounters during their lifetimes as field stars. The few body process is particularly efficient right after the star formation in star clusters. In this article we review the orbital parameters which we expect from the few-body process. First, the solutions of the General Three-Body Problem are described, and then a detailed discussion of the theory relevant to binaries follows. Among other things, Öpik's law of the distribution of semi- major axes of the binaries is derived. In addition, we summarize recent results from the few-body problem of extended bodies which may be relevant to the early dynamical evolution of protostellar clouds.

1. INTRODUCTION

The importance of the few body process in the disruption of small star clusters was first demonstrated by von Hoerner (1960, 1963). The rapid binary formation and subsequent ejection from star clusters was confirmed by van Albada (1968) and Aarseth (1971, 1972, 1974, 1977, 1988). A review of the process is found in Aarseth and Lecar (1975). The binaries may form in a cluster of initially "single" stars, or a certain fraction of stars may be binaries as a result of the star formation process. In either case the three and four body interactions play an important role.

Useful solutions to the General Three-Body problem could not be achieved before the availability of fast computers even though some understanding of the problem was achieved by analytic studies (e.g. Sundman 1912). Another crucial development was the ability to regularize the equations of motion at close encounters which was first achieved by Kustaanheimo & Stiefel (1965). Then the first solution of one particular three-body prob-

J. A. Docobo et al. (eds.), Visual Double Stars: Formation, Dynamics and Evolutionary Tracks, 241–257.

lem (so called Burrau's problem) was fully calculated and illustrated by Szebehely & Peters (1967).

It soon became apparent that the solutions of the three-body problem are extremely sensitive to the initial conditions (Szebehely 1972) and therefore we are dealing with chaotic dynamics. Inspite of this there are many quantities which are stable in statistical sense. In the following we will review some of the properties which are most relevant to the study of wide binary stars. A more comprehensive review is found, for example, in Valtonen (1988) and Valtonen & Mikkola (1991).

2. THE CLASSICAL THREE-BODY PROBLEM

The solutions of the classical three-body problem can be divided in different categories depending on the initial configuration of the three bodies. In the first category the close three-body configuration occurs only once. Typically a star comes from some distance from a binary, interacts with it and the same star or one of the initial binary members receeds away from the remnant binary. In the process the orbital properties of the binary are modified. In case the binary members are the same before and after the encounter we name the event a flyby; otherwise it is called an exchange. In both cases this dynamical process is referred to as a scattering event.

The scattering process was studied extensively by Heggie (1974, 1975a, 1975b, 1977, 1980, 1988) who found a useful analytical expression for the statistical outcome (cross-section) of a scattering event. This expression is derived assuming that the passing star interacts directly only with one binary member, and this assumption leads to a good description of scattering from soft binaries, i.e. when the speed of the approaching star is high in comparison with the orbital speed of the binary (Valtonen & Heggie 1979, Hut & Bahcall 1983, Hut 1984).

The binaries ejected from a star cluster are likely to be hard, i.e. their orbital speeds are greater than the mean speed of stars in the cluster. The statistical distributions of the outcomes of the hard binary – third star scatterings can be derived assuming a temporary bound state (so called resonance) even if it does not actually occur (Mikkola 1986). Therefore we may take over the results of the studies of the resonance states to describe the scattering events of interest to us in star clusters.

Thus we move to the second category of three-body solutions where the trajectories depend sensitively on the initial conditions and the outcome is unpredictable. The system does not disrupt immediately, but there is a period of interplay where typically a star approaches first one star, then another, etc., until finally the system breaks up. After the break-up the system obtains a steady state. It is possible to describe this asymptotic final state using statistical mechanics. The statistical distributions of the orbital properties of the remnant binary and the escaping star are stable and depend only on few parameters. In this sense also the resonance-type three-body problem has been solved, and in the following we describe some of the results.

3. STATISTICAL THEORY OF THE THREE-BODY PROBLEM

In the following we follow the statistical treatment by Monaghan (1976). In the centre of mass frame the energy of the three-body system is

$$E_0 = \frac{1}{2}\frac{m_s m_B}{M}\dot{\mathbf{r}}_s^2 + \frac{1}{2}\frac{m_a m_b}{m_B}\dot{\mathbf{r}}^2 - G\frac{m_a m_b}{r} \tag{1}$$

where bodies of mass m_a and m_b form a binary and the body with mass m_s is the third body which will be considered an escaper. The binary mass is written $m_B \equiv m_a + m_b$ and the total mass of the system $\mathrm{M} \equiv m_B + m_s$. The separation vector between the binary members is $\mathbf{r}$, and the third body is at separation $\mathbf{r}_s$ from the binary centre of mass. G is the gravitational constant.

It is now assumed that the probability of a given end state of the system is proportional to the volume of the phase space. The density of the end states σ (per unit energy) is

$$\sigma = \int \ldots \int \delta\left(\frac{p_s}{2m} + E_B - E_0\right) d\mathbf{r}_s\, d\mathbf{p}_s\, d\mathbf{r}\, d\mathbf{p} \tag{2}$$

where $m = m_s m_B / M$, E_B is the energy of the binary, and $\mathbf{p}$ and $\mathbf{p}_s$ are the canonical momenta associated with $\mathbf{r}$ and $\mathbf{r}_s$.

When Eq. 2 is integrated over $\mathbf{r}_s$ and $\mathbf{p}_s$, we obtain the statistical distributions of the binary. After some algebra (see Monaghan 1976) we obtain

$$\sigma = \left(\frac{4}{3}\pi R^3\right) 8\pi^4 m^{3/2} \left(G m_a m_b\right)^3 \mu^{3/2} \int\int \frac{(E_0 - E_B)^{1/2} e\, de\, dE_B}{\mid E_B \mid^{5/2}}. \tag{3}$$

Here R is the radius of the volume inside which the strong three-body interaction takes place. It is so far an undefined parameter. The reduced mass of the binary is $\mu = m_a m_b / m_B$, and e is the eccentricity of the binary. It is apparent that the binary energy E_B and the binary eccentricity are independently distributed. The eccentricity distribution is

$$f(e) = 2e \tag{4}$$

The binary energy distribution is

$$f(| E_B |) \propto | E_B |^{-n} \tag{5}$$

where $n = 2.5$.

Similarly one may obtain the statistical distributions of the escaper by integrating over the phase space coordinates of the binary. Monaghan (1976) obtains

$$\sigma = \left(\frac{4}{3}\pi R^3\right) \frac{8\pi}{3}^4 G^3 m_a^4\, m_b^4 \left(\frac{m_s}{M}\right)^{3/2} \frac{1}{| E_0 |}. \tag{6}$$

When this is normalized by dividing by the sum of the phase space volumes corresponding to all possible escapers we find

$$f(m_s) = m_s^{-k}/(m_s^{-k} + m_a^{-k} + m_b^{-k}) \tag{7}$$

where $k = 2.5$ (Monaghan 1977, Mikkola and Valtonen 1986).

Numerical experiments have confirmed the eccentricity distribution of Eq. 4 in fully three-dimensional systems; two dimensional systems have a steeper distribution towards $e = 1$ (see Eq. 15 below).

In numerical experiments the binary energy distribution is actually a strong function of the total angular momentum of the system: at zero angular momentum the exponent is $n = 3$, obtains $n = 5.5$ in the middle of the relevant range of angular momenta, and it increases to $n > 10$ at high angular momentum (Saslaw, Valtonen and Aarseth 1974, Mikkola and Valtonen 1986, Mikkola 1994). Heggie (1975a) uses the detail balance argument to derive $n = 4.5$.

It is apparent that the radius of interaction R in Eq. 3 should be inversely proportional to the binary binding energy in order to make the statistical theory comparable to the numerical experiments where the total energy E_0 is practically constant and the initial binary semi-major axis is chosen as a unit distance (Valtonen 1988). Then R should be proportional to this unit, R $\propto E_B^{-1}$, or $n = 5.5$ in Eq. 5.

On the other hand, when we apply the theory to a star cluster situation there is no constant initial binary size. The radius of interaction R will be measured in the distance scale of the cluster of stars, and may be treated as a constant. On the other hand, the total energy $| E_0 |$ of the interacting three-body system ("hard" binary case) now scales with $| E_B |$, as does also $E_0 - E_B$. Therefore the volume of the phase space $\sum$ corresponding to the density of states σ is

$$\Sigma \equiv | E_0 | \sigma \propto \int\int ede\, | E_B |^{-1}\, dE_B \tag{8}$$

which leads to

$$f(E_B) \propto E_B^{-1} \tag{9}$$

or

$$f(a) \propto a^{-1} \tag{10}$$

where a is the semi-major axis of the binary. This is the so called Öpik's law (Öpik 1924).

Also the escaper mass distribution (Eq. 7) is a function of the angular momentum: The exponent obtains $k = 3$ at zero angular momentum and becomes $k < 2$ at high angular momentum (Valtonen 1976, Mikkola and Valtonen 1986).

Once the escape probability as a function of mass is known, it is possible to calculate the mass ratio $Q = m_1/m_2$ of the remaining binary. Using $k = 2.5$ we obtain the Q-distributions shown in Tables 1 and 2. All distributions are normalized to 10 units in the left hand column.

TABLE 1. The distribution of the binary mass ratio Q in three models. Logarithmic scale.

	Q								
	1.0		1.5		3.0		6.0		12.0
Model I		10		15		13		11	
Model II		10		9		4		2	
Model III		10		6		1		0	

TABLE 2. The distribution of the binary mass ratio Q in three models. Linear scale.

	Q										
	1.0		1.1		1.2		1.3		1.4		1.5
Model I		10		8		7		7		6	
Model II		10		8		7		6		5	
Model III		10		8		7		5		4	

The three models are: Model II, the Salpeter mass function $\psi(m) \propto m^{-2.35}$ where m is the stellar mass. Model I, the mass function for light stars, $\psi(m) \propto m^{-1.25} (m \lesssim 1 M_\odot)$; and model III, the mass function for heavy stars, $\psi(m) \propto m^{-3.2}(m \gtrsim 3M_\odot$, Mihalas and Binney 1981). We notice that the Q- distribution becomes flatter for lower mass stars, a trend which was noticed by Abt (1983) in observations.

The lifetime of a few-body system is measured in terms of the crossing time

$$T_{cr} = GM^{5/2} \mid 2E_0 \mid^{-3/2} \tag{11}$$

of the system. The break-up process of a few-body system is like radioactive decay: there is no predictable break-up time but the statistical distribution of lifetimes is a stable quantity. In units of crossing time the lifetime T may be described by

$$f(T) = 0.69 e^{-0.69T/T_{1/2}} \tag{12}$$

where $T_{1/2}$ is the half life of the system. For a zero angular momentum system the half life is

$$T_{1/2} = 4(1 + 4.5/m_r) \tag{13}$$

where m_r is the ratio of the mass of the heaviest body to the mass of the lightest body in the system. High angular momentum systems have somewhat longer lifetimes (Valtonen and Aarseth 1977, Valtonen 1988); typically the range of half lives is $T_{1/2} = 10 - 40$ crossing times (Valtonen 1974).

The break-up of a four-body system is a similar process where the system either breaks up in two binaries or it decays via ejections of two single stars (Mikkola 1983a, 1983b, 1984).

4. THE PLANAR FOUR-BODY PROBLEM

Small planar clusters of protostellar clouds may arise in the fragmentation process of molecular clouds. There may exist triples and quadruples of protostars where the sizes of the bodies are not negligible as compared with the distances between the bodies (Bodenheimer 1995). Collisions between the bodies may become important in the dynamical evolution of the system.

Mikkola, Saarinen and Valtonen (1984) studied the dynamical evolution of finite size bodies in a hierarchically fragmenting disk. In the following we will describe briefly the conclusions from this study as applied to the protostellar clouds. The work was intended for a different application (supermassive bodies in galactic nuclei), but with suitable scaling the results apply also to the few-body systems of protostellar clouds.

The solutions of the classical three-body problem are simple: in the end the system divides in two parts, a binary and a third body. The classical four-body problem is slightly more complicated since the final states are of two different kinds: (1) two binaries, or (2) one binary and two single particles. The collisional four-body problem is even more complex. In the end the number of bodies can vary from one to four and the 2–4 bodies may be found in different combinations of binaries and single bodies.

The collision between two bodies is achieved by modifying the Newtonian force law: Mikkola et al. (1984) use

$$\mathbf{F}_{ij} = \frac{m_i m_j r_{ij}^2}{(\varepsilon + r_{ij}^2)^{2.5}} \mathbf{r}_{ij} - a \frac{\mathbf{W}_{ij}}{1 + \exp\left[100\Delta^2 \left(\frac{r_{ij}^2}{\Delta^2} - \frac{\Delta^2}{r_{ij}^2}\right)\right]} \tag{14}$$

where $\mathbf{F}_{ij}$ is the interparticle force, $\mathbf{W}_{ij}$ is their relative velocity, $\mathbf{r}_{ij}$ their relative position, m_i and m_j are the masses and Δ is the radius of the body. By experimentation it was found that $\varepsilon = 10^{-2}$ and $a = 1$ are suitable parameter values. The force law is practically Newtonian when $r_{ij} > 2\Delta$, but a rapid merger happens at close encounters. The system is thought to be planar and to be composed of equal mass bodies ($m_i = 1, i = 1, \cdots 4$) which form initially two zero eccentricity binaries. The binary orbital radius is chosen as one distance unit. Gravitational constant equals unity.

The final states of the four-body systems were classified as follows:

-1	=	two bodies form a binary system
0	=	unfinished
1	=	two binaries escaping from each other
2	=	a bound triple plus one escaper
3	=	one binary and two escapers

Table 3, column H1 shows the distribution of final states, when the outer orbit of the two binaries about each other is circular and the orbital radius is between 2 and 3 units.

The distribution of the eccentricities of the binaries of category -1 follow the $f(e) = 2e$ distribution from $e = 0$ to $e = 0.7$, and then drops off rapidly when $e > 0.7$ (see Table 4, column H1). The pericentre distances of the binaries concentrate between 2.4 and 3.0 units. The escape speed $v_\infty \simeq 0.25$. The speed unit is such that the relative orbital speed of the initial binaries is $\sqrt{2}$. The escape speed distribution can be derived from Eq. 5 above.

For comparison, experiments were also carried out where the collisions at close encounters were neglected. Column S1 in Table 3 gives the corresponding numbers. Now the eccentricity distribution follows

$$f(e) = \frac{e}{\sqrt{1 - e^2}} \tag{15}$$

TABLE 3. Distribution of final states in the planar four-body experiments

	Set		
Final state	H1	S1	S2
-1	170	–	–
0	194	17	50
1	5	39	104
2	28	54	35
3	0	84	10

which is the result expected from the statistical breakup theory in two dimensions (Monaghan 1976). The escape speeds also follow the statistical theory with the peak of the distribution around $v_\infty = 0.25$ units.

In another set of experiments the outer orbits were made wider and eccentric ($e > 0.5$) such that the pericentre distance of the outer orbit varies uniformly between $q = 2$ and $q = 4$ distance units. In case of collisional systems the most common end result is now of type 1, while only 2–3 % of the cases go to categories 2 and 3. Without collisions the distribution is as given in column S2 of Table 3. The eccentricities of the binaries of the final category 1 follow a distribution which is concentrated between $e = 0.1$ and $e = 0.5$ while in category 3 the eccentricity follows the usual statistical distribution of Eq. 4 (see Table 4, the last two columns). The ejection speed distributions may again be derived from Eq. 5 with the peak at $v_\infty \simeq 0.25$ units.

The processes described above differ from each other by varying eccentricity distribution: it is either very steep towards $e = 1$ or is concentrated towards small values of e.

Earlier studies of the planar binary–binary interaction are found in Saslaw, Valtonen and Aarseth (1974) and Harrington (1974).

5. THE TRAPEZIUM TYPE FOUR-BODY SYSTEMS

The initial binaries may not always lie in a single plane. A trapezium type system of two binaries at a small distance from each may occupy a fully three-dimensional volume of space, or more specifically, the binaries may be randomly oriented relative to each other. This was the case studied by Harrington (1974), Mikkola and Valtonen (1990) and Valtonen et al. (1994).

In the experiments of Mikkola and Valtonen (1990) the radii of the

TABLE 4. The distribution of eccentricities in different dynamical processes.

	H1	S1	S2	
			final state	
e			1	3
0 – 0.1	2	0	5	0
0.1 – 0.2	8	0	29	0
0.2 – 0.3	14	2	40	1
0.3 – 0.4	15	5	46	0
0.4 – 0.5	25	8	31	0
0.5 – 0.6	39	9	17	1
0.6 – 0.7	39	8	19	0
0.7 – 0.8	7	21	14	0
0.8 – 0.9	16	28	3	2
0.9 – 1.0	5	71	4	6

bodies were 0.3 units and the binaries were circular as in the planar problem. However, the binary mass ratio was varied. Also the binaries met each other in a parabolic initial orbit, implying that they were formed in separate systems and came together only thereafter.

Table 5 gives a summary of the different kinds of final configurations in these experiments.

TABLE 5. The distribution of final states in the randomly oriented binary encounters.

Final state	*Binary mass ratio*		
	1:1	2:1	3:1
-1	188	43	13
0	13	14	4
1	151	277	349
2	1	2	0
3	0	1	0
4	125	110	64
5	42	18	12

The additional categories in this table are:

4 = one binary and one escaper
5 = two single bodies escaping from each other.

Valtonen et al. (1994) used a power-law distribution of the masses of the bodies as well as a distribution of the semi-major axes of the binaries. The binaries were in a bound eccentric ($e = 0.9$) orbit relative to each. Also three-body systems were studied. An escape velocity from the cluster was introduced which was smaller than the typical binary speed by a factor of about two. The effective radii of the bodies were about 1 % of the binary orbital radius. The results could now be classified on the basis of escapes from the cluster. The final categories were:

-1 = a binary left in the cluster
0 = unfinished
1 = no bodies left in the cluster, two escapers
2 = one body left in the cluster, one escaper
3 = a binary left in the cluster, two escapers
4 = a binary left in the cluster, one escaper
5 = one body left in the cluster, two escapers
6 = no bodies left in the cluster, three escapers

Table 6 gives some examples of the distribution of final states in four sets of 10 000 examples. The first set S1 only includes three bodies whose masses are uniformly distributed between 1 and 10 (in arbitrary units). The second example is the set S3 where there are four bodies present in the beginning. In the set S7 a power-law distribution of the form $\psi(m) \propto m^{-1.5}$ was used for the four masses in the interval 1 – 10 mass units, while in set S8 the power law was $\psi(m) \propto m^{-3}$ in the same mass interval.
The speed of ejection of the bodies (or binaries) from the cluster is typically a factor of two above the escape speed from the cluster.

6. DISCUSSION

(a) Öpik's law

We saw in Section III that the application of the statistical theory of the three-body break-up to the cluster environment leads to the so called Öpik's law of binary separations, i.e. $f(a) \propto a^{-1}$ where a is the semi-major axis of the binary. Another situation where the same law has been encountered, at least approximately, is the simulation of small groups of finite size bodies. Wiren et al. (1996) start from a group of four bodies in virial equilibrium inside a sphere of radius 1 unit while the bodies have radii typically of 0.2 units. The masses range uniformly from 1 to 4 in some units. The evolution

TABLE 6. Distributions of final states in the randomly oriented binary encounters with a distribution of masses.

	Set			
Final state	S1	S3	S7	S8
-1	1230	159	147	206
0	5960	4972	5210	4737
1	58	40	12	54
2	931	3069	3081	3297
3	0	180	121	38
4	1821	913	661	726
5	0	647	763	921
6	0	21	5	21

of the group is followed until only a binary is left in the group. The final distribution of the projected separations of the binary members is close to the Öpik's law.

Zheng, Valtonen and Valtaoja (1990) and Valtonen and Zheng (1990) simulate the evolution of star clusters with 10 – 50 massive stars, either of equal mass or with a mass distribution of the form $\psi(m) \propto m^{-2}$, together with 1000 massless particles. The stars and particles are placed in virial equilibrium within a sphere of radius 1 pc. After $6 \cdot 10^9 yr$ the cluster is strongly disrupted. Bound star-particle pairs were identified and the distribution of their semi-major axes were plotted. The distribution was found to obey Öpik's law over one decade in a. One may view this as a process by which small mass stars are captured by massive stars to form unequal binaries.

Therefore it is expected that binaries formed in clusters through dynamical interaction, either in collisional or in collisionless systems, should follow Öpik's law. The same statistical theory applies to these different situations.

(b) Eccentricities

In the above experiment of Valtonen and Zheng (1990) the binary eccentricities followed the statistical equilibrium distribution $f(e) = 2e$ very closely. Also the binaries formed in the collisional system of Wiren et al. (1996) give the same distribution. We remember that this is also the result expected from the statistical theory in three-dimensional systems, and should be a good signature of dynamically determined orbital parameters.

However, if the systems are primarily two-dimensional, the eccentricity distribution can be either steeper or shallower than $f(e) = 2e$. In section 4 we discussed the distributions in strongly interacting planar systems. There the distribution is steeper. However, if binaries originally form with zero eccentricity and if they are expelled from the cluster by distant flyby encounters in a planar system, then the eccentricities can remain small, mostly $e < 0.5$. In a strongly collisional system one may also obtain the $f(e) = 2e$ distribution which is truncated at the upper end.

Therefore deviations from the $f(e) = 2e$ distribution should tell us how strongly the initial conditions influence the distributions.

(c) Binary mass ratios

As pointed out in Section 3, the mass ratio should depend sensitively on the initial stellar mass function. Deviations from a universal power-law slope should show up especially at the frequency of the high mass ratios. See Figure 1.

(d) Ejected pairs

In section 5 we discussed the different ways the few-body systems break up and stars or protostellar cloud leave the cluster. It can happen either through ejecting single stars, but quite often by ejecting a binary out of a cluster. This may be an important process by which binary stars find their way from star clusters to the general Galactic field.

(e) Kozai Mechanism

Stars are often found in hierarchical multiple configurations. Also the dynamical simulations produce multiple system. For example, Valtonen and Zheng (1990) find 27 binaries, 12 triples and 26 systems of higher multiplicity per 300 stars among the massive star – test particle pairings. Therefore one should consider long-term evolution in hierarchical triples.

Kozai (1962) showed that in a hierarchical triple system of high relative orbital inclination the secular perturbations of the outer orbit cause a periodic fluctuation in the eccentricity of the inner orbit. In case of perpendicular alignment ($i = 90°$) of the inner and outer orbits the eccentricity $e \to 1$, i.e. the inner binary members collide with each other.

In reality the inclination is seldom exactly $i = 90°$ and the tidal interaction between the stars becomes important when the separation of the stars $r \lesssim 2(R_1 + R_2)$, where R_1 and R_2 are the radii of the two stars (Orlov and Petrova 1996). Thus we may assume that tidal friction causes a rapid decline of the semi-major axis of the inner orbit when close encounters between the stars begin. At the same time the inner orbit becomes again more circular.

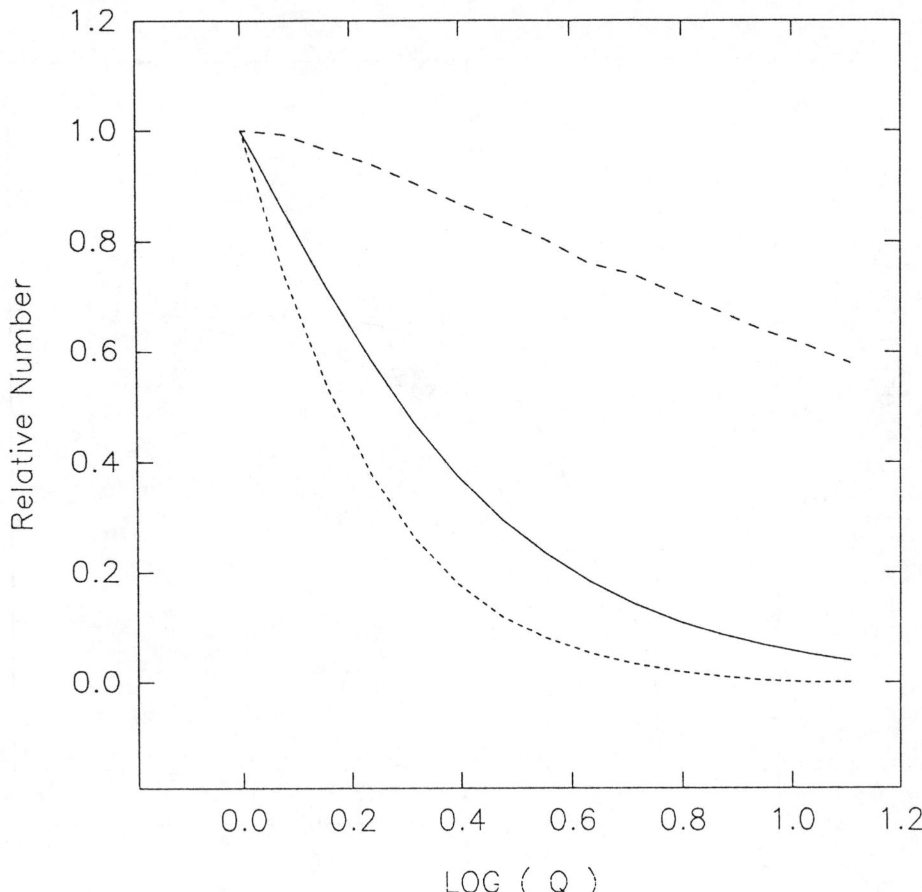

Figure 1. The distribution of the binary mass ratio Q in case of three different initial mass functions for single stars. Solid line: Salpeter mass function. Dotted line: the mass function for heavy stars. Dashed line: the mass function appropriate for low mass stars.

Table 7 gives values of the maximum eccentricity e_{max} as a function of the relative inclination i. We also give a rough estimate of the maximum initial semi-major axis a_{max} of the inner binary which is affected by the Kozai mechanism. The value of a_{max} depends on the separation at which the tidal capture becomes efficient. For the calculation we have assumed the tidal capture distance of 0.03 AU. The orbit calculations were performed by S. Mikkola (paper in preparation).

Figure 2 illustrates how the initial $f(a) \propto a^{-1}$ distribution is modified by tidal encounters (1) if the binaries are isolated, and (2) if they are all

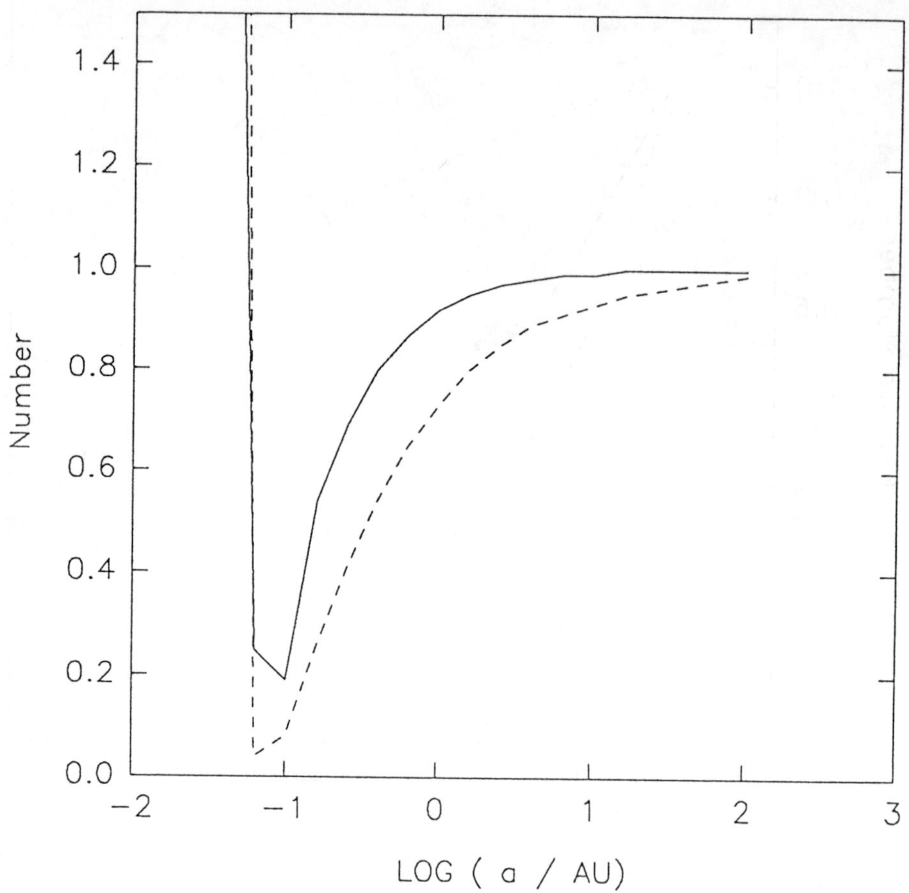

Figure 2. The distribution of semi-major axes a of binaries in the three-body statistical theory including tidal interaction (effective interaction distance 0.03 AU). Solid line: isolated binaries. Dashed line: binaries in hierarchical triple systems (Kozai mechanism and random orientations).

members of triple systems. The Kozai mechanism and a random orientation of relative orbits is assumed. The tidal capture distance is assumed to be 0.03 AU. If the orbital properties are determined at the protostellar phase of the evolution of stars, the tidal capture distance should be increased by 1–2 orders of magnitude and the scale of the abscissa of Fig. 2 should be increased correspondingly. Kozai mechanism does not have time to work at the protostellar stage, and thus this rescaling is only possible in relation to the solid line of Fig. 2.

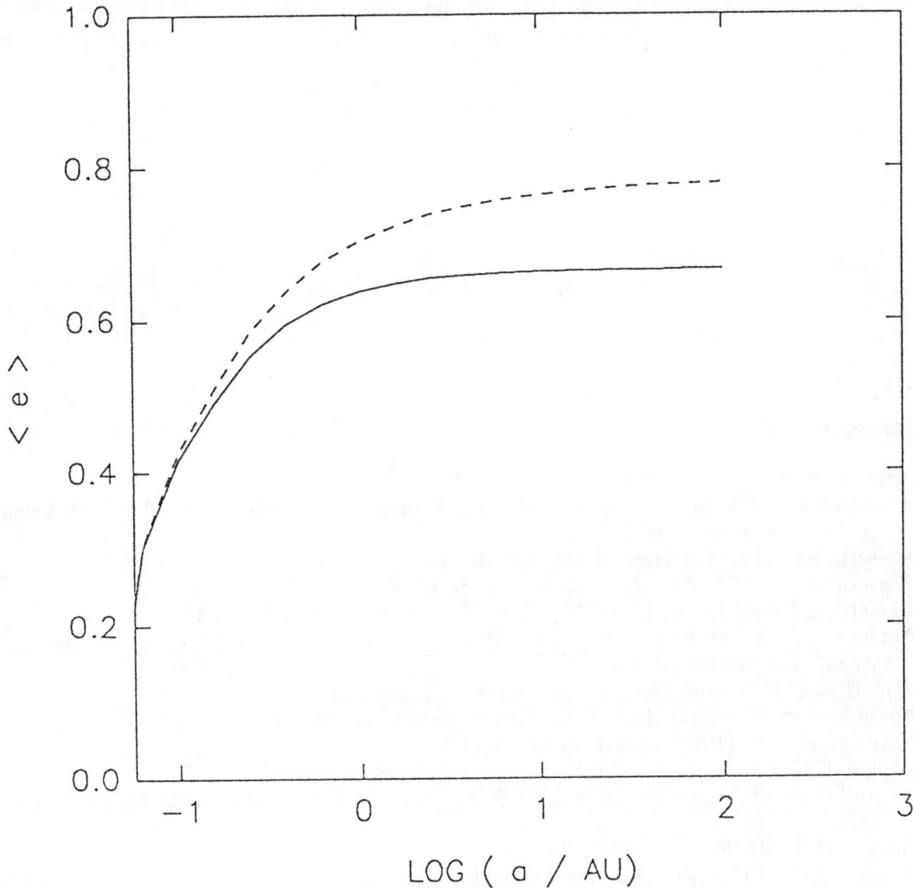

Figure 3. The distribution of the mean binary eccentricity $\langle e \rangle$ as a function of the binary semi-major axis a in the three-body model, including tidal capture (capture distance 0.03 AU). Solid line: three-dimensional systems. Dashed line: two-dimensional systems.

Tidal capture also depopulates the high eccentricity end of the e-distribution of binaries. The average values $< e >= 0.67$ in 3-dimensional systems and $< e >= 0.785$ in 2-dimensional systems are lowered towards small a. Figure 3 illustrates the effect when the tidal interaction distance is 0.03 AU. Kozai mechanism does not influence the latter diagramme very much. The previous remark about rescaling the abscissa is relevant also to this figure.

TABLE 7. The maximum eccentricity $e_{\max}$ for a tidal capture due to Kozai mechanism.

inclination i	e_{max}	a_{max}(AU)
90°	1.0	1
80°	0.975	1.2
70°	0.90	0.3
60°	0.76	0.1
50°	0.56	0.07
40°	0.225	0.04

References

Aarseth, S.J. (1971) *Astrophys. Space Sci.* **13**, 324.

Aarseth S.J. (1972) in Gravitational N-Body Problem, *IAU Colloq. No. 10* , ed. M. Lecar, p. 88. Dordrecht: Reidel.

Aarseth, S.J. (1974) *Astron. Astrophys.* **35**, 237.

Aarseth, S.J. (1977) *Rev. Mex. Astron. Astrofis.* **3**, 199.

Aarseth, S.J. and Lecar, M. (1975) *Ann. Rev. Astron. Astrophys.* **13**, 1.

Aarseth, S.J. (1988), in The Few Body Problem, IAU Colloq. No. 96, ed. M.J. Valtonen, p. 287. Dordrecht: Kluwer.

Abt, H.A. (1983) *Ann. Rev. Astron. Astrophys.* **21**, 343.

Bodenheimer, P. (1995) *Ann. Rev. Astron. Astrophys.* **33**, 199.

Harrington, R.S. (1974) *Celest. Mech.* **9**, 465.

Heggie, D.C. (1974), in The Stability of the Solar System and of Small Stellar Systems, IAU Symp. No 62, ed. Y. Kozai, p. 225. Dordrecht: Reidel.

Heggie, D.C. (1975a), *MNRAS* **173**, 729.

Heggie, D.C. (1975b), in Dynamics of Stellar Systems, IAU Symp. No. 69, ed. A. Hayli, p. 73. Dordrecht: Reidel.

Heggie, D.C. (1977) *Comments Astrophys.* **7**, 43.

Heggie, D.C. (1980), in Globular Clusters, ed. D. Hanes and B. Madore, p. 281. Cambridge: Univ. Press.

Heggie, D.C. (1988) in The Few Body Problem, IAU Colloq. No. 96, edited M.J. Valtonen, page 213, Dordrecht: Kluwer.

Hut, P. and Bahcall, J.N. (1983) *Ap.J.* **268**, 319.

Hut, P. (1984) *Ap.J. Suppl.* **55**, 301.

Kozai, Y. (1962) *Astron. J.* **67**, 591.

Kustaanheimo, P. and Stiefel, E. (1965) *J. Reine Angew. Math.* **218**, 204.

Mihalas, D. and Binney, J. (1981) Galactic Astronomy. Structure and Kinematics, p. 231. San Francisco:Freeman.

Mikkola, S. (1983a) *MNRAS* **203**, 1107.

Mikkola, S. (1983b) *MNRAS* **205**, 733.

Mikkola, S. (1984) *MNRAS* **207**, 115.

Mikkola, S., Saarinen, S. and Valtonen, M.J. (1984) *Astrophys. Space Sci.* **104**, 297.

Mikkola, S. (1986) *MNRAS* **223**, 757.

Mikkola, S. and Valtonen, M.J. (1986) *MNRAS* **223**, 269.

Mikkola, S. and Valtonen, M.J. (1990) *Astrophys.J.* **348**, 412.

Mikkola, S. (1994) *MNRAS* **269**, 127.
Monaghan, J.J. (1976) *MNRAS* **176**, 63.
Monaghan, J.J. (1977) *MNRAS* **179**, 31.
Öpik, E.J. (1924) *Tartu Observatory Publications* **25**.
Orlov, V.V. and Petrova, A.V. (1996) *MNRAS* **281**, 384.
Saslaw, W.C., Valtonen, M.J. and Aarseth, S.J. (1974) *Ap.J.* **190**, 253.
Sundman, K.F. (1912) *Acta Math.* **36**, 105.
Szebehely, V. (1972) *Astron.Astrophys* **77**, 169.
Szebehely, V. and Peters, C.F. (1967) *Astron.J.* **72**, 876.
Valtonen, M.J. (1974) in The Stability of the Solar System and of Small Stellar Systems, IAU Symp. No. 62, ed. Y. Kozai, page 211. Dordrecht: Reidel
Valtonen, M.J. (1976) *Mem. RAS* **80**, 77.
Valtonen, M.J. and Aarseth, S.J. (1977) *Rev. Mex. Astron. Astrofis.* **3**, 163.
Valtonen, M.J. and Heggie, D.C. (1979) *Celest. Mech.* **19**, 53.
Valtonen, M.J. (1988) *Vistas in Astron.* **32**, 23.
Valtonen, M.J. and Zheng, J-Q. (1990) in Nordic-Baltic Astronomy Meeting, Eds. C.-I. Lagerkvist, D. Kiselman and M. Lindgren, p. 353. Uppsala: Univ. Uppsala.
Valtonen, M.J. and Mikkola, S. (1991) *Ann. Rev. Astron. Astrophys.* **29**, 9.
Valtonen, M.J., Mikkola, S., Heinämäki, P. and Valtonen, H. (1994) *Astrophys. J. Suppl.* **95**, 69.
Van Albada, T.S. (1968) *Bull. Astron. Inst. Neth.* **20**, 57.
Von Hoerner, S. (1960) *Z. Astrophys.* **50**, 184.
Von Hoerner, S. (1963) *Z. Astrophys.* **57** , 47.
Wiren, S., Zheng, J.-Q., Valtonen, M.J. and Chernin, A.D. (1996) *Astron.J.* **111**, 160.
Zheng, J.-Q., Valtonen, M.J. and Valtaoja, L. (1990) *Cel. Mech. and Dyn. Astron.* **49**, 265.

THIRD BODY PERTURBATIONS OF DOUBLE STARS

RUDOLF DVORAK
Institute of Astronomy, University of Vienna,
Türkenschanzstr. 17,
A - 1180 VIENNA, AUSTRIA

Abstract. We report on the different results concerning the stability of the hierarchical triple systems where a close binary is accompanied by a third star. There are different possible approaches to answer the question of the stability limits for such triple stars: the most direct investigations can be undertaken in integrating numerically the respective equations of motion for many different initial conditions. It is then difficult to take into account all the important parameters like eccentricities, inclination, phases and masses. Analytical approaches and qualitative methods are more approriate to deal with this problem; the respective results confirm the numerically found results that:
1. for prograde orbits the ratio semimajor axis of the inner orbits to the periastron position of the outer orbit is approximately 3.2
2. for retrograde orbits this ratio is just some 10 percents smaller
3. the results are not sensitive in what concerns the masses involved
4. There is a tendency that the inclinations and eccentricities change slightly the stability limits mentioned above.

1. Introduction

It can be expected that at least one third of the double stars observed are triple systems or contain even four or five members (cf. Batten, 1973, Worley, 1967, Heintz, 1969). This fact stresses the importance to study such multiple systems in detail. In the following I will review on investigations on the dynamics of triple system, a field of growing interest during the last fifty years for obervational and theoretical Astronomy.

In an early work by Wallenquist (1944) the author distinguished 2 major types of triple systems:

- The distant companion types (Dc-systems), which are systems of a close pair and a companion which is at least three times further away than the double star; in the modern nomenclature we call them hierarchical systems.
- The cluster types (Cl-systems) where the distances between the three components are comparable (according to original definition of Wallenquist the third body should be at least closer than three times the separation of the binary); we call them today Trapezium system.

J. A. Docobo et al. (eds.), Visual Double Stars: Formation, Dynamics and Evolutionary Tracks, 259–268.

In the following we will discuss only the gravitational stability of hierarchical systems where we take into account the three stars as mass points with masses m_1, m_2 and m_3 which are of the same order of magnitude (the planetary case will not be taken into account explicitely).

When studying the gravitational three body problem in a Newtonian frame we can express the equations of motion – besides the classical formulation with respect to an origin (inertial system) – in different forms which are more or less useful for the problem in question.

- in the Lagrangian variables the three vectors $\vec{r}_{12}$, $\vec{r}_{23}$ and $\vec{r}_{31}$ are the connecting vectors between the three masses m_1, m_2 and m_3
- in the Jacobi formulation the vector $\vec{R}$ is the connecting vector between the masses m_1 and m_2, the vector $\vec{r}$ is directed from the barycenter between m_1 and m_2 to the third mass m_3.
- in the Delaunay formulation canonical variables are used which are simple functions of the Keplerian elements for the so-called inner and outer orbits of the Jacobi decomposition. The equations of motion are then in the canonical form and obey the Hamiltonian equations.

In principle three different methods can be used to answer the crucial question of stability of such a triple system: the so called qualitative methods allow to draw conclusions of the dynamics of the bodies valid for infinite time, the perturbation approach provides results for a whole bundle of solutions for a certain time scale and the method of numerical integration allows to follow one single solution with very high precision (again only for a limited time span). We will present in this reviewing article the results of all three approaches.

2. Qualitative results

In this section I report on the fundamental research of the following authors: Saari (1974), Marchal and Saari (1975), Szebehely and Zare (1977), Marchal and Bozis (1982), Marchal et al. (1984), Marchal (1990). They suceeded in finding zero velocity curves which separate forbidden regions of motion from possible regions of motion. This is quite similar to the well known Hill's curves of the restricted three body problem; in this model of a dynamical three body problem the third mass m_3 is regarded as possessing an infinitesimal small mass and thus it is not perturbing the circular motion of the primaries m_1 and m_2 (e.g. Szebehely, 1970). In the restricted problem the Jacobian integrals provides the zero velocity curves; in the general three body-problem the product c^2H (where H is the energy integral and c is the angular momentum integral) can be used to distinguish between forbidden and allowed regions of motion for the third mass. The respective plane where such border lines can be drawn is the instantaneous plane connecting all three masses involved.

2.1. A TWO-BODY APPROXIMATION

Denoting by a_1, e_1 and a_2 and e_2 the respective semi-major axis and the eccentricities of the inner and the outer system the Keplerian energy reads as

$$H = -\frac{G}{2}\left(\frac{m_1 m_2}{a_1} + \frac{m_3 \nu}{a_2}\right) \tag{1}$$

with G the gravitational constant and $\nu = (m_1 + m_2)$. The angular momentum integral c expresses as

$$c = \{G[m_3\nu(\frac{a_2(1-e_2^2)}{M} + m_1 m_2(\frac{a_1(1-e_1^2)}{\nu}]\}^{\frac{1}{2}} \tag{2}$$

with M the sum of the three masses involved. An estimation for $m_1 = m_2 = m_3$ leads to

$$H \sim -\frac{Gm^2}{2a_1} \tag{3}$$

and

$$c \sim 2(\frac{Gm^3 a_2}{3})^{\frac{1}{2}} \tag{4}$$

The estimate is valid because $a_2/a_1 \gg 2$; thus in the case of three stars of comparable masses the energy is governed by the close binary and the angular momentum by the outer star's motion. These leads to the approximate expression for

$$c^2 H \sim -\frac{2}{3} G^2 m^5 \frac{a_2}{a_1} \tag{5}$$

This has the following consequencies for hierarchical systems: If the actual value of c^2H for a given triple configuration is smaller than the critical value we have no exchange of bodies (Szebehely and Zare, loc.cit).

After the derivation of two additional equations one can conclude that no exchange can occur if

$$\frac{a_2}{a_1} = \frac{q_2}{a_1} \geq 3.2 \tag{6}$$

where we denote the periastron of m_3 with q_2 (also the eccentricity of the outer orbit is taken into account). The results fit quite well with the ones of Harrington (1972) for prograde motion of the third body.

2.2. ZERO VELOCITY CURVES

Using the notation as it was already introduced we define the two variable lengths the mean quadratic distance ρ and the mean harmonic distance μ (Marchal, loc.cit.)

$$\rho = [\frac{1}{M^*}(m_1 m_2 r_{12}^2 + m_1 m_3 r_{13}^2 + m_2 m_3 r_{23}^2)]^{\frac{1}{2}} \tag{7}$$

$$\frac{1}{\mu} = \frac{1}{M^*}[\frac{m_1 m_2}{r_{12}} + \frac{m_1 m_3}{r_{13}} + \frac{m_2 m_3}{r_{23}}) \tag{8}$$

with $M^* = m_1 m_2 + m_1 m_3 + m_2 m_3$. Furthermore we denote the generalized semi-major axis a, the generalized semi-latus rectum p, the largest mutual distance R and the smallest mutual distance r

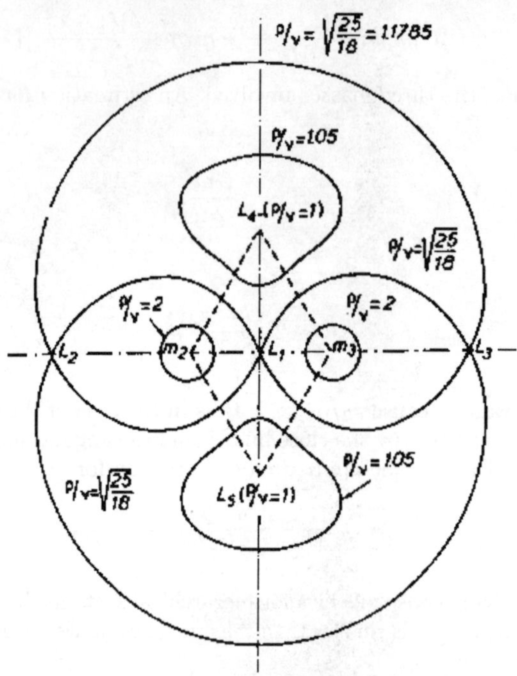

Figure 1. Generalized Hill's zero velocity curves for the three body problem with equal masses. L_n are the respective Lagrangian equilibrium points. The curves of constant ratio ρ/ν define the forbidden regions of motion for the body m_1.

$$a = -\frac{GM^*}{2h}, p = \frac{Mc^2}{GM^{*2}} \tag{9}$$

$$R = \sup r_{ij}, r = \inf r_{ij} \tag{10}$$

It turns out that the following inequality is always fulfilled

$$0 \leq r \leq \mu \leq \rho \leq R \tag{11}$$

The important question is whether the triangle r_{12}, r_{13}, r_{23} is compatible with the known integrals the energy H and the angular momentum c. The estimations led to the following inequality

$$\frac{\rho}{\nu} \leq \frac{\rho}{2a} + \frac{p}{2\rho} \tag{12}$$

Note that the ratio ρ/ν is independent of the scale of the triangle! In the plane defined by the three masses – which is in fact moving in the physical space – we can now draw a map of the ρ/μ= constant lines, where one should keep in mind that the unit of length is the – variable – distance between m_2 and m_3 which is r_{23}. In Fig.1 we show for the special case $m_1 = m_2 = m_3$ these curves. ρ/μ has the value 1 (mininimum) at the Lagrangian points L_4 and L_5 and has saddle points at the collinear Lagrangian points L_1, L_2 and L_3. It can be shown that the ρ/μ= constant lines are the Hill's zero velocity curves in the restricted three-body problem when $m_3 = 0$.

For a given value of ρ (or ν) we can draw such curves: when the energy is negative the generalized semi-major axis is positive and

$$\frac{\rho}{\nu} = \sqrt{\frac{p}{a}}. \tag{13}$$

Then the region of possible motion disconnects into three different ones: around m_2, around m_3 and at larger distances around m_1 and m_2. Thus in this configuration no exchange can occur and the triple systems stays in the hierarchical configuration of a close binary and a third mass far away. Because this stable three-body system has the property that it is scaled with the distance between m_2 and m_3 the absolute distances can become very large or very small. But this fact is not of major importance for a real system as the evolution is rather slowly.

3. Analytical approaches

The first results for the dynamical evolution of triple star systems with the aid of perturbation theory were derived by Harrington (1968) with the von Zeipel method. He could show that using the Delaunay elements in a special form there are no secular trends for the semimajor axes in the hierarchical systems under question. He found only small periodic variations of the inclination and the eccentricities in the observable systems. According to his results no disintegration is to be expected in real systems.

Analytical and also numerical studies were undertaken by Söderhelm (1982, 1984) where he suceeded to include also effects of the eccentricities and the inclinations up to the $3^r d$ order and even the tidal effects. In this paper there is also a classification of real system given which distinguishes 4 different types according to the ratio of their periods.

Using the Hamiltonian form of equations Docobo (1977), Docobo and Prieto (1988) suceeded in solving the problem of hierarchical triple stars up to the third order using a Lie - perturbation technique. In applying it to real start systems Docobo et al. (1992) derived the orbital elements, respectively their time development. There is also a software package available which may be applied to other systems. This work links the problem of theory and observation in a very efficient way.

It is also possible to model a hierarchical triple system by two two-dimensional coupled mappings (Hadjidemetriou and Dvorak, 1996). In this work the initial conditions were chosen along a family of periodic orbits; proceeding along this family the ratio of the dimensions decreases and consequently the coupling becomes stronger. The appearance of instabilities and chaotic motion was studied in a first approach for some orbits: it is evident that for strong coupling instabilities appear and escapes are possible. The study has until now been undertaken for some selected orbits but it is hoped that this approach may provide more complete results in the near future. The mapping methods itself are very effective and fast working: a dynamical system is here replaced by a map on a surface of section. Consequently the orbit in phase space has not to be computed with a small step size, but can be calculated from one surface of section to the other. It can be shown that

TABLE 1. critical values q_2/a_1 for different μ and different values of e_2 with $e_1 = 0$

e_2/μ	0.1	0.5	1.0	5.0	10.0
0.0	2.4	3.0	3.3	6.3	8.0
0.2	2.7	3.3	3.3	6.2	7.8
0.4	2.7	3.2	3.4	6.6	7.8
0.6	2.5	3.3	3.3	6.4	8.4
0.8	2.2	3.4	3.4	5.6	7.6

the so-called symplectic mappings and Hamiltonian systems – as the triple star system is one – are equivalentin their dynamical behaviour in what concerns the principle structure of phase space: periodic orbits, fixed points and chaotic regions. The great advantage is the speed of the method as it produces the results in times some 10^3 faster than a usual integration.

4. Numerical work

The question of stability of triple system and the verifaction using numerical integration techniques was brought up by Harrington (1970, 1972, 1975, 1977). In his series of papers he established with the aid of numerical experiments the following relation for orbitally stable hierarchical triple systems:

$$(\frac{q_2}{a_1}) \leq (\frac{q_2}{a_1})_o\{1 + 0.7\log[\frac{1 + m_3/(m_1 + m_2)}{1.5}]\} + K \qquad (14)$$

where K is 0 for equal masses. This inequality defines the minimum distance of the periastron of the third mass m_3 from the barycenter of the close binary. According to his first results the inclination is not an important parameter; this is not true as it will be shown later. The respective values for prograde and retrograde orbits were found to be 3.5 and 2.75.

Later it was pointed out by Donnison and Mikulkis (1992) that Harrington's stability criterion fits quite well only for the equal-mass case. These authors studied in detail the three-body stability for the circular orbits in the prograde (loc.cit.) and in the retrograde case (Donnison and Mikulskis, 1994). Here they concluded that retrograde orbits are more stable than prograde. Very recently Donnison and Mikulskis (1995) also have taken into account the eccentricities of the orbits and also applied their results to real triple star systems. Introducing the mass parameter $\mu = (m_2 + m_3)/2m_1$ they found after his numerical integrations the stability limits which are shown in table 1.

Other numerical experiments were undertaken by many different authors (cf. Huang and Valtonen (1987), Mikkola and Valtonen (1986)). Valtonen (1988) defined a minimum time of stability for a triple system; he was modelling a parabolic three body encounter and estimated the limiting pericenter distance q in units of the semi-major axis of the primaries which depends on the inclination, the pericenter, the ascending node of the outer orbit as well as the three masses involved. Via the exchange of the energy during the perihelion passage of the third body he was able to give stability boundaries for hierarchical triple systems over at least one pericenter passage. His primarely analytically derived expressions were also confirmed numerically by Huang and Innanen (1983). These q values depending on the inclination were found to be 2.5 ($i = 0^o$ and 20°), 2.375 ($i = 40^o$), 2.125 ($i = 80^o$), 1.25 ($i = 120^o$ and 140°) and 1.125 for retrograde orbits (in units of the semi-major axis of the close binary). It has to be added that longer term stability accumulates energy changes over several pericenter passages and thus to get a stability limit over more such passages should result in q values which are at least 50 percents higher. We then derive values which fit quite well to the results by Harrington.

In a series of three papers Kiseleva et al. (1994a, 1994b) and Eggleton and Kiseleva (1995) established stability borders via the quantity X, which is defined as the ratio of the period of the inner pair to the period of the outere third body. Using extensive numerical integrations where the orbits were integrated up to 10000 periods of the outer body they derived values depending on the mass ratio of the primaries

$$\alpha = log_{10}\frac{m_1}{m_2}, \beta = log_{10}\frac{m_1 + m_2}{m_3} \tag{15}$$

They expressed the initial ratio of the two periods of motion depending both eccentricities, e_{in} and e_{out}, the relative inclination of the two orbital planes, the initial phases and the mass ratio α and β

$$(X_o^{min})^{2/3} = (\frac{mr_{out}}{1 + mr_{out}})^{1/3}\frac{1 + e_{in}}{1 - e_{out}}Y_o^{min} \tag{16}$$

$$Y_o^{min} \sim 1 + \frac{3.7}{\sigma_{out}^{1/3}} + \frac{2.2}{1 + \sigma_{out}^{1/3}} + \frac{1.4}{\sigma_{in}^{1/3}}\frac{\sigma_{out}^{1/3} - 1}{\sigma_{out}^{1/3} + 1} \tag{17}$$

where in this notation we used σ_{in} as the ratio m_1/m_2 and σ_{out} as the mass ratio $(m_1 + m_2)/m_3$, Y_o^{min} is the critical initial ratio q_{out}/Q_{in} where q and Q are the respective periastron and apoastron distances. According to the approximation used in this paper equ. (17) is only dependant on the masses involved, whereas equ. (16) is interesting because it relates observable quantities like periods and eccentricities of the orbits. Although in the papers the dependance on the initial inclinations and phases is give, we concentrate here on the dependance on the eccentricities of the two orbits. The effect of the inclination is such that varying the initial inclinations from 0° to 180° (from prograde to retrograde orbits of the outer star) the orbits became more and more stable; this effect is already well known from other work (cf. Harrington, loc.cit.). Figs. 2 and 3 show the dependance of Y_o^{min}, respectively ΔY for $e_{in} = 0$ and $e_{out} = 0$ with varying the other eccentricity for different values of the mass ratios α and β.

The two figures show additionally the difference between calculated orbits in the "two circular approximation" with initially two orbits on circular orbits and the cases where one of the orbits was set initially to be on a circle whereas the other orbit was an eccentric one. Generally speaking an eccentric inner orbit tends to be more stable than the same system with circular inner and outer orbit or with an outer star in an eccentric orbit. The criteria established in this papers take into account many more properties as the stability criteria used up to now and – as it is shown in their tables in the respective paper (loc.cit) – fit quite well to the real situation for mass ratios usually observed in triple stars systems.

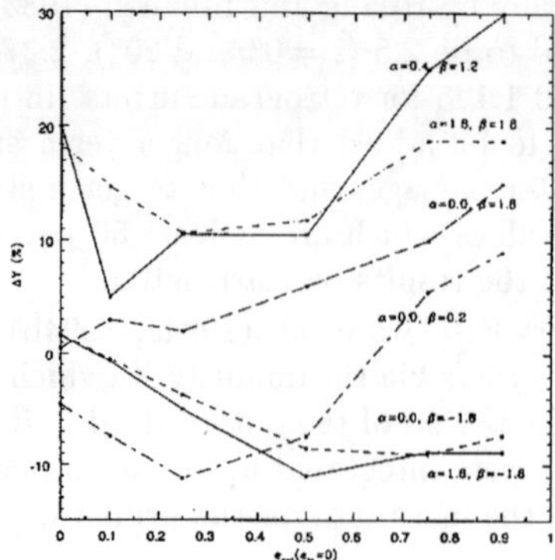

Figure 2. Dependance of the differences ΔY on the outer eccentricity with an initially circular inner orbit for different mass ratios α and β (after Eggleton and Kiselova, 1995)

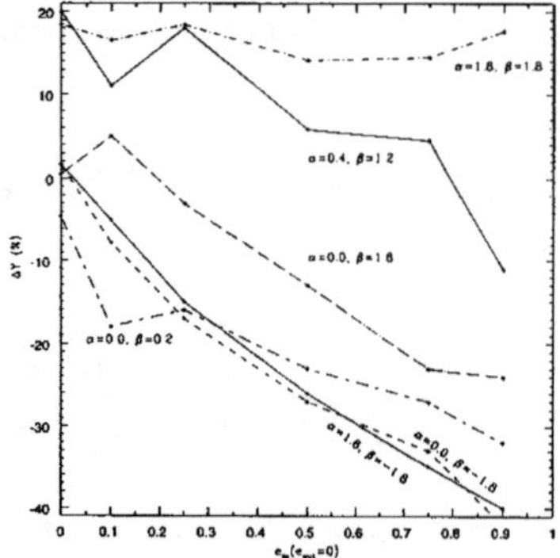

Figure 3. Dependance of the differences ΔY on the inner eccentricity with an initially circular outer orbit for different mass ratios α and β (after Eggleton and Kiselova, 1995)

5. Conclusions

We reported on different papers which are concerned with the stability of the hierarchical triple systems. The progress in this field is very encouraging because we have now many similar results derived with different methods. We know for these dynamical systems the limits of stable motion in the sense such that the system stays in the same configuration of a close binary and third companion. Step by step these stability limits were derived for more and more complex systems:

- all three bodies move in a plane and the inner and outer orbits are regraded to be circular
- the influence of the eccentricities is taken into account
- the full three-dimensional problem is treated
- the influence of the phases of the starting positions are checked
- also the dependance on the different mass ratios m_1/m_2 and m_2/m_3 was investigated

We can finally say that from the early very theoretical work by pure theoreticians through the discovery of the reality of such systems we are now – after 100 years of research in this interesting field which links theory and observation – at a point being able to give a conclusive answer concerning the influence of a third body on a close binary and the stability of the hierarchicle triple star systems.

References

Aarseth, S.J., Anasova, J.P., Orlov, V.V., Szebehely, V.G.: 1994, Close triple approaches and ecsape in the three-body problem, Cel.Mech. and Dynamic.Astron. 60, p. 131 - 137

Abt, Helmut, A.: 1986,: The Ages and Dimensions of Trapezian Systems, AJ 304, p.688-694

Batten, Alan, H.: 1973, Binary and multiple systems of stars, Pergamon Press, 277 p.

Docobo, J.A.: 1977, Aplicación de la teoría de Perturbaciones al estudio de sistema estelares triples, Tesis Doctoral, Universidad de Zaragoza

Docobo, J.A., Prieto, C.: 1988, The Stellar Three-body Problem. Explicit Formulation foir Computing Perturbations, Pub. Obs.Ast R.M. Aller 46

Docobo, J.A., Prieto, C., Ling, J.F.: 1992, Calculation of Perturbations in the Stellar Three-Body Problem in Mc Alister and Hartkopf (eds.) Complementary Approaches to Double and Multiple Star Research, IAU Colloquium 135, ASP Conference Serie, Vol. 32 p. 223 - 224

Donnison, J.R., Mikulskis, D.F.: 1992, Three-body orbital stability criteria for circular orbits, MNRAS 254, p.21 -26

Donnison, J.R., Mikulskis, D.F.: 1994, Three-body orbital stability criteria for circular retrograde orbits, MNRAS 266, p.25 - 30

Donnison, J.R., Mikulskis, D.F.: 1995, The effect of eccentricity on the three-body orbital stability criteria and its importance for triple star systems, MNRAS 272, p. 1 - 10

Eggleton, P. P., Kiseleva, L. G.: 1995, An empirical condition for Stability of Hierarchical triple systems, ApJ 455, p. 640 - 645

Ferrer, S., Osácar, C.: 1992, Harrington's Hamiltonian in the stellar problem of three bodies: Reductions, relative equilibria and bifurcations, Cel.Mech. and Dynamic.Astron. 58, p. 245 - 275

Hadjidemetriou, J., Dvorak, R.: 1996: The hierarchical three body problem as two two-dimensional coupled mappings, in J. Seiradakis (ed), 2nd Hellenic Astronomical meeting, p.571 - 577

Harrington, R.S.: 1968, AJ 73, p. 190 - 194

Harrington, R.S.: 1972, Stability criteria for triple stars, Celest.Mech. 6, p. 322 - 327

Harrington, R.S.: 1975a, Production of triple stars by the dynamical decay of small stellar systems, AJ 80, 12, p. 1081 - 1086

Harrington, R.S.: 1975b, Production of triple stars by the dynamical decay of smaller systems, AJ 80. p. 1081 - 1086

Heintz, W.D.: 1969, Journ. Royal. astr. Soc. Canada 63, p.275

Huang, T.Y., Valtonen, M.J. : 1987, An approximate solution to the energy change of a circular binary i a parabolic three body encounter, MNRAS 1987, p. 333 - 344

Kiseleva, L. G.,Eggleton, P. P., Anasova, J.P.: 1994, A note on the stability of hierarchical triple stars with initially orbits, MNRAS 267, p. 161 - 166

Kiseleva, L. G.,Eggleton, P. P., Orlov, V.V.: 1994, Instability of close triple systems with coplanar initial doubly circular motion, MNRAS 270, p. 036 - 946

Marchal, Ch.,: 1990, The three-body problem, Studies in Astronautics Vol 5, Elsevier, 576 p.

Marchal, Ch., Sari, D.G.: 1975, Hill regions for the general three-body problem, Celest.Mech. 12, p. 115 - 129

Marchal, Ch., Bozis, G.: 1982, Hill stability and distance curves for the general three-body problem, Celest.Mech. 26, p. 311 - 333

Marchal, C., Yoshida, J., Sun, Y.S.: 1984, A test of escape valid even for very small mutual distance I. The acceleration and the escape velocities of the third body. Celest.Mech. 33, p. 193 - 207

Mikkola, S., Valtonen, M.: 1986 MNRAS, 223, p.269

Saari, D.G.: 1974, SIAM, J. Appl. Math, 26 p. 806 - 815

Söderhjelm, S.:1982, Studies of the Stellar Three-body Problem, Astraon.Astrophys. 107, 54 - 60

Söderhjelm, S.:1984, Third - order and tidal effects in the stellar three-body problem, Astron.Astrophys. 141, 232

Szebehely, V., Zare, K.: 1977, Stability of Classical Triplets and their hierarchy, Astron. Astrophys. 58, p. 145 - 152

Szebehely, V.: 1970, Theory of Orbits, Academic Press

Szebehely, V.G.: 1971, Mass effects in the problem of three bodies, Celest.Mech. 6, p. 84 - 107

Valtonen, M. J.: 1988, The general thre- body problem, Vistas in Astronomy Vol. 32, p.23 - 48
Worley, C.E.: 1967, On the Evolution of Double Stars (J. Dommanget, ed.), Commun.Obs.roy, Belgique, Ser. B, no. 17
Wallenquist, A.: 1944, On the apparent distribution and properties of triple and multiple stars, Ann. Uppsala Obs. 1, part 5, 33 p.

NUMERICAL TREATMENT OF SMALL STELLAR SYSTEMS WITH BINARIES

SEPPO MIKKOLA
Tuorla Observatory, University of Turku, 21500 Piikkiö, Finland
E-mail: seppo.mikkola@utu.fi

Abstract. The use of regularization methods based on the Kustaanheimo-Stiefel transformation (KS) is reviewed. The history of the development of such methods is summarized and anecdotal information about the related events is told. Details of the multi-particle regularization methods are given, including the most recent developments.

Key words: few-body systems, numerical methods, regularization

1. INTRODUCTION

In the dynamics of small stellar systems close encounters of two or more stars are common and important leading frequently to formation of binaries (van Albada 1968, Aarseth 1971, 1974, 1977, 1988, Aarseth & Heggie 1976). This was first discovered by von Hoerner (1960, 1963) who attempted to integrate numerically the evolution of few-body systems using the Newtonian form of equations of motion. Loss of precision due to strong interactions prevented detailed numerical studies of such systems until the publication of the two-body regularization method by Kustaanheimo and Stiefel (1965). This was subsequently adopted by other investigators as a tool for treating close encounters. One of the first applications to few-body problem was the accurate solution of the so-called Pythagorean (also known as the Burrau's problem) three-body problem computed by Szebehely & Peters (1967). Also, the KS transformation was used in many-body systems for treating dominant two-body interactions (*e.g.* Aarseth 1972). For a number of years after its invention the KS transformation was not widely known and understood and it only became popular after the publication of the book by Stiefel & Sheifele (1971) which discusses in depth the KS transformation and related analytical and numerical topics.

J. A. Docobo et al. (eds.), Visual Double Stars: Formation, Dynamics and Evolutionary Tracks, 269–288.

Multiparticle regularization techniques were subsequently developed by many authors (Aarseth and Zare 1974, Zare 1974, Heggie 1974, Mikkola 1985, Mikkola and Aarseth 1990, 1993, 1996). The KS treatment has proved itself as probably the best regularization method in practical computations, although others have been published (Lemaitre 1955, Waldvogel 1972). Today there exist KS based computer codes which are capable of handling the most critical dynamical events in stellar systems.

Formation of persistent binaries is a problematic phenomenon in numerical studies of stellar dynamics. Most of the computational effort is wasted in integrating the motion of minutely perturbed binaries. Recently (Mikkola and Aarseth 1996), however, a code has been developed to overcome this problem by means of a method of artificial slow-down of the binaries. This technique loses the phase of the binary (not statistically important because of the chaos), while it retains the secular perturbations correctly and allows a much faster numerical integration.

2. STEPS TOWARDS RELIABLE FEW-BODY SIMULATION

The major stepping stones and events in the development of the numerical tools for the investigation of few-body dynamics, especially strong interactions, are listed below. The list is based entirely on author's personal opinion.

0. **von Hoerner (1960, 1963)** discovered that the formation of binaries in few-body systems is a rule rather than exception. However, von Hoerner was unable to study the process in detail due to computational difficulties: accurate numerical integration of the motion of (eccentric) binaries was too difficult with the numerical tools available at the time.
1. **Kustaanheimo and Stiefel (1965)** published their regularization method (generalization of the celebrated 2-dimensional Levi-Civita transformation), for the (perturbed) two-body motion, which made possible accurate calculations of close encounters and binary motions in clusters.
2. **Aarseth and Zare (1974)** published a three-particle regularization method. The two shortest distances are regularized with the Kustaanheimo-Stiefel (KS) method and the arrangement is appropriately updated when necessary. Also **Zare (1974)** generalized this method by regularizing an arbitrary number of bodies with respect to one dominating body.
3. **Heggie (1974)** realized that it is possible to use an extended phase-space and regularize all the interparticle vectors in the system simultaneously, thus achieving a global regularization.
4. **Mikkola (1985)** was the first to write a concise regularized algorithm for N bodies utilizing Heggie's global method.

5. **Mikkola and Aarseth (1990, 1993)** developed what is known as the **chain method**. This method generalizes the idea of Aarseth & Zare. Forming a chain of interparticle vectors and applying the KS-transformation to each of these, the authors obtained a practical regularized N-body method without extra degrees of freedom.
6. **Mikkola and Aarseth (1996)** further upgraded their method by introducing the idea of **slowing down** tight binaries. This method preserves the secular perturbations, but loses the phase of the binary. Due to the chaos in many-body systems, the phase of a binary should not be statistically important, however.
7. This latest development of **chain & slowdown** appears to be one of the best possibilities to write a computer code for simulations of few-body stellar dynamics (Aarseth 1996).

3. ANECDOTAL

This section is an unusual one in a supposedly scientific paper. However, due to the review-like character of this article I feel it may be fascinating for the reader to know some of the human factors and the role of pure chance in the development and advancement of a scientific idea. With this in mind, I hope that the readers will excuse the somewhat informal style of this section.

The basic transformation (KS) was invented by Paul Kustaanheimo during a conference in 1964. Here the details are given as Jörg Waldvogel (1996, private communication) told them.

> The possibility of generalizing Levi-Civita's (1920) regularization of the Kepler motion to three dimensions was discovered jointly by P. Kustaanheimo and E. Stiefel at a conference on celestial mechanics, organized by Stiefel at the Mathematisches Forschungsinstitut Oberwolfach, Germany, from March 15 to 21, 1964, see the proceedings quoted in Kustaanheimo (1964).
>
> In his talk on spinor notation in the theory of Kepler motion, Kustaanheimo (1964) formally used spinors in Levi-Civita's regularizing transformation and thus laid the grounds for the subsequent discovery of three-dimensional regularization. Stiefel, himself an expert on regularization, immediately recognized the power of these ideas, and in vivid discussions throughout the entire meeting the two scientists worked out the new regularization, later to be referred to as the KS transformation; see Stiefel's own account appended to his obituary in Waldvogel, Kirchgraber, Schwarz & Henrici (1979, p.142).
>
> On the occasion of Kustaanheimo's subsequent stay at Stiefel's institute in Zürich the now classical joint paper (Kustaanheimo and Stiefel 1965) was written, in which the entire theory of three-dimensional regulariz-

> ation is developed, and applications to computing perturbed two-body orbits are described. This development later culminated in a comprehensive book (Stiefel and Scheifele 1971). It may be added that the KS transformation had been known before as the Hopf map, albeit in another context: it was discovered by the topologist Heinz Hopf (1931), as a map from the 3-sphere to the 2-sphere.

This procedure was then adopted by many workers as a tool for treatment of close encounters and binaries in stellar systems (e.g. Szebehely & Peters 1967, Aarseth 1972, Szebehely & Bettis 1972). This was one of the most important steps towards reliable numerical simulations of star clusters.

The next major advancement was made by Aarseth & Zare (1974) (hereafter AZ), Zare (1974), and Heggie (1974). In the words of Sverre Aarseth himself (as he wrote to the author):

> Going back to AZ, the story began when I had to give some lectures at Austin, Texas in autumn 1972. In fact, Szebehely was away on sabbatical so they were able to pay me a small amount. There were 5-6 students at these lectures on basic N-body techniques and also KS since I did have an N-body code with KS already for the 1970 IAU meeting in Cambridge. One of the students was Khalil Zare who was in fact in his last year as an undergraduate! He had been looking at Szebehely's (1967) book concerning the extended phase space. There are quite a few pages (pp. 327-340) on this theme. Being extremely well educated about Hamiltonian dynamics, he was manipulating some equations relating to so-called reduction of the order. He came and showed me his work and we discussed what to do; finally this led to equations of motions without singularities for the two distances. The code was probably written by me who was more experienced in such matters. However, for the integration we used the powerful Runge-Kutta (7/8 order) method of Fehlberg which a staff member at Austin (Dale Bettis) knew well. So we did some test calculations and it was clear it was an excellent method. A little later Zare went on to generalize the formulation to include an arbitrary number of particles that can have regular motions with respect to one reference body. I think Alexander (1986) tried it; nobody else has to my knowledge, but one could envisage a massive black hole application.
>
> The next development occurred in February 1973 on my return to Cambridge. I showed the equations to Douglas (Heggie), who resided in Trinity College (Newton's very own!). Next day (!) he came back with the global formulation for N = 3. Then (or it could have been later) I told him: 'you have regularized the N-body problem globally!' And so it was. But Douglas was only able to write a three-body code.
>
> As you can see from Celes. Mech., Victor Szebehely decided to write a small part introducing the three papers. We certainly did not ask him to

> do this. At some later stage, RK78 was replaced by BS (Bulirsch & Stoer 1966); probably under your influence. By now the method has been used by quite a few people.

In 1981 I (the author) was working with a four-body code and Mauri Valtonen suggested the use of Heggie's global regularization. This looked very complicated and it took considerable time to find a convenient way to use it. However, I believed that 'everybody' had it and thus I had to do it too. At that time I was going to visit Sverre Aarseth in Cambridge and somehow I had (mis)understood that Valtonen had told Aarseth that I had programmed Heggie's global method. This was the most compelling reason why it was necessary to do it! In fact the global regularization code worked first time late in the evening when my flight to Cambridge departed early the next morning! The method was then published much later (Mikkola 1985), largely because Sverre Aarseth urged me to do so!

There is another extraordinary incident related to this method. This occurred in 1987 the day before Good Friday in Princeton. At that time I was a visiting researcher in York University, Toronto, Canada. I had taken a couple of weeks holiday and was cruising around in the USA. On the mentioned day I decided to visit IAS in Princeton, and I had no idea that Sverre Aarseth was there too. Let Sverre tell the story:

> My objective was to formulate a treatment of variable compact clumps (small membership), using Heggie's global method with external perturbations. This formulation was supposed to be included in a tree code being developed by Piet Hut and Steve McMillan (forming the 'root' of the 'tree' as I remember calling it). Douglas was also visiting and I still have his handwritten notes on 'Regularised equations for the perturbed N-body problem' (published later in Heggie (1988)). It states that the notes follow your MNRAS paper (Mikkola 1985). A problem arose in the expressions for the derivatives, which were complicated, and Douglas himself was not able to suggest the relevant terms. So I was stuck on this technical point since I also did not figure it out. I said that there is only one person in the whole world who can resolve this! Next day (the day before Good Friday), you walked in at IAS as a tourist and you had to promise to come back the following day to fix the problem, which you did! I have a listing of this code, which worked fine formally. However, to restrict the order of integration to be analogous to the tree code (leapfrog), many steps were needed. For me, the most important new idea was to treat a clump of variable membership (up to N=4 here), and to consider an arbitrary number of clumps at the same time (like I do KS now, but still using only one chain). So this was in fact an ambitious attempt at introducing automatic decision-making which was useful practice for later developments. For some reason, this code was never implemented in

the tree code. There is a page describing it in my Turku paper (Aarseth, 1988). I often mention this story as an example of having great luck, which has been so important in my life.

The chain method was developed in 1989 when I (the author) went to Cambridge for a two months visit to work with Sverre Aarseth. After doing something else for a week or two I got the idea of a chain (for treatment of four-body subsystems in an N-body code). This was just a generalization of AZ, in a direction other than that of Zare's method for one dominant reference body. The original reaction to my suggestion from Sverre was *'maybe you can write a 2-page note'*; he did not think it to be practical! However, I insisted to write the code and, after the code was running and was observed to be twice as fast as the global method for four-body systems, Aarseth also became interested. We then wrote a full length paper about the chain method concentrating on four-body systems (Mikkola & Aarseth 1990).

The next episode occurred in 1992, when we decided to write a general N-body version of the chain code, including external perturbations. It was deemed to be important for treating strongly interacting subsystems with possibly changing number and membership of bodies. The result was the paper Mikkola & Aarseth (1993).

Finally the invention of the slow-down originally happened in 1990 for an isolated binary, and during my 1995 visit to Cambridge we managed to formulate the Hamiltonian for a system with a slowed-down binary (or many of them) inside a chain. Thus far this (Mikkola & Aarseth 1996) is the latest development in the venture towards a fast reliable few-body code using the KS-regularization.

In the light of all that has been said above it is clear how important the role of Sverre Aarseth has been in the advancement of the regularization methods. He has both participated in the research and constantly encouraged others to work hard towards new developments.

4. KS TRANSFORMATION

The KS-transformations between the 3-dimensional position and momentum **R** and **W** and the corresponding 4-dimensional KS-variables **Q** and **P** may be written

$$\mathbf{R} = \hat{\mathbf{Q}}\mathbf{Q}; \;\; \mathbf{W} = \hat{\mathbf{Q}}\mathbf{P}/(2Q^2) \tag{1}$$

Here $\hat{\mathbf{Q}}$ is the KS-matrix (e.g. Stiefel and Scheifele (1972) p. 24)

$$\hat{\mathbf{Q}} = \begin{pmatrix} Q_1 & -Q_2 & -Q_3 & Q_4 \\ Q_2 & Q_1 & -Q_4 & -Q_3 \\ Q_3 & Q_4 & Q_1 & Q_2 \\ Q_4 & -Q_3 & Q_2 & -Q_1 \end{pmatrix}. \tag{2}$$

With $\mathbf{R} = (X, Y, Z)^t$; $\quad R = |\mathbf{R}|$ we calculate

$$\begin{aligned} u_1 &= \sqrt{\tfrac{1}{2}(R + |X|)} \\ u_2 &= Y/(2u_1) \\ u_3 &= Z/(2u_1) \\ u_4 &= 0 \end{aligned} \tag{3}$$

and the components of **Q** are

$$\mathbf{Q} = \begin{cases} (u_1, u_2, u_3, u_4)^t & ; \quad X \geq 0 \\ (u_2, u_1, u_4, u_3)^t & ; \quad X < 0 \end{cases}. \tag{4}$$

The KS momenta are given by

$$\mathbf{P} = 2\hat{\mathbf{Q}}^t\mathbf{W}. \tag{5}$$

With the introduction of the time transformation

$$\frac{dt}{ds} = R = \mathbf{Q}^2 \tag{6}$$

and the Poincare transform of the Hamiltonian

$$\Gamma = R(H - E), \tag{7}$$

where $H = \frac{1}{2}W^2 - K^2/R$ is the two-body Hamiltonian and $E = H(0)$ is the numerical value of the energy, one obtains in the $\mathbf{P}$, $\mathbf{Q}$, s system

$$\Gamma = \frac{1}{8}\mathbf{P}^2 - K^2 - E\mathbf{Q}^2, \tag{8}$$

i.e. a harmonic oscillator.

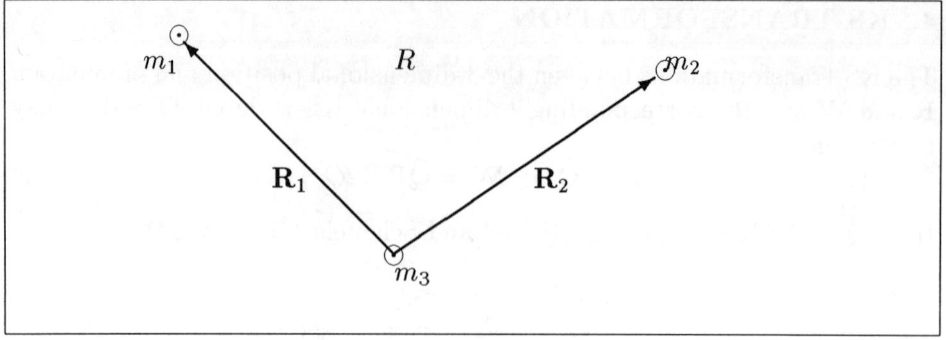

Figure 1. Illustration of the AZ-method. The largest distance $R = |\mathbf{R}_2 - \mathbf{R}_1|$ is chosen to be the non-regularized 'singular' distance.

5. METHOD OF AARSETH AND ZARE

Leaving aside most of the theoretical details we just give the main points of the method. The notation adopted here differs from that in the original paper.

Taking the vectors $\mathbf{R}_1$ and $\mathbf{R}_2$ (see the figure) as new canonical coordinates the generating function takes the form

$$S = \mathbf{W}_1 \cdot \mathbf{R}_1 + \mathbf{W}_2 \cdot \mathbf{R}_2 = \mathbf{W}_1 \cdot (\mathbf{r}_1 - \mathbf{r}_3) + \mathbf{W}_2 \cdot (\mathbf{r}_2 - \mathbf{r}_3) \tag{9}$$

and thus the physical momenta are $\mathbf{p}_1 = \partial S/\partial \mathbf{r}_1 = \mathbf{W}_1$, $\mathbf{p}_2 = \partial S/\partial \mathbf{r}_2 = \mathbf{W}_2$ and $\mathbf{p}_3 = \partial S/\partial \mathbf{r}_3 = -\mathbf{W}_1 - \mathbf{W}_2$.

Application of the KS-transformation to both of the vectors $\mathbf{R}_1$ and $\mathbf{R}_2$ and the momenta $\mathbf{W}_1$ and $\mathbf{W}_2$, accompanied with the time transformation $dt = R_1 R_2 ds = Q_1^2 Q_2^2\, ds$, gives the Hamiltonian in the form

$$\begin{aligned} \Gamma \;=\; & \frac{1}{8\mu_{13}} Q_2^2 \mathbf{P}_1^2 + \frac{1}{8\mu_{23}} Q_1^2 \mathbf{P}_2^2 + \frac{1}{8m_3} \mathbf{P}_1^T \widehat{\mathbf{Q}}_1^T \widehat{\mathbf{Q}}_2 \mathbf{P}_2 \\ & - m_1 m_3 Q_2^2 - m_2 m_3 Q_1^2 - m_1 m_2 Q_1^2 {Q^2}_2/R - E Q_1^2 Q_2^2, \end{aligned} \tag{10}$$

where $\mu_{k3} = m_k m_3/(m_k + m_3)$ and E is the numerical value of the total energy calculated from the initial values. The equations of motion $\mathbf{P}_k' = -\partial\Gamma/\partial \mathbf{Q}_k$, $\mathbf{Q}_k' = \partial \Gamma/\partial \mathbf{P}_k$ derived from this are regular with respect to collisions between the body m_3 and any one of the two bodies m_1 and m_2 (prime denotes differentiation with respect to the new independent variable s). A switching of the reference body is carried out whenever the singular distance R becomes the smallest. This method has proved itself in numerous simulations and is one of the best methods for treating critical three-body encounters. The authors also give additional equations to account for external perturbations. In that case the three-body energy E is no longer constant, but it must be obtained by integration.

6. HEGGIE'S GLOBAL METHOD

For the sake of simplicity we present here only the version of Heggie's theory which is valid in the centre-of-mass coordinate system.

Starting with the Hamiltonian

$$H = \sum_{i=1}^{N} \frac{1}{2m_i} w_i^2 - \sum_{i=1}^{N-1} \sum_{j=i+1}^{N} \frac{m_i m_j}{r_{ij}} \tag{11}$$

Heggie (1974) introduces all the relative vectors $\mathbf{r}_{ij} = \mathbf{r}_j - \mathbf{r}_i$ as new dependent variables by using the generating function

$$S = \sum_{i=1}^{N-1} \sum_{j=i+1}^{N} \mathbf{w}_{ij} \cdot (\mathbf{r}_j - \mathbf{r}_i), \tag{12}$$

where $\mathbf{w}_{ij}$ are the new momenta conjugate to the new coordinates $\mathbf{r}_{ij}$. A proof for the correctness of such an increase of degrees of freedom is also given in Heggie's (1974) article.

After the introduction of the above transformation the old momenta are

$$\mathbf{w}_i = \sum_{j=i+1}^{N} \mathbf{w}_{ij} - \sum_{j=1}^{i-1} \mathbf{w}_{ji} \tag{13}$$

and the new Hamiltonian in the largely extended phase space takes the form

$$\begin{aligned} H(\mathbf{w}_{ij}, \mathbf{r}_{ij}) &= \sum_{i=1}^{N-1} \sum_{j=i+1}^{N} \frac{1}{2\mu_{ij}} \mathbf{w}_{ij}^2 \\ &+ \sum_{i=1}^{N} \frac{1}{m_i} \Big(\sum_{j=i+1}^{N-1} \sum_{k=j+1}^{N} \mathbf{w}_{ij} \cdot \mathbf{w}_{ik} + \sum_{j=1}^{i-2} \sum_{k=j+1}^{i-1} \mathbf{w}_{ji} \cdot \mathbf{w}_{ki} \\ &- \sum_{j=i+1}^{N} \sum_{k=1}^{i-1} \mathbf{w}_{ij} \cdot \mathbf{w}_{ki} \Big) - \sum_{i=1}^{N-1} \sum_{j=i+1}^{N} \frac{m_i m_j}{r_{ij}}. \end{aligned} \tag{14}$$

Application of the KS transformation to each conjugate pair $(\mathbf{w}_{ij},\ \mathbf{r}_{ij})$ together with the time transformation $dt = \prod_{i=1}^{N-1} \prod_{j=i+1}^{N} r_{ij}\, ds$ completes the construction of the globally regular Hamiltonian

$$\begin{aligned} \Gamma &= [H(\mathbf{w}_{ij}, \mathbf{r}_{ij}) - E] \prod_{i=1}^{N-1} \prod_{j=i+1}^{N} r_{ij} \\ &= [H(\frac{\widehat{\mathbf{Q}}_{ij} \mathbf{P}_{ij}}{2Q_{ij}^2}, \widehat{\mathbf{Q}}_{ij} \mathbf{Q}_{ij}) - E] \prod_{i=1}^{N-1} \prod_{j=i+1}^{N} Q_{ij}^2. \end{aligned} \tag{15}$$

The resulting expressions for equations of motion are, however, rather complex and the method was not used for more than three bodies until a more concise formulation was provided by Mikkola (1985).

7. MIKKOLA'S ALGORITHM FOR THE GLOBAL METHOD

Mikkola (1985) published a concise formulation for the computation of the equations of motion for Heggie's global method. The two simple, but essential steps in the reformulation are:

(I) Rename the variables ($\mathbf{w}_{ij}$, $\mathbf{r}_{ij}$) to ($\mathbf{p}_k$, $\mathbf{q}_k$), where $k = (i-1)N - i(i+1)/2 + j$. Also write the kinetic energy $T = \sum_{\mu,\nu=1}^{K} T_{\mu\nu}\mathbf{p}_\mu \cdot \mathbf{p}_\nu$ and the potential energy $U = \sum_{k=1}^{K} \frac{M_k}{q_k}$, where $K = N(N-1)/2$ is the number of $\mathbf{p}$'s and $\mathbf{q}$'s. One easily sees that the constant coefficients $T_{\mu\nu}$ are combinations of inverse masses and M_k's are products of masses.

(II) Instead of the distance product $\prod_{i<j} r_{ij}$ the inverse of the Lagrangian $1/(T+U)$ is adopted as a time transformation function. Originally this selection was based on simplicity of formulation, but it has turned out later that the Lagrangian also improves the numerical stability of the method. [This time transformation was in fact suggested by Szebehely and Zare (1975), but not in connection with KS regularization.]

With the above notations one can write

$$\Gamma = (T - U - E)/(T + U) \tag{16}$$

where

$$T = \sum_{\mu\nu} T_{\mu\nu}\mathbf{P}_\mu^t \widehat{\mathbf{Q}}_\mu^t \widehat{\mathbf{Q}}_\nu \mathbf{P}_\nu/(4Q_\mu^2 Q_\nu^2), \quad U = \sum_k M_k/Q_k^2 \tag{17}$$

and the generation of the equations of motion looks now much more tractable than with the double index formulation. In fact a simple collection of formulae can be readily given:

(i) Calculate for $1 \le i < j \le N$ initial values and auxiliary quantities

$$\begin{aligned}
k &= (i-1)N - i(i+1)/2 + j \\
\mathbf{q}_k &= \mathbf{r}_i - \mathbf{r}_j \\
\mathbf{p}_k &= (m_i\dot{\mathbf{r}}_i - m_j\dot{\mathbf{r}}_j)/N \\
M_k &= m_i m_j \\
a_{ik} &= 1, \quad a_{jk} = -1 \ \text{[other components} = 0] \\
T_{\mu\nu} &= \tfrac{1}{2}\textstyle\sum_{n=1}^{N} a_{n\mu}a_{n\nu}/m_n
\end{aligned} \tag{18}$$

Calculate the initial values for the KS variables $\mathbf{P}_k$ and $\mathbf{Q}_k$ according to the usual formulae. After this the bulk of the numerical effort, the integration of the equations of motion can be carried out using the formulation below. [partial derivatives expressed as subscripts and $\widehat{\mathbf{P}}$ is the analog of $\widehat{\mathbf{Q}}$]

(ii) Given the KS variables $\mathbf{P}_k$ and $\mathbf{Q}_k$ the derivatives $\mathbf{P}'_k$ and $\mathbf{Q}'_k$ are evaluated as:

$$\begin{aligned}
\mathbf{p}_k &= \widehat{\mathbf{Q}}_k \mathbf{P}_k/(2Q_k^2) \\
\mathbf{A}_k &= \textstyle\sum_{\mu=1}^{K} T_{k\mu}\mathbf{p}_\mu \\
d_k &= A_k \cdot \mathbf{p}_k \\
T &= \textstyle\sum_{k=1}^{K} d_k, \quad U = \sum_{k=1}^{K} M_k/Q_k^2 \\
T_{\mathbf{P}_k} &= \widehat{\mathbf{Q}}_k^t \mathbf{A}_k/Q_k^2 \\
T_{\mathbf{Q}_k} &= (\widehat{\mathbf{P}}_k^t \mathbf{A}_k^* - 4d_k\mathbf{Q}_k)/Q_k^2 \\
\mathbf{A}_k^* &= (A_1, A_2, A_3, -A_4)_k^t, \\
U_{\mathbf{Q}_k} &= -2M_k\mathbf{Q}_k/Q_k^4 \\
L &= T+U; \quad H = T-U; \quad \Gamma = (H-E)/L \\
\Gamma_T &= (1-\Gamma)/L; \quad \Gamma_U = -(1+\Gamma)/L.
\end{aligned} \tag{19}$$

The derivatives for the KS vectors $\mathbf{P}$ and $\mathbf{Q}$ and the physical time t are now given by

$$\begin{aligned}
\mathbf{Q}'_k &= \Gamma_T T_{\mathbf{P}_k} \\
\mathbf{P}'_k &= -\Gamma_T T_{\mathbf{Q}_k} - \Gamma_U U_{\mathbf{Q}_k} \\
t' &= 1/L.
\end{aligned} \tag{20}$$

(iii) Finally the transformation from the KS variables to the physical ones are obtained as follows:

For $1 \le i < j \le N,\ k = k(i,j)$ as before, one computes

$$\begin{aligned}
\mathbf{r}_{ij} &= \mathbf{q}_k = \widehat{\mathbf{Q}}_k \mathbf{Q}_k \\
\mathbf{w}_{ij} &= \mathbf{p}_k = \widehat{\mathbf{Q}}_k \mathbf{P}_k/(2Q_k^2) \\
\mathbf{r}_i &= \left(\textstyle\sum_{j=i+1}^{N} m_j \mathbf{r}_{ij} - \sum_{j=1}^{i-1} m_j \mathbf{r}_{ji}\right) / \sum_{n=1}^{N} m_n \\
\dot{\mathbf{r}}_i &= \left(\textstyle\sum_{j=i+1}^{N} \mathbf{w}_{ij} - \sum_{j=1}^{i-1} \mathbf{w}_{ji}\right) / m_i.
\end{aligned} \tag{21}$$

8. THE CHAIN METHOD

8.1. BASIC EQUATIONS

Consider a perturbed N-body system with inertial system coordinates $\mathbf{r}_i$, velocities $\mathbf{v}_i$ and masses m_i $(i = 1, .., N)$. Let $M = \sum m_i$, be the total mass, $\mathbf{r}_0 = \sum m_i \mathbf{r}_i / M$ the centre-of-mass and $\mathbf{v}_0 = \sum m_i \mathbf{v}_i / M$ its velocity. In terms of the cm-coordinates, $\mathbf{q}_i = \mathbf{r}_i - \mathbf{r}_0$, and the corresponding momenta, $\mathbf{p}_i = m_i(\mathbf{v}_i - \mathbf{v}_0)$, $\mathbf{p}_0 = M\mathbf{v}_0$, the (internal N-body) kinetic energy T and the force function U are defined by

$$T = \sum_{i=1}^{N} \frac{1}{2m_i} \mathbf{p}_i^2, \tag{22}$$

$$U = \sum_{1 \le i < j \le N} \frac{m_i m_j}{|\mathbf{q}_i - \mathbf{q}_j|}, \tag{23}$$

while the total Hamiltonian is

$$H_{tot} = T - U + \frac{1}{2M} \mathbf{p}_0^2 - \widetilde{U}, \tag{24}$$

where $\widetilde{U}$ is a perturbing force function. To discuss the regularization of the internal N-body dynamics of the system we drop the centre-of-mass term and the perturbing function. These can be included as discussed later (section 8.3).

Suppose a chain of vectors connecting N bodies has been selected. After relabelling the bodies such that they are labelled 1, 2, .., N along the chain, we may use the generating function

$$S = \sum_{i=1}^{N-1} \mathbf{W}_k \cdot (\mathbf{q}_{k+1} - \mathbf{q}_k) \tag{25}$$

to obtain the old momenta $\mathbf{p}_k = \partial S / \partial \mathbf{q}_k = \mathbf{W}_{k-1} - \mathbf{W}_k;\ k = 2, .., N-1$ $(\mathbf{p}_1 = -\mathbf{W}_1,\ \mathbf{p}_N = \mathbf{W}_{N-1})$ in terms of the new ones. The corresponding chain vectors are defined by $\mathbf{R}_k = \mathbf{q}_{k+1} - \mathbf{q}_k$ and the Hamiltonian may be written

$$\begin{aligned} H = & \sum_{k=1}^{N-1} \frac{1}{2} \left(\frac{1}{m_k} + \frac{1}{m_{k+1}} \right) \mathbf{W}_k^2 - \sum_{k=2}^{N} \frac{1}{m_k} \mathbf{W}_{k-1} \cdot \mathbf{W}_k \\ & - \sum_{k=1}^{N-1} \frac{m_k m_{k+1}}{R_k} - \sum_{1 \le i \le j-2} \frac{m_i m_j}{R_{ij}}. \end{aligned} \tag{26}$$

The non-chained distances can be written in terms of the chain vectors $R_{ij} = |\mathbf{r}_j - \mathbf{r}_i| = |\mathbf{q}_j - \mathbf{q}_i| = |\sum_{i \le k' \le j-1} \mathbf{R}_{k'}|$. Substitution of the KS transformations $\mathbf{R}_k = \widehat{\mathbf{Q}}_k \mathbf{Q}_k$, $\mathbf{W}_k = \widehat{\mathbf{Q}}_k \mathbf{P}_k / (2Q_k^2)$ gives the Hamiltonian in terms of the regularizing variables $\mathbf{Q}_k$, $\mathbf{P}_k$. After applying the time transformation

$$dt = g\, ds, \quad \text{with} \quad g = 1/(T + U), \tag{27}$$

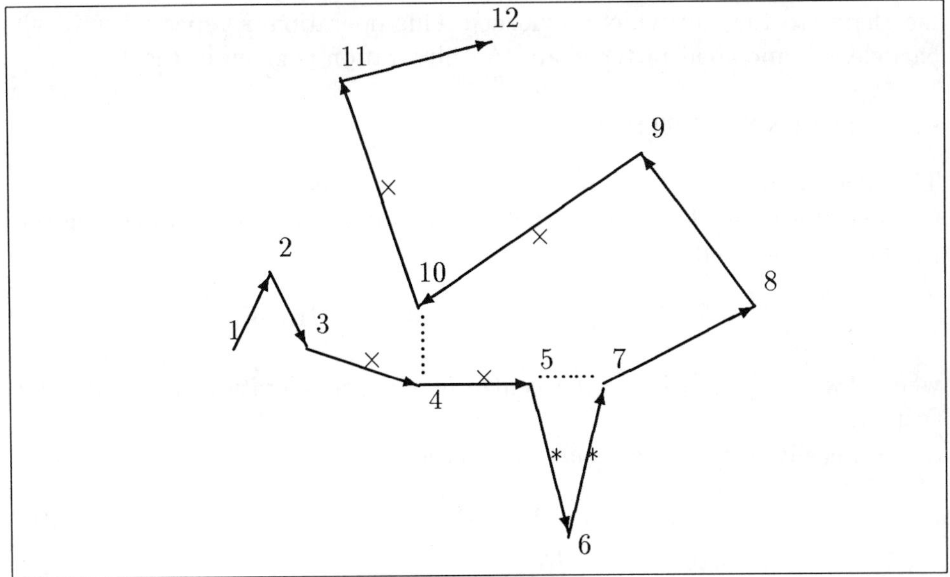

Figure 2. Illustration of the chain and the checking of switching conditions. Distances like $R_{5,7}$ are compared with the smaller of the two distances $R_{5,6}$ and $R_{6,7}$ (marked by *). Interparticle distances like $R_{4,10}$ are compared with the smallest of those in contact with the considered distance (marked by $\times$).

we obtain the regularized Hamiltonian

$$\Gamma = g(H - E) = (T - U - E)/(T + U) \tag{28}$$

in the $(\mathbf{P}, \mathbf{Q}, s)$-system.

The derivation of the equations of motion $\mathbf{P}_k{}' = -\partial\Gamma/\partial\mathbf{Q}_k$, $\mathbf{Q}_k{}' = \partial\Gamma/\partial\mathbf{P}_k$, is now possible when one notes that

$$\frac{\partial}{\partial\mathbf{Q}} = 2\widehat{\mathbf{Q}}^t \frac{\partial}{\partial\mathbf{R}}, \tag{29}$$

which is needed in the formation of the partial derivatives of the non-chained part of the potential. Since the formal structure of the Hamiltonian is essentially the same as in (**??**) the collection of equations resemble largely those of the global method and are omitted here.

8.2. CHAIN SELECTION AND UPDATING

Chain selection begins by finding the shortest interparticle vector. This is taken as the first part of the chain under construction. One proceeds by searching for the particle which is closest to one or the other end of the presently known part of the chain. Once identified, this particle is added to the chain (to the end which is closer). This operation is repeated until all particles are included in the chain. An illustration is given in Fig.1.

8.3. EXTERNAL PERTURBATIONS

The effect of any external force can be formally included by considering (for the sake of argument) the force to be a function of time only and adding the perturbing potential

$$\widetilde{U} = \textstyle\sum_{j=1}^{N} m_j \mathbf{q}_j \cdot (\mathbf{F}_j(t) - \mathbf{F}_0(t)) + M\mathbf{r}_0 \cdot \mathbf{F}_0(t), \tag{30}$$

where $\mathbf{F}_0 = \sum_{j=1}^{N} \frac{m_j}{M} \mathbf{F}_j$ is the centre of mass acceleration in the inertial system.

One easily derives the recursive equations

$$\begin{aligned} \delta\dot{\mathbf{p}}_j &= m_j(\mathbf{F}_j - \mathbf{F}_0);\;\; j = 1,..,N, & (31)\\ \delta\dot{\mathbf{W}}_1 &= -\delta\dot{\mathbf{p}}_1, & (32)\\ \delta\dot{\mathbf{W}}_k &= \delta\dot{\mathbf{W}}_{k-1} - \delta\dot{\mathbf{p}}_k;\;\; k = 2,..,N-2, & (33)\\ \delta\dot{\mathbf{W}}_{N-1} &= \delta\dot{\mathbf{p}}_N, & (34) \end{aligned}$$

where δ denotes the part of the derivative which is due to the perturbation. The corresponding corrections to the derivatives of the KS-momenta are then given by

$$\delta\mathbf{P}_k{}' = 2g\widehat{\mathbf{Q}}_k^t \delta\dot{\mathbf{W}}_k \tag{35}$$

and the rate of change of the internal energy may be written

$$E' = 2\textstyle\sum_{k=1}^{N-1} \mathbf{Q}_k'^t \widehat{\mathbf{Q}}_k^t \delta\dot{\mathbf{W}}_k. \tag{36}$$

For integration of the center-of-mass motion one may use

$$\mathbf{v}_0{}' = g\mathbf{F}_0; \qquad \mathbf{r}_0{}' = g\mathbf{v}_0. \tag{37}$$

9. SLOWDOWN OF TIGHT BINARIES

9.1. GENERAL PRINCIPLE

Let

$$\ddot{\mathbf{r}} = -K^2\mathbf{r}/r^3 + \mathbf{F}, \tag{38}$$

be the equation for the relative motion of the components of a binary. Here $\mathbf{F}$ is a small perturbing acceleration. To solve this numerically for a long time (over very many binary periods), the method of variation of constants has been traditionally adopted. However, in practice this does not give the advantage expected because rapid fluctuations, with a period of the binary motion, normally persist and the integration stepsize can only be extended significantly when the perturbation becomes quite negligible.

A new idea is to slow down the internal (the nearly unperturbed two-body) motion of the binary. Heuristically it is easy to understand that this can be done as long as the binary is only weakly perturbed, and the external force field does not change too fast. Then we get the short-period perturbations wrong, but the secular and long-period terms should be correct to first order at least. The slow-down can be done without referring to element equations, by writing in Cartesian coordinates

$$\begin{aligned} \dot{\mathbf{v}} &= -\kappa^{-1}K^2\mathbf{r}/r^3 + \mathbf{F}, \\ \dot{\mathbf{r}} &= \kappa^{-1}\mathbf{v}, \end{aligned} \tag{39}$$

where κ is the slow-down coefficient. In this new system the slowed-down period is now κ-fold compared to the original one. Thus the numerical integration is speeded up accordingly.

In Hamiltonian formulation we have originally

$$H = \frac{1}{2}p^2 - K^2/r - U, \tag{40}$$

where U is the perturbing force function. The modified formulation may be described by the new Hamiltonian

$$H = \kappa^{-1}(\frac{1}{2}p^2 - K^2/r) - U, \tag{41}$$

which gives the above slowed-down equations of motion. This principle can also be applied more generally in an N-body system, and even in such a complicated situation as we have in the chain regularization.

First separate from the total Hamiltonian H those terms which give the internal interaction of the particular pair of particles forming the isolated weakly perturbed binary (denoted H_B). Second multiply the part H_B by the slow-down factor κ^{-1} to get

$$H_{new} = \kappa^{-1}H_B + (H - H_B) \tag{42}$$

and form the resulting equations of motion as usual in Hamiltonian dynamics. The subsequent numerical solution of these equations is then normally much faster than in the original system. Since the numerical value of the total energy appears in the regularized equations we must also take into account that H_{new} is not a constant, but its value E changes. Considering κ to be a function of time, one may write $\dot{E} = \partial\kappa^{-1}/\partial t\ H_B$. When κ is changed stepwise, its time derivative is a delta function and

$\dot{E} = \delta(t - t_c)(1/\kappa^{new} - 1/\kappa^{old})\, H_B$, where t_c is the moment of change of κ. This can be immediately integrated to get the change of energy

$$\delta E = (1/\kappa^{new} - 1/\kappa^{old})\, (H_B)_{|t=t_c}. \tag{43}$$

The slowed-down part (H_B) must not necessarily represent a Newtonian binary, it may in general be any quasi-periodic part which is only weakly coupled with the rest of the system. Also, it is not necessary that the force field is conservative and derivable from a Hamiltonian.

9.2. APPLICATION TO CHAIN REGULARIZATION

In terms of the chain variables $\mathbf{R}_k$ and $\mathbf{W}_k$ the Hamiltonian can be written in the concise form

$$H = \sum T_{ij} \mathbf{W}_i \cdot \mathbf{W}_j - \sum M_k/R_k - \sum M_{ij}/R_{ij}, \tag{44}$$

where the auxiliary quantities affected by the slow-down procedure are

$$\begin{aligned} T_{kk} &= \frac{1}{2}(1/m_k + 1/m_{k+1}), \\ T_{k\,k+1} &= -1/m_k \qquad k = 1, .., N-1, \\ M_k &= m_k m_{k+1}. \end{aligned} \tag{45}$$

To construct the expression for H_B within the chain we first note that the relevant term in the potential energy is M_b/R_b if b is the index of the distance between the components (m_b, m_{b+1}) of the binary. It is clear that we must select from the kinetic energy those terms having $1/m_b$ or $1/m_{b+1}$ as a factor. The resulting combination of terms, however, also contains the centre-of-mass kinetic energy of the binary. Therefore we must subtract this contribution and add it to the rest of the Hamiltonian.

The centre-of-mass kinetic energy for a pair (m_b, m_{b+1}) is given by $T_{cm} = \frac{1}{2}(\mathbf{W}_{b+1} - \mathbf{W}_{b-1})^2/(m_b + m_{b+1})$. Selecting the relevant terms from the total Hamiltonian, and expanding the above expression for T_{cm} gives the following new formulation for the evaluation of the matrix T_{ij} and the vector M_k in the presence of a slowed-down binary. Instead of the previous expressions we have

$$\begin{aligned} T_{k\,k+1} &= -1/m_k/\kappa_k, \qquad k = 1, 3, 5, .., \leq N \\ T_{k+1\;\,k+2} &= -1/m_{k+1}/\kappa_k, \\ T_{b-1\;\,b+1} &= -(1 - \kappa_b^{-1})/(m_b + m_{b+1}), \\ \delta T_{b-1\;\,b-1} &= \delta T_{b+1\;\,b+1} = -\frac{1}{2} T_{b-1\;\,b+1}, \end{aligned} \tag{46}$$

$$
\begin{aligned}
T_{kk} &= \frac{1}{2}(-T_{k\ k+1} - T_{k+1\ k+2}) + \delta T_{kk}, \\
M_k &= m_k m_{k+1}/\kappa_k \\
\delta E &= -\frac{m_b m_{b+1}}{2a}(1/\kappa_b^{new} - 1/\kappa_b^{old}),
\end{aligned}
$$

where a slow-down coefficient κ_k has been defined for every chain vector, but taken to be $= 1$, except for $k = b$, and a denotes the semi-major axis of the binary.

After expressing the Hamiltonian in terms of the modified T_{ij} and M_k and the application of the KS transformations one gets the equations of motion in the usual way from the new Hamiltonian Γ_{new} by

$$\dot{\mathbf{P}} = -\partial\Gamma_{new}/\partial\mathbf{Q}\,; \qquad \dot{\mathbf{Q}} = \partial\Gamma_{new}/\partial\mathbf{P}. \tag{47}$$

and it is obvious that the algorithm is almost the same as for the original chain method. The only difference is that the auxiliary constants have modified values and there are a few more non-zero components in the matrix T_{ij}.

For decision-making we use an (over)estimate for the relative perturbation due to tidal forces (at the maximum apocentre $2a$) obtained by

$$\gamma = 8\frac{a^3}{m_B}\sum_j m_j/r_{bj}^3, \tag{48}$$

where the summation extends over all other bodies except for the components of the binary, and the distances are computed from one or the other of the binary components since this is satisfactory for decision-making. The expression

$$\kappa = \max(1, (\gamma_0/\gamma)^{\frac{1}{2}}), \tag{49}$$

has been adopted for κ. For the theoretical motivation of this see Mikkola & Aarseth (1996). A boundary value for the relative perturbation of $\gamma_0 \sim 10^{-5}$ may be used, but this depends on the desired accuracy.

10. COMPUTATIONAL ASPECTS

This section discusses practical aspects which are of importance for the implementation of regularization methods.

10.1. ERROR BEHAVIOUR

If the system is integrated in the physical space using $\dot{\mathbf{w}} = -\partial H/\partial\mathbf{r}$; $\dot{\mathbf{r}} = \partial H/\partial\mathbf{w}$, these equations tend to keep the value of H constant. Thus after the occurrence of an error the new value remains constant within the accuracy of the subsequent integration. The regularized equations $\mathbf{P}' = -\partial\Gamma/\partial\mathbf{Q}; \quad \mathbf{Q}' = \partial\Gamma/\partial\mathbf{P}$, tend to keep the quantity Γ constant. On the solution path we have $\Gamma = 0$ theoretically, but errors are unavoidable and this relation does not hold exactly. The physical energy has then the value

$$H = E + \Gamma/g \tag{50}$$

instead of $H = E$. Thus H can be in error by a large amount if g is very small. However, if g increases the system approaches again the correct energy-hypersurface. Specifically if $g = \prod R_k$, and the system disrupts, the error in the physical energy decreases with the increasing distances. However, if a very close multiple encounter occurs then, according to (??), the distance product choice for g forces a large error in H implying unphysical trajectories. The outcome of the interaction may be significantly affected by this, even if the energy finally returns to its correct value. The Lagrangian time transformation $g = 1/L$ does not have these undesirable properties and is thus a recommendable option.

10.2. STEPSIZE CONTROL AND TIME TRANSFORMATIONS

In the actual numerical integration we use a version of the Bulirsch-Stoer (1966) method. This method provides estimates of its local error. It is customary to use a relative error check [$|\delta Y/Y| \leq \epsilon$ for every dependent variable Y] for stepsize control. There are, however, situations in which it is not possible to obtain a given relative accuracy. In such a case the integration often stops by shrinking the stepsize to zero. To avoid this one may use an energy error check instead. The error in the value of the regularized Hamiltonian is

$$\delta\Gamma = \sum(\mathbf{Q}'_k \cdot \delta\mathbf{P}_k - \mathbf{P}'_k \cdot \delta\mathbf{Q}_k). \tag{51}$$

In practice one requires that every individual term in the above expression is, in absolute value, less than the specified tolerance.

As discussed in Mikkola & Aarseth (1990), it is convenient to require that the error in the physical Hamiltonian satisfies

$$|\delta H/L| < \epsilon. \tag{52}$$

This is actually the condition imposed when one uses the Lagrangian time transformation $g = 1/L$ and checks the error in Γ as discussed above. If other time transformations are used then it is necessary to apply an appropriate scaling in the error check (Mikkola & Aarseth 1993).

11. DISCUSSION AND CONCLUSIONS

The KS regularization has proved itself as the most reliable and efficient method available today for treating strong two- or many-body interactions in N-body simulations. The global method of Heggie is, perhaps, the easiest to program and also produces most accurate results during the critical interactions. However, due to the many extra degrees of freedom it also has its problems, especially if the distance ratios in the system grow large (Mikkola and Aarseth 1990). In addition, the efficiency is reduced because the number of equations to be integrated increases rapidly with the number of bodies.

Despite the occasionally slightly lower accuracy, the chain method is a powerful and efficient tool for studying multiple encounters at high precision. However, any such method must inevitably become time-consuming upon the formation of a binary with semi-major axis much smaller than the typical interparticle distance, since the common time-step cannot exceed the Kepler period. Such binaries therefore require special treatment, which is now available in the form of the slow-down procedure, even for the chain method.

In regularization theory it used to be standard practice to employ a time transformation involving the singular distance products. However, such a formulation may exhibit undesirably behaviour as discussed above. Therefore the Lagrangian function, which has the desired property of removing the relevant singularities, is recommended. This simple procedure has proved itself in numerous formulations (*cf.* Mikkola 1983, 1985, Alexander 1986, Mikkola & Aarseth 1990, 1993) and applications.

A computer code, based on the chain-regularization method and containing procedures for the slow-down treatment of weakly perturbed binaries, is now available on request from the author.

Acknowledgments

I am indebted to Sverre Aarseth and Jörg Waldvogel for the information on the 'old times', which they kindly wrote to me on my request.

References

Aarseth, S. J.: 1971, "Binary Evolution in Stellar Systems", *Astrophys. Space Sci.* **13**, 324.

Aarseth, S. J.: 1972, "Direct Integration Methods of the N-body Problem", in *Gravitational N-body Problem*, proceedings of IAU colloquium No. 10, ed. M. Lecar, Reidel, Dordrecht, pp.373–387.

Aarseth, S. J.: 1974, "Dynamical Evolution of Simulated Star Clusterrs. I. Isolated Models", *Astron. Astrophys.*, **35**, 237.

Aarseth, S. J.: 1977, *Rev. Mex. Astron. Astrofis.*, **3**, 199.

Aarseth, S. J.: 1988, "Integration Methods for Small N-Body Systems",p. 287–306, in *The Few Body Problem*, (Valtonen M. J., ed.), Kluwer, Dordrecht, Holland.

Aarseth, S. J. 1996, "Star cluster simulations on HARP",p. 161–170, in *Dynamical Evolution of Star Clusters*, ed. Hut, P. and Makino, J., Kluwer.

Aarseth, S. J. and Zare, K.: 1974,"A Regularization of the Three-Body Problem", *Cel. Mech.*, **10**, 185–205.

Aarseth, S. J. and Heggie, D. C.: 1976, "The Probability of Binary Formation by Three-Body Encounters", *Astron. Astrophys.*, **53**, 259–265.

van Albada, T. S.: 1968, "The evolution of small stellar systems and its implications for the formation of double stars", *Bull. astr. Inst. Neth.*, **20**, 57.

Alexander, M. E.: 1986, "Simulations of binary-single star and binary-binary scattering", *J. Comp. Phys.*, **64**, 195–219.

Bulirsch, R. and Stoer, J.: 1966, "Numerical Treatment of Differential Equations by Extrapolation Methods", *Num. Math.*, **8**, 1–13.

Heggie, D. C.: 1974, "A Global Regularisation of the Gravitational N-Body Problem", *Cel. Mech.*, **10**, 217–241.

Heggie, D. C.: 1988, "The N-Body Problem in Stellar Dynamics", in *Long-Term Dynamical Behaviour of Natural and Artificial N-Body Systems*, (Roy A. E., ed.), 329–347, Kluwer, Dordrecht, Holland.

von Hoerner, S.: 1960, *Z. Astrophys.* **50**, 184.

von Hoerner, S.: 1963, *Z. Astrophys.* **57** , 47.

Hopf, H. 1931, "Uber die Abbildung der dreidimensionalen Sphäre auf die Kugelfläche", *Math. Ann.* **104**. Reprinted in: Selecta Heinz Hopf, 38-63, Springer 1964.

Kustaanheimo, P.: 1964, "Die Spinordarstellung der energetischen Identitäten der Keplerbewegung", p. 333-340. In: E. Stiefel (ed.): *Mathematische Methoden der Himmelsmechanik und Astronautik*, Mathematisches Forschungsinstitut Oberwolfach, Berichte 1. Bibliographisches Institut Mannheim, 1966, 350 pp.

Kustaanheimo, P. and Stiefel, E.: 1965, "Perturbation theory of Kepler motion based on spinor regularization", *J. Reine Angew. Math.*, **218**, 204–219.

Levi-Civita, T.: 1920, "Sur la régularisation du problème des trois corps", *Acta Math.* **42**, 99–144.

Mikkola, S.: 1983, "Encounters of Binaries - I. Equal Energies", *Mon. Not. R. Astr. Soc*, **203**, 1107–1121.

Mikkola, S.: 1985, "A Practical and Regular Formulation of the N-Body Equations", *Mon. Not. R. Astr. Soc*, **215**, 171–177.

Mikkola, S. and Aarseth, S. J.: 1990, "A Chain Regularization Method for the Few-Body Problem", *Cel. Mech.*, **47**, 375-390.

Mikkola, S. and Aarseth, S. J.: 1993, "An Implementation of N-body Chain Regularization", *Celest. Mech. Dyn. Astron.* 57, 439

Mikkola, S. and Aarseth, S. J.: 1996, "A Slow-Down Treatment of Close Binaries", *Celest. Mech. Dyn. Astron.* (to appear).

Stiefel, E. L. and Scheifele, G.: 1971, *Linear and Regular Celestial Mechanics*, Springer, Berlin.

Szebehely, V.: 1967, *Theory of Orbits*, Academic Press, New York.

Szebehely, V. and Bettis, D. G. 1972, in *Gravitational N-body Problem*, proceedings of IAU colloquium No. 10, ed. M. Lecar, Reidel, Dordrecht, pp. 136–147.

Szebehely, V. and Peters, C. F.: 1967, "Complete solution of a general problem of three bodies", *Astron. J.*, **72**, 876.

Szebehely, V. and Zare, K: 1975, "Time transformations in the extended phase-space", *Cel. Mech.*, **11**, 469.

Waldvogel, J.: 1972, "A new regularization of the planar problem of three bodies", *Cel. Mech.* , **6**, 221.

Waldvogel, J., Kirchgraber, U. Schwarz, H. R, Henrici, P.: 1979, "In Memoriam E. Stiefel." *ZAMP* **30**, 133-142.

Zare, K.: 1974, "A Regularization of Multiple Encounters in Gravitational N-body Problems" *Cel. Mech.*, **10**, 207–215.

PERIODIC ORBITS IN THE RESTRICTED THREE BODY PROBLEM WITH RADIATION PRESSURE

A. ELIPE
Grupo de Mecánica Espacial.
Universidad de Zaragoza. 50009 Zaragoza. Spain

AND

M. LARA
Real Observatorio de la Armada. 11110 San Fernando. Spain

Abstract. This paper deals with the Restricted Three Body Problem (RTBP) in which we assume that the primaries are radiation sources and the influence of the radiation pressure on the gravitational forces is considered; in particular, we are interested in finding families of periodic orbits under theses forces.

By means of some modifications to the method of numerical continuation of natural families of periodic orbits, we find several families of periodic orbits, both in two and three dimensions. As starters for our method we use some known periodic orbits in the classical RTBP.

1. Introduction

The restricted three-body problem is one of the most widely studied in celestial mechanics. Its applications span the solar system dynamics, the lunar theory, the motion of spacecrafts, stellar dynamics, etc.. This problem concerns with the motion of a particle of *infinitesimal* mass which is attracted by two primaries which are moving in a non–perturbed Keplerian orbit each around the other.

Nevertheless, in some occasions, the model of three point mass is not sufficient to describe the dynamics of the problem; some additional hypotheses must be introduced and their effects taken into account. In some cases, it is necessary to consider the primaries as rigid bodies and the non sphericity of them gives rise to some differences with respect to the results in the classical RTBP (see for instance the works of Bhatnagar and collaborators [2, 3, 15, 16] and [6, 7, 8]). In other cases, it is considered the so called magnetic-binary problem, that is the motion of a charged particle in the field of two rotating dipoles [9, 13].

Other studies are focussed on the stellar problem, taking the primaries as radiation sources, and the influence of the radiation pressure on the effective gravitational potential is investigated [10, 14, 17]. This is precisely the problem here considered: the motion of a particle in the field of two luminous massive bodies. As it was pointed out by [17], this model may be applied to investigate the accumulation of matter around binaries,

J. A. Docobo et al. (eds.), Visual Double Stars: Formation, Dynamics and Evolutionary Tracks, 289-297.

particularly where the stars are of luminous late type. Even more, new directions are open with the recent discovery of binary pulsars [18, 19, 12].

In this paper, we center our attention in computing families of periodic orbits. In particular, we choose the masses of both primaries to be equal, and only one radiating primary; the reason for choosing both masses equal, is that this case is quite common in binary stars, and besides, this case has been extensively studied for the classic RTBP (see for instance [1] and references therein). The method employed is the numerical continuation with respect to one parameter, namely the radiation pressure coefficient. The algorithm used is the one described in [11], based on the one proposed by Deprit and Henrard [5] that computes the intrinsic variational equations. For computing the families of three dimensional periodic orbits, we use an extension of the one of Deprit and Henrard that will be published elsewhere.

2. Equations of the motion

Let us consider two stars O_1 and O_2 with masses m_1 and m_2 that moves one around the other under a mutual force that is proportional to a function of the mutual distance and in the direction of the line joining them. The force will be specified later on. We will assume that these two stars —called primaries— move on circular orbits around their mutual center of mass O with a constant angular velocity n. Besides, let us consider a point mass O_3, whose mass m_3 is *infinitesimal*, that is to say, it does not affect the motion of the two stars but it is attracted by both primaries.

The potential function acting on O_3 may be written as

$$\mathcal{V} = \mathcal{V}^{13} + \mathcal{V}^{13} = -f\, m_1 m_3 G_1 - f\, m_2 m_3 G_2,$$

with f the gravitational constant and G_i functions depending on the distance $r_i = \|\boldsymbol{x}_i\| = \|\overline{O_i O_3}\|$, that is, $G_i = G_i(r_i)$; these functions will be specified later on.

One of the authors [8] introduced a generic notation for handling this general problem. This is the one that we follow here. For details, see [8]. By introducing the functions

$$g_i = \frac{\partial G_i}{\partial(1/r_i)}, \tag{1}$$

the gradient with respect to an inertial frame is

$$\boldsymbol{\nabla}_{\boldsymbol{X}} G_i = \frac{\partial G_i}{\partial(1/r_i)} \boldsymbol{\nabla}_{\boldsymbol{X}}(1/r_i) = -g_i \frac{\boldsymbol{r}_i}{r_i^3}$$

and hence, the equations of the motion are

$$\ddot{\boldsymbol{r}} = -f\, m_1 g_1 \frac{\boldsymbol{r}_1}{r_1^3} - f\, m_2 g_2 \frac{\boldsymbol{r}_2}{r_2^3}. \tag{2}$$

Let us consider now a synodic frame $Oxyz$, such that the Ox–axis is in the direction of one of the primaries OO_2, the Oz–axis in the normal direction to the orbital plane of the primaries and the Oy–axis the complement to the right-oriented frame. In this reference system, the coordinates of the primaries O_1 and O_2 are $(x_1, 0, 0)$ and $(x_2, 0, 0)$ respectively. Let (x, y, z) be the coordinates of the particle O_3; by virtue of the moving frame theorem, the equations of motion become

$$\begin{aligned}
\ddot{x} - 2n\dot{y} &= n^2 x - f\, m_1 g_1 \frac{x - x_1}{r_1^3} - f\, m_2 g_2 \frac{x - x_2}{r_2^3}, \\
\ddot{y} + 2n\dot{x} &= n^2 y - f\, m_1 g_1 \frac{y}{r_1^3} - f\, m_2 g_2 \frac{y}{r_2^3}, \\
\ddot{z} &= -f\, m_1 g_1 \frac{z}{r_1^3} - f\, m_2 g_2 \frac{z}{r_2^3}.
\end{aligned} \tag{3}$$

As it is usual in the classical RTBP, the units are chosen in such a way that the distance between the primaries, the mean motion of the Keplerian orbit of the primaries and the Gauss constant ($f(m_1 + m_2)$) are equal to one. With these units, and putting $fm_2 = \mu$, equations (3) are

$$\begin{aligned}
\ddot{x} - 2\dot{y} &= x - (1+\mu) g_1 \frac{x + \mu}{r_1^3} - \mu g_2 \frac{x - 1 + \mu}{r_2^3}, \\
\ddot{y} + 2\dot{x} &= y - (1+\mu) g_1 \frac{y}{r_1^3} - \mu g_2 \frac{y}{r_2^3}, \\
\ddot{z} &= -(1+\mu) g_1 \frac{z}{r_1^3} - \mu g_2 \frac{z}{r_2^3},
\end{aligned} \tag{4}$$

with $r_1^2 = (x+\mu)^2 + y^2 + z^2$ and $r_2^2 = (x - 1 + \mu)^2 + y^2 + z^2$.

There is an integral of these equations, the Jacobian constant

$$C = (\dot{x}^2 + \dot{y}^2 + \dot{z}^2) - (x^2 + y^2) - 2(1-\mu)G_1 - 2\mu G_2.$$

The general formulation obtained by the functions g_i (1) has the advantage that the same formula is valid for different types of forces. Thus, the classical case (primaries mass points with only gravitational potential) is recovered for $g_i = 1$. The influence of the non sphericity of some of the primaries (rigid body) is obtained for $g_i = 1 + J_i/r_i^2$. The case we are dealing with in this note —primaries considered mass points taking into consideration the radiation pressure— is formulated by putting $g_i = (1 - \beta_i)$, where β_i is the ratio of the radiation pressure force to the gravitational force [14].

3. Natural families of planar periodic orbits

Let us consider the two degrees of freedom system

$$\begin{aligned}
\ddot{x} &= 2A(x, y; \sigma)\, \dot{y} + W_x(x, y; \sigma), \\
\ddot{y} &= -2A(x, y; \sigma)\, \dot{x} + W_y(x, y; \sigma),
\end{aligned} \tag{5}$$

where A and W are two functions depending on the coordinates and on one parameter σ. This system has the Jacobian integral

$$C = 2W - (\dot{x}^2 + \dot{y}^2).$$

Each manifold $C = -2h$ is determined by the initial conditions. By putting $U = W(x, y; \sigma) + h$, the system (5) together with the Jacobian integral can be written as

$$\begin{aligned}
\ddot{x} &= 2A(x, y; \sigma)\dot{y} + U_x(x, y; \sigma), \\
\ddot{y} &= -2A(x, y; \sigma)\dot{x} + U_y(x, y; \sigma), \\
\mathcal{I} &= \tfrac{1}{2}(\dot{x}^2 + \dot{y}^2) - U(x, y; \sigma) \equiv 0.
\end{aligned} \tag{6}$$

Let us assume that $\boldsymbol{\xi} = \boldsymbol{\xi}(t, \sigma_0)$ —solution of the system (6) for the initial conditions $\boldsymbol{\xi}_0 = (x_0, y_0, \dot{x}_0, \dot{y}_0)$— is periodic with period T_0. Hence,

$$\boldsymbol{\xi}(t; \sigma_0) = \boldsymbol{\xi}(t + T_0; \sigma_0).$$

Let us choose now a variation of the parameter $\sigma = \sigma_0 + \Delta\sigma$. The question now is how to find a new set of initial conditions $\boldsymbol{\xi}_0 + \delta\boldsymbol{\xi}_0$ such that the solution of (6) for these initial conditions be periodic with period $T_0 + \Delta T$, that is,

$$\boldsymbol{\xi}(t; \sigma_0 + \Delta\sigma) = \boldsymbol{\xi}(t + T; \sigma_0 + \Delta\sigma). \tag{7}$$

The Poincaré method of continuity ensures that, under certain conditions, there exist initial conditions

$$\boldsymbol{\xi}_0 + \sum_{k\geq 1} \boldsymbol{\xi}_{0,k} \, (\Delta\sigma)^k \tag{8}$$

such that the solution of the ordinary differential equations (6) is periodic with period

$$T = T_0 + \sum_{k\geq 1} T_{0,k} \, (\Delta\sigma)^k.$$

Under the impossibility of carrying out the infinity terms of the above series, and assuming $\Delta\sigma$ small enough, we consider only the first order approximation for the initial conditions

$$\boldsymbol{\xi}_0 + \Delta\sigma\delta\boldsymbol{\xi}_0; \tag{9}$$

the corrections $\delta\boldsymbol{\xi}_0$ are solutions of the variational system

$$\begin{aligned}
\delta\ddot{x} &= 2A\delta\dot{y} + (U_{xx} + 2A_x\dot{y})\delta x + (U_{xy} + 2A_y\dot{y})\delta y + U_{x\sigma} + 2A_\sigma\dot{y}, \\
\delta\ddot{y} &= -2A\delta\dot{x} + (U_{xy} - 2A_x\dot{x})\delta x + (U_{yy} - 2A_y\dot{x})\delta y + U_{y\sigma} - 2A_\sigma\dot{x}, \\
\delta\mathcal{I} &= \dot{x}\delta\dot{x} + \dot{y}\delta\dot{y} - U_x\delta x - U_y\delta y - U_\sigma \equiv 0.
\end{aligned} \tag{10}$$

Since the system (10) is linear, the general solution is formed by a linear combination of particular solutions that depends on arbitrary integration constants. The desired variations will be determined by computing those values of the integration constants that fulfil the periodicity condition (7). Notice from (7) that the unknown ΔT is an implicit variable; we overcome this inconvenience by expanding (7) up to the first order.

The Cartesian formulation of the variational equations (10) mix secular terms produced by displacements along the orbit with periodic terms. An alternative formulation can be given by using intrinsic coordinates: the tangent p and the normal n displacements obtained from the rotation

$$\begin{aligned}
\delta x &= p\cos\phi - n\sin\phi, \quad \delta\dot{x} = \dot{p}\cos\phi - \dot{n}\sin\phi - \dot{\phi}\,\delta y, \\
\delta y &= p\sin\phi + n\cos\phi, \quad \delta\dot{y} = \dot{p}\sin\phi + \dot{n}\cos\phi + \dot{\phi}\,\delta y.
\end{aligned} \tag{11}$$

When using intrinsic coordinates, the variational differential equations result to be separable [5]: the normal displacements are obtained from the generalized Hill equation

$$\ddot{n} + \Theta n = -2U_\sigma \frac{A + \dot{\phi}}{V} - 2A_\sigma V + U_{y\sigma}\cos\phi - U_{x\sigma}\sin\phi, \tag{12}$$

where

$$V^2 = \dot{x}^2 + \dot{y}^2, \qquad \tan\phi = \dot{y}/\dot{x}, \tag{13}$$

and

$$\Theta = \frac{\ddot{V}}{V} + 2(A + \dot{\phi})^2 + 2A^2 - U_{xx} - U_{yy} - 2V(A_x \sin\phi - A_y \cos\phi). \tag{14}$$

The next step is the computation of the tangent displacement p by integrating the quadrature

$$\frac{d}{dt}\left(\frac{p}{V}\right) = 2\frac{A + \dot{\phi}}{V} n + \frac{U_\sigma}{V^2}. \tag{15}$$

The corrections computed from (9) are tangent approximations, hence the algorithm must consist of two steps: the tangent predictor followed by an isoenergetic corrector (see [5]).

Let us apply the method above exposed to the problem described in section 2. To have planar orbits, we need to make zero the third equation of the system (4). The case for which we compute the families of periodic orbits is the one in which both primaries have equal masses ($\mu = 1/2$), and only one radiating primary (O_2); hence $g_1 = 1$, $g_2 = (1 - \beta)$. The parameter to vary is $\sigma \equiv \beta$. The corresponding variational equations are obtained from the Table 1 (note that in our case, $A = 1$).

TABLE 1. Partial derivatives of the effective potential function U.

$U_x = x - (1-\mu)\frac{x+\mu}{r_1^3} - (1-\beta)\mu\frac{x+\mu-1}{r_2^3}$
$U_y = y\left[1 - \frac{1-\mu}{r_1^3} - (1-\beta)\frac{\mu}{r_2^3}\right], \quad U_z = -z\left[\frac{1-\mu}{r_1^3} + (1-\beta)\frac{\mu}{r_2^3}\right]$
$U_\beta = -\frac{\mu}{r_2}, \quad U_{\beta,x} = \frac{\mu}{r_2^3}(x+\mu-1), \quad U_{\beta,y} = \frac{\mu}{r_2^3}y, \quad U_{\beta,z} = \frac{\mu}{r_2^3}z$
$U_{x,x} = 1 + \frac{1-\mu}{r_1^5}[3(x+\mu)^2 - r_1^2] + (1-\beta)\frac{\mu}{r_2^5}[3(x+\mu-1)^2 - r_2^2]$
$U_{x,y} = U_{y,x} = 3y\left[\frac{1-\mu}{r_1^5}(x+\mu) + (1-\beta)\frac{\mu}{r_2^5}(x+\mu-1)\right]$
$U_{x,z} = U_{z,x} = 3z\left[\frac{1-\mu}{r_1^5}(x+\mu) + (1-\beta)\frac{\mu}{r_2^5}(x+\mu-1)\right]$
$U_{y,y} = 1 + \frac{1-\mu}{r_1^5}[3y^2 - r_1^2] + (1-\beta)\frac{\mu}{r_2^5}[3y^2 - r_2^2]$
$U_{y,z} = U_{z,y} = 3yz\left[\frac{1-\mu}{r_1^5} + (1-\beta)\frac{\mu}{r_2^5}\right], \quad U_{z,z} = \frac{1-\mu}{r_1^5}[3z^2 - r_1^2] + (1-\beta)\frac{\mu}{r_2^5}[3z^2 - r_2^2]$

To begin with, we need an initial periodic orbit. This is chosen from the classical RTBP ($\beta = 0$); with this starter we compute the family for small variations of β. In Figures 1 and 2 we present several periodic orbits corresponding to two different families. The corresponding initial conditions for each orbit —and the value of β— appear in Tables 2 and 3.

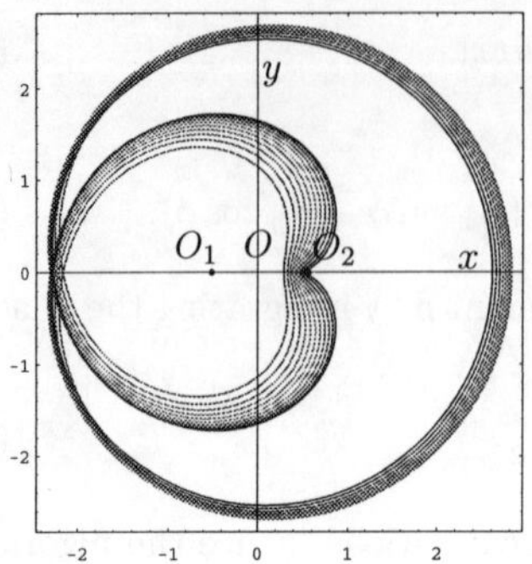

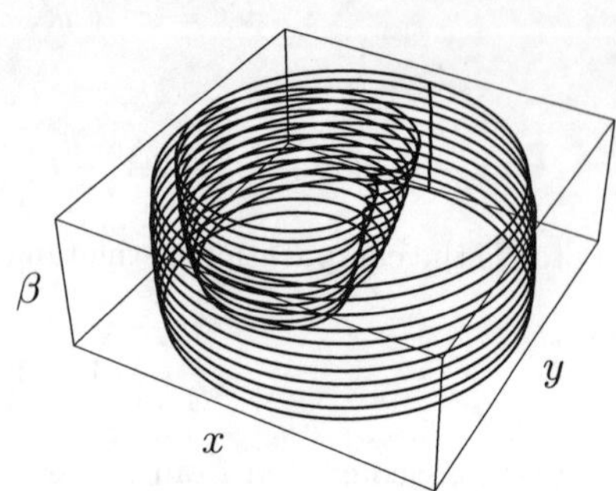

Figure 1. The family of periodic orbits for the Jacobian integral $h = -0.14897239$, and variations of β from $\beta = 0$ till $\beta = 0.986$ and $\delta\beta = 0.1$. The right figure is a spatial representation of the same set of orbits; the vertical axis represents β.

β	T	Trace
0.0	0.1044611108660456E+02	0.3822153997099804E+02
0.1	0.1070172013219846E+02	0.5742991976362558E+02
0.2	0.1091861983180412E+02	0.7877942498646655E+02
0.3	0.1111613216668850E+02	0.1037624867333485E+03
0.4	0.1130400877313352E+02	0.1344319449796839E+03
0.5	0.1148877305694259E+02	0.1742951152990996E+03
0.6	0.1167602660316857E+02	0.2302916965480838E+03
0.7	0.1187195100933765E+02	0.3186898086851346E+03
0.8	0.1208539221579756E+02	0.4888860581472264E+03
0.9	0.1233378082533988E+02	0.9919428359728460E+03
0.986	0.1262371447640814E+02	0.7268460161558314E+04

β	x_0	y_0	$\dot{x}_0$
0.0	0.1807988571131529E+00	0.2658263191705089E+01	0.2745718742061957E+01
0.1	0.1807767909534300E+00	0.2678057761574384E+01	0.2757228143972670E+01
0.2	0.1807995814686417E+00	0.2685745453338287E+01	0.2757616160628099E+01
0.3	0.1807951549726286E+00	0.2685443316541738E+01	0.2750545269799129E+01
0.4	0.1807226536826620E+00	0.2678957662978194E+01	0.2737601543522081E+01
0.5	0.1805502674655450E+00	0.2667173063771944E+01	0.2719537699281033E+01
0.6	0.1802470712670701E+00	0.2650481485311404E+01	0.2696654823880121E+01
0.7	0.1797778243600812E+00	0.2628926901761657E+01	0.2668927197288515E+01
0.8	0.1790970074637502E+00	0.2602209226312452E+01	0.2635996989486309E+01
0.9	0.1781377416068378E+00	0.2569525571175686E+01	0.2597012717347751E+01
0.986	0.1769986552205803E+00	0.2535174972786400E+01	0.2557151270760333E+01

TABLE 2. Several periodic orbits of the family presented in Figure 1 for $h = -0.14897239$.

4. 3-D families

For computing three dimensional families of periodic orbits, we make an extension of the algorithm given by Deprit and Henrard [5]. The main feature of this extension consists in formulating the variational equations in the Frenet frame $(\boldsymbol{t}, \boldsymbol{n}, \boldsymbol{b})$, where $\boldsymbol{t}$ is the tangent, $\boldsymbol{n}$ the normal and $\boldsymbol{b}$ the binormal vectors of the orbit. As it is well known, these vectors are

$$\boldsymbol{t} = \dot{\boldsymbol{x}}/V, \qquad \boldsymbol{n} = \dot{\boldsymbol{t}}/N, \qquad \boldsymbol{b} = \boldsymbol{t} \times \boldsymbol{n},$$

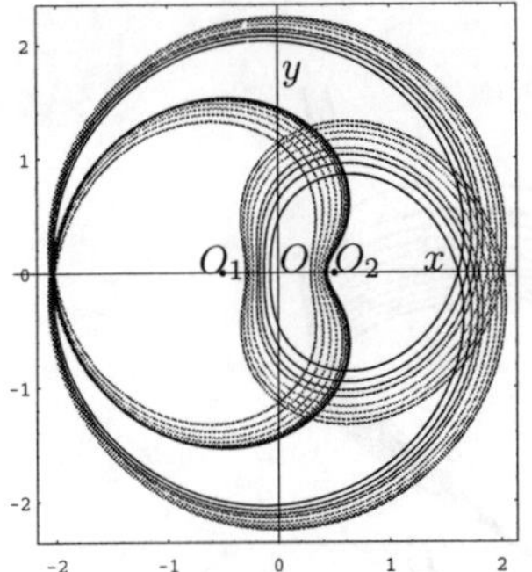

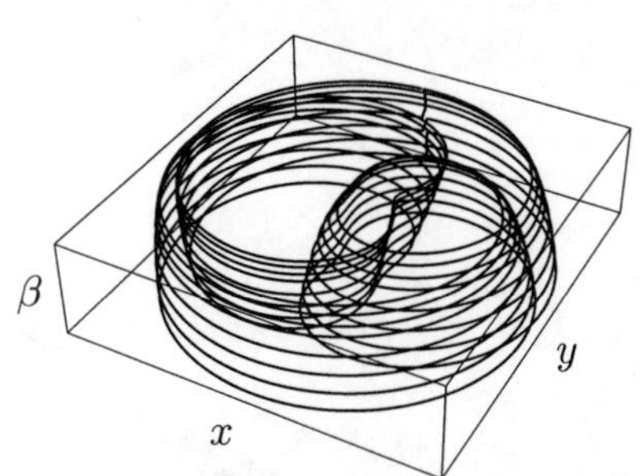

Figure 2. The family of periodic orbits for the Jacobian integral $h = -0.27279945$, and variations of β from $\beta = 0$ till $\beta = 0.537$. The right figure is a spatial representation of the same set of orbits; the vertical axis represents β.

β	T	Trace
0.0	0.1494816481714137E+02	0.6188380629984958E+03
0.099	0.1515637579861915E+02	0.8120723863623102E+03
0.21	0.1533733436551117E+02	0.9804795169153184E+03
0.3	0.1544687308727921E+02	0.1052050814120363E+04
0.399	0.1551583303720730E+02	0.1002491121777919E+04
0.45	0.1551412773178408E+02	0.8793267669154791E+03
0.501	0.1545198506391003E+02	0.6187755693006095E+03
0.537	0.1525584143212497E+02	0.1811784593415472E+03

β	x_0	y_0	$\dot{x}_0$
0.0	-0.2703535202131601E+00	0.2228189299382710E+01	0.2296266524432066E+01
0.099	-0.2690571464753951E+00	0.2217860181527281E+01	0.2280599330513873E+01
0.21	-0.2664917693148786E+00	0.2195799442952172E+01	0.2253077929615262E+01
0.3	-0.2637308598863723E+00	0.2170060236156629E+01	0.2223289994780886E+01
0.399	-0.2599884680685741E+00	0.2131822567937924E+01	0.2181166523493134E+01
0.45	-0.2576830098628738E+00	0.2106121672926874E+01	0.2153905670760668E+01
0.501	-0.2548930500571476E+00	0.2072417468486886E+01	0.2119400522003604E+01
0.537	-0.2518278131800796E+00	0.2031352149271426E+01	0.2079777506305585E+01

TABLE 3. Several periodic orbits of the family presented in Figure 2. The value of $h = -0.27279945$.

with $V^2 = \dot{\boldsymbol{x}} \cdot \dot{\boldsymbol{x}}$ and $N^2 = \dot{\boldsymbol{t}} \cdot \dot{\boldsymbol{t}}$. An intrinsic variation is

$$\boldsymbol{s} = p\,\boldsymbol{t} + q\,\boldsymbol{n} + r\,\boldsymbol{n}.$$

The separability of the tangent variations when formulated in the Frenet frame was demonstrated by [4]. Indeed, given the conservative system

$$\ddot{\boldsymbol{x}} + \mathrm{curl}\boldsymbol{A} \times \dot{\boldsymbol{x}} - \boldsymbol{\nabla}_{\boldsymbol{x}} U = 0,$$

that has the integral

$$C = \dot{\boldsymbol{x}} \cdot \dot{\boldsymbol{x}} - 2U.$$

the intrinsic variations can be computed by solving a coupled system involving only the normal and binormal displacements, and computing the quadrature

$$\frac{d}{dt}\left(\frac{p}{V}\right) = \left(2\,\frac{N}{V} + \frac{\mathrm{curl}\boldsymbol{A} \cdot \boldsymbol{b}}{V}\right) q - \frac{\mathrm{curl}\boldsymbol{A} \cdot \boldsymbol{n}}{V} r + \frac{U_\sigma}{V^2},$$

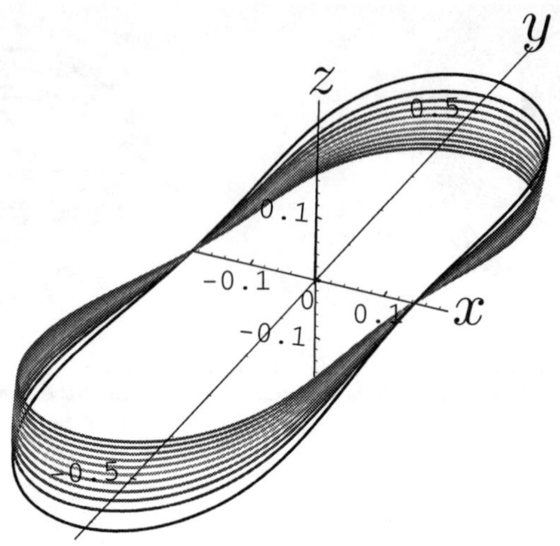

Figure 3. Several 3-D periodic orbits of the family corresponding to a value of the Jacobian constant $h = -1.33815281$.

which shows the sepparability of the tangent displacements. For details, cfr. [4]. Similarly to the two degrees of freedom case, the algorithm consists of a tangent predictor and an isoenergetic corrector.

We now apply the procedure to the spatial motion defined by the system (4). The variational system is obtained from Table 1.

In Figure 3 we present several orbits of a family. Similarly to the planar orbits, the starting periodic orbit for the algorithm is chosen from the classic RTBP. The corresponding initial conditions are showed in Table 4.

5. Acknowledgments

This work has been supported in part by the Ministerio de Educación y Ciencia (DGICYT Projects # PB93-1236-C02-02 and # PB95-0807) .

References

1. Belbruno, E., Llibre, J. and Ollé, M.: 1994, On the families of periodic orbits which bifurcate from the circular Sitnikov motions, *Celestial Mechanics and Dynamical Astronomy* **60**, 99–129.
2. Bhatnagar, K.B. and Chawla, J.M.: 1977. The effect of oblateness of the bigger primary on collinear library points in the restricted problem of three bodies, *Celestial Mechanics* **16**, 129–136.
3. Bhatnagar, K.B. and Hallan, P.P.: 1983. The effect of perturbations in Coriolis and centrifugal forces on the nonlinear stability of equilibrium points in the restricted problem of three bodies, *Celestial Mechanics* **30**, 97–114.
4. Deprit, A.: 1981, Intrinsic variational equations in three dimensions, *Celestial Mechanics* **24**, 185–193.
5. Deprit, A., and Henrard, J.: 1967, Natural families of periodic orbits, *Astron.J.* **72**, 158–172.
6. Elipe, A. and Ferrer, S.: 1985. On the equilibrium solutions in the circular planar restricted three rigid bodies problem, *Celestial Mechanics* **37**, 59–70.
7. Elipe,A. and Arribas, M.: 1986. An extension of Jacobian constant, in *Space Dynamics and Celestial Mechanics*, 53–57, (K.B. Bhatnagar, editor), D. Reidel Publishing Company, Dordrecht.

β	x_0	y_0	z_0
0.0003	0.1638871692679212E+00	-0.2510362681618084E-08	0.7005048919400680E-09
0.0021	0.1653678717992377E+00	-0.2557209555220652E-08	0.6330022239882382E-09
0.0036	0.1665958769975350E+00	-0.2674621274456612E-08	0.5847132335604077E-09
0.0048	0.1675744820870033E+00	-0.3191755856141176E-08	0.6157965335568116E-09
0.0059	0.1684686161461838E+00	-0.1507369601708164E-07	0.2508729056958925E-08
0.0069	0.1692790734294930E+00	-0.2024547203645886E-07	0.2803378619276346E-08
0.0079	0.1700872811034856E+00	-0.1146168880413606E-06	0.1189589651047138E-07
0.0089	0.1708932660742574E+00	-0.1159858110111323E-06	0.5789087144662981E-08

β	$\dot{y}_0$	$\dot{z}_0$	T
0.0003	-0.1303399906260374E+01	0.3637044861399464E+00	3.7173529930360
0.0021	-0.1315310325495506E+01	0.3255891893476048E+00	3.7170517495484
0.0036	-0.1325231216349784E+01	0.2897160756048663E+00	3.7168002723863
0.0048	-0.1333165665236091E+01	0.2572127822288803E+00	3.7165988007882
0.0059	-0.1340437566303580E+01	0.2230901524732224E+00	3.7164138913028
0.0069	-0.1347047571217604E+01	0.1865248622751126E+00	3.7162456025896
0.0079	-0.1353657066609580E+01	0.1404938044559164E+00	3.7160771331593
0.0089	-0.1360266306763544E+01	0.6789365573312622E−01	3.7159084824835

TABLE 4. Several 3-D periodic orbits of the family presented in Figure 3.

8. Elipe, A.:1992. On the restricted three-body problem with generalized forces. *Astrophysics and Space Science*, **188**, 257–269.
9. Kalvouridis, T.J.: 1994. Particle motions around, between or outside the two dipoles of a magnetic binary system, *Astrophysics and Space Science* **213**, 103–112.
10. Kunitsyn, A.L. and Tureshbaev, A.T.: 1985. On the collinear libration points in the photo-gravitational three-body problem, *Celestial Mechanics* **35**, 105–112.
11. Lara, M., Deprit, A., and Elipe, A.: 1995, Numerical continuation of families of frozen orbits in the zonal problem of artificial satellite theory, *Celestial Mechanics and Dynamical Astronomy* **62**, 167–181.
12. Malhotra, R.: 1994. Three-Body effects in the PSR 1257 + 12 planetary system, *Astrophys. J.* **407**, 266–275.
13. Mavraganis, A,G.: 1981. The equilibrium state in the magnetic–binary problem when the more massive primary is an oblate spheroid, *Celestial Mechanics* **23**, 287–293.
14. Schuerman, D.W.: 1980. The restricted three-body problem including radiation pressure, *Astrophysical J.* **238**, 337–342.
15. Sharma, R.K.: 1981. Periodic orbits of the second kind in the restricted three-body problem when the more massive is an oblate spheroid, *Astrophysics and Space Science* **76**,255–258.
16. Sharma, R.K. and Subba Rao, P.V.: 1986. On finite periodic orbits around the equilateral solutions of the planar restricted three-body problem, in *Space Dynamics and Celestial Mechanics*, 71–85, (K.B. Bhatnagar, editor), D. Reidel Publishing Company, Dordrecht.
17. Simmons, J.F.L., McDonald, A.J.C. and Brown, J.C.: 1985. The restricted 3 body problem with radiation pressure, *Celestial Mechanics* **35**, 145–187.
18. Wolszczan, A. and Frail, D.A.: 1992. A planetary system around the millisecond pulsar PSR1257+12, *Nature* **255**, 145–147.
19. Wolszczan, A.: 1994. Confirmation of earth-mass planets orbiting the millisecond pulsar PSR B1257+12, *Science* **264**, 538–542.

DYNAMICS AND EVOLUTIONARY STATUS OF THE YOUNG TRIPLE STELLAR SYSTEM, TY CRA

H. BEUST, P. CORPORON, L. SIESS, M. FORESTINI AND A-M. LAGRANGE
Laboratoire d'Astrophysique de l'Observatoire de Grenoble, B.P. 53, F-38041 Grenoble Cedex 9, France

Abstract. The young star TY CrA was known as a close eclipsing binary, which surprisingly appeared circularized but not synchronized. Recently, a third companion orbiting the close pair was detected. The dynamics of this triple system is here investigated. We show that according to pure 3-body dynamics, the TY CrA central binary should be subject to periodic changes related to the so-called Kozai mechanism in cometary dynamics, which should lead to a rapid collision. Adding tidal effects within that binary preserves its stability. We also show that thanks to the combination of tidal effects and 3-body dynamics, the rotation axes of the components of the central binary may by locked in a particular position (within the orbital plane) which might explain the apparent non-synchronism of the binary. Such a situation would not be stable if the central binary was alone, and if the system was older than $\sim 10^7$ yrs.

1. Introduction

The Herbig star TY CrA has been known as a short period eclipsing, spectroscopic binary [13, 17]. The identification of the radial velocity curves of both components allowed to determine their masses ($3.0\,M_\odot$ and $1.6M_\odot$), orbital characteristics, and rotation velocities [6]. The binary is almost tidally circularized ($e \simeq 0.02$), but the $3.0\,M_\odot$ primary appears surprisingly sub-synchronous [6]. As pointed out by Casey et al. [4], this cannot be explained straightforwardly.

Recently a third component orbiting the original binary system was detected [5, 7]. We recently presented [7] tentative fits of the orbital motion of the tertiary around the center of mass of the binary system, based on heliocentric velocity measurements. All solutions present in fact remarkable common features (see [7]), in particular a high eccentricity ($e' \simeq 0.5$), a tertiary mass $m_3 \simeq 1.2 - 1.4\,M_\odot$, and an inclination of at least 70° with respect to the orbital plane of the eclipsing binary. Such an unusual configuration for a triple system is not surprising and is observed in other hierarchical multiple systems as well [9, 18].

2. Evolutionary status of TY CrA triple system

Although the TY CrA system is thought to be young, its age is controversial. The $3.0\,M_\odot$ primary has a total luminosity of $\sim 90L_\odot$ and effective temperature of 11 000K (see [17]).

J. A. Docobo et al. (eds.), Visual Double Stars: Formation, Dynamics and Evolutionary Tracks, 299–308.

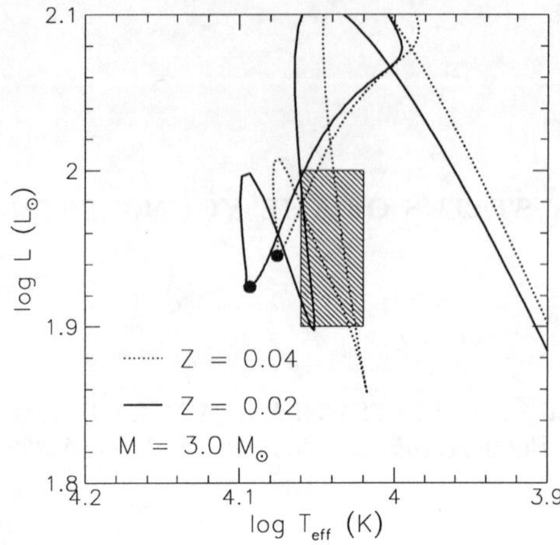

Figure 1. Hertzsprung-Russell diagram showing computed evolution tracks for $M = 3\,M_{\odot}$ and $Z = 0.02$ (solid line) or 0.04 (dashed line), as well as the observed position of the first TY CrA component taking account of the observational uncertainties.

The luminosity of the whole system is dominated by the first component. The location of this component in the Hertzprung-Russell diagram normally allows to determine its total mass m_1 and age t, by comparison with computed evolutionary tracks. However, such theoretical tracks also depends on the metallicity Z of the modeled stars. As TY CrA consists in a young system, it appears reasonable to assume that its metallicity is quite similar or larger than the solar one, namely $Z \geq 0.02$. In Fig. 1, we show $3M_{\odot}$ theoretical tracks (corresponding to $Z = 0.02$ and 0.04) together with the observational box for TY CrA.

We note that the tracks penetrates the box twice (whatever Z), once before the zero-age main sequence is reached and again during the main sequence. This leads to two very different age predictions for the TY CrA system: one very small ($t_{\mathrm{PMS}} \simeq 3\,10^6$ yr or $3.5\,10^6$ yr for $Z = 0.02$ or 0.04), and another considerably larger ($t_{\mathrm{MS}} \simeq 1.5\,10^8$ yr or $\simeq 10^8$ yr for the same Z, respectively).

3. Three body dynamics

Although several non-planar triple stellar systems have been identified, it appears interesting to investigate the dynamics of such an unusual stellar system. Due to three body motion, the two considered orbits (binary system and tertiary) interact and may evolve slowly. Also the question of the stability of the system might be addressed. Finally, this could give clues for understanding the problem of the non-synchronism of the binary.

The interaction between the three components of the TY CrA system can be separated into three body interactions and tidal effects. We will see below that tidal effects have a crucial role for ensuring the stability of the system. However, to make this appear clearly, we develop in this section a purely three body model, ignoring thus tidal effect which will be reintroduced afterwards.

3.1. NUMERICAL STUDY

In order to investigate the dynamics of this triple system, we carried out a numerical integration of the averaged system, taking as input the different orbital fit solutions from [7]. The theoretical background is developed in [3].

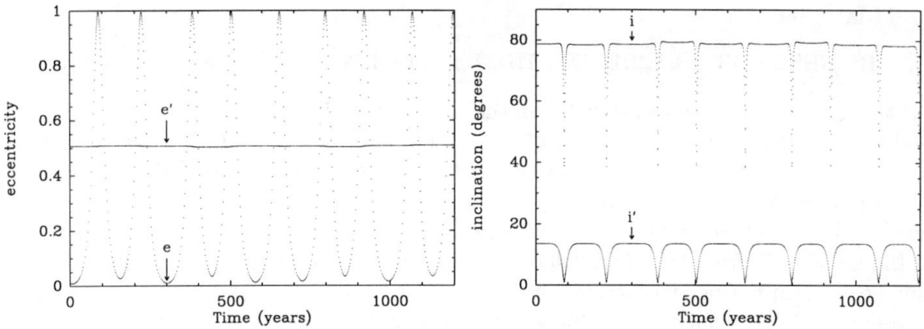

Figure 2. Evolution of the eccentricities and inclinations of the close binary (e, i) and the tertiary orbit (e', i') as a function of time for 1 200 years under the effects of pure 3-body dynamics. A quasi-periodic behavior (confirmed by integration over a longer time scale) is clearly detected; the close binary becomes very eccentric every ~150 years. Conversely, tertiary orbit is very stable. Only very small variations are recorded.

Figure 2 shows an example of temporal evolution. The first obvious conclusion is that the orbit of the binary is much more affected by perturbations than that of the tertiary. Surprisingly, the binary orbit appears to become regularly highly eccentric, while its inclination drops sharply at the same time the eccentricity becomes large. Conversely, the orbit of the tertiary appears much more stable. This general behavior was recorded for all input solutions.

3.2. INTERPRETATION

The recorded dynamical behavior may be understood when comparing the angular momenta of the orbits. It is worth noticing that the angular momentum of the tertiary orbit is significantly larger than that of the orbit of the binary ($\sim$ 5 times more). This ratio is not subject to evolution during the motion. Therefore, the angular momentum of the tertiary is forced to remain permanently close to the global $\vec{C}$, showing that i) the eccentricity of the tertiary should not vary drastically ii) its orbital plane should remain roughly perpendicular to $\vec{C}$. This is obviously confirmed by the numerical integration.

Conversely, the orbit of the binary is expected to be the subject of more drastic changes. The behavior reported above may be described in the frame of the *Kozai mechanism* [16, 2]. This mechanism applies in the Solar System for comets with originally highly inclined orbits with respect of the ecliptic. Under the effect of secular planetary perturbations, the orbit of the comet evolves, but the z-component of the orbital angular momentum of the comet (hereafter J_z) is expected to remain constant. This causes an orbit initially highly inclined and weakly eccentric to potentially evolve to weakly inclined but highly eccentric. This occurs indeed periodically with the precession of the argument of perihelion.

A similar mechanism is at work here on the orbit of the binary. Since the orbit of the tertiary is close to the (OXY) plane perpendicular to the global angular momentum, it is easy to show by rotational invariance of the hamiltonian equations (see [3]) that J_z for the binary orbit should be roughly constant, which is the context of the Kozai mechanism.

4. Tidal effects

4.1. THE NEED FOR CONSIDERING TIDAL EFFECTS

The three body calculations presented in the previous section did not take into account any tidal effect between the components of the clase binary of the TY CrA system. However, this cannot be avoided. It is well known that binaries with orbital periods less than a cut-off period are expected to be circularized that way. In any case, the orbital period of the close binary of the TY CrA system is far less than the cut-off value [14, 20, 19].

Independently from this, Fig. 2 shows obviously that the three-body model of the previous section cannot be satisfactory: the eccentricity of the binary is expected to sometimes almost reach 1 (the exact peak value was about 0.995). Remembering that the semi-major axis is constant, one sees that the periastron distance between the two primaries must become very small whenever the eccentricity is high, showing that the binary should in fact have already coalesced into a single star unless tidal effects prevent the orbit from evolving to high eccentricity values.

4.2. VARIOUS TIDAL MECHANISM

Several mechanisms generate tidal effects in close binaries. They all tend to synchronize the rotation of both components with the orbital motion and to circularize the orbit. In the case when the rotation axes would not be perpendicular to the orbital plane, the tidal effects also act to align both orbital and rotational angular momenta. This is valid for any tidal mechanism [10]. The efficiency of the different mechanism may be compared evaluating characteristic circularization ($t_{\rm circ}$) and synchronization ($t_{\rm sync}$) times for each of them.

The first mechanism invoked is the *equilibrium tide*: due to tidal attraction from its companion, each component is distorted. The analysis of this effect was achieved by Alexander [1], Press et al. (1975), Kopal [15] and Hut [11, 12], in the so-called weak-friction case.

Another tidal mechanism inducing synchronization and circularization known as *dynamical tides* was analyzed by Zahn [23, 24]. This occurs when the non-adiabatic oscillations driven on one component by the perturbing action of its companion are damped by radiative dissipation. This leads to a torque applied to the star, thus coupling orbital motion and rotation.

A final mechanism was discovered by Tassoul & Tassoul ([22] and Refs. therein). They showed that tidally driven meridional currents within each component act for circularization and synchronization. They also gave characteristic times.

4.3. APPLICATION TO TY CrA

Comparing the efficiency of the different tidal mechanism described above means evaluating and comparing the values of the corresponding characteristic times. The formulas giving these times are given in the corresponding references and in [3]. One needs thus values for the various constants appearing in these expressions, once applied to the peculiar case of the close binary of the TY CrA system. Among them we first have the apsidal constants k_j for each component (see [24]).

We thus tabulated these quantities over the lifetime of both components of the binary. More precisely, for each component at a given age, we used the stellar models by Siess et al. [8] to obtain them. The result is illustrated on Fig. 3, where k_2 is plotted as a function

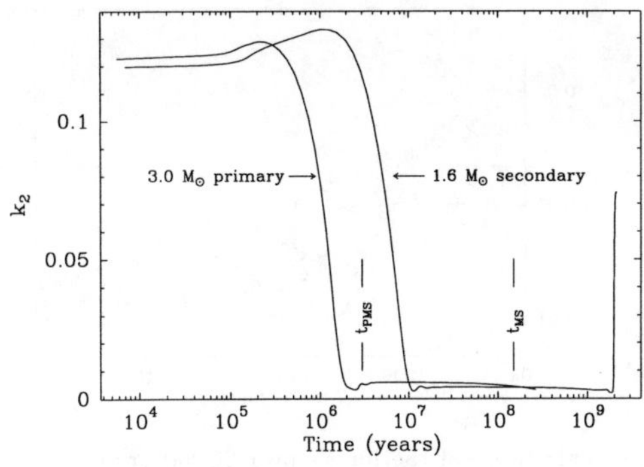

Figure 3. Evolution of the first apsidal constant k_2 as a function of time for the components of the TY CrA central binary, assuming solar metallicity. The temporal location of $t_{\rm MS}$ and $t_{\rm PMS}$ for solar metallicity is indicated.

TABLE 1. Values of characteristic times for the different tidal mechanisms invoked.

Mechanism ($t_{\rm PMS}$: $3\,10^6$ yr)	$t_{\rm circ}$ (yrs)	$(t_{\rm sync})_1$ (yrs)	$(t_{\rm sync})_2$ (yrs)
Equilibrium tide with turbulence [21]	$2.488\,10^{10}$	$1.458\,10^{11}$	$1.175\,10^8$
Dynamical tide [24]	$2.488\,10^{10}$	$6.246\,10^7$	$5.246\,10^8$
Meridional circulation [22]	$2.028\,10^6$	$2.765\,10^4$	$1.993\,10^4$
Mechanism ($t_{\rm MS}$: $1.5\,10^8$ yr)	$t_{\rm circ}$ (yrs)	$(t_{\rm sync})_1$ (yrs)	$(t_{\rm sync})_2$ (yrs)
Equilibrium tide with turbulence [21]	$3.846\,10^{12}$	$1.219\,10^{10}$	$1.193\,10^{11}$
Dynamical tide [24]	$1.134\,10^{10}$	$4.129\,10^7$	$1.926\,10^9$
Meridional circulation [22]	$3.903\,10^6$	$2.001\,10^4$	$4.762\,10^4$

of time for both components. Schematically, for each star, k_2 remains high as long as the star is pre main-sequence, and drops as soon as a radiative core grows. The temporal position of the two possible ages $t_{\rm PMS}$ and $t_{\rm MS}$ mentioned above is indicated (for solar metallicity). We clearly see on the plot the difference between these two models. At $t_{\rm MS}$, both stars are basically on the main sequence, while this is only the case for the $3.0\,M_\odot$ primary at $t_{\rm PMS}$.

Given these results, we thus computed the various characteristic times for all the tidal effects, for both ages. The results are listed in Table 1. We may note that in any case, the most powerful effect is the meridional circulation model by Tassoul & Tassoul [22].

5. Dynamics of the TY CrA system with tidal effects

We carried out a numerical integration of the TY CrA system dynamics, adding the tidal effects to the basic three-body model described above, and taking now into account the evolution of the rotation of the components of the binary. We here only retained the strongest effect, i.e., the meridional circulation model by by Tassoul & Tassoul [22].

If the binary was alone, there would be no reason for both rotation axes not to be perpendicular to the orbital plane. However, the presence of the tertiary component and the non-coplanarity of the orbits may change this simple picture. It is therefore legal to consider that the rotation axes might not be aligned.

For each age ($t_{\rm PMS}$ or $t_{\rm MS}$), several runs with different initial conditions were carried out; they all revealed similar behaviors, showing that the dynamics depends only weakly on the initial conditions. However, significant differences appeared between the behavior

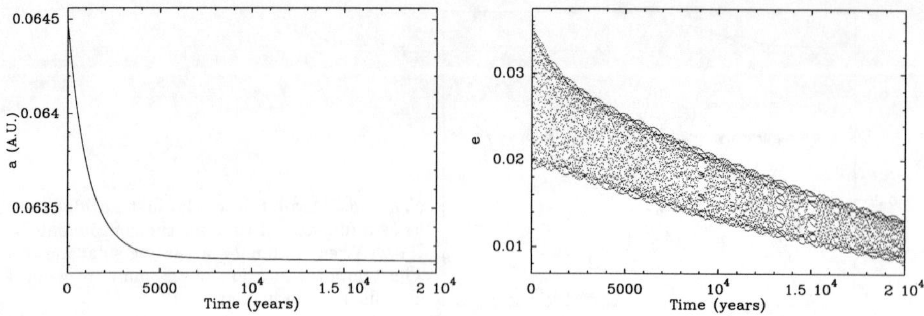

Figure 4. Evolution of the semi-major axis a and eccentricity e of the binary over 20 000 yrs under the effects of dissipative tidal effects, at $t_{\rm PMS}$. An equilibrium value is reached within ~6 000 yrs.

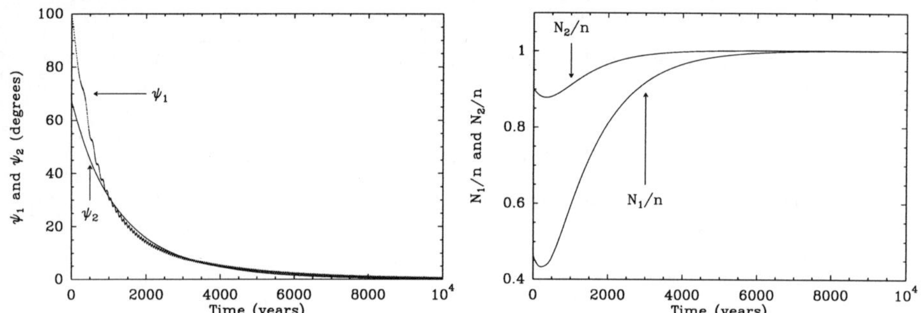

Figure 5. Evolution of the tilt angles ψ_1 and ψ_2 characterizing the alignment of the rotation axes of the binary, under the effects of dissipative tidal effects, at $t_{\rm PMS}$. We note a quick alignment/synchronization process.

recorded for the two models. We present now two typical runs, one for each age.

5.1. THE TYPICAL BEHAVIOR

The dissipative tidal effects act on circularization, synchronization and alignment of the rotation axes with the orbital angular momentum within the close binary. However, such an equilibrium holds theoretically for an isolated binary. It is however worth noticing that a similar equilibrium may be reached for the 3-body TY CrA system. We first show results at $t_{\rm PMS}$. Figure 4 shows the evolution of the semi-major axis a and the eccentricity e of the close binary. We see that after a smooth decrease, we reach a stable equilibrium value. Contrary to Fig. 2, the eccentricity remains small, and slowly decreases; thanks to the tidal effects, no Kozai-like behavior is present. The stability of the TY CrA system appears actually ensured by tidal effects !

Figure 5 describes the evolution of the rotation of the components of the binary. It is convenient here to display the evolution of the tilt angles ψ_i's ($i = 1, 2$) between the rotation axes of each star and the orbital angular momentum of the binary. We see that the rotation axes quickly move to align with the orbital angular momentum of the binary.

Figure 5 also shows the evolution of the rotation velocities N_i of the stars, once divided by the mean orbital motion n (thus, 1 means synchronism). We see that both stars are

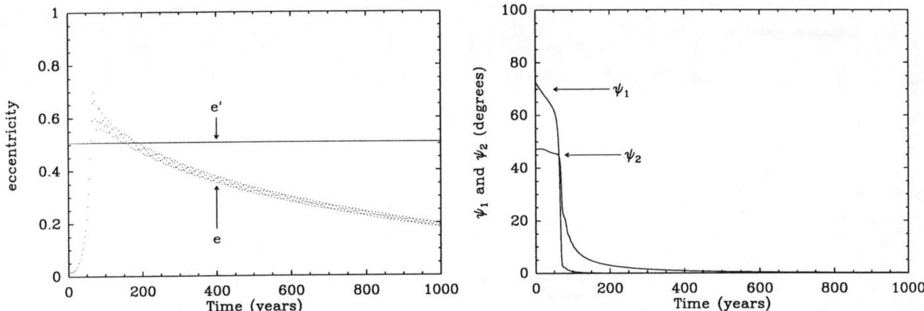

Figure 6. Evolution of the eccentricities of both orbits under the effects of dissipative tidal effects, and of the tilt angles ψ_i's for the components of the binary, at t_{MS}. We note a gradual circularization of the binary following a strong increase of the eccentricity. We also see that the alignment process is much more rapid than at t_{PMS}.

synchronized within ~6 000 yrs.

The behavior described here is characteristic for a tidally interacting binary. Moreover, as expected (see Table 1), the synchronization is achieved earlier than the circularization. It was shown by Hut [11] that this fact is valid for any binary for which the orbital angular momentum largely overcomes the rotational one. However, it is a remarkable fact that this also applies to the TY CrA system, despite the presence of the third companion.

We come now to present results of a similar run at t_{MS}. Figure 6 shows the temporal evolution of the eccentricity e of the binary, and of the tilt angles ψ_i's ($i = 1, 2$). They must be compared to Figs. 4 and 5.

Here, we note a strong initial increase of the eccentricity. Afterwards, it slowly decreases to zero. The sudden increase of the eccentricity is accompanied by a very rapid synchronization and a quick alignment of the axes, which take place much earlier than at t_{PMS}. This may be explained as follows: Thanks to smaller tidal coupling at t_{MS} (see Fig. 3 or Table 1), the 3-body dynamics is strong enough to initiate a full Kozai cycle, making the periastron distance between both components decrease. This leads to a strong increase of the tidal effects. The consequence is a rapid synchronization of the binary. Finally, the equilibrium is reached more quickly than at t_{PMS}, *thanks* to the 3-body dynamics, and *although* the tidal effects are globally smaller (Table 1).

5.2. A PECULIAR BEHAVIOR

Testing various initial configurations of the rotation axes reveals that the behavior described above is quite generic. We come now to describe the results of a specific run which might give clues for understanding the subsynchronosm problem of the primary. We first recall that in all our runs, once the orientation of the rotations axes are chosen, the initial rotation velocities is fixed to fit the observed present $v\sin(i)$'s of the stars (8 km s^{-1} for the primary and 35 km for the secondary [4, 6, 7]). In the following run, the initial orientation of the 3.0 $M_\odot$ primary was nearly pole-on; thus, its initial rotation velocity was high (super-synchronous), contrary to all other possible orientations.

Figure 7 shows the results of this run at t_{PMS}: we first note that the same equilibrium as before is obtained, but somewhat later, the tilt angle ψ_1 of the 3.0 $M_\odot$ primary remaining quite a long time close to 90° before dropping to zero.

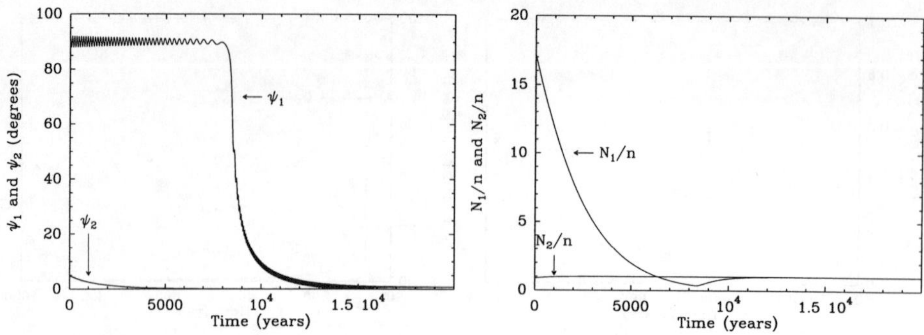

Figure 7. Evolution of the tilt angles ψ_1 and ψ_2 and rotation velocties (N_1/n, N_2/n) over 20 000 yrs, under the effects of dissipative tidal effects, at $t_{\rm PMS}$ with an initially pole-on primary. The 3.0 $M_\odot$ primary remains quite a long time at $\psi_1 = 90°$, while the 1.6 $M_\odot$ secondary is quickly aligned. This remains true even when the initial value of ψ_2 is high. We see that the 3.0 $M_\odot$ primary is initially highly super-synchronous, despite a very low $v\sin(i)$. This is due to the pole-on orientation of its rotation axis.

It thus seems that a marginally stable equilibrium $\psi_1 = 90°$ exists when the rotation of the primary is high enough. $\psi_1 = 90°$ means that the rotation axis of the 3.0 $M_\odot$ primary lies in the orbital plane of the binary. Such a marginal equilibrium might apply for the today situation of the TY CrA system, and thus explain why the binary does not appear synchronized. As the binary is an eclipsing one, $\psi_1 = 90°$ is compatible with a pole-on orientation of the rotation axis of the primary with respect to the line of sight. Before concluding anything, we must examine the origin of this marginal equilibrium. First, we may note that it is not observed at $t_{\rm MS}$ (see [3]). In that case, the tilt angles ψ_i's drop to zero much more quickly than at $t_{\rm PMS}$, within less than 500 yrs.

5.3. INTERPRETATION

In order to understand this peculiar behavior, one has to describe the variations of ψ_1. The details of the calculations are given in [3]. Schematically, to lowest order, we may write:

$$\frac{{\rm d}\psi_1}{{\rm dt}} = \left(\frac{{\rm d}\psi_1}{{\rm dt}}\right)_{\rm tidal} - \sin\eta_1 \sin i \left(\frac{{\rm d}\Omega}{{\rm dt}}\right)_{3-{\rm body}}, \tag{1}$$

where η_1 is a rotation angle around the angular momentum of the binary, defining (with ψ_1) the direction of the rotation axis of the primary. Here $({\rm d}\psi_1/{\rm dt})_{\rm tidal}$ stands for the tidal contribution, while the second term is explicitely due to the three body interaction, by means of the precession velocity $({\rm d}\Omega/{\rm dt})_{3-{\rm body}}$ of the orbital plane of the binary. This velocity is roughly constant (see [3]). Once applied to the TY CrA system, it appears that that $({\rm d}\psi_1/{\rm dt})_{\rm tidal} \ll ({\rm d}\Omega/{\rm dt})_{3-{\rm body}}$. Hence, by virtue of Eq. (1), a stable point for ψ_1 implies $\sin\eta_1 \simeq 0$. It is then easy to derive that as soon as $N_1/n \gtrsim 0.5$, the only stable point is closer to 90° than 0.2°. In fact, contrary to an isolated binary, the equilibrium is controlled by the 3-body term of Eq. (1) which dominates the tidal term.

6. Long-term evolution of the TY CrA system

The theory described in the previous section appears indeed well suited to explain the apparent non-synchronism of the central binary of TY CrA. The 3-body dynamics enhances

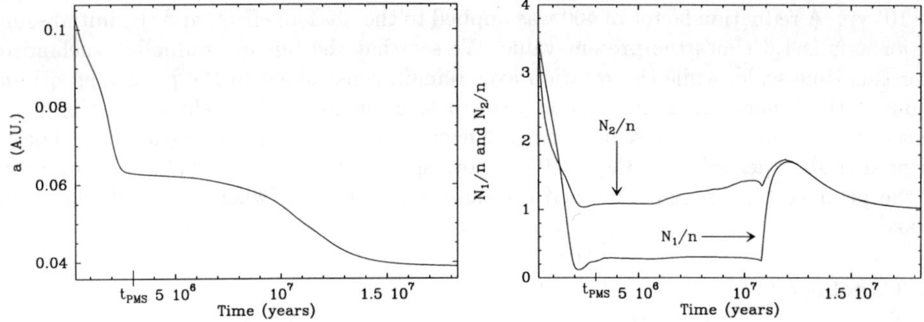

Figure 8. Long-term evolution of the semi-major axis and rotation velocities of the central binary under tidal effects, with reduced Tassoul constants, for 1.8 10^7 yrs. The semi-major axis regularly decreases. The value at $t_{\rm PMS}$ corresponds roughly to the present value. Both stars are initially supersynchronous, but a gradual slow-down process is detected. The stars quicly synchronize after 10^7 yrs. Note that the primary remains significantly subsynchronous for a long time.

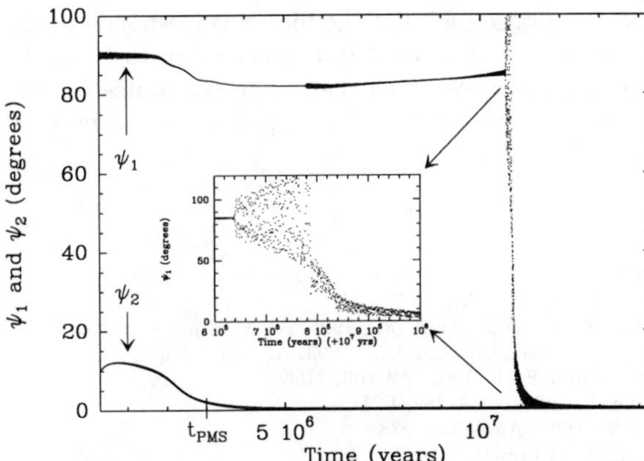

Figure 9. Same as Fig. 8, but for the tilt angles ψ_i $(i = 1, 2)$. The rotation axis of the primary remains constrained to the precessing orbital plane of the binary for $\sim 1.1\,10^7$ yrs. The central inner plot is an enlargement of the transition epoch for ψ_1 between $\sim 90°$ and $\sim 0°$. The alignment is achieved within a few 10^5 yrs.

a pseudo-equilibrium constraining the rotation axis of the primary to the orbital plane of the binary. However, we see from Fig. 7 that the duration of this phase should not exceed $\sim$ 10 000 years. Even if the age of the TY CrA system is unknown, this is a very short duration. In fact, if we assume the values given by Table 1 for the characteristic times of the meridional circulation model by Tassoul & Tassoul [22], there is no way to explain a non-synchronism of the binary after at most 10^5 years. One has then to admit than these constants are significantly smaller in the present case. This might be justified by the fact that all these quantities highly depend on the internal viscosity of the star, which is in any case very poorly known.

In order to test this theory, we decided to investigate the dynamics of TY CrA over a longer time-scale (a few 10^7 yrs), with significantly reduced Tassoul constants. Compared to the previous runs, we had to take into account the fact that the internal structure of the stars may evolve over such a time-scale. This is why we decided to introduce the evolution of all concerned quantities at every time during the integration, using the models by Siess et al. [8]. Our aim was mainly to investigate the long-term stability of the marginal $\psi_1 = 90°$ equilibrium, so that we gave this initial value to ψ_1.

Figures 8 and 9 show the result of an integration with such initial conditions for

$1.4\,10^7$ yrs. A reduction factor of 400 was applied to the Tassoul effect, and the initial semi-major axis is 1.6 times the present value. We see that the binary gradually synchonizes over this time-scale, while the rotation axes remain constrained to the precessing orbital plane of the binary for a few million years before aligning. This behavior is similar to that of Fig. 7, but the time-scale is much longer ($\sim 10^7$ yrs). This time-scale must not be surprising, since according to Fig. 3, 10^7 yrs corresponds to the time when both components of the primary have roughly reached the main sequence, i.e., when the tidal effects are weaker.

7. Conclusion

The long-term run shows that the survival of the marginal equilibrium characterized by $\psi_1 \simeq 90°$ is able to last several 10^6 yrs long, provided the tidal effects are not too strong. If we assume that the present age of TY CrA is $t_{\rm PMS} \simeq 3\,10^6$ yrs, this situation could still be valid today. However, it is unrealistic to imagin that the marginal equilibrium could still survive at $t_{\rm MS}$.

We may therefore suggest that i) the present age of TY CrA is probably $t_{\rm PMS}$ rather than $t_{\rm MS}$ and ii) the present status of the rotation axes of the binary corresponds to the marginal equilibrium. This requires the Tassoul & Tassoul tidal mechanism to be in the case of TY CrA significantly smaller than usually. We think that this is not in itself a very strong constraint, In any case, the stability of the system first appears ensured by tidal effects.

References

1. Alexander M.E., 1973, Ap&SS 23, 459
2. Bailey M.E., Chambers J.E., Hahn G., 1992, A&A 257, 315
3. Beust H., Corporon P., Siess L., Forestini M., Lagrange A.-M., 1996, A&A, in press
4. Casey B., Mathieu R.D., Suntzeff N., Lee C.-W., Cardelli J.A., 1993, AJ 105, 2276
5. Casey B., Mathieu R.D., Suntzeff N., Walter F.M., 1995, AJ 109, 2156
6. Corporon P., Lagrange A.-M., Bouvier J., 1994, A&A 282, L21
7. Corporon P., Lagrange A.-M., Beust H., 1996, A&A 310, 228
8. Siess L., Forestini M., Dougados C., 1996, in preparation
9. Hale A., 1994, AJ 107, 306
10. Hut P., 1980, A&A 92, 167
11. Hut P., 1981, A&A 99, 126
12. Hut P., 1982, A&A 110, 37
13. Kardopolov V.I., Sahanionok V.V., Phylipjev G.K., 1981, Perem. Zvezdy 21, 589
14. Koch R.H., Hvrinak B.J., 1981, AJ 86, 438
15. Kopal Z., 1978, Dynamics of close binary systems, Reidel, Dordrecht
16. Kozai Y., 1962, AJ 67, 591
17. Lagrange A.-M., Corporon P., Bouvier J., 1993, A&A 274, 785
18. Lestrade J.F., Phillips R.B., Hodges M.W., Preston R.A., 1993, ApJ, 410, 808
19. Mathieu R.D., Mazeh T., 1988, ApJ 326, 256
20. Mayor M., Mermilliod, J.-C., 1984, *Observational Tests os the Stellar Evolution Theory*, eds. A. Maerder, A. Renzini, Reidel, Dordrecht, p. 411
21. Press W.H., Wiita P.J., Smarr L.L., 1975, ApJ 202, L135
22. Tassoul J.-L., Tassoul M., 1992, ApJ 395, 259
23. Zahn J.-P., 1975, A&A 41, 329
24. Zahn J.-P., 1977, A&A 57, 383

CROSSING AREAS OF QUASIPERIODICITY: A NEW TECHNIQUE

J.SEIMENIS
University of the Aegean, Samos, Greece
and
Research Group of Astronomy and Space Geodesy.
Universitat Politecnica de Catalunya, Barcelona, Spain

Abstract. In this paper we develop the method of rational approximations in order to describe in an "analytical" way one- and two-dimensional periodic orbits of a logarithmic potential. During that "experiment" we were specifically interested if we can describe quasiperiodic orbits, in the same way. Moreover a very interesting question was if we can follow the change from periodicity to quasiperiodicity and vise-versa, using rational forms. Our results give us the right to defend that both periodic and quasiperiodic orbits is possible to be described by the same simple rational formulae.

1. Introduction

An important question in nonlinear galactic dynamics is to ask how we can describe "analytically" the families of periodic orbits in a galactic-type model. Furthermore, the description of quasiperiodic orbits, using "analytical" formulae, is a more complicated matter.

This paper contains the development of the method of rational approximations in the case of periodic and quasiperiodic orbits of the two-dimensional galactic model

$$V(x, y) = \ln(x^2 + \frac{y^2}{U^2} + C^2) \tag{1}$$

The above logarithmic potential (1) is of special importance for galactic dynamics, because is a model for a barred galaxy with a bulge of radius C (U is the axisymmetric ratio of the equipotential curves). More specifically, we try to parametrize the change from periodicity to quasiperiodicity, using simple rational forms.

The results are very interesting, in the view that the major research works on this field up to now, are made by usual numerical methods (e.g. Runge-Kutta method) only. Therefore, our method can be considered as a new numerical technique in the field of the orbit theory.

The first idea of the method was proposed by [Prendergast 1982]. That time was developed for second order nonlinear ordinary differential equations and was applied to the Duffing problem, the van der Pol oscillator and other similar problems of one degree of freedom. Later additional mathematical proofs were found by [Wood 1984], related with

J. A. Docobo et al. (eds.), Visual Double Stars: Formation, Dynamics and Evolutionary Tracks, 309–312.

the convergence of the method and the comparison with other well-known methods on the solution of nonlinear ordinary differential equations (e.g. single Fourier series, Galerkin approximation, fourth-order finite-difference method).

The method proposes the description of the solution of the non-linear differential equation(s) of motion in the form

$$x_i = \frac{N_i}{D}$$

where N_i and D are Fourier series in t (D starts always with 1 and $i = 1, 2, 3$ depending on the degrees of freedom we have - here we take $i = 2$).

The form of the series are: $\sum A_k \cos(k\omega t)$ or $\sum A_k \sin(k\omega t)$ for N_i and $1+\sum D_l \cos(l\omega t)$ for D (ω : the frequency).

The chosen form for N_i and D is connected with the shape of the orbits we want to describe and of course with the initial conditions. The series are truncated at a given order. In our study we use only one or two terms - the accuracy is quite good - and that made our formulae very simple.

The above method was applied by the author in several papers (e.g. see [Seimenis 1989 ; Contopoulos and Seimenis 1990]). [Contopoulos and Seimenis 1990] applied this method in order to find simple families of periodic orbits in the two-dimensional logarithmic potential (1). The importance of this potential is shown on the big number of papers on that. Among these are works of [Richstone 1982], [Levison and Richstone 1986] and [Miralda-Escude and Schwarzschild 1989]. Understanding the orbital structure of this potential is vital to understand the dynamics of galaxies.

Up to now there was an open question if it is possible to describe quasiperiodic orbits with simple formulae. Therefore in this paper we give an answer in that question, using the method of rational approximations and a new technique based on analytical and numerical parts.

2. Equations and expressions

The potential (1) has the following equations of motion:

$$x'' + \frac{2U^2x}{U^2x^2 + y^2 + C^2U^2} = 0, \qquad y'' + \frac{2y}{U^2x^2 + y^2 + C^2U^2} = 0, \tag{2}$$

where C and U depend on the specific model we have (C is the radius and U the axisymmetric ratio of the equipotential curves - in this study we used values obtained from the works of [Richstone 1982; Levison and Richstone 1986; Miralda-Escude and Schwarzschild 1989; Contopoulos and Seimenis 1990). We replace the forms

$$x = \frac{N}{D} \quad \text{and} \quad y = \frac{M}{D}$$

as well as their derivatives in the above equations of motion (2) and we have the "new" equations:

$$\begin{aligned} &(N''D^2 - 2N'D'D - ND''D + 2ND'^2)\cdot \\ &\quad \cdot(U^2N^2 + M^2 + C^2U^2D^2) + 2U^2ND^4 = 0, \\ &(M''D^2 - 2M'D'D - MD''D + 2MD'^2)\cdot \\ &\quad \cdot(U^2N^2 + M^2 + C^2U^2D^2) + 2MD^4 = 0. \end{aligned} \tag{3}$$

Then we have to solve in an approximating way the nonlinear system of the equations (3). In order to do this we propose the expressions:

$$\begin{aligned} N &= a + A\cos(m\omega t) \\ M &= G\cos(n\omega t) \qquad \text{or} \qquad G\sin(n\omega t) \\ D &= 1 + B\cos(n\omega t) \end{aligned}$$

The above coefficients a, A, B and G as well as the frequency ω are computed by solving a nonlinear algebraic system (see Appendix). The form of M depends on the family of periodic orbits which we want to represent, e.g. if we want a family that starts from a point on x-axis we have to take for $M = G\sin(n\omega t)$ because if $t = 0$ we obtain $M = 0$ and therefore $y = 0$.

Therefore searching for periodic orbits of any resonance $m : n$ with an initial point out of the axes x, y we propose to represent the approximate solution of x and y with the formulae:

$$x = \frac{a + A\cos(m\omega t)}{1 + B\cos(n\omega t)} \qquad y = \frac{G\cos(n\omega t)}{1 + B\cos(n\omega t)} \tag{4}$$

3. The technique

Since the expressions (4) are generic we studied the case $n = 1$ and various m. After the solution of the nonlinear system for the computation of the coefficients a, A, B, D and the frequency ω (see Appendix) we found that in the range of $m \in [0, 1]$ we have three different families of periodic orbits. It is obvious that for $m = 0$ (n always is 1) we have an one-axis orbit, the known straight-line family. For $m = \frac{1}{2}$, we have the resonance 1:2 and the anti-banana family. For $m = 1$ we have the resonance 1:1 and the well-known loop family (see [Seimenis 1995]).

Our interest was turned to the other values of m (other than $0, 1/2, 1$). So, our technique was to run the value of m from 0 to 1 and to observe what happened in any step. Running the value of m is similar like to run the value of a parameter of our problem. Therefore if we keep the energy constant, running the value of m corresponds with running the values of a, A, B, G and ω .

In that study we have the presence of a quasiperiodic orbit (the phenomenon from 0 to 0.05 has a continuity, and this continuity is observed for all m values). Few points before $m = 1$ at $m = 0.95$ we have similar behaviour (we have a box orbit, which in a sense is of similar behaviour of the previous one, if you look at the family of periodic orbits represented by the value $m = 1$). Going more close to $m = 1$, at $m = 0.997$ we observed a smooth return to periodicity which is obtained, as was expected, at the resonance 1:1 [Seimenis 1995].

4. Conclusions

In this paper we propose a new technique to represent families of periodic and quasiperiodic orbits. It is very interested to know that using the method of rational approximations we have representations of periodic and quasiperiodic orbits by the *same* simple formulae. In fact we cross areas of quasiperiodicity by using our formulae for periodic(!) orbits and when we must find the periodicity, we find it ! This is a new technique, and we believe that in a next step will be used as a method of finding the points of bifurcation and the changes of the stability.

References

1. G. Contopoulos and J. Seimenis, "Application of the Prendergast method to a logarithmic potential", *Astronomy and Astrophysics* **227** (1990) 49-53.
2. H. Levison and D. Richstone, "Dynamical Models of a sample of population II stars", *Astrophysical Journal* **308** (1986) 627-634.
3. J. Miralda-Escude and M. Schwarzschild, "On the orbit structure of the logarithmic potential", *Astrophysical Journal* **339** (1989) 752-762.
4. K. Prendergast, "Rational approximation for non-linear ordinary differential equations", pp. 369-371, in *The Riemann problem, complete integrability and arithmetic applications*, Lecture Notes in Mathematics **925**, D. Chudnovsky and G. Chudnovsky (Eds.), Springer, New York, 1982.
5. D. Richstone,"Scale-free models of galaxies II. A complete survey of orbits", *Astrophysical Journal* **252** (1982) 496-507.
6. J. Seimenis,"A new theoretical method for nonlinear dynamical systems", *Physics Letters* **A139** (1989) 151-155.
7. J. Seimenis, "The method of rational approximations: Periodic orbits and quasiperiodicity", *Reports of Mathematical Physics* **36**,2/3 (1995) 433-438.
8. D. Wood,"Rational-Fourier-series Approximations for Periodic Boundary-value Problems", *IMA Journal of Applied Mathematics* **33** (1984) 229-244.

Appendix

For the computation of the coefficients a, A, G, B and the frequency ω we have to follow the next steps:

(1) We insert the formulae (4) in the equations (3) and we obtain two truncated Fourier series equal to zero.

(2) In order to have the values for a, A, G, B and ω we must solve a non-linear algebraic system of five equations. We choose these equations among those we produced in the first step (by setting equal to zero all coefficients of the truncated Fourier series). For this purpose we use only the five equations of the lower order. The remaining terms are error terms. The nonlinear system can be solved by using the normal Newton-Raphson method. There is also possibility to replace some of the equations of the system with equations of initial conditions or the equation of the energy. For instance, the equation of the energy is produced by replacing the formulae (4) in the Hamiltonian of the model:

$$H = \frac{1}{2}(x'^2 + y'^2) + \ln(x^2 + \frac{y^2}{U^2} + C^2) = h,$$

where h is the value of the energy. If you put $t = 0$ you obtain an equation which is a function of a, A, G, B and ω and therefore can be used as an equation of our nonlinear system. The replacement of the highest order equation of the system by the energy-equation is suggested, because in that case our solutions conserve in the best way the energy integral. This last equation, which is produced by replacing the rational formulae in the Hamiltonian, is in an approximate way *the energy integral.* Replacement of one more equation by an equation of initial condition is under consideration and depends on the nonlinearity of the system (e.g. if the system has highly nonlinear terms of the unknown quantities we do not replace them, because we shall obtain solutions with no good accuracy).

APPLICATION OF THE GAUSS' METHOD TO THE STELLAR THREE BODY PROBLEM

A. ABAD
Grupo de Mecánica Espacial. Universidad de Zaragoza. 50009 Zaragoza. Spain

AND

F. BELIZÓN
Real Observatorio de la Armada. 11110 San Fernando. Spain

Abstract. In this paper, we deal with the stellar three body problem, that is, one star is far away from the other two stars. The outer orbit is assumed to be Keplerian. To analyze the effect of the distant star on the orbit of the close stars, we use the Gauss method; this method consist of replacing the gravitational attraction of the third star by the gravitational attraction of an infinitesimal non–homogeneous elliptic ring. We obtain the force vector for the Gauss method in terms of elliptic integrals. Finally, we compare the results obtained by this model with the classical third body model.

1. Introduction

Usually, the triple stellar systems present a hierarchical distribution in which the third star P is located far away from the other two stars S and O. This characteristic allows to assume the outer orbit as a Keplerian orbit. In this paper we propose the application to this problem of the Gauss method. In order to compute the secular variation of an orbit due to the influence of a third body, Gauss replaced the attraction of the third body P by the attraction of an infinitesimal elliptic ring (wire) whose mass element is proportional to the interval of time needed to describe the line element.

The expression of the force of attraction of the elliptic ring, given by Gauss, is apparently purely analytical. However, it is expressed in the frame of the principal axes of a cone with vertex S and with section the orbit of P. The transformation matrix of the reference frame is numeric, and consequently, the application of this expression cannot be but numeric. Hill [5] was the first to apply this expression to compute the secular perturbations produced by Venus on Mercury.

Halphen [6] applied the Weiersstrassian elliptic functions for obtaining a new expression of the force of attraction of the Gauss' elliptic ring in terms of hypergeometric functions. This new method was improved by Goriachev [8], who used it to compute the secular variations of Ceres. Musen [9] took again the method, rewrote it in terms of vectors and tensors and applied it to the long–range solar and lunar effects on the orbit of an artificial satellite. The algorithm derived by Musen does not present any difficulty if one contents

J. A. Docobo et al. (eds.), Visual Double Stars: Formation, Dynamics and Evolutionary Tracks, 313–320.

himself with numerical values of the perturbation effects, but it presents almost insuperable complications for a symbolic treatment.

A new approach to this problem has been made by Roth [10] under the restriction that the orbit of P be circular; under this assumption, he obtained completely analytical expressions to describe the effects of P. He applied them to a spacecraft orbiting a Galilean satellite, although the solution of this problem was achieved by means of a semi-analytical method.

Recently, we extended the result of Roth and we obtained [1], by means of expansions in power series of the eccentricity, an analytical expression for the potential of the ring when the eccentricity of the orbit described by P is small. However, in general this is not valid when applied to the stellar problem, since the outer orbit is not necessarily almost circular. In this paper we develop an analytical method to obtain no the potential, but the force function for any value of the eccentricity.

2. Force of attraction of an elliptic ring

Let us consider three stars O, S and P, such that both S and P move around O, and the orbit of P (with mass M) is assumed to be Keplerian (see figure 1).

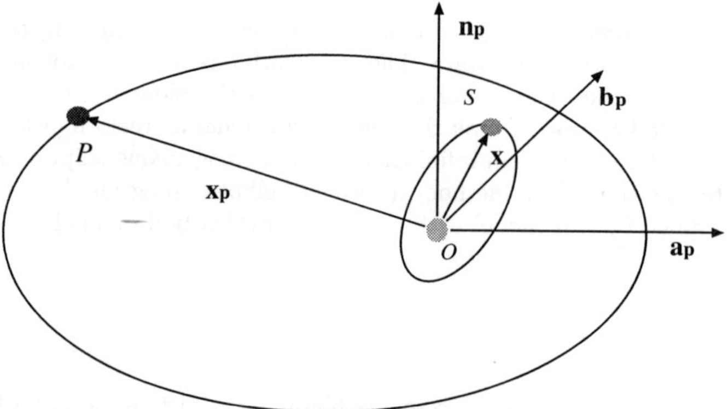

Figure 1. Third body model

The force of attraction $\boldsymbol{F}$ that P produces on S is expressed as

$$\boldsymbol{F}(\boldsymbol{x}, \boldsymbol{x}_p) = GM\left(\frac{\boldsymbol{x}_p - \boldsymbol{x}}{\|\boldsymbol{x} - \boldsymbol{x}_p\|^3} - \frac{\boldsymbol{x}_p}{\|\boldsymbol{x}_p\|^3}\right), \tag{1}$$

where G stands for the gravitational constant, and the vectors $\boldsymbol{x}$ and $\boldsymbol{x}_p$ represent, respectively, the position vector of S and P relative to O.

The originality proposed by Gauss [7] consists in replacing the previous force (1) by the attraction of an infinitesimal elliptic ring (see figure 2) whose density is $dm = M d\ell/2\pi$, with ℓ the mean anomaly of the orbit of P.

The attraction force created by this ring is

$$\boldsymbol{F}(\boldsymbol{x}) = \frac{GM}{2\pi}\int_0^{2\pi} \frac{\boldsymbol{x}_p - \boldsymbol{x}}{\|\boldsymbol{x} - \boldsymbol{x}_p\|^3}\, d\ell. \tag{2}$$

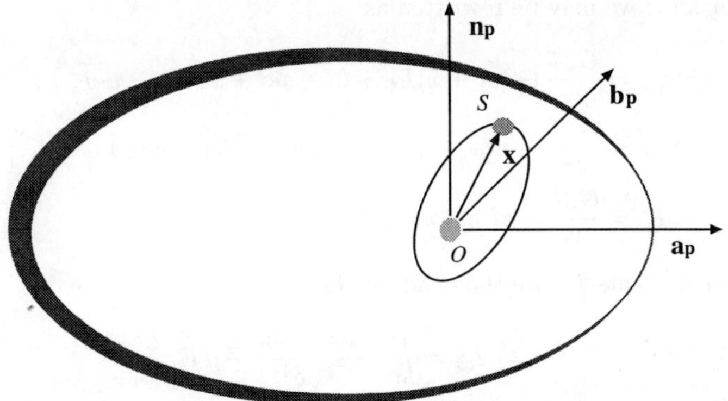

Figure 2. Gauss' model

It can be easily verified, see. e.g. [9], that the second term of (1) becomes 0.

To carry out this quadrature (2) in a symbolic mode, we adopt as reference frame the apsidal system, $\mathcal{S}_p = (\boldsymbol{a}_p, \boldsymbol{b}_p, \boldsymbol{n}_p)$, of the orbit of P around O, that is to say, the vector $\boldsymbol{a}_p$ is the unit vector in the direction of the periastrum of the orbit of P; the vector $\boldsymbol{n}_p$ is the unit vector in the direction of the angular momentum, and $\boldsymbol{n}_p = \boldsymbol{n}_p \times \boldsymbol{a}_p$. In this reference frame, the vectors $\boldsymbol{x}$ and $\boldsymbol{x}_p$ are

$$\begin{aligned} \boldsymbol{x} &= x\,\boldsymbol{a}_p + y\,\boldsymbol{b}_p + z\,\boldsymbol{n}_p\,, \\ \boldsymbol{x}_p &= a(\cos E - e)\,\boldsymbol{a}_p + a\sqrt{1-e^2}\sin E\,\boldsymbol{b}_p\,, \end{aligned}$$

where a and e are, respectively, the semi–major axis and the eccentricity of the orbit of P, and E is the eccentric anomaly. Taking into account the classical differential relation between mean and eccentric anomalies, $d\,\ell = (1 - e\cos E)\,d\,E$, we easily deduce from (2) the expression of the force (1) in this apsidal reference frame, $\boldsymbol{F} = F_x\,\boldsymbol{a}_p + F_y\,\boldsymbol{b}_p + F_z\,\boldsymbol{n}_p$:

$$\begin{aligned} F_x &= \frac{GM}{2\pi}\int_0^{2\pi}(x - a\cos E + ae)(1 - e\cos E)/\delta\,d\,E\,, \\ F_y &= \frac{GM}{2\pi}\int_0^{2\pi}(y - a\sqrt{1-e^2}\sin E)(1 - e\cos E)/\delta\,d\,E\,, \qquad (3) \\ F_z &= \frac{GM}{2\pi}\int_0^{2\pi} z\,(1 - e\cos E)/\delta\,d\,E\,, \end{aligned}$$

where $\delta = \sqrt{(x - a\cos E + ae)^2 + (y - a\sqrt{1-e^2}\sin E)^2 + z^2}$ is the mutual distance between S and P.

Gauss, Halphen and Musen integrated (2) in the frame defined by the principal axes of inertia of the cone with vertex S and section the ellipse defined by the orbit of P. The obtaining of the rotation matrix that relates both frames introduces numerical complications in this method; however they can be avoided by integrating directly the quadratures (3). This is the way followed in this note. With this approach, we obtain an analytical expression of the force; this method is faster and simpler than the classical ways of obtaining the force.

Equations (3) we may be rewritten as

$$\begin{aligned} F_x &= \frac{GM}{2\pi}\left[-(ae+x)I_{00} + (a+ae^2+xe)I_{01} - aeI_{20}\right] ,\\ F_y &= \frac{GM}{2\pi}\left[-yI_{00} + yeI_{10} + a\sqrt{1-e^2}I_{01} - ae\sqrt{1-e^2}I_{11}\right] ,\\ F_z &= \frac{GM}{2\pi}\left[-zI_{00} + zeI_{10}\right] , \end{aligned} \tag{4}$$

where the coefficients I_{cs} are the quadratures

$$I_{cs} = \int_0^{2\pi} \frac{\cos^c E \sin^s E}{\delta^3} dE. \tag{5}$$

3. Integration of I_{cs}

The quadrature I_{cs} will be computed by means of two changes of variable that leads these integrals to a linear combination of the complete elliptic integrals of the first and the second kind.

The first change we make is the classical transformation to rationalize combinations of circular functions

$$t = \tan\frac{E}{2} ,$$

With this change, the expression of the distance δ can be written as

$$\delta^2 = \frac{p(t)}{(1+t^2)^2} ,$$

where $p(t)$ is a polynomial of degree four, whose coefficients depend on x, y, z, a, e. The interval of integration $[0, 2\pi]$ is converted into the interval $(-\infty, \infty)$. Once the change is made, we have

$$I_{cs} = \sum_{n=0}^{5} \alpha_n T_n , \tag{6}$$

with α_n integer number, and T_n the quadrature

$$T_n = \int_{-\infty}^{\infty} \frac{t^n\, dt}{(\sqrt{p(t)})^3}. \tag{7}$$

The transformed polynomial $p(t)$ must be positive for any value of t for δ is a distance; hence the four roots of the polynomial must be complex. Let us denote by $a_1 \pm b_1 i, a_2 \pm b_2 i$ the four roots.

The second change of variable (borrowed from Tricomi [11]) is

$$t = a_2 + b_2 \tan\left(\phi + \frac{\theta_1+\theta_2}{2}\right) , \tag{8}$$

where

$$\theta_1 = \arctan\left(\frac{b_2+b_1}{a_2-a_1}\right), \qquad \theta_2 = \arctan\left(\frac{b_2-b_1}{a_2-a_1}\right).$$

With this change, the quadrature T_n is converted into

$$T_n = \sum_{m=0}^{4} \beta_{nm}(x, y, z, a, e) J_m, \qquad \text{with} \qquad J_m = \int_{-\frac{\pi}{2}}^{\frac{\pi}{2}} \frac{\cos^m \phi \sin^{4-m} \phi \, d\phi}{(1 - k^2 \sin^2 \phi)^{3/2}} \tag{9}$$

and

$$k^2 = \frac{4\sqrt{(a_2 - a_1)^2 + (b_2 + b_1)^2}\,\sqrt{(a_2 - a_1)^2 + (b_2 - b_1)^2}}{(\sqrt{(a_2 - a_1)^2 + (b_2 + b_1)^2} + \sqrt{(a_2 - a_1)^2 + (b_2 - b_1)^2})^2}\,.$$

Finally, by using the relations given in [4], the expression of J_m may be written as

$$J_m = \gamma_k(k)\, K(k) + \gamma_e(k)\, E(k)\,, \tag{10}$$

where $K(k)$ and $E(k)$ represent respectively the complete elliptic integrals of the first and the second kind.

Merging both changes of variable, we express, finally, the integrals I_{cs} as a linear combination of $K(k)$ and $E(k)$ with coefficients functions of x, y, z, a, e in the form

$$I_{cs} = \Phi_k(x, y, z, a, e)\, K(k) + \Phi_e(x, y, z, a, e)\, E(k)\,. \tag{11}$$

The expressions of $\Phi_k(x, y, z, a, e)$ and $\Phi_e(x, y, z, a, e)$ have been obtained automatically by using a set of packages written in *Mathematica* by Abad and Deprit and described in [2]. More details about this integration will be presented in [3].

The expressions of (11) combined with (4) can be used to write an efficient FORTRAN program to evaluate the force of attraction of the elliptic ring; this program has been employed in the numerical experiments described in the next section.

4. Numerical experiments

To check the efficiency of the model of Gauss with respect to the third body perturbation we integrate the perturbed two body problem equations

$$\ddot{\boldsymbol{x}} = -\frac{\mu}{\|\boldsymbol{x}\|^3}\boldsymbol{x} + \boldsymbol{F} \tag{12}$$

using the force models given by (1) and (2). In both cases we use a Runge-Kutta method of order 8 and 13 stages.

From this set of the orbital elements of S and P

$$\begin{array}{llllll} a_s = 1, & e_s = 0.1, & i_s = 30°, & \Omega_s = 50°, & \omega_s = 100°, & T_s = 0, \\ a = 20, & e = 0, & i = 0°, & \Omega = 80°, & \omega = 20°, & T = 0, \end{array}$$

we computed the initial position and velocity vectors.

In order to make a better analysis of the evolution of the orbit of S with both models we choose the semi–major axis of the inner orbit as the unit of distance and its period as the unit of time. With this assumption and supposing the the three stars with the same mass, we get $\mu = 4\pi^2$ and $GM = 2\pi^2$. The period of the outer orbit is equal to 126.49 units of time.

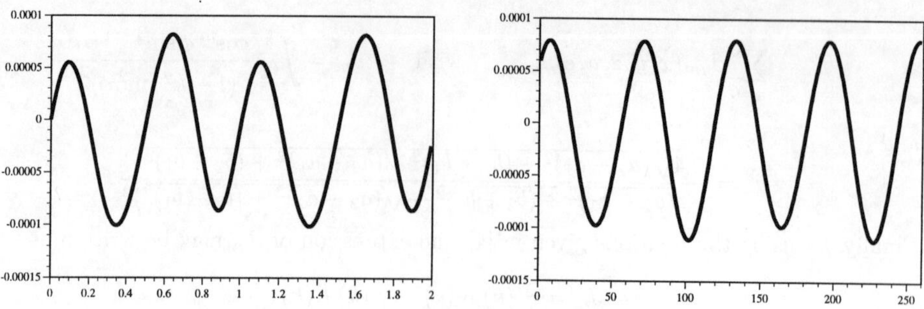

Figure 3. Difference of semi–major axis versus time

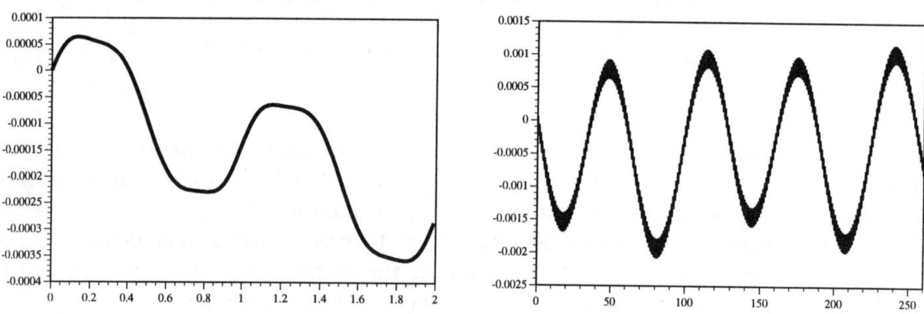

Figure 4. Difference of eccentricities versus time

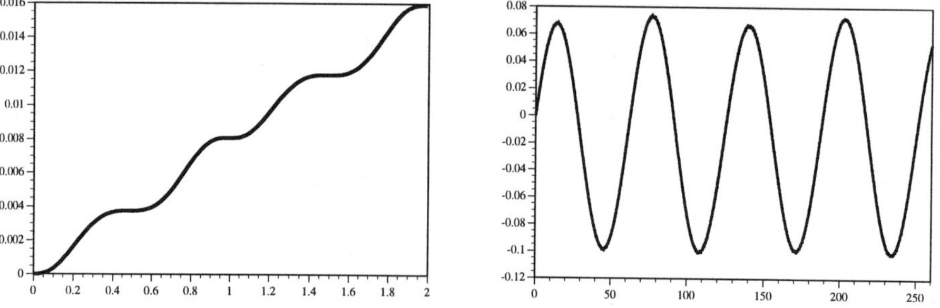

Figure 5. Difference of inclinations versus time

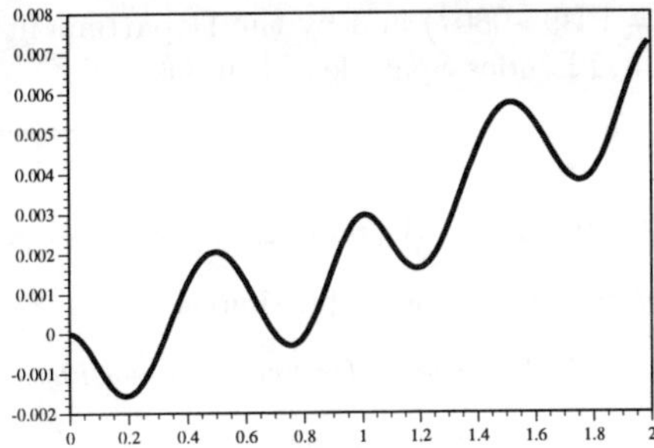

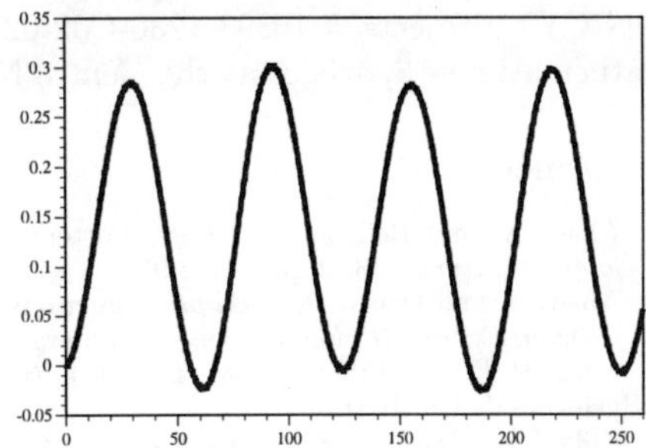

Figure 6. Difference of longitudes of the node versus time

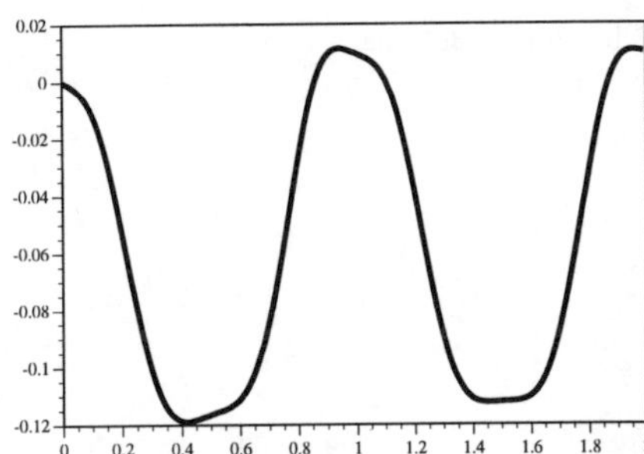

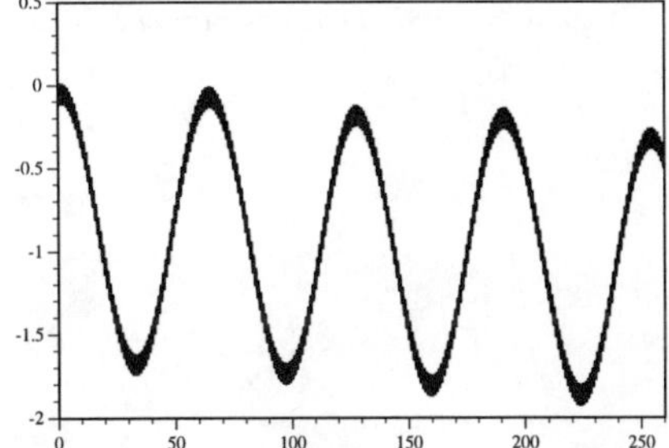

Figure 7. Difference of arguments of periastrum versus time

After the integration we obtain the evolution of the orbital elements of the inner orbit. The figures 3 to 7 show the difference between the orbital elements obtained by both methods. Left column figures represent two periods of the inner orbit, the right column figures represent two periods of the outer orbit. The differences of angular variables are expressed in degrees.

All orbital elements (with the exception of the semi–major axis) present periodic differences with the same period that the outer orbit. They present also short periodic variations with smaller amplitude. The amplitude of the variations on the semi–major axis presents also the short periodicity but the amplitude of these differences is maintained along the two periods of the outer orbit. The right part of the figure 3 is not representative because it has been obtained by taking only one point at each unit of time (recall that the unit of time has been chosen to be one period of the inner orbit). We present it only for the sake of showing that the differences maintain the same amplitude along the long period.

The magnitude of the differences between both models permits to say that the Gauss model can be used to study the long term evolution of the stellar triple systems.

5. Acknowledgments

Seminars and discussions with Professor André Deprit were very useful. Comments from an anonymous referee improved the manuscript. Thanks to Pablo Abad for drawing the figures

1 and 2. This work has been supported in part by the Ministerio de Educación y Ciencia (DGICYT Projects # PB93-1236-C02-02 and # PB95-0807) and by the Département de Mathématiques Spatiales at the Centre National d'Etudes Spatiales (Toulouse).

References

1. Abad, A. and Belizón, F. : 1996, Pottential of attraction for an elliptic ring. *Mechanics Research Communications*, **23**, 2, pp 111–116.
2. Abad, A. and Deprit, A. : *Elliptic Functions and Integrals by machine.* in preparation.
3. Belizón, F.: *Ph. D. dissertation.* in preparation.
4. Byrd, P. F. and Friedman, M. D.: 1954, *Handbook of Elliptic Integrals for Engineers and Physicist.* Springer–Verlag, Berlin.
5. Hill, G.W.: 1882, *Astron. Papers for the preparation of the American Ephemeris*, **1**.
6. Halphen, G. H.: 1888, *Traité des fonctions elliptiques et leurs applications.* Gauthier Villars, Paris.
7. Gauss, K. F.: 1818, *Collected Works*, **3**, p. 331
8. Goriachev N.N.: 1937, "On the method of Halphen of the computation of secular perturbations" (in Russian), pp 1–115, University of Tomsk
9. Musen P.: 1963, *Rewiews of Geophysics*, **1**, 1, pp 85-122
10. Roth J.: 1983, *Acta Astronautica*, **10,** 5–6 , pp 301-307
11. Tricomi, S.: 1937, *Funzioni Ellitiche.* Nicola Zanichelli Editore. Bologna.

THE STROBOSCOPIC METHOD APPLIED TO THE STUDY OF ZETA CANCRI

J. F. LING[1] AND J. A. DOCOBO[1]
Observatorio Astronómico "Ramón María Aller"
P.O. Box 197; Universidad de Santiago de Compostela, Spain
[1] *Dpto. Matemática Aplicada. Universidade de Santiago. Spain.*

1. The ζ Cancri System

The well-known multiple star system ζ Cancri = WDS 08122N1739 = ADS 6650 = STF 1196 was discovered by Sir W. Herschel on November 21, 1781.

It consist of four stars which form a hierarchical system. The two stars A and B form a visual binary having an orbit with a period aproximatly of 60 years.

The star C has got an invisible companion c which orbits around for about 17 years.

The pair Cc moves round the centre of masses of the pair AB with a period of more than thousand years at a distance of about 7".

Components	Period aprox.	Magnitudes	Epectral type	Orbit
A - B	59^y	5.6 - 6.0	F8 V	(1) Inner
C - c	17^y	6.2	G0 V	
AB - Cc	1115^y			(2)Outer

TABLE 1.

In the table we also show the magnitudes and the spectral type of the components of the system.

J. A. Docobo et al. (eds.), Visual Double Stars: Formation, Dynamics and Evolutionary Tracks, 321–329.

For the dynamical studies of this system we have taken into account some assumptions previously mentioned by Harrington et al. (1992) and Heintz (1996)

As first approximation the binary Cc, formed by two stars with very similar masses, was treated as a single object. So the system was considered as a triple star A, B and Cc. We denoted as INNER ORBIT (or Orbit One) those of the pair AB and OUTER ORBIT (or Orbit Two) those of the Cc-AB.

As the outer orbit exhibits only a very slow mouvement we have assumed it to be a keplerian orbit. This simplification has already been suggested by other authors (Brown, 1936-37; Docobo, 1977).

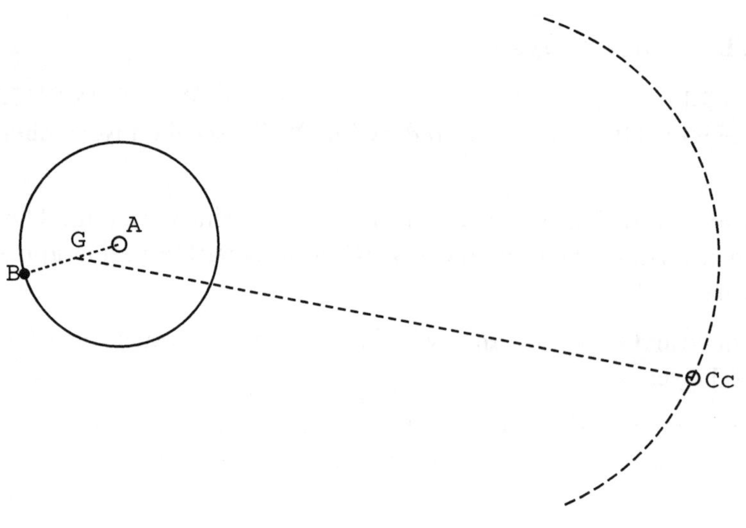

Figure 1.

2. THE STROBOSCOPIC METHOD

To study the perturbations of the inner orbit AB we have used a semianalytical method called the STROBOSCOPIC METHOD (Roth, 1973, 1979), which allows a qualitative study to be made of the perturbations in a rapid form and with it we can work directly in the classical orbital elements.

For these reason the Lagrange equations of motion must be transformed, conveniently, in a close form using as independet fast angular variable the eccentric anomaly of the inner orbit, E_1.

$$
\begin{aligned}
\frac{da_1}{dE_1} &= \frac{\mu'_1}{\mu_1} 2a_1^2 \frac{\partial R}{\partial E_1}\gamma_1 \\
\frac{de_1}{dE_1} &= \frac{\mu'_1}{\mu_1}\frac{a_1\sqrt{1-e_1^2}}{e_1}\left(\sqrt{1-e_1^2}\frac{\partial R}{\partial E_1} - (1-e_1\cos E_1)\frac{\partial R}{\partial \omega_1}\right)\gamma_1 \\
\frac{di_1}{dE_1} &= \frac{\mu'_1}{\mu_1}\frac{a_1}{\sqrt{1-e_1^2}\sin i_1}(1-e_1\cos E_1)\left(\cos i_1\frac{\partial R}{\partial \omega_1} - \frac{\partial R}{\partial \Omega_1}\right)\gamma_1 \\
\frac{d\Omega_1}{dE_1} &= \frac{\mu'_1}{\mu_1}\frac{a_1}{\sqrt{1-e_1^2}\sin i_1}(1-e_1\cos E_1)\frac{\partial R}{\partial i_1}\gamma_1 \\
\frac{d\omega_1}{dE_1} &= \frac{\mu'_1}{\mu_1}a_1(1-e_1\cos E_1)\left(\frac{\sqrt{1-e_1^2}}{e_1}\frac{\partial R}{\partial e_1} - \frac{\cot i_1}{\sqrt{1-e_1^2}}\frac{\partial R}{\partial i_1}\right)\gamma_1 \\
\frac{dt}{dE_1} &= \frac{a_1^{\frac{3}{2}}}{\mu_1^{\frac{1}{2}}}(1-e_1\cos E_1)\gamma_1
\end{aligned}
\tag{1}
$$

We must note that the perturbation function R can be expanded in powers of the small constant ϵ, the ratio of the semimajor axes of the orbits at some given instant. Due to observational limitations it is sufficient to use only the first nonzero perturbing term, which is the term in ϵ^3.

For the integration we start from a periastron passage $T_k^{(0)}$ were the initial values $\vec{x}_k^{(0)} = (P_1^{(0)}, t_1^{(0)}, e_1^{(0)} a_1^{(0)}, i_1^{(0)}, \Omega_1^{(0)}, \omega_1^{(0)}) = cte$ are known. Then we make the integration from periastron to periastron and we can obtain, at the end of the revolution k, the perturbed orbital elements.

$$\vec{x}_k = \vec{x}_k^{(0)} + \sum_{i=1}^{3} \Delta\vec{x}_k^i(2\pi)$$

$$t_k = t_k^{(0)} + \sum_{i=1}^{3} \Delta t_k^i(2\pi) + T_k^{(0)}$$

In the periastron of the next revolution we reinitialiced the elements as follows.

$$
\begin{aligned}
\vec{x}_{k+1}^{(0)} &= \vec{x}_k \\
t_{k+1}^{(0)} &= t_k
\end{aligned}
\tag{2}
$$

And then we repeat de proccess.

So with this method the solution comes from making succesive aproximations during the fundamental period of the fast variable.

It is possible to obtain the perturbed elements in any time, the detailed study of the topic was published by us in 1991 (Ling, 1991) and 1995 (Ling et al., 1995).

3. APPLICATION TO ζ CANCRI

3.1. INITIAL ORBITAL ELEMENTS

In order to initialice the calcul procedure for applying the stroboscopic method we have used the most recent orbital elements and masses published for this system. These correspond to those calculated by Heintz in 1996 (Heintz, 1996), which are showed in table 2. Previous orbital elements were obtained by Makemson (1933) and Gasteyer (1954).

Orbital elements of W. D. Heintz (1996)	
Inner orbit (A-B)	Outer orbit (Cc-AB)
$P_1 = 59^y$	$P_2 = 1115^y$
$T_1 = 1989.23$	$T_2 = 1970$
$e_1 = 0.325$	$e_2 = 0.24$
$a_1 = 0''880$	$a_2 = 7''70$
$i_1 = 165^0$	$i_2 = 146^o$
$\Omega_1 = 30^04$	$\Omega_2 = 74^o2$
$\omega_1 = 204^0$	$\omega_2 = 345^o5$
Masses: $m_A = 1.11\odot$, $m_B = 1.00\odot$, $m_{Cc} = 1.92\odot$	

TABLE 2.

Althoug this system was discovered in 1781, as menction previously, it was negleted up to 1825 when the sistematics observations first began and have continued to the present day. So we have chosen as the starting point for stroboscopic calculation the periastron epoch immediately previous to the first measurement itself, that is to say the epoch of 1810.045 . The initial orbital elements of the inner orbit were obtained in such a way to minimize the difference between Heintz's values for the osculating epoch

1989.23 and the values calculated by the stroboscopic method starting in 1810.045. The initial elements so obtained are showed in TABLE 3.

INITIAL ELEMENTS OF THE INNER ORBIT (A-B)
$P_1^{(0)} = 59^y50$
$T_1^{(0)} = 1810.045$
$e_1^{(0)} = 0.293$
$a_1^{(0)} = 0''880$
$i_1^{(0)} = 166^o5$
$\Omega_1^{(0)} = 30^o0$
$\omega_1^{(0)} = 202^o4$

TABLE 3.

3.2. STUDY OF PERTURBATIONS

The perturbations study were made for each inner orbit element separately for 100 instants that correspond to multiples of 2π in the fast variable E_1, that is to say, the time of periastron passage of the inner orbit. The initial periastron K=0 corresponds to the epoch 1810.045 and the last one K=100 that of 7769.333. This supposes a time interval of more than 6000 years and can be considered sufficient for a long-term study of the system.

We also compare the results with a numerical integration method, precisely a Runge-Kutta-Verner method that appears in the subroutines of IMSL library (1982).

Both results are represented graphycally in the following figures (2-10). The solid line represents the stroboscopic results and the dashed line the numerical ones.

Even though the value of the semimajor axis a_1, obtained using the stroboscopic method is the same in all periastrons, it varies periodically between every two; which means that short-term perturbations only exist in this element.

In other figures we can see that there is a good relationship between the values found by the two methods.

In reference to the time of periastron passage, the values are reduced to one revolution.

Up to here we have presented the study on the elements referred to the apparent plane, but we have also considered the angular elements that result from the relation to the outer orbital plane called the mutual elements, that is to say the mutual inclination, mutual node and mutual longitude of periastron.

Acknowledgements

We would like to pay spetial tribute to the late Dr. R. S. Harrington, who dedicated a lot of his time to the study of the stelar three-body problem.

This study was carried out as part of the research projects PB92-1074 and XUGA 24301B96 financed by the Spanish D.G.I.C.Y.T. and the XUNTA DE GALICIA (Spain) respectively.

References

Brown, E. W.: (1936-37) *Monthly Notices Roy, Astron. Soc.*, **97**, 25
Docobo, J. A.: (1977) Aplicación de la Teoría de Perturbaciones al Estudio de Sistemas Estelares Triples. *Ph. D.*, University of Zaragoza. Spain
Gasteyer, C.: (1954) *Aston. J.*, **59**, (1219), 243-250
Harrington, R. S.; Douglass, G. G. and Worley, C. E.: (1992) *ASP Conference Series*, **32**, 321-322
Heintz, W. D.: (1996) *Aston. J.*, **111**, (1), 408-411
IMSL Library.: (1982) Problem-Solving Software System For Mathematical Statistical FORTRAN Programming *IMSL Lib-0009*, **Ed. 9**, revised.
Ling, J. F.: (1991) *Astrophys. and Space Science*, **185**, 51-61
Ling, J. F.; Docobo, J. A. and Abad, A. J..: (1995) *Aston. J.*, **110**, (2), 875-879
Makemson, M.: (1933) *Aston. J.*, **42**, (17), 153-157
Roth, E. A.: (1973) *Celes. Mech.*, **8**, 245
Roth, E. A.: (1979) *J. Appl. Math. Phys.*, **30**, 315

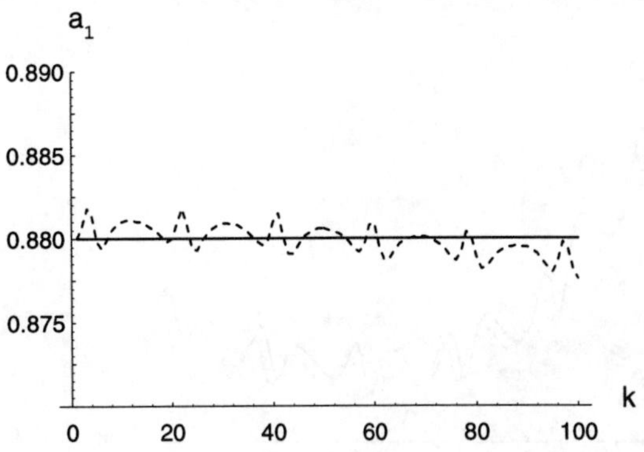

Figure 2. Semimajor axis

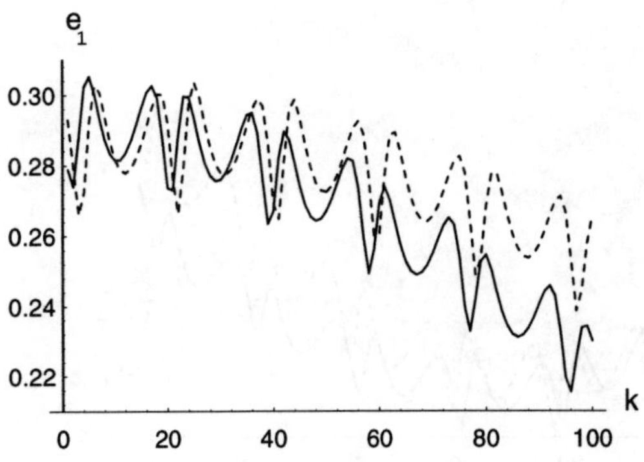

Figure 3. Eccentricity

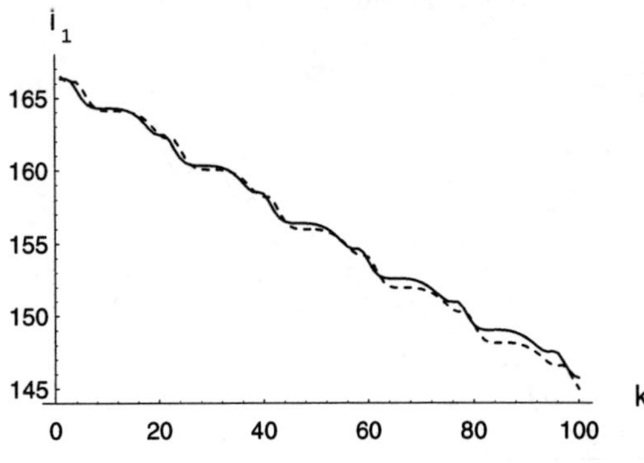

Figure 4. Inclination

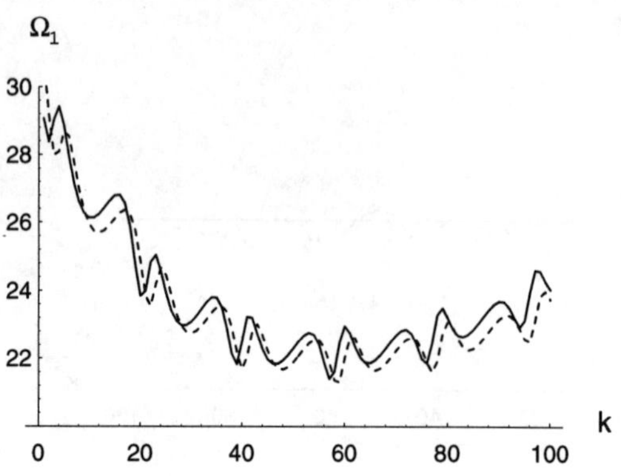

Figure 5. Node

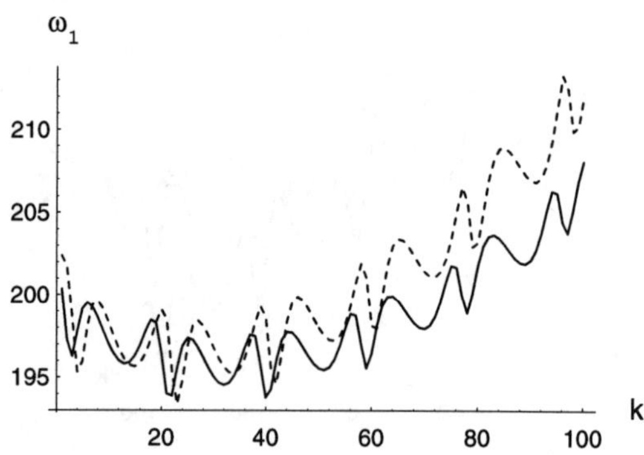

Figure 6. Longitude of periastron

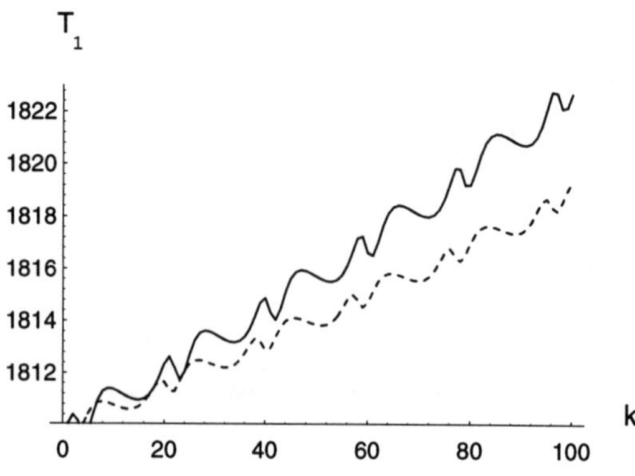

Figure 7. Periastron epoch

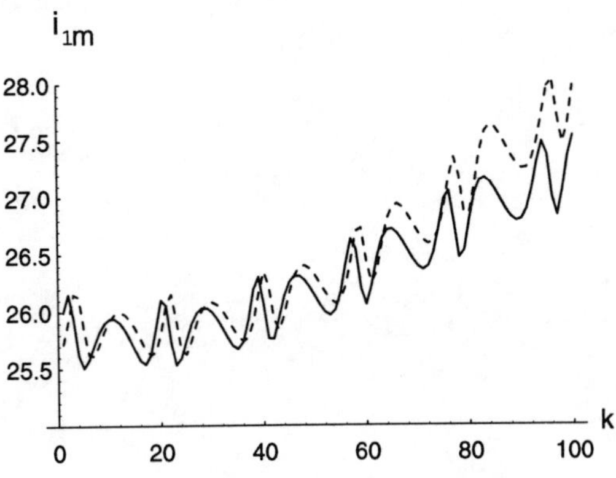

Figure 8. Mutual inclination

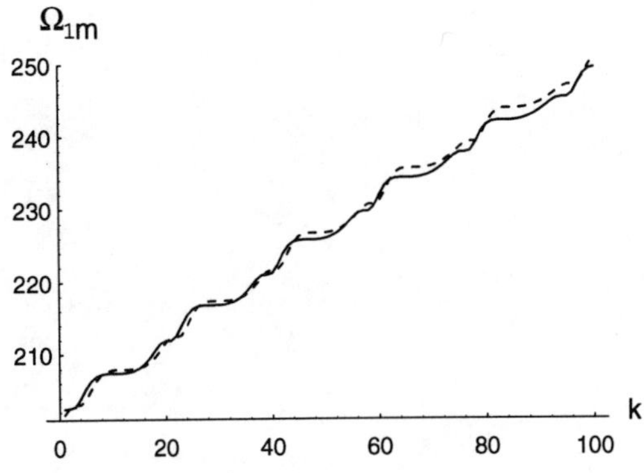

Figure 9. Mutual Node

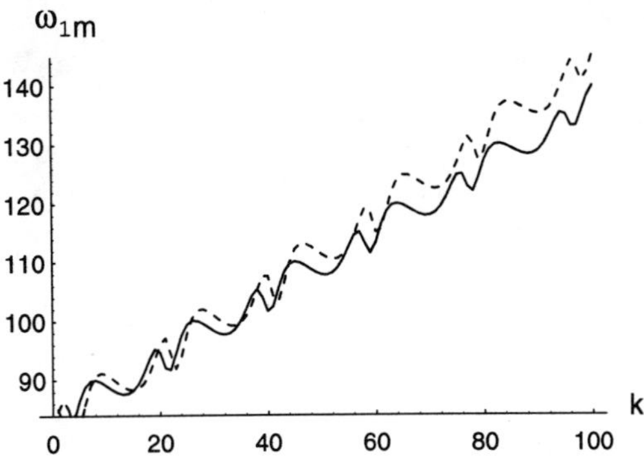

Figure 10. Mutual longitude of periastron

ON THE TWO-BODY PROBLEM WITH SLOWLY DECREASING MASS

C. PRIETO[1] AND J. A. DOCOBO[2]
Observatorio Astronómico "Ramón María Aller"
P.O. Box 197; Universidade de Santiago de Compostela, Spain
[1] *Dpto. Matemática Aplicada. Universidade de Vigo. Spain.*
[2] *Dpto. Matemática Aplicada. Universidade de Santiago. Spain.*

Abstract. We intend to present two approximate analytic solutions of the two-body problem with slowly decreasing mass which are obtained through the integration of the Hamilton equations.

The law of mass variation used was $\dot{m} = -\alpha m^n$ on the first case, and $\dot{m}_i = \alpha_i m_i^{n(i)}; i = 1, 2$ on the second which gives rise to a perturbed Keplerian problem dependent on one and two small parameters respectively.

Practical appplications are included.

1. INTRODUCTION

During the second half of the 19th century, M. Ch. Dufour (Dufour, 1866) and Th. V. Oppolzer (Oppolzer, 1884) were the first to examine the astronomical phenomenon of variable mass, more specifically in relation to the Earth-Moon system altered by an increase in mass due to the impact of meteorites.

Opplozer's paper inspired H. Gylden (Gylden 1884) set out the solution to the system of differential equations which describes the two-body motion when the masses are subject to variations by a known function of time.

In 1893 F. Mestschersky, (Mestschersky 1893, 1902, 1949) was the first to point out a specific case of this problem which is integrable taking into consideration a law of mass variation

$$\mu(t) = \frac{1}{a + \alpha t}$$

where a and α are constants, and the transformation

$$\xi = \frac{x}{a + \alpha t}; \quad \eta = \frac{y}{a + \alpha t}; \quad \tau = \frac{-1}{a(a + \alpha t)} \tag{1}$$

It is of vital importance to have this exact solution available when studying the nature of other solutions both numerical as well as analytic, since they can be compared with it.

J. A. Docobo et al. (eds.), Visual Double Stars: Formation, Dynamics and Evolutionary Tracks, 331–339.

In his later studies, Mestschersky (1902, 1949) demonstrates that transformation (1) is still valid for a more general case when

$$\mu(t) = \frac{1}{\sqrt{\alpha+\beta t+\gamma t^2}}$$

$$\alpha, \quad \beta, \quad \gamma; \quad \text{constants}$$

and also establishes the equation which characterizes the absolute movement of variable mass points

$$m_i \dot{\vec{v_i}} = \vec{F_i} + \dot{m}_i \left(\vec{u_i} - \vec{v_i}\right) \tag{2}$$

where $\vec{u_i}$ is the absolute speed of the mass centre of particles dm_i just before the moment of union with m_i, or of the particles separated in the precise instant subsequent to separation from m_i; $\vec{v_i}$ is the absolute speed of the point m_i. before union with or after separation from the particles dm_i and $\vec{F_i}$ is the force acting on m_i.

Up until then and except in its very origin, the two-body problem with variable mass had been considered a strictly mathematical problem, and therefore no physical application had been used. It could be said that in 1924, Jeans was the first to pose this as an astrophysical problem basing his studies on the theories of Eddington (1924) on the relationship between luminosity and star mass by generalizing a law of mass loss

$$\dot{m} = -\alpha m^n \tag{3}$$

where a and n are real numbers, the first of them positive proximate to zero and n varying between 1.4 and 4.4, law that is called Eddington-Jeans's law.

Note that if in (3) $n = 2$ we have Mestschersky's first integrable case, and if $n = 3$ the second one.

In 1963, J. D. Hadjidemetriou used the Lagrange equations corresponding to the two-body problem with isotropic mass loss considering the problem as one of two-bodies subject to the perturbation $\vec{\gamma} = \frac{-1}{2}\frac{\dot{m}}{m}\vec{v}$.

Taking into account also the law of mass variation $\dot{m} = -\alpha m^n$, Hadjidemetriou numerically integrated the equations for the elliptic case using a Runge-Kutta method.

In a later study, Hadjidemetriou (1966) describes the equations of motion in terms of eccentric anomaly and develops the derivatives de/dE, $d\omega/dE$, dm/dE in series in which they appear in independent terms, as sines and cosines of E. Hence he integrated the equations analytically in successive orders.

Among other considerations the author also contemplates the possibility that each of the components of a binary system loses mass according to a different law.

Since F. Mestschersky's initial study a great many researchers have dedicated much of their time to this problem, one of the classics of Celestial Mechanics, which has become known as the Gylden-Mestschersky problem.

An ample bibliography can be found in the published works of J. Dommanget (1963), E. N. Polyakova (Polyakova 1994) and C. Prieto (Prieto 1995).

The specific case which results in a slow isotropic mass loss has also been the focus of exhaustiv studies carried out by researchers such as, L. Vander Laan (1972), F. Verhulst (1969(1), 1969(2)), F. Verhulst and W. Eckhaus (1970), etc.

1.1. SLOWLY DECREASING MASS

It has been proved that in most cases stellar mass loss occurs very slowly which justifies the fact that the factor α. in relation (3) is a small parameter.

More specifically, it is presently accepted that for hot stars, annual mass loss ranges between 10^{-5} and 10^{-7} solar masses. Examples are given in Table 1

Star	Annual Mass Loss (in solar masses)
P Cyg	(20-30) $\times 10^{-6}$
λ Cep	(4.0-5.1) $\times 10^{-6}$
9 Sgr	4.0 $\times 10^{-6}$
ϵ Ori	(3.1-3.3) $\times 10^{-6}$
ξ Ori	(1.9-2.3) $\times 10^{-6}$
HD 48099	(0.63-0.64) $\times 10^{-6}$
HD 42088	(0.13-0.20) $\times 10^{-6}$

TABLE 1.

For colder stars a decrease in mass is estimated to be between 10^{-7} and 10^{-14} solar masses for supergiant K5 and dwarf stars, respectively. As regards the Sun, the latest estimates indicate that each year it loses in the order of $(2.2 \pm 0.6) \times 10^{-14}$ of its mass.

By applying these considerations to the case of a double star with no transfer of mass among its components we have a real case of the two-body problem with slowly decreasing mass.

This problem can be treated as a perturbed Keplerian motion in the sense that if at any given moment the mass remained constant, then the relative course would correspond to the two-body problem (osculating orbit). In this case and in accordance with the method of variation of the constants, we may assume that even though the relative orbit is not yet conical globally speaking, it is so at each instant, although now the elements which define it are temporal functions.

2. THE HAMILTONIAN PROBLEM.

As in the two-body problem we can obtain the Hamiltonian function associated to a two-body problem with variable mass:

$$F = \frac{1}{2}\dot{r}^2 - \frac{Gm(t)}{r}$$

we will choose the unit of time so that for G = 1 and therefore $\mu = m$

As we we have already seen, the loss of mass (3) depends on a small parameter α so it seems reasonable that we could find a development of the Hamilton function in a small parameter related to α which we could use to integrate the equations up to different orders of perturbation.

We know that for this kind of integration Delauny's variables are very usefull. For this reason we write the Hamiltonian in that variables

$$F = F(L, G, \ell; t) = -\frac{1}{2}\frac{m^2}{L^2} + \frac{\dot{m}}{m} Le \sin E \tag{4}$$

an expression already obtained by Deprit (1983).

So, the canonical equations will be

$$\begin{aligned}
\frac{d\ell}{dt} &= \frac{\partial F}{\partial L} = \frac{m^2}{L^3} + \frac{\dot{m}}{m}\left(e + \frac{G^2}{L^2 e}\right)\sin E \\
\frac{dg}{dt} &= \frac{\partial F}{\partial G} = -\frac{\dot{m}}{m}\frac{G}{eL}\sin E \\
\frac{dL}{dt} &= -\frac{\partial F}{\partial \ell} = -\frac{\dot{m}}{m}\frac{eL\cos E}{(1-e\cos E)} \\
\frac{dG}{dt} &= -\frac{\partial F}{\partial g} = 0
\end{aligned} \tag{5}$$

The equations of motion can be analytically integrated until the second order of perturbation using the Deprit's method (Deprit 1963). To do so would firstly require the development of the Hamilton function (4) in a small parameter power ϵ so that for $\epsilon = 0$ the Keplerian case would result. It seems reasonable that this parameter should be related to the coefficient α which appears in the law of mass variation because if $\alpha = 0 \Rightarrow \dot{m} = 0$ and we have the Two-Body problem. We will simply choose ϵ as the nondimensioned value of α.

This choice also justifies the application of the method as far as the second order only, since α is very small in higher orders and would therefore not contribute significantly.

Let us now consider for the function $m(t)$ a development in the Taylor series of the form:

$$m = m_0 + \sum_{p=1}^{\infty} \frac{1}{p!} m_0^{(p)} (t - t_0)^p \tag{6}$$

m_0 being the value of the mass in a certain initial instant t_0 and where $m_0^{(p)}$ represents the pth derivative of the function m with respect to t evaluated as $t = t_0$.

If this expression is substituted in (4) and certain operations effected, the result is

$$F = F_0 + \frac{\epsilon}{1!} F_1 + \frac{\epsilon^2}{2!} F_2 + \cdots$$

where the F_i terms are, up to the second order:

$$\begin{aligned}
F_0 &= -\frac{m_0^2}{2L^2} \\
F_1 &= \frac{m_0^{n+1}}{L^2}(t - t_0) - eLm_0^{n-1}\sin E \\
F_2 &= 2eLm_0^{2n-2}(n-1)(\sin E)(t - t_0) - \frac{m_0^{2n}(n+1)}{L^2}(t - t_0)^2
\end{aligned}$$

As is well known the Deprit's method is based on the application of the Lie transformation to the initial canonical system with the aim of obtaining other systems which are simpler to integrate.

If we applied this integration up to second order we obtain the generating function of the canonical transformation as:

$$W = \epsilon W_1 + \frac{\epsilon^2}{2!} W_2$$

being

$$W_1 = -e^* L^{*4} m_0^{n-3} \left[\frac{e^*}{2} + \cos E^* - \frac{e^*}{2}\cos(2E^*)\right]$$

and

$$\begin{aligned} W_2 &= L^{*4} m_0^{2(n-3)} \Big[\tfrac{E^* L^*}{2} \Big[e^* L^{*2}(2n-3) - 2G^{*2} \Big] + \tfrac{e^* L^*}{4} \Big[4G^{*2} - \\ &-8L^{*2}(n+1) - 2e^{*2} L^{*2}(n-3) \Big] \sin E^* - \\ &-e^{*2} \left(n \tfrac{L^{*3}+1}{4} + L^{*3} + 1 \right) \sin(2E^*) + \\ &\tfrac{L^{*3} e^{*3}(n+1)}{6} \sin(3E^*) \; + 2e^*(n+1) m_0^2 (t-t_0) \cos E^* - \\ &-\tfrac{e^{*2}}{2}(n+1) m_0^2 (t-t_0) \cos(2E^*) - \tfrac{e^{*3}}{2}(n+1) \sin E^* \cos(2E^*) \Big] \end{aligned}$$

On the other hand, since the new Hamilton function:

$$F^* = F_0^* + \epsilon F_1^* + \frac{\epsilon^2}{2!} F_2^* = -\frac{m_0^2}{2L^{*2}} - \epsilon \frac{m_0^{n+1}}{L^{*2}}(t-t_0) - \frac{\epsilon^2}{2!} \frac{m_0^{2n}(n+1)}{L^{*2}}(t-t_0)^2$$

only depends on L^* and time, the canonical equations can be integrated into the form:

$$\begin{aligned} \tfrac{dG^*}{dt} &= -\tfrac{\partial F^*}{\partial g^*} = 0 \Rightarrow G^* = cte \\ \tfrac{dL^*}{dt} &= -\tfrac{\partial F^*}{\partial \ell^*} = 0 \Rightarrow L^* = cte \\ \tfrac{dg^*}{dt} &= \tfrac{\partial F^*}{\partial G^*} = 0 \Rightarrow g^* = cte \end{aligned}$$

$$\frac{d\ell^*}{dt} = \frac{\partial F^*}{\partial L^*} = \frac{m_0^2}{L^{*3}} - \frac{2\epsilon m_0^{n+1}}{L^{*3}}(t-t_0) + \frac{\epsilon^2 m_0^{2n}(n+1)}{L^{*3}}(t-t_0)^2 \tag{7}$$

$$\ell^* = A^*{}_0 + A^*{}_1 t + A^*{}_2 t^2 + A^*{}_3 t^3$$

where

$$\begin{aligned} A^*{}_1 &= \tfrac{m_0^2 + 2\epsilon m_0^{n+1} t_0 + \epsilon^2 m_0^{2n} t_0^2 (n+1)}{L^{*3}} \\ A^*{}_2 &= -\tfrac{\epsilon m_0^{n-1}}{L^{*3}} \left(m_0^2 + \epsilon m_0^{n+1} t_0 (n+1) \right) \\ A^*{}_3 &= \tfrac{\epsilon^2 m_0^{2n}(n+1)}{3L^{*3}} \end{aligned} \tag{8}$$

and $A^*{}_0$ is an integration constant.

2.1. PRACTICAL APPLICATION

As a test of our analytic solution we have considered the case of a double star in which the total mass loss is of the order of that of the Sun's, that is we will use $\alpha = 0.35 \times 10^{-14} \odot/u.t.$ from the Eddington-Jeans law as E. L. Schatzman and F. Praderie (Schatzman y Praderie 1993) suggest. Likewise, we will use $n = 2$, having chosen this value to enable us to avail of the Mestschersky exact solution. The initial values taken for the mass and the orbital elements were the following: $m(t_0) = 1$, $a(t_0) = 1$, $e(t_0) = 0.5$, $\omega(t_0) = 0$ y $f(t_0) = 0$.

From a certain time $t_0 = 0$, in an interval of 600.000 units of time (u.t.), we calculated the values of the orbital elements in 1.200.000 instants each two separated 0.5 u.t., an interval which would correspond to the Sun-Earth system approximately 100.000 years away.

With the aim of making the development of (6) convergent, the algorithm developed was reinitiated at each stage, that is, each 0.5 units of time, although we tested different values between 0.1 and 0.9 to see if there was any influence on the solution and have demostrated that there is no significant effect. For the purpose of comparison, Table 2 gives the results corresponding to $t - t_0$ equals 0.2, 0.5 and 0.8 with the exact Mestschersky solution.

To apply the Mestsersky solution we have used the relation $\tau = \frac{t}{1+\alpha t}$ instead of the expression in (1), since we have proved any function of the form $\tau = \frac{\pm 1}{\alpha(1+\alpha t)} + C$ is valid as a time change. In particular, if we put $C = \frac{1}{\alpha}$ it gives the above result.

	a	e
Exact Solution	1.0000000095492905	0.49999999999999381
$\Delta t = 0.2$	1.0000000094711963	0.49999999994142251
$\Delta t = 0.5$	1.0000000095180446	0.49999999997655867
$\Delta t = 0.8$	1.0000000095297594	0.49999999998534476
	ω	f
Exact Solution	$1.7747582837255322 \times 10^{-16}$	600600.5485507190485773
$\Delta t = 0.2$	$1.2280071297662721 \times 10^{-16}$	600600.5473249171422396
$\Delta t = 0.5$	$1.2288285088346467 \times 10^{-16}$	600600.5472071655169355
$\Delta t = 0.8$	$1.2292317044210630 \times 10^{-16}$	600600.5471493967954257

TABLE 2. Values for orbital elements in the last instant

Likewise, we integrated the equations of motion using a Runge-Kutta method of the eighth-order for systems of first-order ordinary differential equations. As a comment on the results obtained we can state that by comparing the analytic solution and the numerical solution with the exact, ours gives smaller differences in the final instant, as is shown in Table 3 where A denotes the analytic, N the numerical and M the exact Mestsersky solution.

All these calculations were carried out on a FUJITSU vectorial computer model VP 2400/10 in the Supercomputation Center of Galicia (CESGA), the calculation times being 59^s12 for the exact solution, 1^m05^s33 for our analytic solution and $2^m04^s12$for the numerical method.

	a	e	ω	f
A-M	10^{-9}	10^{-10}	10^{-16}	10^{-3}
N-M	10^{-7}	10^{-9}	10^{-15}	10^{-2}

TABLE 3. Differences in the values obtained for each orbital element.

3. THE BIPARAMETRIC CASE.

Let us consider now a law of variation similar to (3), but instead of globally expressing the mass loss, this is to be taken independently for each of the two-bodies, that is

$$\dot{m_i} = -\alpha_i {m_i}^{n(i)} \quad i = 1, 2$$

where n_i is represented by $n(i)$ so as to facilitate formula writing.

This law of variation had already been proposed by J. D. Hadjidemetriou (Hadjidemetriou 1963) and its application solves a perturbed two-body problem dependent on two small parameters.

From a practical point of view, the fact that the components of numerous binary systems present different physical characteristics (mass, magnitude, spectral type, etc) it may be of benefit to avail of laws of mass loss like those considered in this present study, when studying the evolution of the system.

As in the uniparametric problem, we now express function (4) in a series of powers of two small parameters $\epsilon_i (i = 1, 2)$ which are related to mass loss in each of the bodies. We simply choose ϵ_i as the adimensioned value of $\alpha_i (i = 1, 2)$.

$$\begin{aligned} F &= F_0 + \epsilon_2 F_{01} + \epsilon_1 F_{10} = -\frac{(m_{10}+m_{20})^2}{2L^2} + \\ &+ \epsilon_1 \left[\frac{\left(m_{10}^{n(1)+1} + m_{10}^{n(1)} m_{20} \right)}{L^2} (t - t_0) - \frac{Lem_{10}^{n(1)} \sin E}{m_{10}+m_{20}} \right] \\ &+ \epsilon_2 \left[\frac{\left(m_{20}^{n(2)+1} + m_{20}^{n(2)} m_{10} \right)}{L^2} (t - t_0) - \frac{Lem_{20}^{n(2)} \sin E}{m_{10}+m_{20}} \right] \end{aligned}$$

So, the corresponding equations of the movement are:

$$\frac{d\ell}{dt} = \frac{\partial F}{\partial L}; \qquad \frac{dg}{dt} = \frac{\partial F}{\partial G}$$

$$\frac{dL}{dt} = -\frac{\partial F}{\partial \ell}; \quad \frac{dG}{dt} = -\frac{\partial F}{\partial g} = 0$$

We have now a canonical system dependent on two small parameters, which may be analytically integrated to a certain order of perturbation using a biparametric method. The method which we have used to integrate the system is that developed by A.J. Abad and J. Ribera (Ribera 1981; Abad & Ribera 1984) which is based on the concept of the Lie transformations (Hori 1966; Deprit 1969).

The integration technique essentially consists in obtaining the generatrix function $W_\epsilon = \epsilon_1 W^1 + \epsilon_2 W^2$ from a completely canonical transformation which transforms the canonical system into another which is more manageable. To do so, we supose that the components W^1 and W^2 of the generatrix function can likewise be developed in a series of powers of $\epsilon = (\epsilon_1, \epsilon_2)$.

Let us consider now, as in the uniparametric case, an infinitesimal canonical transformation which passes us from one set of variables (L, G, ℓ, g) to another (L^*, G^*, ℓ^*, g^*), so that the new Hamiltonian function is independent from ℓ^* . More specifically, we choose

$$F_0^* + \epsilon_1 F_{10}^* + \epsilon_2 F_{01}^* = -\frac{(m_{10}+m_{20})^2}{2L^{*2}} +$$

$$+\epsilon_1 \left[\frac{\left(m_{10}^{n(1)+1} + m_{10}^{n(1)} m_{20}\right)}{L^{*2}}(t-t_0) - \frac{L^* e^* m_{10}^{n(1)} \sin E^*}{m_{10}+m_{20}} \right] +$$

$$+\epsilon_2 \left[\frac{\left(m_{20}^{n(2)+1} + m_{20}^{n(2)} m_{10}\right)}{L^{*2}}(t-t_0) - \frac{L^* e^* m_{20}^{n(2)} \sin E^*}{m_{10}+m_{20}} \right]$$

In this conditions and with the equations of method we can obtain W^1 and W^2

$$W^1 = \frac{L^{*4} e^* m_{10}^{n(1)} \cos E^*}{(m_{10}+m_{20})^3}\left(1 - \frac{e^* \cos E^*}{2}\right)$$

$$W^2 = \frac{L^{*4} e^* m_{20}^{n(2)} \cos E^*}{(m_{10}+m_{20})^3}\left(1 - \frac{e^* \cos E^*}{2}\right)$$

So that, the change of variables, in the first order, is given by

$$L = L^* - \frac{\partial W}{\partial \ell^*}; \quad G = G^* - \frac{\partial W}{\partial g^*} = G^*$$

$$\ell = \ell^* + \frac{\partial W}{\partial L^*}; \quad g = g^* + \frac{\partial W}{\partial G^*}$$

the and the canonical equations in the new variables can be integrated in the following way

$$\frac{dG^*}{dt} = -\frac{\partial F^*}{\partial g^*} = 0 \Rightarrow G^* = cte$$

$$\frac{dL^*}{dt} = -\frac{\partial F^*}{\partial \ell^*} = 0 \Rightarrow L^* = cte$$

$$\frac{dg^*}{dt} = \frac{\partial F^*}{\partial G^*} = 0 \Rightarrow g^* = cte$$

$$\frac{d\ell^*}{dt} = \frac{\partial F^*}{\partial L^*} = \frac{(m_{10}+m_{20})^2}{L^{*3}} - \epsilon_1 \frac{2m_{10}^{n(1)+1} + 2m_{10}^{n(1)} m_{20}}{L^{*3}}(t-t_0) +$$

$$+\epsilon_2 \frac{-2m_{10} m_{20}^{n(2)} + 2m_{20}^{n(2)+1}}{L^{*3}}(t-t_0)$$

$$\ell^* = A^*{}_0 + A^*{}_1 t + A^*{}_2 t^2$$

where

$$A^*{}_1 = \frac{(m_{10}+m_{20})^2 + 2(m_{10}+m_{20}) t_0 (\epsilon_1 m_{10}^{n(1)} + \epsilon_2 m_{20}^{n(2)})}{L^{*3}}$$

$$A^*{}_2 = -\frac{(m_{10}+m_{20})(\epsilon_1 m_{10}^{n(1)+1} + \epsilon_2 m_{20}^{n(2)} m_{10})}{m_{10} L^{*3}}$$

and $A^*{}_0$ are an integration constant.

Note that in this case the expression ℓ^* is only a polynomial of the second order and not of the third, as in the uniparametric case, since here we have only developed as far as the first order, not the second as in the other case.

Acknowledgements

This study was carried out as part of the research projects PB92-1074 and XUGA 24301B96 financed by the Spanish D.G.I.C.Y.T. and the XUNTA DE GALICIA (Spain) respectively.

References

Abad, A. J. and Ribera, J.:1984. *Publicaciones Seminario Matemático García de Galdeano, Serie II, Sección 1*, **3**. Univ. de Zaragoza. Spain
Deprit, A.:1963. *Celest. Mech.*, **31** 1-22
Deprit, A.:1983. *Celest. Mech.*, **31** 1-22
Dommanget, V.:1963. *Ann. Obs. R. Belg.*, **9** 213
Dufour, M. Ch.:1866. *Comptes Rendus Hebdomadaires de L'Accademie de Sciences*, 840-842
Eddington, A. S.:1924 *Mont. Not. R. Astron. Soc.*, **84** 308-332
Gyldén, H.:1884. *Astron. Nachr.*, **2593-94** 1-6
Hadjidemetriou, J.:1963. *Icarus*, **2** 440-451
Hadjidemetriou, J.:1966. *Icarus*, **5** 34-46
Jeans, J. H.:1924. *Mont. Not. R. Astron. Soc.*, **85,1** 2-11
Jeans, J. H.:1925. *Mont. Not. R. Astron. Soc.*, **85,9** 912-914
Mestschersky, F.:1893. *Astron. Nachr.*, **3153** 8-9
Mestschersky, F.:1902. *Astron. Nachr.*, **3807** 229-240
Mestschersky, F.:1949. *Moscow, Gostekhizdat*
Oppolzer, Th. V.:1884. *Astron. Nachr.*, **2573** 67-72
Polyakhova, E. N.:1994. Astron. Rep., 38(2), 283-291
Prieto, C.:1995. *PhD Publicacións do Departamento de Matemática Aplicada* **Num. 1**, I.S.B.N.: 84-89581-00-2. D.L.: C-1714-95. Spain
Ribera, J.:1981. *PhD*. Univ. Zaragoza. Spain
Schatzman, E. L. y Praderie, F.:1993. *Springer-Verlag*, Berlin
Van Der Laan, L. and Verhulst, F.:1972. *Celest. Mech.*, **6** 343-351
Verhulst, F.:1969. *Bull. Astron. Inst. Neth.*, **20** 215-221
Verhulst, F.:1969. *Bull.* 5th *ICNO Conference, Kiev*, 158-168
Verhulst, F. and Eckhaus, W.:1970. *Int. J. Non-Linear Mechanics*, **5** 617-624

ON THE NUMERICAL INTEGRATION OF TWO BODY PROBLEM WITH VARIABLE MASS

C. CALVO
Grupo de Mecánica Espacial. Universidad de Zaragoza, Spain.

AND

M. PALACIOS
Dpto. Matemática Aplicada (CPS) and Grupo de Mecánica Espacial. Universidad de Zaragoza, Spain.

Abstract. Using the computational advantages offered by the formulation of orbital problems with respect to the ideal frame with or without regularization, the numerical integration of the two-body variable-mass problem is developed. An isotropic slow mass change model in several cases is considered, because this is the situation with more analytical difficulties, although physically is quite well substantiated in celestial mechanics problems. Jeans' law of mass loss is taken into account and the evolution of orbital elements is obtained in both the uniparametric and biparametric problems.

1. Introduction

In the nineteenth century, Meshcherskii (1897) [2] formulated the two-body problem with decreasing mass in terms like these: assume that mass is ejected isotropically from the two-body system at very high velocities and is lost to the system, how will the orbits change with time and what will be the variation in quantities such as the energy and the angular momentum. For isotropic variation one means uniform ejection, without collisions and continuously in all directions.

¿From that year onwards, the differential equations describing the problem with appropriate initial conditions has been the subject of many investigations; during the last decades [5], the problem again became of importance in the context of some astrophysical theories. No general approach to the problem is known and this is due to the particular difficulties of the problem, which is non-linear, non-autonomous and should be solved for all relevant election of functions $m(t)$ and all possible initial values. Moreover, the equations contain singularities.

The equations of the relative motion

$$\ddot{\boldsymbol{r}} = -\frac{\mathcal{G}m(t)}{r^3}\,\boldsymbol{r} \tag{1}$$

J. A. Docobo et al. (eds.), Visual Double Stars: Formation, Dynamics and Evolutionary Tracks, 341–346.

gives [7], after some manipulations, the following equation for the relative distance, r:

$$\ddot{r} = -\frac{\mathcal{G}\, m(t)}{r^2} + \frac{G^2}{r^3},$$

where G stands for the norm of the angular momentum. Taking into account the last equation, the variation of the energy can be given by

$$\frac{dE}{dt} = -\mathcal{G}\,\frac{\dot{m}}{r}$$

so the energy of the system is a monotonically increasing function of time during the process of ejection of mass. This property illustrates the importance of the process of loss of mass in double star systems, as it may cause disruption of binary systems and at the same time change the velocity distribution of stars in the galaxy. Important efforts have been done in connection to the effects of galactic mass loss in galactic dynamics [4].

Recently, by means of analytical methods, Prieto [6] study the type of variation produced on the orbital elements when the motion of a star with respect to another is considered to be perturbed by the supposed isotropic loss of mass and make a qualitative analysis of the evolution of the classical orbital elements. It also makes in nine cases a comparison between the analytical and a numerical solution considering that the loss of mass is independent in both bodies.

Hadjidemetriou [1] establishes that the isotropic loss of mass is a problem equivalent to a two-body problem perturbed by $\boldsymbol{\gamma} = -\dot{m}/(2\,m)\,\boldsymbol{v}$, where $\boldsymbol{v}$ is the velocity vector.

The purpose of this paper is the numerical integration of the two body problem with decreasing masses perturbed by tidal effect, radiation pressure or a third body, considering the biparametric problem, that is, both bodies change their masses independently. In order to gain some digits in the numerical solution we pose the problem in several sets of variables, mainly Cartesian coordinates, projective coordinates and regularized projective coordinates. Some figures are presented showing the behavior of the numerical solutions when compared with the analytical one.

2. Several Formulations for Orbital Dynamics

Let us remember several formulations for the equations of the motion of a two-body problem to which is superimposed a perturbing force $\boldsymbol{F}$, in such a way that the dynamical equations show as clearly as possible the dynamical properties of the solutions. This type of formulation provides great efficiency when integrated numerically [3].

All the following frames have the same origin at the center of mass of the system. First at all, we will consider the *orbital frame* $\mathcal{O}$ defined by three orthonormal vectors $(\boldsymbol{u}, \boldsymbol{v}, \boldsymbol{n})$ as follow: $\boldsymbol{u}$ is the radial vector, $\boldsymbol{n}$ is the instantaneous normal vector to the orbit and $(\boldsymbol{u}, \boldsymbol{v})$ is the instantaneous orbital plane.

If we select the orbital frame at an initial time t_0, we obtain a fixed orthonormal reference system that will be named *departure frame* and denoted by $\mathcal{D}$. The formulation of the relative motion of the mass $\mu_1(t)$ around the mass $\mu_2(t)$ expressed in Cartesian coordinates with respect to the departure frame will be called **cd model** and is given by the differential system:

$$\left.\frac{d^2\boldsymbol{r}}{dt^2}\right|_{\mathcal{D}} = -\frac{\mu}{r^3}\boldsymbol{r} + \boldsymbol{F}\,, \tag{2}$$

where $\mu = \mathcal{G}(\mu_1(t) + \mu_2(t))$.

Let us remark that the orbital frame is the departure frame rotated by a vector $\boldsymbol{w} = (G/r^2)\,\boldsymbol{n} + (r/G)\,(\boldsymbol{F}\cdot\boldsymbol{n})\,\boldsymbol{u}$. If we take the second term of this angular velocity vector, $\boldsymbol{w}_{\mathcal{I}} = (r/G)\,(\boldsymbol{F}\cdot\boldsymbol{n})\,\boldsymbol{u}$, we may consider a new frame $\mathcal{I}$, the *ideal frame*, as the departure frame rotated by $\boldsymbol{w}_{\mathcal{I}}$. It is defined by the orthonormal vectors $(\boldsymbol{u}_{\mathcal{I}}, \boldsymbol{v}_{\mathcal{I}}, \boldsymbol{n}_{\mathcal{I}})$ with $\boldsymbol{n} = \boldsymbol{n}_{\mathcal{I}}$. Consequently the fundamental plane is the same in both orbital and ideal frames and it coincides with the orbital plane. Hansen called it ideal because the velocity is identical in the departure and the ideal frames.

Actually, the motion may be decomposed in two non independent parts: the motion on the orbital plane and a slow rotation of the orbital plane, which will be represented by means of a quaternion q_{di}. The formulation of the relative motion described by equation (2) expressed in polar coordinates with respect to the ideal frame using quaternions will be called **piq model**, and it is given by the differential system:

$$\begin{aligned}
\dot{q}_{di} &= -\frac{r}{2G}(\boldsymbol{F}\cdot\boldsymbol{n}_{\mathcal{I}})\,(0, \cos\theta, \operatorname{sen}\theta, 0)\,q_{di}, \\
\ddot{r} &= \frac{G^2}{r^3} - \frac{\mu}{r^2} + (\boldsymbol{F}\cdot\boldsymbol{u}_{\mathcal{I}})\cos\theta + (\boldsymbol{F}\cdot\boldsymbol{v}_{\mathcal{I}})\operatorname{sen}\theta, \\
\dot{G} &= r\left((\boldsymbol{F}\cdot\boldsymbol{v}_{\mathcal{I}})\cos\theta - (\boldsymbol{F}\cdot\boldsymbol{u}_{\mathcal{I}})\operatorname{sen}\theta\right), \\
\dot{\theta} &= \frac{G}{r^2},
\end{aligned} \tag{3}$$

where θ is the angle between $\boldsymbol{u}_{\mathcal{I}}$ and $\boldsymbol{u}$.

Looking at (3) it seems quite natural to replace the independent variable t and the distance r by the dimensionless variable θ and the quantity ρ, such that

$$\rho = \frac{1}{r} \quad \text{and} \quad r^2 d\theta = G dt.$$

Rescaling the force to $\boldsymbol{F}^* = \dfrac{r^3}{G^2}\boldsymbol{F}$, we obtain the differential system:

$$\begin{aligned}
q'_{di} &= -\frac{1}{2}(\boldsymbol{F}^*\cdot\boldsymbol{n}_{\mathcal{I}})\,(0, \cos\theta, \operatorname{sen}\theta, 0)\,q_{di}, \\
\rho'' &= -\rho'\left(-(\boldsymbol{F}^*\cdot\boldsymbol{u}_{\mathcal{I}})\operatorname{sen}\theta + (\boldsymbol{F}^*\cdot\boldsymbol{v}_{\mathcal{I}})\cos\theta\right) \\
&\quad -\rho\left(1 + (\boldsymbol{F}^*\cdot\boldsymbol{u}_{\mathcal{I}})\cos\theta + (\boldsymbol{F}^*\cdot\boldsymbol{v}_{\mathcal{I}})\operatorname{sen}\theta\right) + \frac{\mu}{G^2}, \\
G' &= G\left(-(\boldsymbol{F}^*\cdot\boldsymbol{u}_{\mathcal{I}})\operatorname{sen}\theta + (\boldsymbol{F}^*\cdot\boldsymbol{v}_{\mathcal{I}})\cos\theta\right), \\
t' &= \frac{1}{G\rho^2},
\end{aligned} \tag{4}$$

where derivatives are taken with respect the new independent variable θ; these equations represent the orbital dynamics of (2) and (3) expressed in regularized polar coordinates with respect to the ideal frame using quaternions; this formulation of the equations of the motion will be called **pirq model**.

In order to study the motion of a binary system with isotropic loss of mass we incorporated to the three models *cd, piq* and *pirq* the so called Eddintong-Jeans law expressed as

$$\frac{d\mu(t)}{dt} = -\alpha\mu(t)^n\,, \quad 1.4 < n < 4.4$$

where α is a constant, in each component, i. e., the loss of mass in each body is considered to be independent, what composes the biparametric model of the problem.

3. Numerical results

INPAID is a FORTRAN software package that we have implemented to numerically propagate orbital problems established as first order initial value problem considering the three models mentioned in the preceding section.

Two numerical propagators have been included in it: a) a Runge-Kutta of order 8 with 13 stages and constant step-size; b) a predictor-corrector in the $P(EC)^m E$ mode, with k stages (k= 6, 8, 10) and order k+2. The predictor is an explicit Adams-Bashforth and the corrector is an optimal AS:S with stability properties and order k+2.

From our own previous experiments [3] we deduce that the 'piq' and 'pirq' models have a better performance and give two more digits of accuracy than the 'cd' model; besides, the 'pirq' model shows slightly better results for eccentric orbits because of the regularization.

As all step-by-step numerical method, reaching more than a few hundred revolutions within the required accuracy is very expensive, if not impossible, because of the accumulated truncation error and excessive CPU time consuming (small efficiency).

In the context of the two-body variable-mass problem we have performed several numerical tests. First at all, we have considered that the only disturbing force is due to the isotropic loss of mass and that it is conducted by the Meshcherskii particular case of the Eddintong-Jeans law that makes (1) integrable, that is, we take $\boldsymbol{F} = 0$ and $d\,\mu(t)/d\,t = -\alpha\,\mu(t)^2$ with $\alpha = 0.000005$.

In order to describe approximatively the motion of a binary star we took the following initial conditions: semi-major axis $a = 2a.u.$, several values of eccentricity e (0.1, 0.3, 0.5 and 0.7, in different cases), argument of pericenter $\omega = 0$, true anomaly $f = 0$.

In the figures presented here we have constrained ourselves to the Runge-Kutta method with an integration step-size of $h = 1$ month and $h = 4$ months. All the figures represent the error in semi-major axis when compared with the analytical solution. Similar results have been obtained for all the orbital elements.

Figure 1 and Figure 2 (left) show that 'cd' model gives a worse performance than 'piq' and 'pirq' model and the last improves the others. This behaviour is similar for other eccentricities and step-sizes.

Figure 2(right) and Figure 3 show that a high eccentricity is not a major difficulty if the 'pirq' model is used; only 2 digits have been lost when the 'big' step-size of 4 months has been taken.

4. Conclusions

The formulation of the two-body variable-mass problem following the schemes of our 'piq' and 'pirq' models is a good way to analyze numerically the behaviour of dynamical systems of those characteristics. In particular, the 'pirq' model is highly recommended for not very small eccentricities.

Acknowledgments

This work has been supported in part by the Spanish Ministerio de Educación y Ciencia (DGICYT Project # PB95-0807).

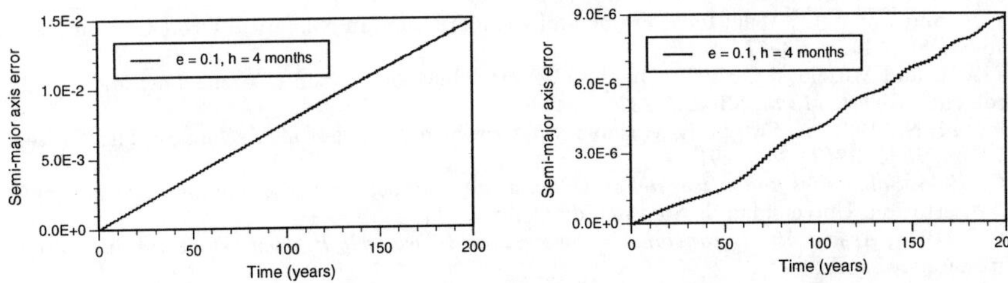

Figure 1. Analytical vs. numerical cd and piq model solutions.

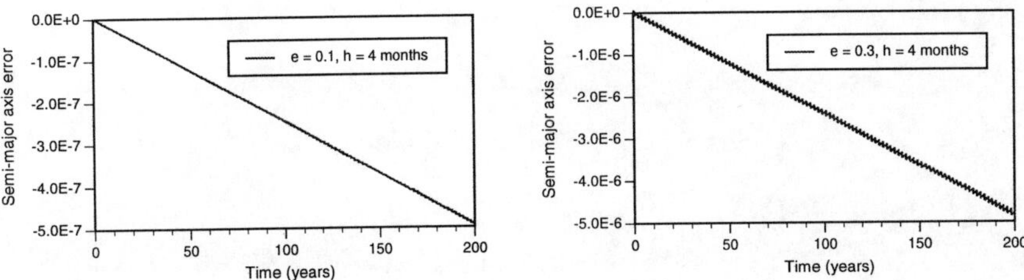

Figure 2. Analytical vs. numerical pirq model solutions.

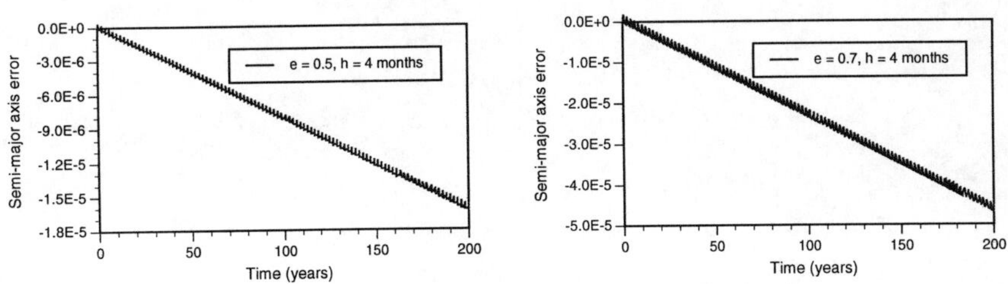

Figure 3. Analytical vs. numerical pirq model solutions.

References

1. Hadjidemetriou, J.D.: 1963, Two-body Problem with Variable Mass: a new Approach. *Icarus*, **2**, 440.
2. Meshcherskii, F.: 1902, Ueber die Integration der Bewegungsgleichungen in Probleme zweier Körper von veränderlicher Masse. *Astron. Nachr.*, **3807**, 229–240.
3. Palacios, M. and Calvo, C.: 1996, Ideal Frame and Regularization in Numerical Orbit Computation. *J. Astronaut. Sci.*, **44**, 1, 63–77.
4. Plastino, A.R. and Muzzio, J.C.: 1992, On the Use and Abuse of Newton's Second Law for Variable Mass Problems. *Celest. Mech.*, **53**, 227–232.
5. Polyakhova, E.N.: 1994, A Two-body Variable-Mass Problem in Celestial Mechanics: The Current State, *Astron. Rep.*. **38(2)**, 283–291.
6. Prieto, C.: 1995, *Soluciones analmticas del problema de dos cuerpos con masa lentamente decreciente.* Doctoral disertation. Universidad de Santiago de Compostela.
7. Verhulst, F.: 1973, *Asymptotic Expansions in the Perturbed Two-body Problem.* Doctoral disertation. Rijksuniversiteit, Utrecht.

PERTURBED GYLDEN SYSTEMS AND TIME–DEPENDENT DELAUNAY–LIKE TRANSFORMATIONS

LUIS FLORÍA
Grupo de Mecánica Celeste I.
Dept. de Matemática Aplicada a la Ingeniería,
E. T. S. de Ingenieros Industriales. Universidad de Valladolid.
E – 47 011 Valladolid, Spain.

Abstract. We study the possibilities and limitations of the application of generalized Delaunay–like transformations (in the 6–dimensional phase space) and TR–like mappings (in the 8–dimensional, extended phase space) to perturbed two–body problems with a time–varying Keplerian parameter $\mu(t)$, that is, to Gylden–type systems. For the sake of theoretical completeness, both negative– and positive–energy motion (with nonstationary coupling parameter) are, in principle, considered. Our developments are intended to introduce canonical variables parallelling the classical ones of Delaunay and the Delaunay–Similar variables of Scheifele.

Key words: Gylden systems, canonical transformations, Delaunay–like map, TR– transformation, perturbed elliptic– and hyperbolic–type motion.

1. Introduction

As a first approximation in the study of other more complex problems involving variable mass or variable gravitational constant, one can consider time–dependent Kepler systems.

To deal with perturbed Gylden systems, say perturbed Keplerian systems with time–varying gravitational parameter $\mu(t)$, we analyze the possibility of introducing certain canonical sets of *generalized* Delaunay (GD) and Delaunay–Similar (GDS) variables. In particular, such variables are intended to account for some perturbation effects.

Variants of generalized *Delaunay–like and Delaunay–Scheifele–like mappings* (Floría 1994a,c and 1996), which normalize certain classes of (time–independent) perturbed Keplerian systems and introduce sets of *generalized* Delaunay and Delaunay–Similar variables *incorporating some perturbing effects* into their definitions, both in the *elliptic* and in the *hyperbolic* perturbed case, are now applied to the treatment of certain time–dependent problems.

The present study aims to extend previous results (most of them originally obtained within the framework of the analytical treatment of orbital motion of artificial Earth satellites) due to Bond & Broucke (1980), Bond & Janin (1981), Deprit (1981a,b and

J. A. Docobo et al. (eds.), Visual Double Stars: Formation, Dynamics and Evolutionary Tracks, 347–356.

1983), Hori (1961), Scheifele (1970) and Scheifele & Graf (1974), whose detailed reference and scope can also be found in some of our papers (Floría 1993, 1994a,b,c, 1995 and 1996).

The derivation of the classical elliptic Delaunay elements by means of a *Delaunay mapping* (see, e. g. Deprit 1981b, §2, pp. 115–118) was adapted to the construction of classical hyperbolic Delaunay elements (Floría 1995), generalized to the introduction of a certain kind of perturbed elliptic elements (Floría 1994a and 1996), and also modified for the derivation of perturbed hyperbolic elements (Floría 1996) with the help of what we call a *Delaunay–like transformation.* In particular, this notion generalizes the quasi–Delaunay map used by Deprit (1981b, §4, p. 126, Formula [37]) to reduce the class of what he called the *quasi–Keplerian systems* (*ibid.*, p. 124, Formula [32].)

On the other hand, on the basis of a TR–transformation (Deprit 1981a), we proposed (Floría 1994b) a *systematic* derivation of classical Delaunay–Similar canonical elements for the Kepler problem, and considered (Floría 1994c) a special extension capable of contracting certain perturbed Keplerian problems onto a standard Keplerian system.

For the treatment of *Gylden* systems, Deprit (1983, §3), Salmassi (1985, §2) and Prieto (1995, Chapter 2, §2.3.1) have employed time–dependent Delaunay mappings.

In this paper, the said concept of a *time–dependent Delaunay–like transformation* will be extended to the construction of a kind of perturbed Delaunay variables, both in the case of perturbed bounded and hyperbolic–type orbital motion. For convenience, we have avoided the use of universal–like functions. The derivation will be presented under a unified form, covering both types of two–body motion, and the transformation will *absorb* some perturbing effects due to the *non–Keplerian* terms occurring in the Hamiltonian systems under consideration. In particular, the kind of perturbation (as taken into account in this research) which turns out to be compatible with a Keplerian–like description of motion is just of the type corresponding to the *Deprit–like potentials* (Deprit 1981b, p. 138.) In this respect, notice that our generalization of previous classical approach and results is intended to be *consistent* with the intrinsic dynamical structure of Keplerian motion.

As for the Delaunay–Similar sets, for convenience and brevity, we restrict ourselves to studying the basic type of variables, without including perturbations into their definitions. Derived in extended phase space, they will employ a pseudo–time (say, an *anomaly*) as the independent parameter, which will be a certain true–like anomaly. The reparametrization of motion will be accomplished by means of a differential transformation of the time variable, generalizing the classical Sundman transformation.

This way of proceeding takes advantage of the use of the *Homogeneous Canonical Formalism*: introduce the physical time t as an additional canonical coordinate whose conjugate momentum is the nagative of the total energy of the dynamical system. Extended phase–space formulation facilitates the introduction of new independent variables different from the physical time t. Additional details can be found, e.g. in Stiefel & Scheifele (1971), Chapter VIII (and, in particular, §30 and §34), and Chapter X, §37.

2. Time–Dependent Delaunay–like Maps and Gylden Problems

Consider the 6–dimensional phase space coordinatized by the so–called *Hill–Whittaker polar nodal variables* (Deprit 1981b, §2), denoted by the symbols $(r, \theta, \nu; p_r, p_\theta, p_\nu)$. Unlike other sets of variables and orbital elements, they are applicable to *any* two–body conic–section orbit. Let t represent the physical time, the inclination being given by $\cos I = p_\nu / p_\theta$.

Let $\mathcal{H}_0$ denote the Hamiltonian corresponding to a *Gylden system* (Deprit 1983; Deprit *et al.* 1989; Prieto 1995). For our purposes, e.g. to take into account the possible *nonsphericity of the gravitational field*, the perturbed Gylden problem under consideration will be given by a Hamiltonian $\mathcal{H} = \mathcal{H}_0 + V$, namely

$$\mathcal{H} \equiv \mathcal{H}_0\,(r\,;\,p_r\,,\,p_\theta\,;\,\mu(t)\,) + V\,(r\,;\,p_\theta\,,\,p_\nu\,;\,\mu(t)\,;\,\varepsilon\,)\,, \tag{1}$$

$$\mathcal{H}_0 = \frac{1}{2}\left[p_r^2 + \frac{p_\theta^2}{r^2}\right] - \frac{\mu(t)}{r}\,, \quad V \equiv \sum_{j=0}^{2} \frac{1}{r^j}\, V_j\,(p_\theta\,,\,p_\nu\,;\,\mu(t)\,;\,\varepsilon)\,. \tag{2}$$

For special cases, the V_j might be taken as

$$V_j \equiv V_j\,(p_\theta\,,\,p_\nu\,;\,\mu(t)\,;\,\varepsilon\,) = \sum_{l=1}^{n_j} \varepsilon^{\,l}\, \mathcal{V}_{(j)\,,\,l}\,(p_\theta\,,\,p_\nu\,;\,\mu(t))\,, \tag{3}$$

and the (small) parameter ε formalizes a perturbation of the basic Gylden problem. However, *the specific functional form of the V_j will not be (in principle) significant for the subsequent developments*, although (at a certain stage of our study) we shall have to analyze and characterize the *compatibility* between the perturbation and a model of Keplerian dynamical structure.

We introduce generalized Delaunay–like variables by means of a *time–dependent canonical transformation* derived from a *generating function* S:

$$(\,r\,,\,\theta\,,\,\nu\,;\,p_r\,,\,p_\theta\,,\,p_\nu\,;\,\mu(t)\,) \;\xrightarrow{\;S\;}\; (\,l_D\,,\,g_D\,,\,h_D\,;\,L_D\,,\,G_D\,,\,H_D\,)$$

$$S \equiv S\,(r,\theta,\nu\,;\,L_D,G_D,H_D\,;\,\mu(t)) = \theta G_D + \nu H_D + \int_{r_0}^{r} \sqrt{Q}\,dr, \tag{4}$$

$$\begin{aligned} Q &\equiv Q_{GD}\,(\,r\,;\,L_D\,,\,G_D\,,\,H_D\,;\,\mu(t)\,;\,\varepsilon\,) \\ &= \frac{2\,\mu(t)}{r} \mp \frac{\mu^2(t)}{L_D^2} - \frac{G_D^2}{r^2} - 2\,V\,(\,r\,;\,G_D\,,\,H_D\,;\,\mu(t)\,;\,\varepsilon\,)\,. \end{aligned} \tag{5}$$

Throughout this paper, the following **sign convention** is to be taken into account: the *upper* sign will refer to *bounded (elliptic-type)* motion, while the *lower* sign will allude to *positive–energy (hyperbolic–type)* orbits. The subscript D means *Delaunay–like*, irrespective of the nature of the orbit. The quantity r_0 will be the *lowest positive root* of the r–equation

$$Q\,(r;L_D,G_D,H_D;\mu(t);\varepsilon) = 0 \Rightarrow r_0 = r_0\,(L_D,G_D,H_D;\mu(t);\varepsilon)\,. \tag{6}$$

It will be shown that the transformation derived from S requires the evaluation of certain quadratures that, for convenience in writing, we denote

$$I_0 = \int_{r_0}^{r} \frac{d\,r}{\sqrt{Q}}\,, \quad I_1 = \int_{r_0}^{r} \frac{d\,r}{r\,\sqrt{Q}}\,, \quad I_2 = -\int_{r_0}^{r} \frac{d\,r}{r^{\,2}\,\sqrt{Q}}\,. \tag{7}$$

The first of them is related to the *integral form of the Kepler equation*, and requires [Formula (17)] the introduction of an eccentric–like anomaly u, in terms of which we also calculate I_1. In its turn, I_2 is solved by means of the *polar equation* [see Eqs. (21) and (22) below] *of a conic–section* in terms of plane polar coordinates in the orbital plane, which introduces a true–like anomaly angle f.

The transformation equations generated by (4) and (5) are

$$p_r = \frac{\partial S}{\partial r} = \sqrt{Q}\,, \quad p_\theta = \frac{\partial S}{\partial \theta} = G_D\,, \quad p_\nu = \frac{\partial S}{\partial \nu} = H_D\,, \tag{8}$$

$$l_D = \frac{\partial S}{\partial L_D} = \pm\frac{\mu^2(t)}{L_D^3}\, I_0\,, \tag{9}$$

$$g_D = \frac{\partial S}{\partial G_D} = \theta - \frac{\partial V_0}{\partial G_D} I_0 - \frac{\partial V_1}{\partial G_D} I_1 + \left[G_D + \frac{\partial V_2}{\partial G_D}\right] I_2\,, \tag{10}$$

$$h_D = \frac{\partial S}{\partial H_D} = \nu - \frac{\partial V_0}{\partial H_D} I_0 - \frac{\partial V_1}{\partial H_D} I_1 + \frac{\partial V_2}{\partial H_D} I_2\,. \tag{11}$$

Now, if we rewrite the function Q in the form

$$\begin{aligned} Q &= \frac{2\,\mu(t)}{r} \mp \frac{\mu^2(t)}{L_D^2} - \frac{G_D^2}{r^2} - \sum_{j=0}^{2} \frac{2}{r^j}\, V_j\,(G_D\,, H_D\,; \mu(t)\,;\,\varepsilon) \\ &= \sum_{j=0}^{2} \frac{1}{r^j}\, A_j\,(L_G\,, G_D\,, H_D\,; \mu(t)\,;\,\varepsilon)\,, \end{aligned} \tag{12}$$

we can *recognize and characterize* the type of perturbation compatible with a model of Keplerian dynamical structure. Indeed, consider a *hypothetical* Kepler problem with coupling parameter $\mu^* = \mu - V_1$ in which the semi–major axis and the energy–like terms are related to A_0, and the (modified) angular momentum is given by $A_2 = G_D^2 + 2V_2$. We conclude that, in order to be allowed to proceed in the same way as when dealing with the classical Delaunay mapping (Deprit 1981, §2; Floría 1994a, 1996), and the results to be consistent with a Keplerian–like description of motion for a fictitious orbit incorporating the perturbations due to the V_j–terms, one must have $V_0 \equiv 0$ and $V_1 \equiv 0$, the only *admissible* perturbation being then of the form

$$V\,(r\,; p_\theta\,, p_\nu\,; \mu(t); \varepsilon) = \frac{1}{r^2}\, V_2\,(p_\theta\,, p_\nu\,; \mu(t); \varepsilon)\,, \tag{13}$$

although additional refinements will be required at a later stage, with the subsequent simplifications in the generating relations derived from S. Accordingly, for consistency of the said pattern, in what follows we shall restrict ourselves to considering this special case of perturbing function. Under these circumstances, to determine the quadratures and complete the transformation equations, we introduce some *subsidiary quantities* $a \equiv a\,(L_D; \mu(t))$, $e \equiv e\,(L_D, G_D, H_D; \mu(t); \varepsilon)$, $p \equiv p\,(G_D, H_D; \mu(t); \varepsilon)$, $\Gamma \equiv \Gamma\,(G_D, H_D; \mu(t); \varepsilon)$ through the formulae

$$L_D = \pm\sqrt{\mu(t)\,a}\,, \quad \pm\mu(t)\,a\left(1 - e^2\right) = G_D^2 + 2\,V_2 \equiv \Gamma^2, \tag{14}$$

$$p = \frac{\Gamma^2}{\mu(t)} = \pm\, a\left(1 - e^2\right), \qquad e^2 = 1 \mp \frac{\Gamma^2}{L_D^2}\,, \tag{15}$$

and obtain a decomposition of Q in the form

$$Q_{\mp} = \pm \frac{\mu a e^2}{r^2}\left[1 - \left(\frac{a \mp r}{a e}\right)^2\right], \qquad r_0 = \pm a(1 - e). \tag{16}$$

Define an *auxiliary integration variable* $u \equiv u(r; L_D, G_D, H_D; \mu(t); \varepsilon)$, of an eccentric–like anomaly type, by the relation

$$\frac{a \mp r}{a e} = \mathcal{C}_{\mp}(u) \Rightarrow r = a\mathcal{R}_{\mp}(u), \; \mathcal{R}_{\mp}(u) = \pm(1 - e\mathcal{C}_{\mp}(u)), \tag{17}$$

with the following meaning for the functions of u:

$$\mathcal{C}_{-}(u) = \cos E, \qquad \mathcal{S}_{-}(u) = \sin E,$$

$$\mathcal{C}_{+}(u) = \cosh F, \qquad \mathcal{S}_{+}(u) = \sinh F,$$

the arguments E and F referring to the usual notation for an (elliptic and hyperbolic, respectively) eccentric–like anomaly.

Then, the expression for Q and the quadrature I_0 result in

$$Q_{\mp} = \frac{\mu e^2 \mathcal{S}^2_{\mp}(u)}{a(1 - e\mathcal{C}_{\mp}(u))^2}, \tag{18}$$

$$I_0 = \pm\sqrt{\frac{a^3}{\mu}}\,(u - e\mathcal{S}_{\mp}(u)) = \frac{L_D^3}{\mu^2}(u - e\mathcal{S}_{\mp}(u)), \tag{19}$$

which converts the Delaunay equation for l_D into the *Kepler equation*

$$l_D = \Phi_{\mp}(u) = \pm(u - e\mathcal{S}_{\mp}(u)). \tag{20}$$

Another *auxiliary integration variable* $f \equiv f(r; L_D, G_D, H_D; \mu(t); \varepsilon)$, of a true–like anomaly type, is now introduced to obtain the quadrature I_2:

$$r = \frac{p}{1 + e\cos f} \Longrightarrow Q_{\mp} = \frac{\pm\mu e^2 \sin^2 f}{a(1 - e^2)} = \frac{\mu e^2}{p}\sin^2 f, \tag{21}$$

$$\Longrightarrow I_2 = -f/\Gamma. \tag{22}$$

In addition to this, we consider a kind of mean motion n such that

$$\mu(t) = n^2 a^3 \Longrightarrow n(L_D; \mu(t)) = \pm\mu^2(t)/L_D^3. \tag{23}$$

These elements allow us to *complete the set of transformation formulae*:

$$l_D = \Phi_{\mp}(u) = \pm(u - e\mathcal{S}_{\mp}(u)) = n\left(t - t_{\text{pericentre}}\right), \tag{24}$$

$$g_D = \theta - \frac{\partial\Gamma}{\partial G_D} f, \qquad \frac{\partial\Gamma}{\partial G_D} = \frac{1}{\Gamma}\left[G_D + \frac{\partial V_2}{\partial G_D}\right], \tag{25}$$

$$h_D = \nu - \frac{\partial\Gamma}{\partial H_D} f, \qquad \frac{\partial\Gamma}{\partial H_D} = \frac{1}{\Gamma}\frac{\partial V_2}{\partial H_D}, \tag{26}$$

along with the equations for the conjugate momenta

$$L_D = \pm\sqrt{\mu a}\,, \quad G_D = p_\theta\,, \quad H_D = G_D \cos I = p_\nu\,. \tag{27}$$

The original Hamiltonian (1) must be expressed in the modified Delaunay–like variables. From the general theory of canonical transformations defined by a time–dependent generating function (Wintner 1947, §46; Schneider 1984, §§1.5.3.4), as in Deprit (1983, §3) or Salmassi (1985):

$$\mathcal{H}^*_\mp \equiv \mathcal{H}^*_\mp\,(l_D, -, -; L_D, G_D, H_D; t; \varepsilon) = \mp\frac{\mu^2(t)}{2\,L_D^2} + \frac{\partial S}{\partial t}. \tag{28}$$

¿From (4) and (5), along with the simplification (13),

$$\frac{\partial S}{\partial t} = \mu'\,I_1 \mp \frac{\mu\,\mu'}{L_D^2}\,I_0 + \mu'\,\frac{\partial V_2}{\partial \mu}\,I_2\,, \quad \text{with} \quad \mu' \equiv \frac{d\,\mu(t)}{d\,t}\,. \tag{29}$$

The quadratures I_0 and I_2 are known from Formulae (19) and (22), but I_1 is to be determined. Moreover, all these formulae should be expressed in terms of the new Delaunay–like canonical variables. To this end, as done by Salmassi (1985), conditions allowing application of Fourier analysis of the two–body problem (see e.g. Brouwer & Clemence 1961, Chapter II; Wintner 1947, Chapter IV; Schneider 1984, §§4.2.5; Schneider 1992, §§3.6.5) should be established.

As an illustration, at this stage we concentrate on the consideration of *elliptic–type motion*. In such a case,

$$I_1 = E\sqrt{a/\mu} = L\,E/\mu\,, \tag{30}$$

and therefore

$$\begin{aligned}\frac{\partial S}{\partial t} &= \mu'\left[\frac{L}{\mu}\,E - \frac{L}{\mu}\,(E - e\sin E) - \frac{1}{\Gamma}\frac{\partial V_2}{\partial \mu}\,f\right]\\ &= \frac{\mu'}{\mu}\,L\,e\,\sin E - \mu'\,\frac{\partial \Gamma}{\partial \mu}\,f = \frac{\mu'}{\mu}\,r\,p_r - \mu'\,\frac{\partial \Gamma}{\partial \mu}\,f\,.\end{aligned} \tag{31}$$

As in Salmassi (1985), to express these formulae in Delaunay–like variables, we can resort to results similar to those of the Fourier analysis of the Kepler motion (e. g. the classical Fourier series expansions in terms of anomalies with coefficients given by Bessel functions involving the eccentricity). In particular, as a simple way of legitimating the process, this approach would require the eccentricity–like quantity e, as given in (15) by

$$e^2 = 1 - \Gamma^2\,(G_D, H_D; \mu(t); \varepsilon)\,/L_D^2\,,$$

to be independent of time t. Accordingly, the function Γ, and therefore the perturbation V_2, should also be time–independent: $V_2 \equiv V_2\,(p_\theta, p_\nu; \varepsilon)$.

In the case of *hyperbolic–type motion*, the quadrature I_1 is solved in terms of the hyperbolic eccentric–like anomaly F, and then Fourier analysis of *non–periodic* Keplerian orbits (Bucerius 1947; Bucerius & Schneider 1966, vol. I, §5.5, pp. 50–54; Schneider 1984, §§4.2.5, pp. 180–183; Schneider 1992, §§3.6.5, pp. 87–92) should be adapted and applied: instead of expansions of the elliptic motion into Fourier series, we should have to deal with *Fourier–integral representations* in which the coefficient–function can be expressed by means of a Hankel function depending on e. This process would introduce Hankel functions to express F in terms of nt. Application of this device also requires the *time–independent* nature of e and, therefore, of the perturbation V_2.

Mathematical foundations of the theory of Fourier integrals can be found e.g. in the classical books by Carslaw (1950) or Titchmarsh (1967).

Account of these details will be given elsewhere.

3. TR–like Transformations and Gylden Systems

The main concern of this section lies in the analysis of the possibility of defining canonical variables of a Delaunay–Similar (DS) type which might be applied to the reduction of the *negative–energy* Gylden problem.

As presented here, the construction of a *generic*, unspecified set of DS–like variables is based on a *variant* of the TR–mapping approach of Deprit (1981a). The unified, compact treatment carried out in the present section hinges on the use of certain *unspecified functions of the momenta* belonging to the new set of canonical variables created by the transformation.

We apply Deprit's technique toward enlarging Scheifele's construction of elliptic Delaunay Similar elements. The first step consists in defining a particular class of time–dependent generating functions which involves an arbitrary function. Another arbitrary function enters at the second step where it is shown how one might transform the independent variable. In Floría (1994b), appropriate selection of these arbitrary functions yields back the special cases considered by Scheifele and others (1970, 1974), Deprit (1981a), Bond & Broucke (1980), or Bond & Janin (1981). Remember that these authors worked out their transformations in order to find integrable approximations of non integrable Hamiltonian systems (namely: the main oblatenes perturbation problem in Artificial Satellite Theory).

Now we consider the 8–dimensional, extended phase space coordinatized by the enlarged canonical set of the Hill–Whittaker polar nodal variables, with $x_0 \equiv t$, namely: $(r, \theta, \nu, x_0; p_r, p_\theta, p_\nu, p_0)$. Here p_0 is the negative of the time–varying energy (per unit of mass) of the Gylden system. We define a canonical transformation,

$$(r, \theta, \nu, x_0; p_r, p_\theta, p_\nu, p_0) \xrightarrow{S} (\varphi, l, g, h; \Phi, L, G, H),$$

to a new set of eight unspecified DS–like variables, which is accomplished by means of a *generating function*

$$S(r,\theta,\nu,x_0;\Phi,L,G,H) = \theta G + \nu H + x_0 L + \int_{r_0}^{r} \sqrt{Q}\, dr, \tag{32}$$

where r_0 is any simple zero of the function Q (usually, the lowest positive root of the r–equation $Q \equiv Q_{DS}(r, x_0; \Phi, L, G, H) = 0$), and the function appearing under the radical sign is given by the expression

$$Q \equiv Q_{DS}(r, x_0; \Phi, L, G, H) = \frac{2\mu(x_0)}{r} - 2L - \frac{\gamma^2}{r^2}, \tag{33}$$

where $\gamma \equiv \gamma(\Phi, L, G, H)$ is a generic, still *unspecified* function of the new momenta. Its unspecified nature gives us the clue to the following general development. At a later stage, its expression will be chosen so as to produce a further simplification on the Hamiltonian of the Gyldem problem.

The generating relations derived from S yield

$$p_r = \frac{\partial S}{\partial r} = \sqrt{Q} \; , \qquad \varphi = \frac{\partial S}{\partial \Phi} = \gamma \frac{\partial \gamma}{\partial \Phi} I_2 \; ,$$

$$p_\theta = \frac{\partial S}{\partial \theta} = G \; , \qquad g = \frac{\partial S}{\partial G} = \theta + \gamma \frac{\partial \gamma}{\partial G} I_2 \; ,$$

$$p_\nu = \frac{\partial S}{\partial \nu} = H \; , \qquad h = \frac{\partial S}{\partial H} = \nu + \gamma \frac{\partial \gamma}{\partial H} I_2 \; ,$$

$$p_0 = \frac{\partial S}{\partial x_0} = L + \mu'(x_0)\, I_1 \; , \qquad l = \frac{\partial S}{\partial L} = x_0 - I_0 + \gamma \frac{\partial \gamma}{\partial L} I_2 \; ,$$

the integrals I_j being, *formally*, the same as in (7). Their evaluation will be facilitated by the introduction of *subsidiary* quantities a, e and p,

$$a = \frac{\mu}{2L} , \quad \gamma^2 = \mu a (1 - e^2) = \mu p , \tag{34}$$

$$p = \frac{\gamma^2}{\mu} = a(1 - e^2) , \quad e^2 = 1 - \frac{2L}{\mu} p = 1 - \frac{2L\gamma^2}{\mu^2} , \tag{35}$$

which allows one to factorize Q_{DS} in the form

$$Q = \frac{\mu}{a r^2} \left[(a e)^2 - (a - r)^2 \right] = \frac{\mu a e^2}{r^2} \left[1 - \left(\frac{a - r}{a e} \right)^2 \right] . \tag{36}$$

The auxiliary integration variables E and f, defined by the relations

$$\frac{a - r}{a e} = \cos E , \quad r = a(1 - e \cos E) , \quad r = \frac{p}{1 + e \cos f} , \tag{37}$$

lead to the final expression for the quadratures

$$I_0 = \sqrt{\frac{a^3}{\mu}} (E - e \sin E) = \frac{\mu}{(2L)^{3/2}} (E - e \sin E) , \tag{38}$$

$$I_1 = E \sqrt{a/\mu} = E/\sqrt{2L} , \qquad I_2 = -f/\gamma . \tag{39}$$

As a consequence, after the required calculations, one obtains the following table of formulae for the coordinates of the new DS set:

$$\varphi = -\frac{\partial \gamma}{\partial \Phi} f , \qquad g = \theta - \frac{\partial \gamma}{\partial G} f ,$$

$$l = x_0 - \frac{\mu(x_0)}{(2L)^{3/2}} (E - e \sin E) - \frac{\partial \gamma}{\partial L} f , \qquad h = \nu - \frac{\partial \gamma}{\partial H} f ,$$

while

$$p_0 = L + \mu'(x_0)\, E/\sqrt{2L} , \quad p_r^2 = Q , \quad p_\theta = G , \quad p_\nu = H .$$

Notice that the functional form of these expresions gives rise to *implicit function and inversion* problems.

4. Application to a Gylden System

In what follows the subscript h refers to the *homogeneous canonical formalism* in the extended phase space. The application of the preceding transformation to the homogeneous Hamiltonian

$$\mathcal{H}_{0,h} \equiv \mathcal{H}_h = \frac{1}{2}\left(p_r^2 + \frac{p_\theta^2}{r^2}\right) - \frac{\mu(x_0)}{r} + p_0 , \tag{40}$$

of a conventional *Gylden* problem, brings it into an expression

$$\widetilde{\mathcal{H}}_{0,h} \equiv \widetilde{\mathcal{H}}_h = \frac{1}{2r^2}\left(G^2 - \gamma^2\right) + \frac{E}{\sqrt{2L}}\,\mu' . \tag{41}$$

It can be simplified via a reparametrizing transformation $t \longrightarrow \tau$ introducing a new *pseudo–time* τ with the help of a differential relation $dt = \widetilde{f}\,d\tau$, where the function $\widetilde{f}$ can be chosen proportional to a power of the distance, $\widetilde{f} = k_\alpha\, r^\alpha$. In this case, with $\mathcal{G}$ a function of the new canonical variables,

$$\widetilde{f} = r^2/\mathcal{G} , \quad dt = \left(r^2/\mathcal{G}\right) d\tau . \tag{42}$$

The transformed Hamiltonian, with τ as the new time parameter, reads

$$\mathcal{K}_h \equiv \mathcal{K}_{0,h} = \widetilde{f}\,\widetilde{\mathcal{H}}_h = \frac{1}{2\,\mathcal{G}}\left(G^2 - \gamma^2\right) + \frac{E\,r^2}{\mathcal{G}\,\sqrt{2L}}\,\mu' . \tag{43}$$

This Hamiltonian should be expressed in terms of the new canonical variables of a DS type. For this purpose (since γ and $\mathcal{G}$ are *per se* functions of the new variables), implicit function results, inversion theorems and Fourier analysis of the two–body problem (as at the end of Section 2) must be invoked again to develop the second term of the right–hand side in (43).

According to Floría (1994b), certain specifications of γ and $\mathcal{G}$ (as functions of *canonical momenta*) lead to simple forms for the first term in (43). Following the scheme of Floría (1994c) with diverse choices of γ and $\mathcal{G}$, the above developments are readily adapted to the case of perturbed motion, with perturbations similar to those given in (2). Under these circumstances, as expected, the functional form of the resulting terms will be more involved.

Finally, to deal with *hyperbolic–type motion*, the just mentioned results (with appropriate modifications of γ and intermediate calculations similar to those required while studying Delaunay–like mappings in 6–dimensional phase space) can be adapted to a DS formulation of positive–energy motion.

Acknowledgements

The author is gratefully indebted to Prof. Manfred Schneider (Technische Universität München) for providing a copy of the article by Bucerius (1947) and information about the book by Bucerius and Schneider (1966, 1967). A copy of the Ph. D. Thesis by C.

Prieto (1995) was provided by Dr. Alberto Abad and Dr. Antonio Elipe (Universidad de Zaragoza). This research has been partially supported by the DGICYT (Ministerio de Educación y Cultura) of Spain, under Grant PB 95–0807.

References

Bond, V. and Broucke, R. (1980) Analytical Satellite Theory in Extended Phase Space, *Celestial Mechanics*, **21**, pp. 357–360.

Bond, V. R. and Janin, G. (1981) Canonical Orbital Elements in Terms of an Arbitrary Independent Variable, *Celestial Mechanics*, **23**, pp. 159–172.

Brouwer, D. and Clemence, G. M. (1961) *Methods of Celestial Mechanics*, Academic Press, New York and London.

Bucerius, H. (1947) Zur FOURIER–Analyse der Lösungen des Zweikörperproblems, *Astronomische Nachrichten*, **275**, pp. 193–202.

Bucerius, H. and Schneider M. (1966, 1967) *Himmelsmechanik* (I und II). Bibliographisches Institut Hochschultaschenbücher, Mannheim

Carslaw, H. S. (1950) *Introduction to the Theory of Fourier's Series and Integrals* (Third Edition, revised and enlarged). Dover, New York (republication of the edition published by Macmillan and Company in 1930).

Deprit, A. (1981a) A Note Concerning the TR–Transformation, *Celestial Mechanics*, **23**, pp. 299–305.

Deprit, A. (1981b) The Elimination of the Parallax in Satellite Theory, *Celestial Mechanics*, **24**, pp. 111–153.

Deprit, A. (1983) The Secular Acceleration in Gylden's Problem, *Celestial Mechanics*, **31**, pp. 1–22.

Deprit, A., Miller, B. and Williams, C. A. (1989) Gylden Systems: Rotation of Pericenters, *Astrophysics and Space Science*, **159**, pp. 239–270.

Floría, L. (1993) On the Definition of a Family of Sets of Canonical Elements for Hyperbolic Motion, *Comptes rendus de l'Académie des Sciences de Paris*, Série **II**, **316**, pp. 1369–1374.

Floría, L. (1994a) Delaunay–like Transformations. In: K. Kurzyńska, F. Barlier, P.K. Seidelmann & I. Wytrzyszczak (Editors), *Dynamics and Astrometry of Natural and Artificial Celestial Bodies*, pp. 175–180. Astronomical Observatory of A. Mickiewicz University, Poznań, Poland.

Floría, L. (1994b) On the Definition of the Delaunay–Similar Canonical Variables of Scheifele, *Mechanics Research Communications*, **21**, pp. 409–414.

Floría, L. (1994c) A Canonical Reduction of a Class of Perturbed Two–Body Problems, *Revista de la Academia de Ciencias Excactas, Físico–Qímicas y Naturales de Zaragoza*, **49**, pp. 77–92.

Floría, L. (1995) A Simple Derivation of the Hyperbolic Delaunay Variables, *Astronomical Journal*, **110**, pp. 940–942.

Floría, L. (1996) Canonical Elements for Perturbed Orbital Motion Using an Arbitrary Independent Variable (Paper AAS 96–136), *Advances in the Astronautical Sciences*, **93**. (In press.)

Hori, G.–i. (1961) The Motion of a Hyperbolic Artificial Satellite around the Oblate Earth, *Astronomical Journal*, **66**, pp. 258–263.

Prieto, C. (1995) *Soluciones analíticas del problema de dos cuerpos con masa lentamente decreciente.* (Doctoral Dissertation.) Universidad de Santiago de Compostela.

Salmassi, M. (1985) Second Order Adiabatic Invariants Associated with the Two–Body Problem with Slowly Varying Mass, *Celestial Mechanics*, **37**, pp. 359–369.

Scheifele, G. (1970) Généralisation des éléments de Delaunay en Mécanique Céleste. Application au mouvement d'un satellite artificiel, *Comptes rendus de l'Académie des Sciencies de Paris*, Série **A**, **271**, pp. 729–732.

Scheifele, G. and Graf, O. (1974) *Analytical Satellite Theories Based on a New Set of Canonical Elements.* AIAA Paper No. 74–838. AIAA Mechanics and Control of Flight Conference, August 1974, Anaheim, California.

Schneider, M. (1984) *Himmelsmechanik* (2., verbesserte Auflage). Bibliographisches Institut Wissenschaftsverlag, Mannheim–Wien–Zürich.

Schneider, M. (1992) *Himmelsmechanik (Band I: Grundlagen, Determinierung).* Bibliographisches Institut Wissenschaftsverlag, Mannheim–Leipzig–Wien–Zürich.

Stiefel, E. L. and Scheifele, G. (1971) *Linear and Regular Celestial Mechanics*. Springer–Verlag, Berlin–Heidelberg–New York.

Titchmarsh, E. C. (1967) *Introduction to the Theory of Fourier Integrals*. (Second Edition). Oxford University Press. (Reprinted, with corrections, from the Second Edition, 1948).

Wintner, A. (1947) *The Analytical Foundations of Celestial Mechanics* (Second printing from the 1941 edition). Princeton University Press, Princeton, New Jersey.

A DIRECT GEOMETRICAL METHOD FOR DETERMINATION OF THE ELLIPTIC ORBIT OF A BINARY STAR USING ITS PROJECTION ON THE CELESTIAL SPHERE

A. A. KISSELEV
Main Astronomical Observatory of RAS
196140, 64 Pulkovskoe sh., St.Petersburg, Russia

Abstract. A classical problem of determination of a true elliptic orbit of a visual double star is solved in a quite new way. The key-formula of the method (2) determine the relative range of the main semiaxes of the true ellipse, if the positions of main star's projection onto a plane of the apparent ellipse is known. Then the angles of mutual orientation of both apparent and true ellipses may be obtained by some simple formulae using only the values of the major and minor semiaxes of two ellipses. The given algorithm is useful to analyse the modern speckle interferometric observations of very close binaries which allow to draw the ellipses of apparent orbits even in 5–10 years of observations.

We consider the classical problem of determination of the elements of the true elliptical orbit of a visual double star on the basis of optical observations. The geometrical construction of this problem is well known (Fig. 1). We assume that the true and the apparent ellipses coincide with their centres which serve us, as the origin of the polar coordinate system. We assume also that the apparent elliptical orbit of the binary is well ascertained. Thus, we regard as known quantities the four decisive parameters to the problem. These parameters are the following:

a, b — the semiaxes of the apparent ellipse; r_0, α — the polar coordinates of the point Σ' in the plane of apparent ellipse; Σ' — the projection of the main star on the plane of apparent ellipse; α — the position angle of $O\Sigma'$ relative to the direction of the semi-major axes of apparent ellipse.

The data above let us to determine the two main dynamical parameters of the true orbit:

– firstly — the eccentricity e :

$$e = \left(\frac{r_0}{r_\Pi}\right) = \left(\frac{r_0}{ab\sqrt{a^2 \sin^2 \alpha + b^2 \cos^2 \alpha}}\right) \tag{1}$$

here: r_Π is the radius $O\Pi'$, where the point Π' is the projection of the true periastron (point Π) on the plane of the apparent ellipse;

– secondly, we can obtain the range of semi-major axes A :

$$A^4(1-e^2) - A^2(a^2+b^2-r_0^2) + a^2b^2 = 0 \tag{2}$$

J. A. Docobo et al. (eds.), Visual Double Stars: Formation, Dynamics and Evolutionary Tracks, 357–359.

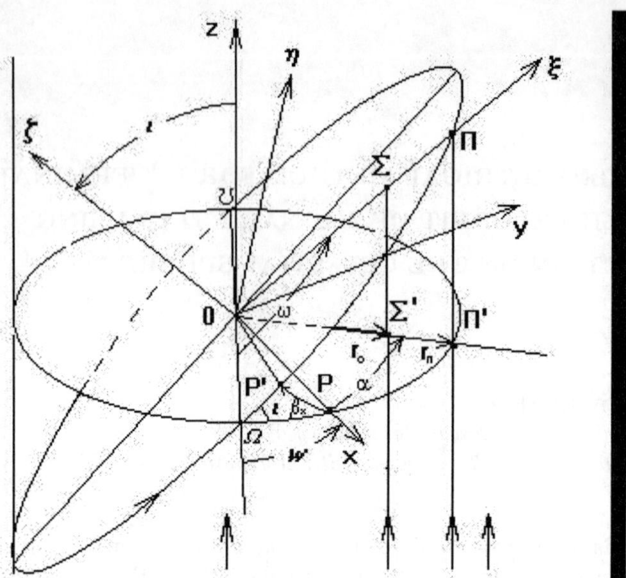

Figure 1. The relative position of apparent and true ellipses. Here : $O\Omega$ — node line; $\Omega\Pi'$ — apparent ellipse; $\Omega\Pi$ — true ellipse; Σ, Σ' — primary star and its projection onto the plane Oxy; $O\Pi, O\Pi'$ — major semiaxes of true ellipse and its projection onto the plane $O\xi\eta$; W, i, ω — angles of the orientation of true ellipse relatively apparent one.

The equation (2) has a fundamental meaning for the problem as whole. The details of its deducing have been considered by the author (A. A. Kisselev, 1990). Now, we dispose all the data to find the orientation of true ellipse relative to apparent one. Let us introduce two rectangular coordinate system (Fig. 1): $Oxyz$, where x and y are in the plane of apparent ellipse (system A) and $O\xi\eta\zeta$, where ξ and η are in the plane of the true ellipse (system T). We assume that the directions of the axes Ox and $O\xi$ coincide with the semi-major axes of the apparent and the true ellipses, respectively. The transformation of unit vectors from system A to system T may be done by using the rotational matrices Z, X, Z and the Euler's angles W, i, ω:

$$\overline{r} = Z(\omega)X(i)Z(-W)\overline{r}_{X,Y,Z}. \tag{3}$$

The Euler's angles of rotation have the following meaning:

- W is the arc from the node line $O\Omega$ to the direction Ox;
- i is the inclination;
- ω is the arc from the node line $O\Omega$ to the direction $O\xi$ (longitude of periastron).

The detailed analysis of transformation (3) (A. A. Kisselev, 1990) allows to prove the formulae which determine the Euler's angles:

$$\sin^2 W = \left(\frac{b^2}{a^2 - b^2}\frac{(a^2 - B^2)(A^2 - a^2)}{(A^2B^2 - a^2b^2)}\right), \tag{4}$$

$$\cos^2 i = \left(\frac{a^2b^2}{A^2B^2}\right), \tag{5}$$

$$\sin^2 \omega = \left(\frac{B^2}{A^2 - B^2} \frac{(A^2 - b^2)(A^2 - a^2)}{(A^2B^2 - a^2b^2)} \right), \tag{6}$$

where we denote $B^2 = A^2(1 - e^2)$ according to (1) and (2). The usual astronomical orientation of the orbit under consideration may be determined if the position angle of semi-major axis of apparent orbit Θ_{ab} is given. Thus we have :

$$\Omega = \Theta - W.$$

Here, we have to notice that the angles in (3) are counted from Ox direction to Oy, and position angles in Astronomy of double stars are counted clockwise (from the North direction).

Example. To illustrate the effectiveness of the algorithm, we applied it for computing the orbit of the close binary $51 Tau$, by using speckle interferometric observations made with the 6-meter Telescope (Balega, 1988). The direct measurements of the drawing of the apparent ellipse published in the issue let us to estimate the parameters of the apparent ellipse as follow :

$$a = 0.1290'', \quad b = 0.0734'', \quad r_0 = 0.0212'', \quad \alpha = 11.6^o, \quad \Theta_{ab} \simeq -15^o.$$

These data allow to calculate by formulae (1),(2),(4)–(6) the parameters of the true ellipse. They are presented in the Table 1.

TABLE 1. Comparison of the orbit elements obtained by different methods.

	This method	Balega method
A	$0.1293''$	$0.129''$
e	0.1710	0.171
i	$54.7^o(125.3^o)$	125.9^o
ω	$20.2^o(200.2^o)$	160.9^o
Ω	$-15.5^o(164.5^o)$	170.0^o
B	$0.1274''$	
W	$0.5^o(180.5^o)$	

It seems that the agreement is fairly good. The discrepancy in ω is avoided if we know the direction of orbital motion.

References

1.- A. A. Kisselev and others (1990). Astronomical and geodetic investigation: Nearby binary and multiple stars, **pp. 38–43**, Sverdlovsk, (in Russian).

2.- J. J. Balega, Ju. Ju. Balega (1988), *Letters to Astron. J.*, **Vol. no. 10**, **pp. 927–930**, (in Russian).

ON REDUCTION OF THE NONSTATIONARY TWO–BODY PROBLEM TO OSCILLATOR FORM

IGNACIO APARICIO AND LUIS FLORÍA
Grupo de Mecánica Celeste I
Dept. de Matemática Aplicada a la Ingeniería
E. T. S. de Ingenieros Industriales. Universidad de Valladolid
E – 47 011 Valladolid, Spain.

Abstract. We deal with a perturbed Gylden system in homogeneous canonical formulation. After a simple extension of the Burdet–Ferrándiz (BF) transformation to an enlarged phase space, we convert the canonical equations of motion (derived from the transformed Hamiltonian of the problem) into a set of four second–order differential equations for the position–like variables. Attempts to establish analogies with conservative cases suggest that, although linearization cannot be assured, regularization of the equations of motion is attainable in a slightly modified BF formulation.

Key words: regularization, time–varying Keplerian parameter, perturbed Gylden Hamiltonian, BF linearizing transformation, pseudo–time.

1. Introduction

Within the framework of Linear and Regular Celestial Mechanics (Stiefel & Scheifele 1971; Deprit *et al.* 1994), we analyze the possibility of reducing the equations of motion of the two–body problem with time–dependent Keplerian parameter into the form of perturbed linear oscillators.

We concentrate on *Gylden systems* (Deprit 1983; Deprit *et al.* 1989), and confront a Burdet–Ferrándiz (BF) linearizing transformation with a time–dependent two–body problem. In conservative cases, Ferrándiz & Fernández–Ferreirós (1991) obtained exact linearization of the equations of motion governing a class of perturbed Kepler problems in which the perturbing potential is given by a slighty simpler expression than that studied here.

Starting from a Newtonian, non–canonical approach to the equations of motion, *Burdet* (1969) studied the linearization of the Kepler problem, in particular by the so–called *focal method.* The type of *focal* variables used by this author, which can be conceived as homogeneous Cartesian coordinates in a projective space, are of special interest in the present paper. The *focal method* was developed in terms of a set of four coordinates: the direction cosines of the position vector and the reciprocal of the distance, the independent

J. A. Docobo et al. (eds.), Visual Double Stars: Formation, Dynamics and Evolutionary Tracks, 361–366.

variable being a kind of true anomaly. These projective coordinates were made *canonical* by *Ferrándiz* (1988, 1991, and references therein; also Deprit *et al.* 1994, §§4.4), who completed them with the respective conjugate momenta, obtaining eight redundant variables of focal type, the so–called *BF variables*, in terms of which the equations of motion are similar to those of Burdet: *the BF variables reduce the pure Kepler problem to four harmonic oscillators.*

Clarification on the *canonical* or *weakly canonical* character of the BF transformation, and other related questions, can be found in the article by Deprit, Elipe & Ferrer (1994). In addition to this, the regularizing transformations of the independent variable are usually suggested by that of *Sundman*, by considering the reparametrizing function proportional to a power of the distance. As for stabilization of the equations of motion, the introduction of first–integrals and constraints (of a geometrical or mechanical character) into the equations sometimes helps to achieve this goal.

We investigate the two–body problem with a time–varying Keplerian parameter $\mu(t)$. Linearization of the equations of motion derived from a perturbed Gylden Hamiltonian, formulated in *homogeneous canonical formalism*, will be analyzed. From the onset, we opt for extended phase–space formulation by slightly modifying the approach by Ferrándiz and Fernández–Ferreirós (1991): our version of the BF–transformation includes time t, and we will follow a *variant* of the *focal method canonical treatment* of Ferrándiz & Fernández–Ferreirós (1991), with a *true–like anomaly* as the new independent variable. We shall arrive at *regularized equations of motion* resembling those of a perturbed 4–dimensional oscillator.

2. Statement of the problem

¿From the onset, we consider an extended, 8–dimensional phase space coordinatized by the enlarged canonical set of the *Hill–Whittaker polar nodal* variables, $(r, \theta, \nu, t; R, \Theta, N, T)$ where t stands for the physical time and the momentum T (canonically conjugate to t) is the negative of the total energy of the system. The *inclination* I of the instantaneous orbital plane of motion is given by the relation

$$\cos I = N/\Theta . \tag{1}$$

Remember also that these canonical variables are universally applicable to any two–body conic–section orbit.

We start from a rather general perturbation model formulated in these variables: the perturbing part will be proportional to a power of the inverse of the distance r, the coefficient being some function of time and canonical momenta that can also be expanded in powers of a small parameter ε.

Let $\mathcal{H}_0$ denote the standard Gylden Hamiltonian. In homogeneous canonical formulation, we consider a perturbed Gylden system given by

$$\begin{aligned} \mathcal{H}_h &\equiv \mathcal{H}_h(r, -, -, t; R, \Theta, N, T; \varepsilon) \\ &= \mathcal{H}_0(r, \mu(t); R, \Theta) + \frac{1}{r^j} U\left(t; \Theta^2, N, T; \varepsilon\right) + T, \end{aligned} \tag{2}$$

$$\mathcal{H}_0 = \frac{1}{2}\left[R^2 + \frac{\Theta^2}{r^2}\right] - \frac{\mu(t)}{r}, \tag{3}$$

where $\mu(t)$ stands for the time–dependent Keplerian parameter, and the *perturbation* obeys the expression (truncated at the desired power of ε)

$$U \equiv U_j \left(t\,;\,\Theta^2\,,\,N\,,\,T\,;\,\varepsilon\right) \;=\; \sum_{l\geq 1} \varepsilon^{\,l}\, U_{j\,,l}\left(t\,;\,\Theta^2\,,\,N\,,\,T\right)\,, \tag{4}$$

understanding that ε is a dimensionless measure of smallness, introduced to separate the small terms according to their relative size.

If necessary, as in Ferrándiz & Fernández–Ferreirós (1991) §2, thanks to relation (1), the functional dependence in $U_j\,(t\,;\,\Theta^2\,,\,N\,,\,T\,;\,\varepsilon)$ might also be rewritten in the form $U_j^*\,(t\,;\,\Theta^2\,,\,I\,,\,T\,;\,\varepsilon)$.

The form of these Hamiltonians, intended as approximations to account for the possible *nonsphericity of the gravitational field* due to an oblate central body, is originally based on that corresponding to some *radial intermediaries* (Deprit 1981) for the *Main Problem* in Artificial Satellite Theory. According to some previous knowledge on *exact* linearization of this type of perturbation in the conservative case (Burdet 1969; Ferrándiz & Fernández–Ferreirós 1991), we restrict ourselves to dealing with the *exponent* $j = 2$.

In *extended Cartesian variables* $(Q_0, \mathbf{Q}, P_0, \mathbf{P})$, Hamiltonian (2) reads

$$\mathcal{H} = \frac{1}{2}\,\|\,\mathbf{P}\,\|^{\,2} \;-\; \frac{\mu(Q_0)}{r} \;+\; \frac{1}{r^j}\,U\left(Q_0\,;\,\Theta^2\,,\,N\,,P_0\,;\,\varepsilon\right) \;+\; P_0, \tag{5}$$

where $\mathbf{Q} = (Q_1, Q_2, Q_3)$ denotes the position vector of the particle, r is the radial distance, and the vector $\mathbf{P} = (P_1, P_2, P_3)$ contains the respective conjugate momenta. To this set we append the pair of canonically conjugate variables $(Q_0 = t, P_0 = T)$. The momenta N and Θ take on the form

$$N = Q_1\,P_2 \;-\; Q_2\,P_1\,, \qquad \Theta^2 \;=\; \|\,\mathbf{Q}\,\times\,\mathbf{P}\,\|^{\,2}, \tag{6}$$

where $\times$ stands for the usual cross product in space $\mathrm{I\!R}^3$.

3. A simple extension of the BF mapping. Time transformation

In this section we deal with a transformation of the dependent variables and a change of the time parameter, and their effect on Hamiltonian (5).

As in Ferrándiz & Fernández Ferreirós (1991), we perform a *modified BF transformation* (see also Deprit *et al.* 1994, §§4.4). The mapping is *weakly canonical* (Deprit *et al.*, 1994), increases the number of variables and introduces the set $(q_0, \mathbf{q}, q_4, p_0, \mathbf{p}, p_4) \equiv (q_0, q_1, q_2, q_3, q_4, p_0, p_1, p_2, p_3, p_4)$ by means of the defining relations:

$$\begin{aligned} Q_i \;&=\; \frac{q_i}{q_4}\,, \qquad P_i \;=\; p_i\,q_4 \;-\; q_i\,\frac{q_4}{\|\,\mathbf{q}\,\|^{\,2}}\,\left\{(\mathbf{q}\mid\mathbf{p}) + q_4 p_4\right\}\,, \quad i\;=\;1\,,2\,,3\,, \\ Q_0 \;&=\; q_0\,, \qquad P_0 \;=\; p_0\,, \end{aligned}$$

where we have introduced the abbreviations $\mathbf{q} \equiv (q_1, q_2, q_3)\,,\; \mathbf{p} \equiv (p_1, p_2, p_3)$, and $(-\mid-)$ denotes the standard inner product in a real vector space $\mathrm{I\!R}^n$.

Under this mapping, the homogeneous Hamiltonian (5) becomes

$$\widetilde{\mathcal{H}} \;=\; \frac{1}{2}\,\|\,\mathbf{p}\,\|^{\,2}\,q_4^2 \;+\; \frac{1}{2}\,\frac{q_4^4\,p_4^2}{\|\,\mathbf{q}\,\|^{\,2}} \;-\; \frac{1}{2}\,\frac{(\mathbf{q}\mid\mathbf{p})^2\,q_4^2}{\|\,\mathbf{q}\,\|^{\,2}} \;-\; \mu\,(q_0)\,\frac{q_4}{\|\,\mathbf{q}\,\|}$$

$$+ \frac{q_4^j}{\|\mathbf{q}\|^j} U\left(q_0 ; c^2, N, p_0; \varepsilon\right) + p_0, \qquad \text{where} \tag{7}$$

$$c^2 = \|\mathbf{q}\|^2 \|\mathbf{p}\|^2 - (\mathbf{q} \mid \mathbf{p})^2, \tag{8}$$

$$\cos I = N/c, \quad N = q_1 p_2 - q_2 p_1 . \tag{9}$$

Thanks to the extended phase–space formulation, we carry out a change of time parameter, from t to a new fictitious time s, by reparametrizing with the help of a *generalized Sundman transformation*, the new independent variable (proportional to a true anomaly) being defined by

$$t \longrightarrow s : \quad dt/ds = \widetilde{f} = \|\mathbf{q}\|^2 / q_4^2 = r^2 . \tag{10}$$

Bearing in mind the above relations (8) and (9), the *new homogeneous Hamiltonian*, with s as the independent variable, will take on the form

$$\mathcal{K} = \widetilde{\mathcal{H}} \widetilde{f} = \frac{1}{2} c^2 + \frac{1}{2} p_4^2 q_4^2 - \mu(q_0) \frac{\|\mathbf{q}\|}{q_4}$$

$$+ \frac{q_4^{j-2}}{\|\mathbf{q}\|^{j-2}} U\left(q_0 ; c^2, N, p_0 ; \varepsilon\right) + \frac{\|\mathbf{q}\|^2}{q_4^2} p_0 . \tag{11}$$

4. Canonical equations from the transformed Hamiltonian

In what follows, as in Ferrándiz & Fernández–Ferreirós 1991, pp. 5–6 use will be made of the abbreviations and first–integrals and constaints

$$a = 1 + 2 \frac{\partial U}{\partial c^2}, \quad b = \frac{\partial U}{\partial N}, \quad \Gamma = \frac{\partial U}{\partial q_0} . \tag{12}$$

$$\|\mathbf{q}\|^2 = 1, \quad (\mathbf{q} \mid \mathbf{p}) + q_4 p_4 = 0 . \tag{13}$$

The canonical equations of motion generated by $\mathcal{K}$, namely

$$q'_\alpha \equiv \frac{d q_\alpha}{d s} = \frac{\partial \mathcal{K}}{\partial p_\alpha}, \quad p'_\alpha \equiv \frac{d p_\alpha}{d s} = - \frac{\partial \mathcal{K}}{\partial q_\alpha}, \quad \alpha = 0, ..., 4, \tag{14}$$

can be presented (for the special case when $j = 2$) in the form

$$\begin{aligned}
q'_0 &= q_4^{-2}, \qquad q'_4 = p_4 q_4^2 \\
q'_1 &= a \{p_1 - (\mathbf{q} \mid \mathbf{p}) q_1\} - b q_2, \\
q'_2 &= a \{p_2 - (\mathbf{q} \mid \mathbf{p}) q_2\} + b q_1, \\
q'_3 &= a \{p_1 - (\mathbf{q} \mid \mathbf{p}) q_3\}, \\
p'_0 &= (d\mu(q_0)/d q_0) q_4^{-1} - \Gamma, \qquad p'_4 = -p_4^2 q_4 - \mu(q_0) q_4^{-2} + 2 p_0 q_4^{-3}, \\
p'_1 &= -a \left\{\|\mathbf{p}\|^2 q_1 - (\mathbf{q} \mid \mathbf{p}) p_1\right\} + \mu(q_0) q_1 q_4^{-1} - b p_2 - 2 p_0 q_1 q_4^{-2}, \\
p'_2 &= -a \left\{\|\mathbf{p}\|^2 q_2 - (\mathbf{q} \mid \mathbf{p}) p_2\right\} + \mu(q_0) q_2 q_4^{-1} + b p_1 - 2 p_0 q_2 q_4^{-2}, \\
p'_3 &= -a \left\{\|\mathbf{p}\|^2 q_3 - (\mathbf{q} \mid \mathbf{p}) p_3\right\} + \mu(q_0) q_3 q_4^{-1} - 2 p_0 q_3 q_4^{-2}.
\end{aligned}$$

5. Second–order equations for the position–like variables

Next, by forming the s–derivatives of the first five equations, and taking into account the expressions for the p'_α and the above notations and constraints, we obtain a set of second–order differential equations for the q_α:

$$q_0'' = -2\,q_4'\,q_4^{-3}, \quad q_4'' = -\left[c^2 + 2\,U\left(q_0\,;\,c^2,\,N,\,p_0\,;\,\varepsilon\right)\right] q_4 + \mu\,(q_0)\,,$$

$$q_1'' = -\left(a^2 c^2 - b^2\right) q_1 - 2\,b\,q_2' - \frac{ab' - a'b}{a}\,q_2 + \frac{a'}{a}\,q_1'\,,$$

$$q_2'' = -\left(a^2 c^2 - b^2\right) q_2 + 2\,b\,q_1' + \frac{ab' - a'b}{a}\,q_1 + \frac{a'}{a}\,q_2'\,,$$

$$q_3'' = -a^2 c^2 q_3 + \frac{a'}{a}\,q_3'\,.$$

Observe that these equations are *regular*, their form resembling that of harmonic equations. In particular, for a pure Gylden system (say, $U \equiv 0$) there results $a \equiv 1$, $b \equiv 0$ and $\Gamma \equiv 0$, and the equations of motion read

$$q_k'' + c^2 q_k = 0\,, \quad k = 1, 2, 3 \text{ (three uncoupled harmonic oscillators)}\,,$$
$$q_4'' + c^2 q_4 = \mu\,(q_0)\,.$$

As a final remark, notice that *perturbed* Gylden problems might also be treated by applying variants of some other transformation approach, for instance of the type developed by Berkovich (1979, 1980, 1981, 1982) or by other researchers. In this respect, a general review of the variable–mass two–body problem has been published by Polyakhova (1994).

Future work will report on other aspects of our treatment.

Acknowledgements

The second author (L. Floría) acknowledges partial financial support received from the DGICYT of Spain under Grant PB 95–0807.

References

Berkovich, L. M. (1979) Method of Exact Linearization of Nonlinear Autonomous Differential Equations of Seconf Order, *Journal of Applied Mathematics and Mechanics (PMM)*, **43**, pp. 673–683.

Berkovich, L. M. (1980) Transformation of the Gil'den–Meshcherskii Problem to Its Stationary Form and the Laws of Mass Change, *Journal of Applied Mathematics and Mechanics (PMM)*, **44**, pp. 246–249.

Berkovič, L. M. (1981) Gylden–Meščerskii Problem, *Celestial Mechanics*, **24**, pp. 407–429.

Berkovich, L. M. (1982) On Integrability of the Gil'den–Meshcherskii Problem, *Journal of Applied Mathematics and Mechanics (PMM)*, **46**, pp. 132–133.

Burdet, C. A. (1969) Le mouvement Képlérien et les oscillateurs harmoniques, *Journal für die reine und angewandte Mathematik*, **238**, pp. 71–84.

Deprit, A. (1981) The Elimination of the Parallax in Satellite Theory, *Celestial Mechanics*, **24**, pp. 111–153.

Deprit, A. (1983) The Secular Acceleration in Gylden's Problem, *Celestial Mechanics*, **31**, pp. 1–22.

Deprit, A., Elipe, A. and Ferrer, S. (1994) Linearization: Laplace vs. Stiefel, *Celestial Mechanics* **58**, pp. 151–201.

Deprit, A., Miller, B. and Williams, C. A. (1989) Gylden Systems: Rotation of Pericenters, *Astrophysics and Space Science*, **159**, pp. 239–270.

Ferrándiz, J. M. (1988) A General Canonical Transformation Increasing the Number of Variables with Application to the Two–Body Problem, *Celestial Mechanics*, **41**, pp. 343– 357.

Ferrándiz, J. M. and Fernández–Ferreirós, A. (1991) Exact Linearization of Non–Planar Intermediary Orbits in the Satellite Theory, *Celestial Mechanics*, **52**, pp. 1–12.

Polyakhova, E. N. (1994) A Two–Body Variable–Mass Problem in Celestial Mechanics: The Current State, *Astronomy Reports*, **38**, pp. 283–291.

Stiefel, E. L. and Scheifele, G. (1971) *Linear and Regular Celestial Mechanics*. Springer–Verlag, Berlin–Heidelberg–New York.

ORBITS FOR TWO SOUTHERN DOUBLE STARS

D.M.D. JASINTA
Bosscha Observatory, Lembang 40391, Bandung Indonesia
Department of Astronomy, Institut Teknologi Bandung,
Jl. Ganesha 10, Bandung 40132, Indonesia

1. Introduction

The paper presents new orbit for two southern visual double stars BU 271 AB (ADS 14847 AB, CCDM 21198-2621) and BU 1004 AB (CCDM 04021-3429). Each orbit was determined by means of the Thiele-van den Bos method using three fundamental points and an areal constant value. A programme developed by Nys (1983) involving some conditions of validity of the method proposed by S. Arend and J. Dommanget was applied. Dynamical parallax and mass of each component was computed by means of the Baize & Romani (1946) method. Residual of observations were presented as well as the ephemerides.

For each star, the position angles have been reduced to the equinox 2000.0. The residual of observations obtained were for the epoch of the observations.

2. The orbits

2.1. BU 271 AB (ADS 14847 AB, CCDM 21198-2621)

The 110 years observation only covers 45° of arc, but a Keplerian motion in the apparent orbit is well marked. The secondary had moved away from the primary, reached its longest distance and is now approaching the primary.

J. A. Docobo et al. (eds.), Visual Double Stars: Formation, Dynamics and Evolutionary Tracks, 367–372.

Using the three fundamental points and the areal constant value:

t	θ_{2000}	ρ
1880.00	230°0	2″300
1930.00	246.0	3.500
1985.00	266.0	2.225
Areal constant		-0.0540837

and after an emphirical correction of $\Delta\Omega = +1^o$, a set of orbital element was yielded:

P(yr)	261.619			A	0″0471
T	1845.991	i	74°08	F	−0″2356
a	3″010	Ω_{2000}	59°09	B	0″2386
e	0.424	ω	57°57	G	−0″2749

The physical parameters and the ephemerides are given in Table 1, while the residual of observations in Table 3.

2.2. BU 1004 AB (CCDM 04021-3429)

The observations span over 100 years and cover an arc of 70°. In about 20 years, the secondary will reach the periastron.

The orbit was obtained using the following three fundamental points:

t	θ_{2000}	ρ
1881.85	154°8	1″790
1935.00	123.0	1.680
1980.00	90.5	1.435

and areal constant value -0.0315358.
Additional emphirical corrections: $\Delta T = +2$ yr and $\Delta\Omega = -1^o$ have been adopted to yield the following orbital parameter:

P(yr)	281.831			A	$0\overset{''}{.}9875$
T	2018.824	i	$124\overset{\circ}{.}77$	F	$-0\overset{''}{.}8492$
a	$1\overset{''}{.}853$	Ω_{2000}	$59\overset{\circ}{.}01$	B	$-0\overset{''}{.}3963$
e	0.692	ω	$85\overset{\circ}{.}35$	G	$-1\overset{''}{.}6429$

The dynamical parallax, mass of each component, and the ephemerides are presented in Table 2, while Table 4 gives the residuals of observations.

Acknowledgements

The author acknowledges with appreciation Dr. C.E. Worley for communicating the observational data of the systems and the Director of Leids Kerkhoven-Bosscha Fonds for the conference visit grant to attend the meeting.

References

Baize, P., Romani, L. (1946) *Ann. Astrophys.*, **9**, 13
Nys, O. (1983) *Comm. Obs. R. Bel.*, **128**, 3

TABLE 1. Dynamical parallax, masses, and ephemerides of BU 271 AB

π	$0''070$	$\mathcal{M}_{AB}$	$1.160\ \mathcal{M}_{\odot}$
		$\mathcal{M}_{A}$	$0.794\ \mathcal{M}_{\odot}$
		$\mathcal{M}_{B}$	$0.367\ \mathcal{M}_{\odot}$

t	θ_{2000}	ρ
1997.00	277°2	1″716
1999.00	279.4	1.647
2001.00	281.8	1.580
2003.00	284.3	1.514
2005.00	287.1	1.450
2007.00	290.2	1.388
2009.00	293.6	1.330
2011.00	297.2	1.275
2013.00	301.1	1.225
2015.00	305.4	1.180
2017.00	310.0	1.141

TABLE 2. Dynamical parallax, masses, and ephemerides of BU 1004 AB

π	$0''007$	$\mathcal{M}_{AB}$	$4.406\ \mathcal{M}_{\odot}$
		$\mathcal{M}_{A}$	$2.373\ \mathcal{M}_{\odot}$
		$\mathcal{M}_{B}$	$2.034\ \mathcal{M}_{\odot}$

t	θ_{2000}	ρ
1997.00	119°0	0″966
1999.00	115.0	0.945
2001.00	110.9	0.920
2003.00	106.5	0.891
2005.00	101.7	0.856
2007.00	96.6	0.814
2009.00	90.8	0.763
2011.00	84.0	0.699
2013.00	75.7	0.621
2015.00	64.8	0.531
2017.00	49.2	0.436

TABLE 3. Residual of observations of BU 271 AB

t	θ_t	ρ	n	Obs.	$(O-C)_\theta$	$(O-C)_\rho$
1874.73	250°	1″8	1	BU	22.7	-0.11
1876.66	278.0	2.23	1	STN	49.7	0.17
1876.72	226.6	2.20	1	HWE	-1.8	0.13
1877.66	234.2	2.07	1	HWE	5.1	-0.07
1879.68	224.7		1	STN	-5.6	
1879.69	231.3	2.34	1	STN	1.0	0.06
1886.78	237.0	2.77	1	MLF	3.1	0.07
1891.52	238.3	2.65	1	BU	2.6	-0.27
1891.54	236.7	2.72	2	BU	1.0	-0.20
1897.16	241.6	3.16	5	SEE	4.0	0.03
1898.75	239.5	2.91	4	DOO	1.4	-0.27
1898.84	236.3	3.37	1	BU	-1.8	0.19
1899.75	237.4	3.21	2	BAR	-1.0	0.01
1901.76	238.8	3.12	3	DOO	-0.2	-0.14
1909.64	242.0	3.13	1	OL	0.8	-0.28
1909.67	243.4	3.08	3	OL	2.2	-0.33
1910.04	243.9	3.47	3	BU	2.6	0.05
1916.14	247.2	3.05	4	LV	4.4	-0.44
1918.76	245.7	3.51	1	LV	2.2	0.01
1919.50	244.3	3.51	1	OL	0.6	0.00
1924.70	247.4	3.48	4	VOU	2.4	-0.04
1925.65	249.3	3.33	1	I	4.1	-0.18
1925.73	246.5	2.49	1	OL	1.3	-1.02
1926.94	244.3	3.49	3	FOX	-1.2	-0.02
1928.52	247.4	3.63	2	JSP	1.5	0.12
1929.48	244.8	3.40	3	OL	-1.4	-0.10
1929.73	247.0	3.47	4	WAL	0.8	-0.03
1929.80	246.0	3.55	3	FEN	-0.2	0.05
1932.72	247.9	3.49	3	WRH	0.9	0.01
1933.43	245.6	3.25	3	BRT	-1.6	-0.23
1936.59	250.6	3.42	4	SMW	2.7	-0.03
1939.05	251.0	3.47	4	B	2.4	0.05
1943.51	252.5	3.08	4	VOU	2.7	-0.28
1954.06	254.6	3.17	3	B	1.8	0.01
1954.806	254.68	3.245	6	VAD	1.6	0.103
1959.65	255.0	2.98	3	B	0.4	-0.05
1962.70	257.4	2.73	2	KNP	1.7	-0.22
1964.73	257.4	2.87	3	KNP	1.0	-0.02
1966.81	258.2	2.91	2	NBG	1.0	0.08
1968.537	259.6	2.75	1	MRO	1.7	-0.03
1970.513	261.6	2.75	1	WOR	2.9	0.03
1970.79	263.2	2.82	3	KLE	4.4	0.11
1975.721	260.5	2.54	3	WOR	-0.5	-0.01
1976.456	263.1	2.67	1	VAD	1.8	0.15
1976.852	261.5	2.72	1	HLN	-0.0	0.21
1977.633	263.6	2.62	1	VAD	1.7	0.14
1985.450	266.40	2.242	1	PAN	0.1	0.036
1986.460	266.52	2.161	1	PAN	-0.5	-0.008

TABLE 4. Residual of observations of BU 1004 AB

t	θ_t	ρ	n	Obs.	$(O-C)_\theta$	$(O-C)_\rho$
1881.85	154°1	1″79	2	BU	1.2	-0.00
1895.08	144.8	1.54	3	SLR	-1.3	-0.23
1897.72	144.2	1.77	1	SEE	-0.4	0.01
1898.84	143.4	1.97	2	A	-0.5	0.21
1900.80	138.2	1.76	3	DOO	-4.6	0.00
1901.93	137		3	DOB	-5.1	
1902.98	141.2	1.90	3	SCT	-0.3	0.15
1906.06	141.2	1.78	5	OL	1.5	0.03
1910.19	137.9	1.80	3	OL	0.6	0.06
1912.00	135.3	1.65	3	COX	-0.9	-0.09
1913.98	136.3	2.04	2	VDS	1.3	0.31
1915.09	134.5	1.77	4	VOU	0.1	0.04
1915.76	133.3	1.71	4	DOO	-0.7	-0.02
1919.91	131.3	1.87	3	DAW	-0.2	0.15
1921.08	130.7	2.04	3	DAW	-0.1	0.32
1921.44	129.4	1.81	4	VOU	-1.2	0.09
1921.98	130.1	1.85	3	DAW	-0.1	0.13
1922.93	129.7	1.88	2	DAW	0.1	0.17
1927.53	128.5	1.72	2	FIN	1.7	0.02
1928.77	127.5	1.67	4	BRU	1.4	-0.03
1928.81	126.4	1.74	5	WAL	0.4	0.04
1929.05	126.6	1.76	2	OL	0.7	0.06
1931.67	125.0	1.53	1	ALD	0.7	-0.16
1935.52	122.4	1.60	4	B	0.6	-0.08
1936.80	121.2	1.58	4	SMW	0.2	-0.10
1942.65	117.5	1.70	3	VOU	0.2	0.04
1943.24	115.2	1.50	1	HIR	-1.7	-0.16
1944.14	116.4	1.67	2	HIR	0.1	0.01
1945.04	114.8	1.44	3	HIR	-0.9	-0.22
1945.92	115.4	1.69	3	WOY	0.3	0.04
1945.95	114.2	1.69	4	GTB	-0.9	0.04
1946.71	134.0	1.80	4	WOH	19.4	0.15
1947.08	114.3	1.69	3	SMW	-0.1	0.04
1949.05	113.8	1.59	3	B	0.7	-0.05
1957.669	107.35	1.460	1	THE	0.1	-0.145
1959.03	107.1	1.39	4	B	0.8	-0.21
1959.03	106.0	1.48	3	KNP	-0.3	-0.12
1963.00	104.0	1.44	3	KNP	0.6	-0.14
1964.80	103.2	1.42	4	B	1.1	-0.15
1966.04	102.2	1.56	2	KNP	1.0	0.00
1975.14	96.1	1.60	3	HLN	1.9	0.11
1975.731	94.5	1.40	4	WOR	0.8	-0.09
1976.794	93.6	1.60	2	HLN	0.7	0.12
1976.853	93.4	1.43	2	HLN	0.6	-0.05
1977.29	94.0	1.46	2	WRH	1.5	-0.01
1981.17	90.6	1.32	3	WRH	1.4	-0.11
1982.662	87.8	1.45	2	ARY	-0.1	0.04
1986.854	84.5	1.33	4	SCA	0.6	-0.03

THE RESTRICTED 2+2 BODY PROBLEM: THE PERMISSIBLE AREAS OF MOTION OF THE MINOR BODIES CLOSE TO THE COLLINEAR EQUILIBRIA OF THEIR CENTER OF MASS

T.J.KALVOURIDIS AND A. G. MAVRAGANIS
National Technical Univ. of Athens, Dep. of Engineering Sci., Section of Mechanics, 5 Heroes of Polytechnion Ave. GR 157 73, Greece

The $2+2$ body problem stated by Whipple in the middle of 80's gave rise to the study of a specific case of the $2+\nu$ body systems, that before then focused almost exclusively to the investigation of the case which corresponds to $\nu = 1$.

The most of the interest that has been paid in this problem concerns the study of the equilibrium positions. In our contribution we deal with the areas of the possible motion of the infinitesimal masses when their center of mass remain in equilibrium on the syzygies' axis.

As it is known, the restricted $2+2$ body problem consists of two attractive primaries (i.e spherical bodies) moving on circular orbits around their center of mass and two infinitesimal masses S_i, $(i = 1,2)$ which, acted upon by the gravitational attraction of the primaries and their Newtonian interaction, move on the plane of the primaries revolution.

Under the above assumptions, the dimensionless equations of motion in the synodic coordinate system have the form (Whipple, 1984),

$$\left.\begin{aligned} \ddot{x}_i - 2\dot{y}_i &= \frac{1}{\mu_i}\frac{\partial T}{\partial x_i}, \\ \ddot{y}_i + 2\dot{x}_i &= \frac{1}{\mu_i}\frac{\partial T}{\partial y_i}, \end{aligned}\right\} \qquad i = 1,2 \tag{1}$$

with

$$T = \sum_{j=1}^{2} \mu_j \left[\tfrac{1}{2}(x_j^2 + y_j^2) + \frac{1-\mu}{r_{1j}} + \frac{\mu}{r_{2j}} + \frac{\mu_{3-j}}{\rho} \right],$$

where

$$\begin{aligned} r_{1j}^2 &= (x_j - \mu)^2 + y_j^2, \\ r_{2j}^2 &= (x_j + 1 - \mu)^2 + y_j^2, \end{aligned} \qquad j = 1,2$$

and

$$\rho^2 = (x_2 - x_1)^2 + (y_2 - y_1)^2, \qquad \mu_i = \frac{m_i}{M_1 + M_2},$$

with μ_i $(i = 1,2)$ the reduced masses of S_i and M_i the masses of the primaries.

J. A. Docobo et al. (eds.), Visual Double Stars: Formation, Dynamics and Evolutionary Tracks, 373–376.

These equations admit an integral of motion, that is the Jacobi's integral whose general expression takes, after some arrangement, the form

$$\frac{1}{2}\sum_{j=1}^{2}\mu_j(\dot{x}_j^2+\dot{y}_j^2)=T-C, \tag{2}$$

where C is a constant.

By this integral we can check the accuracy of any numerical elaboration of equations (1) and find the areas where the motion of each minor body is permissible.

We emphasize here the advantage of using in all computational sequences the definition of the center of mass of the two minor bodies given by the relations,

$$\sum_{j=1}^{2}\mu_j x_j=\left(\sum_{j=1}^{2}\mu_j\right)x_c, \qquad \sum_{j=1}^{2}\mu_j y_j=\left(\sum_{j=1}^{2}\mu_j\right)y_c. \tag{3}$$

When this point occupies an equilibrium position, that is to say, when

$$\dot{x}_c^e=\ddot{x}_c^e=0, \qquad \dot{y}_c^e=\ddot{y}_c^e=0,$$

it coincides, in linear approximation, with the Lagrangian points of the restricted 3-body problem i.e $x_c^e=x_{L_j}$, $y_c^e=y_{L_j}$ j=1,2,..,5 (Whipple, 1984).

Let us now consider that both bodies Si, $(i=1,2)$, start moving from the $x-$ axis, i.e. with initial conditions $x_{j0}\neq 0$, $y_{j0}=0$, $\dot{x}_{j0}=0$, $\dot{x}_{j0}\neq 0$, $(i=1,2)$. Then, in view of equations (3) we obtain for the integral of motion an expression of the form,

$$\tfrac{1}{2}\sum_{j=1}^{2}\mu_j(\dot{x}_{j0}^2+\dot{y}_{j0}^2)=f(x_{10};C)\geq 0, \tag{4}$$

which may be used to determine the areas within the body S_1 is permitted to move. Obviously, to these areas we must associate those for the motion of S_2 which are obtained easily through the first of equations (3). The bottom of these areas show the positions where the two minor bodies come to equilibrium close to the Lagrangian points L_i, $(i=1,2,3)$.

According to Whipple only these positions are admissible as collinear equilibrium solutions of the problem. Other points satisfying the equations of equilibrium but lying "apart" from the Lagrangian points must be excluded as making the interaction between the two minor bodies negligible. As an application we give in the figures below, the areas of motion of the body S_1 (non-shaded areas) for a system with $\nu_1=\nu_2=10^{-12}$, $\nu=0.2$ used as example in the referenced paper. The respective areas of S_2 coincide in this case with those of S_1 .

C
(10⁻¹¹)
0.355239358
0.355239348
0.355239338
x10
-1.27125
-1.27120
-1.27115
-1.27110
L1
-1.27100
-1.27095
-1.27090

(a)

C
(10⁻¹¹)
0.380465367
0.380465357
0.380465347
0.380465337
x10
-0.43825
-0.43820
-0.43815
-0.43810
L2
-0.43805
-0.43800
-0.43795

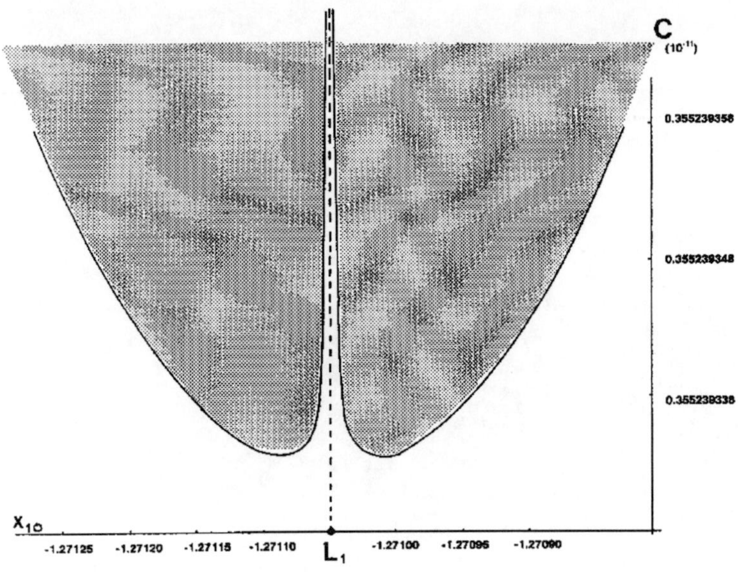

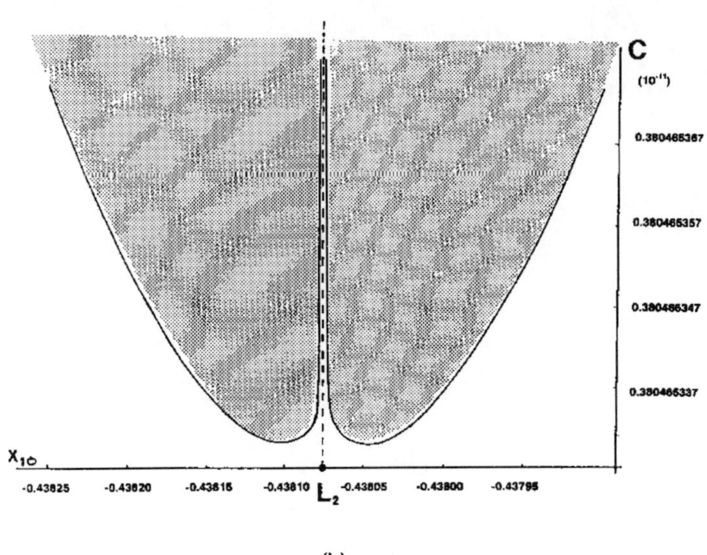

(b)

VISUAL DOUBLE STARS ORBITS OBTAINED BY APPARENT MOTION PARAMETERS METHOD AT PULKOVO

A.A. KISSELEV, O.V. KIYAEVA, L.G. ROMANENKO
Pulkovo observatory

Abstract. The orbital elements and masses of 12 wide visual double stars are determined by the Apparent Motion Parameters (AMP) method. These binaries are the following: ADS 48, 5983 (Delta Gem), 7251, 8002, 10329, 10386, 10759 (Psi Dra), 11061 (40/41 Dra), 11632, 12169, 12815 (16 Cyg) and 14636 (61 Cyg). Eight orbits are deduced by the first time. The observational basis of the investigation consists of: (1) 20-30 years long series of high accuracy positions obtained by photographic observations with the 26-inch refractor at Pulkovo, (2) accurate relative radial velocity of the components observed by the correlation method and (3) trigonometric parallaxes. These data provide the orbit calculations using AMP-method on the basis of very short arc of apparent tracks. The peculiarities of the AMP-method are considered.

The observation of visual double stars is a traditional work at Pulkovo observatory. After the Second World War the observations of double stars were renewed with a help of the 26-inch Zeiss refractor (f=10.4 m) using the photographic technique. The main purpose of this work is to accumulate long-term series of uniform accurate relative positions of the components (i.e. obtained by one telescope at the same site) to provide the reliable observational basis for kinematics and dynamics investigations of the star systems in the neighborhood of the Sun. As a theoretical basis for the investigations we worked out the Apparent Motion Parameters (AMP) method [1].

To use the AMP-method we have to enlarge the observational basis by adding to the data the relative radial velocities of the components of the binaries and their trigonometric parallaxes. These data were obtained by special organized spectroscopic observations at the Zelentchouk astrophysical observatory [2] and in Crimea [3].

Now, we have accurate 20-30 years series of relative positions and also radial velocities of some dozens of double stars in the neighborhood of the Sun. Most of these binaries are wide pairs with a period of revolution beyond 10000 years, although their orbits are not available yet.

In this paper, we consider the results of orbit determination by using the AMP-method for 12 selected visual double stars, including 4 well-known stars - ADS 48, 7251, 11632 and 14636 (61 Cyg), which orbits have already been determined by classical methods. The common data for these stars are given in Table 1. We notice that all stars are physical pairs which are located closer than 25 parsec from the Sun (except ADS 11061, which

J. A. Docobo et al. (eds.), Visual Double Stars: Formation, Dynamics and Evolutionary Tracks, 377–382.

TABLE 1. Common Data of wide Binaries investigated at Pulkovo.

N ADS (Name)	α_{1950} δ_{1950}	ρ [″]	m_A m_B	SP_A SP_B	μ_{xA} μ_{xB} [″/yr]	μ_{yA} μ_{yB} [″/yr]	π_{tr} ±err [″]
48	$00^h03^m\!.0$	6.0	8.3	K6 V	+0.875	−0.131	0.094
	45°32′		8.3	M0 V	+0.875	−0.183	± 4
5983	07 17.1	6.0	3.2	F0 IV	−0.016	−0.016	0.061
(δGem)	22 05		8.2	K6 V			± 4
7251	09 11.0	17.8	7.6	M0 V	−1.558	+0.568	0.166
	52 54		7.7	M0 V	−1.553	+0.672	± 4
8002	10 57.0	5.3	8.5	K0 V	−0.188	−0.069	0.070
	25 42		9.0	K5 V	−0.188	−0.069	
10329	17 02.7	12.1	8.0	K5 V	−0.371	+0.242	0.040
	59 29		9.2	M0 V	−0.393	+0.261	± 6
10386	17 09.1	22.2	8.8	K8 V	+0.070	−0.117	0.049
	54 33		9.3	K8 V	+0.089	−0.101	± 5
10759	17 42.8	30.1	4.6	F5 IV	+0.013	−0.270	0.060
(ψDra)	72 11		5.8	F8 V	+0.042	−0.289	± 8
11061	18 03.8	19.1	5.8	F5 V	+0.032	+0.127	0.021
(40/41Dra)	80 00		6.2	F5 V	+0.038	+0.131	± 4
11632	18 42.5	14.0	8.2	M4 V	−1.333	+1.869	0.284
	59 30		8.7	M5 V	−1.366	+1.819	± 3
12169	19 10.8	8.0	6.0	G5 V	−0.225	+0.622	0.042
	49 45		6.5	G5 V	−0.183	+0.627	± 4
12815	19 40.5	39.3	6.0	G2 V	−0.168	−0.153	0.047
(16 Cyg)	50 24		6.2	G5 V	−0.147	−0.141	± 7
14636	21 04.7	28.2	5.2	K5 V	+4.135	+3.250	0.296
(61 Cyg)	38 30		6.0	K7 V	+4.122	+3.112	± 6

trigonometric parallax was determined in Pulkovo [4]). The values of the apparent motion parameters (AMP) obtained from Pulkovo observations are presented in Table 2.

Here, ΔT and T denote, respectively, the time interval and its middle, used to compute the AMP; ρ, θ - apparent distance and position angle of the component B relative to A; μ, ψ - the apparent motion B relative to A and its position angle; ρ_c - the radius of the curvature of the apparent short arc observed during time interval ΔT.

We succeeded in computing the value of ρ_c only for six binaries, with display appreciable rotation; in other cases, the curvature of observed arc was not detected. In the last columns, there are the data for relative radial velocities of the components $\Delta v_r = v_{rB} - v_{rA}$ and the sum of their masses. The last values were estimated by the mass-luminosity relation in the most cases, or by using the ancient remote observations [5] (these cases are marked with one asterisk), or by using the dynamical criterium, see below (these cases are marked

TABLE 2. The Apparent Motion Parameters of visual binaries derived at Pulkovo.

N ADS	T ΔT	ρ [″]	θ [°]	μ [″]	ψ [°]	ρ_c [″]	ΔV_r [km/s]	M_{A+B} $M_\odot$
48	1980 22	5.967 ± 2	173.45 ± 1	0.0460 ± 2	250.8 ± 3	3.3 ±5	−3.2 ± 2	0.72* ± 8
5983 (δGem)	1980 18	6.107 ± 5	221.06 ± 4	0.0246 ± 10	347 ± 3	5.5	−0.4	2.5
7251	1972 21	17.828 ± 3	84.82 ± 1	0.0928 ± 4	185.6 ± 3	17.4 ±1.0	+0.5 ± 2	1.0
8002	1980 18	5.227 ± 6	103.02 ± 6	0.0299 ± 8	190 ± 2	3.0	−0.5 ± 3	1.0
10329	1910 163	11.86 ± 8	50.89 ± 10	0.0167 ± 5	334 ± 5	–	+0.7 ± 7	1.2
10386	1971 33	22.231 ± 6	134.12 ± 4	0.0087 ± 8	34 ± 2	–	−1.0 ± 2	1.0
10759 (ψDra)	1986 161	30.15 ± 1	15.36 ± 6	0.0138 ± 11	130 ± 3	–	+3.4 ± 4	5.0 **
11061	1985 22	19.055 ± 3	231.46 ± 3	0.0107 ± 9	19 ± 9	–	+0.4 ± 2	5.0 **
11632	1922 28	16.950 ± 2	153.178 ± 2	0.0648 ± 1	262.52 ± 5	3.1 ±1	+2.6 ± 3	0.68* ± 2
12169	1974 25	7.974 ± 5	209.94 ± 3	0.0252 ± 4	73.0 ± 9	–	−0.3 ± 3	1.8
12815 (16 Cyg)	1976 33	39.262 ± 3	133.52 ± 1	0.0137 ± 3	108.1 ±1.3	–	−0.9 ± 3	2.0
14636 (61 Cyg)	1966 18	28.211 ± 5	143.13 ± 1	0.1331 ± 9	195.5 ± 6	25.9	+1.1 ± 1	1.3

* – according to remote observations, ** – dynamical conditions

with two asterisks).

The data from Table 2 were used to compute the orbital elements for six binaries by the AMP-method in the usual way [1]. The results of these computations are presented in Table 3. Here we have to point out that the orbital elements of binaries ADS 5983 and 8002 should be considered as preliminary due to uncertain values of ρ_c in the first case and π_{tr} in the second one. Otherwise, the orbit of the nearby double star ADS 11632 we appreciate as fairly reliable. To compute this orbit we have used the ancient observations from Hopmann's list [6] as basic data to obtain the AMP and the Pulkovo observations as remote ones. This choice is explained by the fact that the curvature of the observed arc of the apparent orbit in 1910-1935 was essential greater than in 1970.

We emphasize that values of positional angle of the ascending node Ω and inclination i as computed by AMP-method are true and unequivocal. They correspond to the direction

TABLE 3. The AMP-orbits of visual double stars.

N ADS	a [″]	P [yr]	e	ω [°]	i [°]	Ω [°]	T_p [yr]	π_{tr} [″]	M_{A+B} [$M_\odot$]
48	13.3 ±5.5	1982 ±820	0.55 ±16	164 ±21	125 ±3	357 ±7	1967 ±780	0.094 ± 4	0.72 ± 8
5983 (δGem)	6.0	622	0.75	64	118	359	1645	0.061 ± 5	2.5
7251	17.1 ±1.7	1048 ±150	0.24 ± 8	57 ±10	162 ± 3	149 ±15	2250 ±170	0.166 ± 5	1.0
8002	4.2	397	0.64	283	138	351	1767	0.070	1.0
11632	17.5 ±2.1	585 ±110	0.35 ± 4	264 ±18	110 ± 1	142 ± 1	1778 ± 14	0.284 ± 3	0.68 ± 2
14636 (61Cyg)	24.4 ± 7	659 ±17	0.48 ± 3	146 ± 3	126 ± 1	176 ± 1	1697 ± 12	0.296 ± 6	1.3

of the orbital moment vector $\vec{q} = \vec{r} \times \vec{v}$ relative to astrocentric coordinates $O\xi\eta\zeta$, where the $O\xi$ and $O\eta$ axes are directed to the East and the North in the tangential plane.

To compute the orbits for the remaining six double stars we cannot use the key-formula (1) of the AMP-method, because of lack of the ρ_c values, which allow to determine r —the distance between components A and B in space. But we can estimate the value of r by using the simple geometrical and dynamical consideration (2). The corresponding formulae are the following:

$$r^3 = \pm k^2 \frac{\rho\rho_c}{\mu^2} sin(\psi - \theta) \quad [AU^3], \tag{1}$$

$$\frac{\rho}{\pi_{tr}} \leq r \leq \frac{2k^2}{v^2}, \tag{2}$$

$$v^2 = \left(\frac{\mu}{\pi_{tr}}\right)^2 + \left(\frac{v_r}{4.74}\right)^2 \quad [AU^2/yr^2], \tag{3}$$

where $k^2 = 4\pi^2(M_A + M_B)$ $[AU^3/yr^2]$ and M_A, M_B are expressed in units of solar mass.

The expression (2) ascertains the lower and the upper limits for the value of r. The left hand part of the inequality means that the vector's projection in the plane is shorter as its true length, and the right hand part affirms the premise that the binary considered is a physical pair, that is to say, the relative motion of the components is elliptical.

Thus, the condition (2) allows us to adopt some suitable values of r which agree with the defined APM and allows to obtain the realistic orbits, satisfying the observations. To compute these orbits we have chosen three variants of r :

1. $r = \frac{\rho}{\pi_{tr}} = r_{min} = r_{\xi\eta}$ $\quad \beta = 0°$
2. $r = -\frac{4}{\pi} r_{min} = -1.273 r_{\xi\eta}$ $\quad \beta = -38°$
3. $r = +\frac{4}{\pi} r_{min} = +1.273 r_{\xi\eta}$ $\quad \beta = +38°$

TABLE 4. The probable orbits of wide visual double stars derived by AMP-method.

N_{ADS}	β	r	a	P	e	ω	i	$\overrightarrow{Q}$		$\overrightarrow{P}$	
γ	[°]	[AU]	[AU]	[$year$]		[°]	[°]	$l°$	$b°$	$l°$	$b°$
7251	−38	137	156	1900	0.28	353	142	297	−74	275	+15
+11°	00	107	96	940	0.22	133	169	331	−39	230	−14
	+38	137	156	1900	0.12	87	139	340	− 2	289	+86
true	−17	112	103	1048	0.24	57	162	325	−55	234	0
10329	−38	377	843	22000	0.55	302	44	88	− 7	188	−60
+19°	00	296	380	6800	0.30	303	19	70	+25	36	−60
	+38	377	843	22000	0.63	23	39	38	+56	56	−32
10386	−38	578	616	15000	0.38	124	57	92	−20	21	+45
−49°	00	454	389	7700	0.20	332	49	113	− 4	203	+5
	+38	578	616	15000	0.56	247	52	136	+15	204	−55
10759	−38	640	4862	152000	0.93	51	107	28	−54	125	− 6
+72°	00	502	943	13000	0.48	22	106	43	−64	159	−12
	+38	640	4862	152000	0.90	337	104	89	−70	210	−15
11061	−38	1155	2658	61000	0.84	347	130	342	−72	206	−13
+10°	00	907	1178	18000	0.84	138	162	276	−19	182	− 9
	+38	1155	2658	61000	0.72	128	120	261	+24	162	+20
12169	−38	242	316	4200	0.55	332	52	43	−19	185	−68
−5°	00	190	184	1870	0.73	318	8	80	+10	158	−49
	+38	242	316	4200	0.65	171	47	129	+38	132	−54
12815	−38	1064	3266	132000	0.96	129	38	69	−22	32	+63
−34°	00	835	1219	30000	0.78	57	58	138	− 7	57	+51
	+38	1064	3266	132000	0.70	106	73	156	+ 1	66	− 2
14636	−38	121	87	711	0.42	144	124	313	+38	111	+49
+27°	00	95	63	434	0.70	198	147	295	+11	182	+64
	+38	121	87	711	0.76	291	142	273	−31	216	+43
true	+37	119	83	658	0.48	146	126	309	+38	114	+51

Where β is the angle between the vector $\overrightarrow{r}$ and the $O\xi\eta$ plane in the astrocentric system. The cases 2 and 3 correspond to the most probable values (statistically) of the vector $\overrightarrow{r}$ for the given projection in a plane $r_{\xi\eta}$ if the random distribution of $\overrightarrow{r}$ is assumed. The results of the orbital elements, computed by the three variants of r, for eight wide pairs are presented in Table 4.

For two of these stars, ADS 7251 and ADS 14636 (61 Cyg), the orbital elements computed by AMP-method using the data from Table 2 are also given. These orbits in the Table 4 are noticed by the symbol "true".

In the last two columns of the table, the coordinates l and b appear, which define the orientation of orbital momentum (unit vector $\overrightarrow{Q}$) and periastron (unit vector $\overrightarrow{P}$) in the galactic system.

As we can see, the computed orbit's variants differ very essentially in dynamical sense, but not essentially in geometrical sense. Therefore, we can use these results in the statistical investigations of wide binaries orbits orientation in the neighborhood of the Sun. The comparison of true and probable orbits proves our suggestion in two cases (ADS 7251 and 14636). The present limited investigation shows also that the observed distribution of $\overrightarrow{Q}$ and $\overrightarrow{P}$ for a dozen wide double stars may be regarded as random.

In the conclusion, we pay attention to the fact that the expression (2) provides the possibility to verify the adopted sum of mass in definition of dynamic constant (k^2). In our case, the sum of mass for stars ADS 10759 and 11632, according to the mass-luminosity relation, would be equal to 3.0 and 2.6 and the corresponding value r_{max} are for ADS 10759 $r_{max} = 2k^2/v^2 = 410AU$ and for ADS 11061 $r_{max} = 768AU$. These values are less than the geometrical ones $r_{min} = \rho/\pi_{tr}$ (equal to 502 AU and 907 AU respectively). This contradiction allows us to suppose that there are hidden masses in both systems. Therefore, we have adopted the masses equal to 5 for each of the binaries, as it is indicated in Table 2. In one case, this suggestion was independently proved by A.Tokovinin [7], who found that both components of the ADS 11061 were spectral double stars.

The authors hope that this work opens the way to dynamical investigations of wide double stars.

References

[1] A.A. Kisselev, O.V. Kiyaeva (1980) The method of Apparent Motion Parameters, used to determine the orbit elements of a visual double star on the basis of short arc observations, *Astron. J. USSR*, **Vol. no. 57**,p. 1227–1241 (in Russian).

[2] L.G. Romanenko, E.L. Chentsov (1994) Determination of relative radial velocities of the components of visual binary stars from observations with the 6-meter telescope, *Astron. J. USSR*, **Vol. no. 71**,p. 278–281 (in Russian).

[3] A.A.Tokovinin (1994) Radial velocities of the components of wide visual double stars, *Astron. J. USSR*, **Vol. no. 71**,p. 293–296 (in Russian).

[4] A.A. Kisselev *et al.* (1994) Trigonometric parallaxes of 12 visual double star , *Izvestia GAO at Pulkovo*, **No. 208**, p. 9–17 (in Russian).

[5] O.V. Kiyaeva (1983) Using old observations for correction of the visual double star orbit, *Astron. J. USSR*, **Vol. no. 60**,p. 1208–1216 (in Russian).

[6] J. Hopmann (1955) Der Doppelstern ADS 11632, *Mitt. Univ. Sternwarte Wien*, **Vol. no. 7**, p. 67–82.

[7] A.A. Tokovinin (1995) The multiple system ADS 11061, *Pis'ma Astron. J. USSR*, **Vol. no. 21**, p. 286–293 (in Russian).

A NEW METHOD USED TO RE-VISIT THE VISUAL ORBIT OF THE SPECTROSCOPIC TRIPLE SYSTEM η ORIONIS A

D. POURBAIX
Department of Astronomy
University of Florida
Gainesville FL 32611-2055, U.S.A.
e-mail: pourbaix@astro.ufl.edu

AND

P. LAMPENS
Koninklijke Sterrenwacht van België
Ringlaan 3, B-1180 Brussels, Belgium

Abstract. Relatively few close binaries are simultaneously resolved by spectroscopic and visual techniques. However, they are important because they allow the estimation of fundamental stellar data such as masses and distances even without recourse to physical hypotheses. In addition, if eclipses also occur in the system, the dimensions of the components can be estimated.

The primary component of the bright multiple star η Orionis, consisting of a spectroscopic triple system, with A_c orbiting around the eclipsing pair A_{ab} in less than 10 years, is such a system. An extensive discussion of new spectroscopic data for the eclipsing pair A_{ab} has recently been published (DK96). We hereby will focus our attention on the third component, the interferometrically observed A_c companion.

We used a new method for the estimation of the elements of the visual orbit with *no* a priori information on the orbital parameters (PD94) and an automated Lehmann-Filhés procedure for determining the elements of the spectroscopic orbit. The results of both computations are being illustrated and discussed.

Satellite data forthcoming from the HIPPARCOS mission may help in determining this visual orbit, only partially covered by ground-based observations, more stringently. High-angular resolution space missions, capable of directly resolving a large fraction of the known spectroscopic binary systems, are expected to contribute substantially in that respect.

1. Introduction

η Orionis consists visually of three early-type stars ABC, the brightest of which form a pair at a separation of 1.6". The primary A is a spectroscopic triple system A_{abc}, studied by Zizka and Beardsley (1981). They determined elements for the short-period and the long-period orbits after removal of respectively the long-period and the short-period variations.

J. A. Docobo et al. (eds.), Visual Double Stars: Formation, Dynamics and Evolutionary Tracks, 383–388.

TABLE 1. Visual orbit of η Ori A

Parameter	Value	Stand. devia.	Convent. param.	Value
A	0.0132	0.0084	a (")	0.047
B	−0.0265	0.0092	i (°)	101.43
F	−0.0291	0.0132	Ω (°)	131.16
G	0.0247	0.0171	ω (°)	232.83
s	1.10	0.44	$e(s^2/(1+s^2))$	0.55 ± 0.20
P (yr)	9.356	0.17		
T	1983.91	0.99		
a^3/P^2	1.21x10^{-6}			

Note: A, B, F, G are the Thiele-Innes constants from which the conventional parameters are derived. s is directly related to eccentricity.

The short-period spectroscopic orbit of the eclipsing pair A_{ab} ($\approx$ 8 days) has been very recently re-investigated by De Mey et al. (1996). The third component A_c orbits the eclipsing pair in less than 10 years. It has also been observed by speckle interferometric techniques (CHARA).

2. Computational methods

We used a nonlinear least-squares adjustment in the computations of both the speckle and the spectroscopic orbits. This requires the a priori knowledge of a sufficiently good starting point, i.e. an approximate solution. The way to produce such a starting point automatically is different for the two programmes: the one for the visual orbit determination uses the method of *Simulated Annealing* (**Ptolemy**, Pourbaix (1994) and references therein) while for the spectroscopic orbit determination, a slightly modified version of the Lehmann-Filhés has been implemented (**Spectros**, Pourbaix (1997)). Direct minimisation follows then: both programmes make use of the Broyden-Fletcher-Goldstarb-Shanno algorithm (Dennis and Schnabel, 1996). The final solution is the one that minimises the 'distance' between the observations and the computed data points. Statistical quantities such as correlation matrices, confidence limits and graphical outputs are also produced.

3. Results

3.1. THE VISUAL ORBIT

The orbital solution determined from 1000 runs of the programme **Ptolemy** is presented in Table 1 and illustrated in Figure 1. Table 2 shows the values of the observables computed with the estimated orbital parameters at the epochs of the observations and the corresponding residuals. The residuals in ρ and in $\rho\ \Delta\theta$ are generally within the errors quoted by CHARA (McAlister et al.,1984; 1988). The hereby adopted solution furthermore confirms that the system was not observable on 1980.726, when the separation was smaller than 0.03", i.e. below the resolution limit of a 4m telescope ($\rho_c = 0.028$").

TABLE 2. Ephemeris and residuals computed with the elements of Table 1. An asterisk denotes 180° reversal of the published angle (col. 3).

Obs	Date	θ_{2000} (°)	$(O-C)_\theta$ (°)	ρ (")	$(O-C)_\rho$ (")
1	1975.956	132.781	1.152	0.045	−0.001
2	1976.858	127.925	−1.297	0.057	−0.001
3	1976.860	127.913	0.615	0.057	0.000
4	1976.923	127.636	−1.709	0.057	0.003
5	1977.087	126.929	0.597	0.058	−0.005
6	1977.734	124.214	−0.691	0.058	0.001
7	1977.742	124.179	−1.156	0.058	−0.000
8	1978.751	119.442	−0.325	0.051	−0.007
9	1979.036	117.796	0.820	0.048	0.005
10	1979.771	112.160	6.252	0.038	0.004
11	1979.774	112.134	−0.622	0.038	−0.001
12	1980.016	109.511	−3.901	0.035	−0.003
13	1985.854	129.603*	1.275	0.054	0.002
14	1986.889	125.056*	1.917	0.059	0.001
15	1988.255	118.615*	−2.550	0.050	0.002

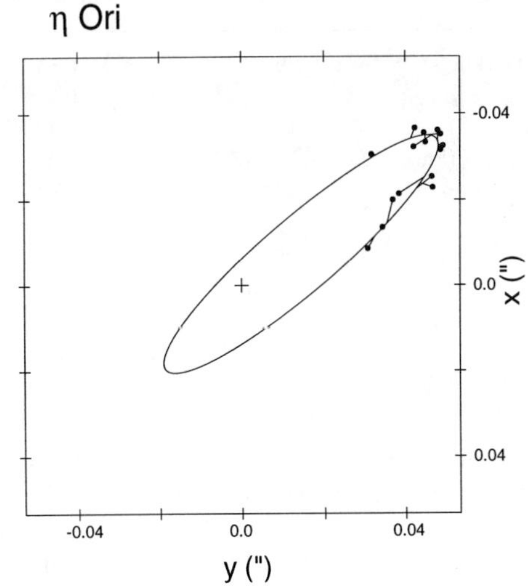

Figure 1. Visual orbit of η Orionis $A_{ab,c}$ based on speckle data

3.2. THE SPECTROSCOPIC ORBIT

The elements of the long-period spectroscopic orbit, derived from the published data (radial velocities for A_{ab} corrected for the ≈ 8 days orbit (Table III, ZB81)) and computed

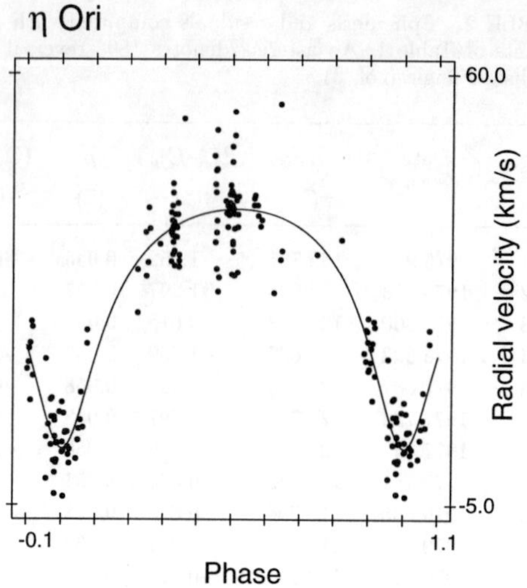

Figure 2. Spectroscopic orbit of η Orionis A_{ab} after removal of the short-period orbit

TABLE 3. Spectroscopic orbit of η Ori A

Parameter	Value	Stand. deviation
V_0 (km/s)	29.16	0.55
e	0.42	0.04
ω (°)	177.4	6.6
$a_{ab} \sin i$ (km)	759x10^6 = 5.07 AU	24x10^6
P (yr)	9.246	0.023
T	1906.71	0.13
$T' = T + 8\text{x}P$	1980.67	0.31
K	18.03	
σ_{O-C}(km/s)	4.98	

with **Spectros**, are given in Table 3. One can verify that the presented solution, though computed in a different way, closely corresponds to the one proposed by these authors. Figure 2 illustrates the fit to the computed radial-velocity curve: the mean residual error is ± 5km/s. The few data in the phase interval 0.6 to 0.9 still impede a definitive orbit calculation.

4. Discussion

Common orbital elements of the solutions in Tables 1 and 3 should be confronted in order to assess the reality of the errors and, if consistent, combined to better constrain the various parameters of the solution.

The orbital elements such as P, T, e and ω can be directly compared: the period and the eccentricity agree within the quoted uncertainties. The differences in time of periastron passage, $\Delta T' \approx 3$ yrs, and in longitude of periastron, $\Delta\omega \approx 50°$, are much larger, indicating that uncertainties on some elements of the speckle orbit are still underestimated as can be expected for a pair that is unresolved at periastron. Combination of the remaining elements provides a direct determination of the masses and the distance. Taking $i = 101°$ from the visual orbit, one derives the true separation of the pair A_{ab} from the spectroscopy: $a_{ab} = 5.17$ AU. De Mey et al. (1996) derived $F(M) = 1.53\mathcal{M}_\odot$ and $\mathcal{M}_{\rm ab} = 21.6\mathcal{M}_\odot$. From this, one computes the mass of the third component $\mathcal{M}_{\rm c} = 12.3\mathcal{M}_\odot$ and $\mathcal{M}_{\rm ab}/\mathcal{M}_{\rm c} = 2.3$. Since the visual semi-major axis equals the sum of the spectroscopically derived a_{ab} and a_c with respect to the centre of mass, one has:

$$a_{tot} = a_{ab} + a_c = a_{ab}(1 + \frac{\mathcal{M}_{\rm ab}}{\mathcal{M}_{\rm c}}) = 3.3a_{ab}.$$

So, the true linear separation of the speckle pair $A_{ab,c}$ equals 17.05 AU. Confrontation with the apparent semi-axis major derived from the speckle orbit gives a distance of 363 pc or $\pi = 0.0028''$. This is in excellent agreement with a previous photometric calibration (330pc, Waelkens et al. 1988) as well as with the mean photometric parallax found from Walraven photometry for subgroup 1a of the Orion association (380pc, Brown 1995) to which η Orionis belongs. We further derive a total mass of 33.9 $\mathcal{M}_\odot$ for the triple system and mass ratios of

$$\frac{\mathcal{M}_{\rm a}}{\mathcal{M}_{\rm c}} = 0.89$$

and

$$\frac{\mathcal{M}_{\rm b}}{\mathcal{M}_{\rm c}} = 0.86.$$

Though a better agreement between different parallax determinations is found than in previous works (WL88), this should only be considered as indicative. Indeed, with the large errors found on both orbit computations this remains a risky thing to do. More recent data and a combined determination of the spectroscopic-visual orbit are needed and will allow to constrain the information on masses and distance of this very interesting system at a much more satisfactory level.

5. Future projects and acknowledgement

We plan to merge the two methods into one combined least-squares method for a better constrained determination of the orbital solutions for spectroscopic-visual binaries such as η Orionis. We thank Dr. H. Eichhorn for helpful comments.

References

Brown, A., 1995, Ph.D. Thesis, Leiden University, unpublished.
De Mey, K., Aerts, C., Waelkens, C. and Van Winckel, H., 1996, AA **310**, 164 (DK96).
Dennis Jr., J.E. and Schnabel, R.B, 1996, in: Numerical Methods for Unconstrained Optimization and Nonlinear Equation (2nd ed.), SIAM, Philadelphia.
McAlister, H.A., 1984, CHARA Contribution No.1.
McAlister, H.A., Hartkopf, W.I. 1988, CHARA Contribution No.2.
Pourbaix, D., 1994, AA **290**, 682 (PD94).
Pourbaix, D., 1997, in preparation.

Waelkens, C. and Lampens, P., 1988, AA **194**, 143 (WL88).
Zizka, E.R. and Beardsley, W.R., AJ **86**, 1944 (ZB81).

THE REEXAMINATION OF GLIESE 623A ORBITAL MOTION ON THE BASIS OF AUTOMATIC MEASUREMENTS

N.A.SHAKHT, E.V.POLYAKOV AND V.B.RAFALSKY
Pulkovo Observatory,
196140, St.Petersburg, Russia
e-mail:shakht@gao.spb.su

keywords: stars,dark components,orbit

1. INTRODUCTION

The principal programmes realized by means of Pulkovo 26-inch refractor [F =10.4 m, D=65 cm, M=19."81 in mm] are determinations of trigonometric parallaxes , the observations of double and multiple stars and the study of the motion of the selected stars with the suspected unseen companions. These programmes have been initiated by the Chief of Photographic Astrometry Department Prof.A.N.Deutsch.

The well-known objects such as 61 Cygni, δ Gem,Lalande 21185, Gliese 623 belong to last programme and have the long-term sets of the observations. The part of these objects must have componentss which can posess small masses close to substellar or planetary ones.
The stars for the programme of objects with unseen components have been chosen due to the geographical position ($\varphi = +59^o46'.0$, $\lambda = -02^h01^m.3$) and seeing conditions at Pulkovo.

Pulkovo programme of stars with unseen components contains single and multiple stars of late spectral classes , a part of them are red dwarfes with small masses. This programme is described by **Kisselev** *at al.*, **1992**.

Thanks to the specific conditions we have the possibility to observe these stars close to a meridian only in the limited interval of time. This circumstance exerts influence on the weight and the precision of parallaxes determinations for these objects. That is way we often prefer to exclude the parallax displacement by means of its certain catalogue's value from the observational positions of star and then we use the corrected positions for a study of the motion free from parallax.

J. A. Docobo et al. (eds.), Visual Double Stars: Formation, Dynamics and Evolutionary Tracks, 389–394.

It is known that the star Gliese 623 (AC48^{o}1595/1589) [R.A.= $16^h22^m.6$ Decl.= $+48^o28'.0$(1950.0), $10^m.3$, dM3] **Gliese, 1969** is a binary (Gliese 623 A and B) . The second component of this system is an optically invisible one and its mass is close to the substellar one. The first astrometric study of the motion of Gliese 623 A has been made by **Lippincott and Borgman, 1978** (hereafter **LB**) on the basis of observations at Sproul Observatory in 1938-1977. The first estimation of the lower limit of the mass of a dark companion equaled 0.06-0.08 solar mass has been made by these authors too. Then the other methods have been applied for improvement of the parameters of photocentric orbit and estimation of the mass of Gliese 623 B ,see , for instance, **Marcy and Moore,** 1989,(hereafter **MM**)
The numerous estimations of the low limit of this mass yield it in the range 0.067 – 0.087 solar masses if the primary has the mass equaled 0.31 solar masses according to MM. Nevertheless some difficulties remain in the determination of parallax and hence in the estimations of mass of the component.

2. THE OBSERVATIONS OF GLIESE 623 AT PULKOVO

We present the results of astrometric observations over the interval 1979-1995 years . The following Pulkovo astronomers taken part in these observations : A.A.Kisselev (23 plates), N.A.Shakht (21 plates), O.A.Kalinichenko (19 plates), T.P.Kis seleva (16 plates), O.P.Bykov (12 plates) and some others. During these observations the plates NP-27 were used.

The all number of the obtained plates is about one hundred with 6 exposures on each plate in average. We are observing this star during 3 months in course of spring season. The relative positions of the star have been determined in the system of reference stars. These stars with magnitudes and co lours near to the ob ject have been chosed by means of Palomar Atlas.

In spite of some variations of seeing conditions and also some modifications in ocular part of telescope we keeped the constant system consisted of 5 stars for 93 plates which we are analysing here.

3. MEASUREMENTS AND TREATMENT OF DATA

The first measurements of all plates have been made with semi-automatic Zeisse measuring machine "Ascorecord"and now we repeated this measurements by means of automatic complex "Fantasy". The description of measurements, technique of treatment and the information about reference stars are given in the report (**Shakht at al.**, this issue).

4. A DETERMINATIONS OF THE PROPER MOTION AND THE ESTIMATION OF THE PARALLAX OF GLIESE 623

For the study of the motion of this object the following equations are used:

$$\Delta X_j = C_x + \mu_x(t_j - t_0) + P_x\pi_x$$

(1)

$$\Delta Y_j = C_y + \mu_y(t_j - t_0) + P_y\pi_y$$

where $\Delta X_j, \Delta Y_j$ are positions of object on each plate with respect to standard one, C_x, C_y- the constants which are due to the errors of standard plate , t_j- the moment of observations, t_0- the moment of observations of the standard plate equaled 1987.31, μ_{xy}- the relative proper motion, P_{xy}- parallactic factors, π_{xy}- the values of the relative parallax.

The most of parallactic factors P_{xy} have only positive sign. It is corresponding to narrow interval of time accessible for the observations of this star. In consequence of it we don't aim at the exact determinations of the parallax and known equations have been solved on the first study only to estimate the precision of our series and any possible correlations between unknowns. Finally we have obtained

μ_x = + 1".1460 ±0."0009, μ_y =-0."4462 ±0."0009 (m.e.)

π_x=+ 0."128 ±0."019, π_y= +0."136 ±0".018

It should be noted that although the weight of the parallax is not considerable owing to geographical positions of Pulkovo, our data gives the values of π_{xy}close to catalogue's one 0."138 (Gliese,1969.0) and near to π_{abs} equaled 0."134 according to LB.

On the following stage of the treatment we solved the systems of equations where only values of constants and proper motions were as unknowns and the left parts $\Delta X^{'}$,$\Delta Y^{'}$ represented relative positions of star corrected by means of π_{abs} LB. Also in this case the values of proper motions have been obtained similar to above-mentioned ones. The mean annual residuals R_x,R_y for mean moment of observations and their mean errors ϵ_x, ϵ_y in milliarcsec are given in the Table 1, where N is a number of plates.

Table 1. The mean annual residuals

Moment	N	R_x	ϵ_x	R_y	ϵ_y
1979.246	4	-18	14	-62	10
1980.243	8	+30	5	+32	7
1981.246	4	-64	8	+48	10
1982.287	5	-28	5	-36	7
1983.248	7	+18	3	+ 8	11
1984.303	6	+44	6	+64	7
1985.303	6	-42	6	+22	7
1986.227	5	-20	5	-42	6
1987.249	7	+22	7	+ 2	7
1988.303	5	+18	5	+58	5
1989.271	5	-46	10	+ 2	10
1990.304	9	-16	7	-26	5
1991.412	4	+32	10	-10	4
1992.362	4	-32	14	+38	14
1993.311	5	-40	11	-52	11
1994.321	2	+20	4	-18	10
1995.339	7	+48	6	+28	7

5. DETERMINATION OF THE PHOTOCENTRIC ORBIT

On the basis of the residuals R_x,R_y we have estimated the dynamical elements of the preliminary photocentric orbit P,e and T_o by means of graphical methods by **Shain,1936** and by **Van de Kamp,1981** and rectangular coordinates of the ellipse of the photo centric orbit have been calculated in the units of the great se miaxis :

$$x = cosE - e; \quad y = sinE(1 - e^2)^{1/2} \tag{2}$$

where E – the excentrical anomaly. Then the following equations have been solved:

$$\Delta X_j^{'} = C_x + (t_j - t_0)\mu_x + x(B) + y(G) \tag{3}$$

$$\Delta Y_j^{'} = C_y + (t_j - t_0)\mu_y + x(A) + y(F)$$

where $(B), (G), (A), (F)$ – the Thiele-Innes constants obtained with respect to photocenter.

We have made many versions of the solutions for the systems (3) with the assumed values of dynamic elements P, e, T_0 in following ranges correspondly: $3^y.70 - 3^y.80$ years, $0.50 - 0.58$ and 1984.1-1984.5 for different systems of the reference stars and for different intervals of observations.

Finally we adopted the following dynamical elements which coincide with the results obtained by means of visual measurements:

$$P = 3^y.76 \pm 0^y.10, \quad e = 0.51, \quad T_0 = 1984.3$$

. because the solution of the equations (3) with these elements gives the minimum of the error of the unit weight in X,Y coordinates, The use of the dynamical elements of LB for our series gives the same precision but in this case the constants obtained from (3) yield the geometrical elements which have a great difference from ones obtained by other authors. With our dynamical elements we have determined the constants (B), $(G), (A), (F)$ and then we have determined the great semiaxis α which is equal to $0.''053 \pm 0.''006$ and i, ω, Ω which are equal to 141°, 289°, 149° . All these values don't differ from our previous results (see, **Shakht, 1995**) with the exception of ω and Ω which were equaled to $265^o, 126^o$. Taking into account our value of α, the parallax π (**LB**) and the mass of the main star equaled 0.31 solar masses (**MM**) we have estimated the low limit of the mass of the satellite as 0.09 ± 0.03 solar masses.
The error of P is obtained by graphically estimation, the values of i change in the ranges $\pm 5^o$, and the values of ω and Ω have the most dependence on the adopted dynamical elements, different reference systems and intervals of observations. Their values may have the scatter more than 10°.

6. CONCLUSION

Our analysis has shown that automatic measurements give the results similar to the results obtained by means of visual measurements and we have not found noticeable systematic differences in the positions of star.

The value of proper motion, parallax and great semiaxis of photocenter are close to our previous results founded on the basis of visual measurements. Nevertheless we obtained some difference in geometrical elements and intend to improve this preliminary orbit.

This study is founded on the classical technique with the solution of the equations (3). This method is applied since in this case we have difficulties for determination of the projection of the focus for an apparent ellipse. But this method needs to a control because of probable correlations between unknowns in (3) , and thus some other methods may be use for the correction of this photocentric orbit.

7. REFERENCES

Gliese W.(1969), Catalogue of Nearby Stars, *Veröffent. Astr.Rech. Inst.Heidelberg,* **Vol 22** *p.p.3-116*

Kamp van de,(1981),Stellar Path,*in series"Astrophysics and Space Science Library,* **Vol.85** *" Dortrecht ,Holland, p.p.1-255*

Kisselev A.A.,Kiyaeva O.V. and Shakht N.A.,(1992), Astrometry with the Long-Focus Telescope at Pulkovo *in series "Problems of the Study of the Univers",* **v.13**,*St-Petersburg p.p.142-165 (in Russian)*

Lippincott S.L.and Borgman E.R.,(1978),GC 20986 a New Unresolved Astrometric Binary,*PASP* **v.90,No 534**,*p.p.226-229*

Marcy G.W. and Moore D.,(1989),The Extremely Low Mass Companion to Gliese 623, *Ap.J.* **No341**,*p.p.961-967*

Shain G.A.,(1936),Double Stars, *in the textbook "The Astroph. and Stellar Astronomy"Leningrad-Moscow, p.p.255-258(in Russian)*

Shakht N.A.,Polyakov E.V.,Rafalsky V.B.The Automatic Machine "Fantasy" Employment For the Measurements of the Stars With Dark Companions, *this issue*

Shakht N.A.,(1995),The Observations of Gliese 623 and Some Other Objects with Suspected Unseen Components, *Proc.Coll.* **No166 IAU**,*p.359.*

CORRECTION OF VISUAL BINARY STARS ORBITS WITH A PRECISE NUMBER OF OBSERVATIONS OR NORMAL POINTS

E. VIÑUALES AND R. CID
Grupo de Mecánica Espacial. Universidad de Zaragoza
50009 ZARAGOZA. SPAIN

Abstract. R. Cid (1958) established that four position angles θ and three angular distances ρ are the necessary and sufficient number of data for computing the orbit of a visual double star. Later on, M. Liso (1952) obtained the method for computing the orbit with only one distance and six position angles. Based on these computation methods, that have a tendency to give a larger importance to position angles, we build some correction methods that minimize the observation-computation differences. Now, considering the obtained results,these processes are applied to several orbits with the aim of analyze which of these methods are the best for each type of orbit.

1. The Cid's Method for determination of orbits in double stars

This method utilizes only three complete observations (ρ_i, θ_i, t_i), $(2 \leq i \leq 4)$ and one incomplete one (θ_1, t_1).

Thiele's equation for the epochs (t_i, t_j) with $(1 < i < j < 4)$, allows to write six equations that, by convenient division among them, give

$$\begin{aligned} \frac{nt_{12} - F(W-V)}{nt_{14} - F(W)} &= N = \frac{\rho_2 \sin\theta_{12}}{\rho_4 \sin\theta_{14}}, \\ \frac{nt_{13} - F(W-U)}{nt_{14} - F(W)} &= Q = \frac{\rho_3 \sin\theta_{13}}{\rho_4 \sin\theta_{14}}, \\ \frac{nt_{23} - F(V-U)}{nt_{24} - F(V)} &= R = \frac{\rho_3 \sin\theta_{23}}{\rho_4 \sin\theta_{24}}, \\ \frac{nt_{23} - F(V-U)}{nt_{34} - F(U)} &= S = \frac{\rho_2 \sin\theta_{23}}{\rho_4 \sin\theta_{34}}, \end{aligned} \tag{1}$$

with $E_{ij} = E_j - E_i$, $U = E_{34}$, $V = E_{24}$, $W = E_{14}$, $F(X) = X - \sin X$, $\Phi(XY) = F(X) - F(Y) - F(X-Y)$, $\theta_{ij} = \theta_j - \theta_i$.

The quantities N, Q, R, S, verify the relation $QS = RS + RN$.

J. A. Docobo et al. (eds.), Visual Double Stars: Formation, Dynamics and Evolutionary Tracks, 395–399.

Making use of the notations

$$h = \frac{t_{13} - Qt_{14}}{Qt_{12} - Nt_{13}}, \qquad k = \frac{t_{12} - Nt_{14}}{Qt_{12} - Nt_{13}},$$
$$p = \frac{Rt_{23} - RSt_{34}}{Rt_{24} - St_{34}}, \qquad q = \frac{St_{23} - RSt_{34}}{Rt_{24} - St_{34}}.$$

the system (1) is composed by the independent equations

$$\begin{aligned} &hF(W-V) - kF(W-U) + F(W) = 0, \\ &F(V-U) - pF(V) + qF(U) = 0, \\ &\Phi(VU) - (1-N)\Phi(WU) + (1-Q)\Phi(WV) = 0. \end{aligned} \tag{2}$$

Thus, the problem is reduced to solve this *fundamental system.*

With the aim of improving approximate solutions of the system (2) U_0, V_0, W_0, we consider u, v, w, as unknown errors and define $U = U_0 + u$, $V = V_0 + v$, $W = W_0 + w$, that will satisfy exactly the system (2). In order to calculate the values u, v, w, we make the classical approach $\sin u \simeq u, \sin v \simeq v, \sin w \simeq w$, and $\cos u \simeq \cos v \simeq \cos w \simeq 1$.

With the variables U, V, W, well-determined, we can calculate $C(= a^2\sqrt{1-e^2}\cos I)$, n and P. The remaining orbital elements can be evaluated by the process followed in the Thiele-Innes method.

2. The Liso's Method for the computation and improvement of orbits in visual binary stars

M. Liso (1962) decided in his method, based upon the Cid one, to omit some distance in favor of position angles. Thus, he considered as observational data, one distance (ρ_1) and six positional angles (θ_i) ($1 \le i \le 6$). Defining the quantities U, V, W, X, Y, by the differences $U = E_2 - E_1$, $V = E_3 - E_1$, $W = E_4 - E_1$, $X = E_5 - E_1$, $Y = E_6 - E_1$, and denoting for the sake of simplicity $[E_{ik}] = nt_{ik} - F(E_{ik})$, and representing by α, β, γ, the cross ratio of four points $(\theta_1, \theta_2, \theta_3, \theta_i)$, $(i = 4, 5, 6)$,

$$\alpha = \frac{[V][W-U]}{[V-U][W]}, \qquad \beta = \frac{[V][X-U]}{[V-U][X]}, \qquad \gamma = \frac{[V][Y-U]}{[V-U][Y]}. \tag{3}$$

Adding to these equations (3) the ones that result by permuting the order in the cross ratio of four points, we can obtain other tree equations

$$\begin{aligned} &[U]\Phi(W,V) - [V]\Phi(W,U) + [W]\phi(V,U) = 0, \\ &[U]\Phi(X,V) - [V]\Phi(X,U) + [X]\phi(V,U) = 0, \\ &[U]\Phi(Y,V) - [V]\Phi(Y,U) + [Y]\Phi(V,U) = 0. \end{aligned} \tag{4}$$

Equations (3) and (4) form the *fundamental system* in the unknowns U, V, W, X, Y, n.

After solving the system for determining the orbital elements, we can proceed in similar way described in the Cid's method.

3. A variant of the Thiele-Innes method for correction of visual binary stars orbits

A correction method based upon the Cid or Liso methods would be the most convenient way to correct the orbit found by these methods. The way followed consists in improving the inherent variables of the methods, rather than the orbital elements themselves. Once the inherent variables have been improved, we would apply again the orbit determination method.

Since the variables in both methods are differences of eccentric anomalies (with the exception of n), the construction of a correction method based on these methods result very difficult. Instead, we applied for the correction a variant of the Thiele-Innes method developed by one of the authors (E. Viñuales, 1994).

This variant consists of considering the expression

$$\tan\theta = \frac{bX + gY}{X + fY}, \tag{5}$$

where $b = B/A$, $f = F/A$, $g = G/A$, $X = \cos E - e$, $Y = \sqrt{1-e^2}\sin E$ and A, B, F, G are the Innes constants. By differentiating equation (5) and after some algebraic manipulations we get the equation

$$\begin{aligned} \Delta\theta &= \frac{A}{\rho}\cos\theta(X\Delta b + Y\Delta g) - \frac{A}{\rho}Y\sin\theta\Delta f \\ &\quad - \frac{\cos I}{\rho^2}(P\Delta e + nQ\Delta T + R\Delta n), \end{aligned} \tag{6}$$

where $N = 1 - e^2 - eX$ and P, Q, R are the expressions

$$P = -Y\left[1 + \frac{X^2+Y^2}{N(1-e^2)}\right], \quad Q = \frac{Y^2 + X(1-e^2)(X+e)}{N\sqrt{1-e^2}}, \quad R = -Q(t-T).$$

Proceeding in a similar way, the corresponding equation for the distances may be written as

$$\begin{aligned} \Delta\rho &= [(A\cos\theta + B\sin\theta)P' + (F\cos\theta + G\sin\theta)P'']\,\Delta e \\ &\quad + [(A\cos\theta + B\sin\theta)R' + (F\cos\theta + G\sin\theta)R'']\,\Delta n \\ &\quad + [(A\cos\theta + B\sin\theta)nQ' + (F\cos\theta + G\sin\theta)nQ'']\,\Delta T \\ &\quad + \cos\theta\,(X\Delta A + Y\Delta F) + \sin\theta\,(X\Delta B + Y\Delta G)\,. \end{aligned} \tag{7}$$

where

$$P' = -\left[1 + \frac{2Y^2}{(1-e^2)N}\right], \qquad P'' = \frac{Y(2X+eN)}{(1-e^2)N},$$

$$Q' = \frac{Y}{\sqrt{1-e^2}N}, \qquad Q'' = -\frac{(X+e)\sqrt{1-e^2}}{N},$$

$$R' = -Q'(t-T), \qquad R'' = -Q''(t-T).$$

TABLE 1. Observations and differences observation-calculus

t	θ	ρ	n	Observer	$\Delta\theta$	$\Delta\rho$
1918.71	47.0	2.11	19	Ol 10 Miller 9 pg	1.4	−0.05
1919.78	37.2	1.88	10	Ol Miller 5 pg	1.7	−0.13
1920.64	25.5	1.85	22	Bar 10 Bab 7 A 3 Fox 2	−0.6	−0.03
1921.68	13.8	1.63	22	Bar 13 A 4 Bab 3 Ol 2	0.7	−0.11
1922.70	357.6	1.58	26	Bar 18 Bab 7 Ol 1	−0.4	−0.03
1923.72	340.6	1.52	8	VB 4 A 2 Bab Ol 1	0.0	0.01
1924.77	320.9	1.45	16	Bab 7 4 A B VB 4	0.5	0.02
1925.75	299.3	1.51	5	VB 3 GΣ 2	−0.7	0.10
1926.64	279.6	1.47	3	VB	−2.0	0.04
1927.67	262.5	1.53	4	VB 3 GΣ 1	0.9	0.03
1928.69	244.2	1.64	3	VB	0.1	0.02
1929.72	228.8	1.72	9	VB Kpr 3 A 2 GΣ 1	−0.2	−0.03
1930.57	219.5	1.89	7	VB 4 Kpr 3	1.1	0.00
1931.68	206.2	1.98	7	VB 4 Sim 3	−0.2	−0.07
1932.79	195.0	2.25	6	Sim GΣ 3	−1.4	0.03
1933.54	191.4	2.34	6	VB 4 Kpr 2	0.9	0.01
1934.74	183.6	2.53	6	VB 3 Rabe 2 Bz 1	1.6	0.03
1937.73	165.3	2.88	10	Kpr 1 Rabe 3 Bz 6	0.6	0.02
1938.89	158.7	3.00	8	Bz Rabe 4	−0.3	0.03
1939.80	153.9	3.06	7	Rabe 4 Bz 3	−0.9	0.01
1940.66	149.9	3.21	9	Rabe 5 Bz 4	−1.1	0.10

Finally, by solving the system formed by (6) and (7), by the method of least squares we have the corrections of the orbital elements e, n, T and the Innes constants which will allow to find the corrections of the rest of the orbital elements.

We have applied the above methods to the binary ADS 15972. The orbital elements obtained are: $P = 44^a.73$, $T = 1970.34$, $e = 0.41$, $a = 2''\!.39$, $\Omega = 153^\circ\!.6$, $\omega = 211^\circ\!.1$, $I = 169^\circ\!.9$ and the following table shows differences observation-calculus.

4. Acknowledgments

We are very grateful to Drs. Docobo and Ling for providing us recent observations of the pair ADS:15972. This work has been supported in part by the Ministerio de Educación y Ciencia (DGICYT Project # PB95-0807)

References

Cid, R.:1952, Método de Mejora de Orbitas de Estrellas Dobles Visuales, *Urania*, **no.232**.

Cid, R.:1958, On the Necessary and Sufficient Observations for Determination of Elliptic Orbits in Double Stars, *Urania*, **no.232**.

Cid, R.:1960, Método de cálculo de órbitas elípticas en estrellas dobles visuales y aplicación al par ADS 13169, *Revista de la academia de Ciencias*, Universidad de Zaragoza. **Serie 3, Vol.XV**.

Liso Puente, M.:1962, Métodos de cálculo y mejora de órbitas de binarias visuales *Seminario Matemático U. Zaragoza.*

Viñuales, E.:1994, Generalización de Métodos de calculo y corrección de órbitas de estrellas dobles variables *Seminario Matemático Garcia de Galdeano*, Universidad de Zaragoza. **Serie 2,no.46**.

TABLE 2. Observations and differences observation-calculus (cont.)

t	θ	ρ	n	Observer	$\Delta\theta$	$\Delta\rho$
1941.82	146.7	3.17	14	Rabe 6 Dur Bz 3 VB 2	0.4	−0.01
1942.87	142.1	3.24	12	Rabe 7 Bz 3 Dur 2	0.0	0.0
1943.87	139.0	3.25	12	Rabe Bz VB 4	0.8	−0.03
1945.80	133.3	3.43	7	Bz 4 VB 3	2.3	0.09
1946.76	125.9	3.25	5	Rabe	−1.4	−0.10
1948.72	121.6	3.35	13	Rabe 6 VB 4 Bz 3	1.4	0.0
1949.78	115.2	3.27	10	Rabe 8 Mark 2	−1.0	-0.06
1950.78	110.1	3.03	5	Rabe	−2..4	−0.27
1951.81	107.2	3.06	14	Rabe 6 Bz 5 Mark 3	−1.4	−0.21
1953.51	99.5	3.15	19	Rabe 16 Dju 3	−2.4	−0.03
1955.45	92.1	3.00	21	Rabe 16 Wor 5	−1.6	−0.06
1956.82	87.7	2.82	8	Wor 5 C 3	0.2	−0.12
1957.69	83.5	2.74	10	Wor 4 C B 3	0.3	−0.12
1958.81	78.7	2.61	4	C 3 Wor 1	1.3	−0.14
1959.78	72.2	2.59	16	Wor 8 hz 4	0.3	−0.05
1960.92	66.8	2.42	8	Wor 5 C 3	2.0	−0.08
1961.74	58.9	2.33	8	Wor B 4	−0.2	−0.68
1962.66	52.8	2.15	15	B 8 Wor 4 C 3	0.6	−0.12
1963.57	44.7	2.06	4	Wor	0.2	−0.08
1964.70	35.8	2.10	4	Wor	2.2	0.11
1965.80	20.9	1.88	3	Wor	0.0	0.05
1966.79	8.0	1.66	5	Wor	0.2	−0.03
1967.84	351.3	1.60	4	Wor	0.0	0.03
1968.74	335.5	1.50	4	Wor	0.2	0.02
1969.79	314.1	1.52	6	Wor	−0.3	0.10
1970.73	295.5	1.52	6	Wor	0.6	0.11
1971.66	276.2	1.47	4	Wor	0.4	0.02
1972.75	256.2	1.51	8	Wor hz 4	0.8	−0.02
1973.74	238.8	1.66	9	Wor 6 Dur 3	−0.3	0.00
1974.79	222.6	1.79	5	Wor	−1.9	−0.01
1975.65	212.7	1.92	4	Wor	−1.6	−0.01
1975.83	211.73	1.96	6	(F) HEI	−0.7	0.00
1976.786	202.54	2.19	6	(F) HEI	−0.3	0.08
1977.67	195.26	2.3	6	(F) HEI	0.0	0.05
1978.64	187.6	2.4	3	HEI	−0.0	0.01
1978.76	175.3	2.47	10	(F) HEI	0.2	0.07
1979.83	179.85	2.54	3	(F) HEI	0.2	0.00
1980.59	175.3	2.73	2	HEI	0.3	0.08
1980.82	173.29	2.71	5	(F) HEI	−0.2	0.04
1981.74	168.65	2.81	5	(F) HEI	0.2	0.03
1981.85	166.9	2.68	1	ZUL	−0.9	−0.10
1982.60	166.9	3.	2	HEI	2.9	0.13
1984.62	154.5	3.09	2	HEI	0.0	0.04
1984.82	155.53	3.14	2	(F) HEI	1.9	0.07
1985.51	151.61	3.1	2	(F) HEI	1.0	−0.01
1987.90	141.9	3.64	2	MUL	1.0	0.38
1989.71	132.5	3.56	2	MUL	−1.4	0.24

SECTION IV

ASTROMETRIC DATA FROM SATELLITES

HIPPARCOS CONTRIBUTION TO BINARY EVOLUTION RESEARCH

J. DOMMANGET
Observatoire Royal de Belgique

Abstract. The Hipparcos space mission is one of the most fascinating experiment in astrometry never realised. Its importance from a general point of view has not to be recalled but it may be interesting to here examine the part it might have for double star astronomy and more precisely for our research on the evolution of the binaries.

In this introductory lecture, we first recall the theories that have been proposed in the past and how we to-day may consider the problem on the basis of our present knowledge from a somewhat personal point of view. In that respect we shall here consider the histogram of the real separations of the components, the period-eccentricity relation, the mass-ratio distribution and the HR diagram for binaries. The evolution by secular mass-loss is finally considered with its different possible consequences.

Finally the Hipparcos contribution to binary star evolution is discussed on the basis of the different output of the satellite, as the positions (absolute and relative), the parallaxes, the proper motions, the photometric data and the discovery of new systems.

¿From a personal point of view, it appears that beside the parallaxes (the original duty of the mission), the most important result of the satellite is the discovery of new systems due to the general survey of the sky by the mission as well as the photometric data that will be provided for the individual components of the close pairs.

1. GENERAL COMMENTS

The origin and the evolution of the Universe have always been a subject of concern. The same happened of course for each of its constitutive elements as for example for the double and multiple stars as soon as their existence was discovered. That stars exist is already a subject of question for the human beings but that many of them are systems of a few objects close together is a remarkable feature that had to excite the curiosity.

This curiosity was strengthened when it was discovered that there exist not only visual pairs, but spectroscopic and photometric binaries and that finally all these systems seem to belong to only one category of objects showing component separations from a few hundredths, to some thousands of astronomical units and that the distinction between the three different sub-categories lies more in our technical difficulties to observe them than in a deep real physical difference.

It is thus evident that a general theory on the evolution of double and multiple systems should take into account all categories of systems: close pairs (spectroscopic and

J. A. Docobo et al. (eds.), Visual Double Stars: Formation, Dynamics and Evolutionary Tracks, 403–427.

photometric binaries), medium wide ones (visual), very wide ones (common proper motion systems) and also particular objects as triple and multiple systems, trapezium type systems, systems with variable components, with white dwarfs etc...

The evolution of the binaries has thus been proposed as the subject for many meetings: colloquium, symposiums, workshops, etc. but every time the center of interest of the discussed matter shifted to the close pairs. When one goes through all researches on binary evolution, one finds that the very great majority of the papers concern these last systems, I mean spectroscopic and photometric binaries. This seems to be due to the fact that such systems are rapidly revolving and rapidly offer a lot of information on parameters as orbital elements, masses, radii, chemical composition, etc. of their components.

At the contrary, the medium wide systems and the wide pairs do not lead to the same interest because orbital elements are only obtained after much more extended periods of observations and the physical parameters of their components need specific independent programs. As a consequence, the progress in their knowledge is much slower and can thus not lead to the same amount of various researches.

Already at the IAU Colloquium "On the Evolution of Double Stars" held at Uccle (Belgium) (1967) - *the first meeting specially devoted to the evolution of the binaries* - K. Aa. STRAND remarked in his general introductory lecture, that: " *judging from the contributed papers, a great deal of the discussions will center on the evolution of close binaries either presently on or off the main sequence*". Fifteen years later at the IAU Colloquium $n^0$69 held at Bamberg on close pairs, M. PLAVEC (1982) remembered that this Uccle Colloquium has finally been the first symposium of a long series of such meetings on close double systems.

And this situation did not changed since then. The same happened at all meetings on double stars. As a consequence, our knowledge on the evolution of the binaries remains more confined in the field of the close pairs and all its improvement seems to return to them without leading to many particular information about the genesis and the evolution of the medium wide and large systems: the visual double stars. It appears alike the knowledge of the formation and the evolution of the close pairs is self-satisfactory.

This is regrettable. Not only because it is not logical to try to understand the existence and evolution of a whole, by only showing some interest for one of his subsets, but also because the visual binaries are the only systems which are free of any complicated mechanism encountered in the case of close pairs, as mass exchange, tidal effects, Roche lobes, common envelopes, etc. They are the purest application of the two body problem and the simplicity of their dynamical equations makes them the easiest objects to handle in this field.

It is thus evident that the researches about the evolution of the close pairs and those about the visual ones show a completely different approach and our knowledge on the evolution of the visual binaries may not expect very much from that on the evolution of the close systems: techniques, environment and data are of such a different nature that only specific researches on the visual pairs may help in finding something about their origin and their evolution.

I think thus that if we want to make some progress in our investigation, we have to direct more efficient efforts to the medium wide and wide pairs that means to the most simple systems, that means to those who do not show any other mechanism as the orbital Keplerian law because their evolution is certainly more easy to detect, to follow and to understand under that circumstances.

Fortunately, at the present symposium, we must state that probably for the first time the papers on visual systems are prevailing.

It seems to me evident that any discovery in the field of visual binaries will lead to an important contribution to the evolution of binaries in general and thus of close pairs and finally also to the evolution of the single stars. It is in that direction that I will try to give here a synthesis of what may be expected in their field especially on the basis of the Hipparcos awaited results.

2. OUR KNOWLEDGE ON THE ORIGIN AND THE EVOLUTION OF THE VISUAL BINARIES

To explain the origin and the evolution of the binaries, a few different processes have been proposed. They all may be classified into three general categories of which some may be divided in a few sub-categories showing little differences. The fundamental principles of these three categories are: a mutual capture at the time of a common approach of two independent stars, a common origin mechanism and an independent nuclei formation.

2.1. THE CAPTURE BY MUTUAL APPROACH OR ENCOUNTER THEORY

In case of a *distant approach* of two independent stars, created *in different places in space very far from each other*, no sufficient important mechanism seems to exist to transform the original hyperbolic relative orbit into an elliptic one at the first periastron passage, both stars returning thus to the interstellar medium. At the contrary, in case of a *close approach*, it may happen, as first proposed by J. STONEY (1867), that by "collisions" between the two stars (without common destruction!... that means by grazing collision), due to the consecutive dissipation of the orbital energy, an elliptic orbit results and at each successive periastron passages the period shortens and the periastron distance increases. When the periastron distance is sufficiently large, the orbital elements cease to evolve, the orbital eccentricity remaining relatively high. Thus, as seen by Stoney, this theory needs a complementary mechanism to explain the binaries having low orbital eccentricity.

On the other hand, in case of some other particular less dramatic circumstances (dense medium surrounding the stars and breaking the relative motion, chocks with surrounding planets, tidal disruption, etc.) it may be also that at each periastron passage, the eccentricity is decreasing and the orbit shortening, meaning that finally the high eccentric orbit are the youngest ones and the low eccentric orbits, the eldest ones... Some orbital transform due to other approaches by surrounding stars has also been considered.

Supported by Lord KELVIN (1891), M. J. BOSSLER (1928) etc. and later, on the basis of a study of the distribution of the angle between the orbital planes and the galactic one, by O. J. SCHMIDT (1945), this process has first been rejected by T. J. J. SEE (1893). It is not often retained to-day as the principal phenomenon to consider for creating binaries, because a mutual close approach of two stars is very scarce and as a consequence this phenomenon cannot explain the very high observed frequency of binaries. It also needs the consideration of any consecutive evolutionary phase after formation of the system, to explain the wide distribution of the eccentricities.

2.2. THE COMMON ORIGIN THEORY (FISSION OR SCISSION)

This theory very early appeared as the most probable formation process. It easily explains the existence of the close pairs with all their characteristics but unfortunately does not immediately explain the existence of the wide pairs, thus the visual systems, without a complementary mechanism of evolution of the orbital elements.

The starting idea lies in researches conducted by C. MAC-LAURIN (1741) who showed that a rotating homogeneous mass of fluid gives raise to an ellipsoid of revolution as it was illustrated one century later by a very simple experiment proposed by M. PLATEAU (1843). In a letter to the french Academy of Science in 1834, C. G. J. JACOBI (see: J. LIOUVILLE, 1834) demonstrated the stability of such an ellipsoid even if it has three unequal axis and is rotating around its shortest one. H. POINCARE (1911) investigated the stability conditions of a Jacobi's ellipsoid and explained under which conditions such a body may break up into two comparable, though unequal masses (fission theory). A very complete list of papers concerning Jacobi's figure of equilibrium for a rotating mass of fluid is given by W. M. HICKS (1882) and by G. H. DARWIN (1886).

In his papers of 1893 - referring to papers of G. H. DARWIN (1886; 1887) - and of 1896, T. J. J. SEE estimates that it is easy to show that the binaries may only have been formed from a unique nebula whatever the ultimate process may be. But by discussing this theory, he expresses (1896) the opinion that the binaries originated with a small orbital eccentricity and that the: "*high eccentricities .. have been developed by the secular action of tidal friction*" (p.5). But he cannot explain the very large orbital pairs because when the orbit becomes larger, the tidal effect at periastron vanishes. In 1897 H. G. DARWIN by developing the *theory of the bodily tides* (see: H. G. DARWIN, 1911) showed the importance of the tidal effects in the formation of a close binary. In his address to H. POINCARE when delivering on presenting to him the Gold Medal of the Royal Astronomical Society, G. H. DARWIN (1900) gave a clear history of the contribution of the awarded in solving the tidal problems and thus in clarifying our knowledge on the birth and evolution of the binaries. Some other complementary researches were conducted later by him on the equilibrium of a rotating mass of liquid, that may be found in H. G. DARWIN (1910) where all his previous papers on the subject are collected.

In a detailed study, H. N.RUSSELL (1910) considering different initial situations, shows how two masses revolving in contact may evolve. Applying his theory to a sample of double stars principally from the Burnham Double Star Catalogue he concludes that "*it is more reasonable to suppose that binary stars have originated by fission*".

It is thus not surprising that P. BAIZE (1932) believed that the sequence: scission - cepheides - spectroscopic and photometric binaries - visual binaries - common proper motion pairs should be considered as the real process of evolution.

J. H. JEANS (1919) expressed his favor for the scission theory but invoked disturbing effects from the galaxy of stars to explain the large systems and their large eccentricities. Later (1924) he replaced this hypothesis by that of a mass-loss that we will discuss later (Section 4.-). In a general synthesis on cosmogony, (1928, p.303) he finally expressed the opinion that there may be two kinds of double star formation: *by scission* on one side and *by independent condensations* in the parent nebula as here explained in Subsection 2.3.

For completeness it is interesting to mention the critical work of R. A. LYTTLETON (1954) concerning the first point of view of JEANS about the scission theory.

2.3. THE INDEPENDENT NUCLEI FORMATION OR FRAGMENTATION THEORY

At the contrary of this two mechanisms of binary formation, the present theory proposes to reject as much as possible, the need of any consecutive evolutionary phase after the genesis of the system and to admit the formation of the components in their present relative situations, so that the systems just after formation, would approximately have the same characteristics as they presently show. That means that the distribution of the stars and that of the components of the binaries and multiple systems in space did not fundamentally change since their formation.

F. R. MOULTON (1906) is perhaps the first one to have suggested this possibility when he writes: "*It is perhaps suggestive that there are many nebulae which seem to have broken in parts, and which appear to be capable of developing into several separate stars*" (p.526). Three years later (1909), he showed that the fission theory could only have occurred if the original mass was *in the nebulous state*. He also expressed the idea that both aggregation and dissipation of matter should be simultaneously considered. He may thus be considered as the first author to imagine a form of mass-loss.

Unfortunately, it appears here also that this theory do not perhaps fit with all the observed characteristics of the binaries. For example, to explain the period-eccentricity correlation that we will discuss later, C. M. VARSAVSKY (1962) estimated necessary to introduce a friction phenomenon of the newly created nuclei inside the original cloud and more specifically a periastron drag force by the atmosphere of the primary. F. HOYLE (1953) in a study of the fragmentation of gas clouds into galaxies and stars, as well as G. P. KUIPER (1955) and W. HEINTZ (1969) do not need to invoke any subsequent evolutionary phase as mass ejection or magnetic fields. Confirming this opinion, J. P. ARCORAGI & alt. (1991) have shown on the basis of a "three-dimensional hydrodynamical simulation" that the fragmentation of a uniform, isothermal elongated molecular cloud slowly rotating around an axis perpendicular to its elongation, may practically lead to any kind of binary or multiple star. We also need to mention first the recent interesting and extended researches by A. DUQUENNOY & M. MAYOR (1991a; 1991b) from which it follows that the fragmentation is a possible mechanism of binary formation, but also the Proceedings edited by the same authors (1992) of papers presented at a workshop in Bettmeralp (Switserland), where one should remark the general synthesis by C. J. CLARKE on the various possibilities to create close binaries from a common original cloud.

A recent discussion of the fragmentation theory has also been given by A. P. BOSS (1992).

Under this item one should not forget the formation of binaries by the evolution of small clusters which looses their components. This would explain not only the binaries but also the triple, quadruple etc. systems.

2.4. ARE THERE ONE KIND OR TWO KINDS OF BINARY FORMATION MECHANISMS?

Facing the incapability of any of the above theories to explain both types of binaries, close pairs and very wide ones, T. J. J. SEE (1910) in another important research on the evolution of stellar systems discussed both last theories in length and concluded *first* that amongst them "*we seem finally reduced, by exclusion, to the first process as the law of nature*" (p.126), but *further*, after having considered also a possible rupture of nebulous masses rotating in equilibrium under the pressure and attraction of their parts (independent nuclei formation), that "*in the actual universe both processes are at work together*" (p.132). On the basis of an extended statistical research and the comparison of the luminosity curves for respectively the single stars and the components of close pairs on one side and wide pairs on the other, E. ÖPIK (1923; 1924) comes to the conclusion that the capture theory has to be rejected but that the fission and the independent nuclei processes have to be retained respectively for the close and the wide systems. We just have seen that J. H. JEANS (1928) came to the same conclusion.

One decade later, F. R. MOULTON (1931) considered that the capture, no more than the fission theory, was sufficiently verified by theoretical or observational evidences to retain them as the *unique* process of binary formation whereas A. BLEKSLEY, (1934) estimated that the binaries may have different origins and thus that close pairs and visual ones are different kinds of binaries.

G. P. KUIPER (1935) believes in a unique group of binaries (p.121) and rejects the fission theory because it "*is entirely unable to account for the origin of binaries wider than a few astronomical units*" (p.149). He confirmed somewhat later (1955) his opinion of the formation of *all* double stars by independent condensations.

Applying to the problem of the origin of the double stars, the theory concerning the birth of twins, N. BONEFF (1951) found also that, the capture theory being rejected, "*les couples serrés ont été formées d'après la théorie du dédoublement d'une masse unique, tandis que les couples éloignées - d'aprèrs la théorie des noyaux séparés*" (p.199).

Of course in the different statistical diagrams, we observe a gap between both categories of objects, for instance in the distribution diagram of the real component separation or better of the semi-axis major in the region from 0,3 to 7 a.u.(see: Figure 3). But nobody may presently assure that this gap is real or that it is not a selection effect due to the use of different observation techniques which do not sufficiently overlap.

One argument against this conception of the problem is that it would be surprising that this gap - if real - would just fit with the "technical observational" gap. On the other hand, if this is so, both mechanisms would lead to two different categories of systems showing their own characteristics that should be recognised in any sample of binaries covering the entire observation material. Now V. TRIMBLE (1974, 1978) in a first research on spectroscopic binaries, finds that at the time their components arrive at the main sequence, one observes two different binary production mechanisms, one (fission of a protostar?) which systematically yields mass ratios near unity and another (capture? formation by independent nuclei that became gravitationally bound?) which generally gives unequal masses. On the basis of the distribution of their periods, H. A. ABT & S. G. LEVY (1976, 1978) estimate that there is a proof for two distinctive kinds of binaries amongst solar type stars: those with periods less than 100 years are born by fission, the others are pairs of protostars that contracted separately. This limit would be of 10 years for B2-B5 (IV, V) absorption and emission stars.

Later, V. TRIMBLE (1986) extending her research to the visual binaries, confirmed her previous opinion that there are two distinct kinds of binaries, the frequency diagrams of the mass ratios for visual binaries and for spectroscopic ones being *distinctly different* (p.247). Considering the mass-ratio distribution of a sample of spectroscopic binaries, T. MAZEH & D. GOLDBERG (1992) found also differences between short-period and long-period binaries.

3. STATISTICAL RESEARCHES ON BINARIES

¿From the theoretical point of view, it is always possible to imagine any kind of evolution by considering well chosen hypothesis and by using our extended astrophysical knowledge in stellar formation and evolution. But one must be careful when starting in this way because one is never sure that what we think definitely admitted is real. See for instance the spectacular results recently obtained by the NASA with the space probe launched by GALILEO in the atmosphere of Jupiter. Many theories will have to be rejected and it is not sure that even our knowledge on the formation of the planets and of the solar system will not have to be revised at least to some extend. And Jupiter is a very nearby object! How far are the stars..!

But from the practical point of view one may have a different approach. It is sure that whatever the mechanism of formation and of evolution may have been, it should have left some traces and produced visible effects on the characteristics of the binaries observed to-day. It is thus not surprising that since the early time of double star astronomy, many astronomers have been temptated by the establishment of frequency diagrams of the physical, dynamical and astrometric parameters of the binaries and also of correlations that could exist between them. We will see that of all these diagrams, the histogram of the real separations, the period-eccentricity correlation, the distribution of the mass-ratios (correctly estimated to some extend by the difference in magnitude) and the H-R diagram are the most important for organizing our researches on binaries evolution.

3.1. THE HISTOGRAM OF THE REAL SEPARATIONS

From the four diagrams we just mentioned, the first one seems to be the most interesting because it concerns a parameter which may be known for many binaries, not only for the visual pairs (when their distances are known by one or another way) but even for the spectroscopic ones. And more especially if one limits the material to the nearby stars. At the contrary, periods and eccentricities are only known for orbital pairs which restricts much more the available material; mass-ratios are not well defined by direct observation, due to the generally undetermined position of the center of mass even when the orbits are known.

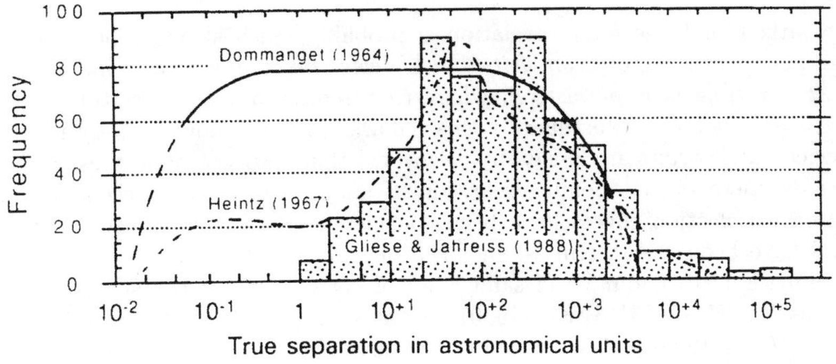

Figure 1.

The shape of this histogram may be of a great signification as we will see it later. It has been the subject of many researches. We shall retain here three results respectively established on one side by J. DOMMANGET (1970) in 1964 and by W. HEINTZ (1967, 1969) one year later, on the basis of very wide materials, and on the other by W. GLIESE & H. JAHREISS (1988) using only all known binaries in the surrounding of the sun to some 25 parsecs. They are given in Figure 1 - (J. DOMMANGET & P. LAMPENS, 1993) and show differences for separations inferior to a few tens of an AU expressing the insufficiency of the statistical material. Concerning spectral types or any other physical characteristics it appeared from a research conducted by H. ABT (1988), that the frequency of the separation does not vary significantly along the main sequence, but that the most frequent separations and the limiting separations vary markedly along the main sequence. But M. A. GIANNUZZI (1989) estimates that these results may be partially due to selection effects and at the contrary - in agreement with previous results obtained by J. L. HALBWACHS (1983, 1986) - she showed on the basis of Gliese & Jahreiss' catalogue of nearby stars that the distribution of the component separations is entirely independent from the spectral type of the primary component.

3.2. THE PERIOD-ECCENTRICITY CORRELATION

In addition to the distribution diagrams of some physical and dynamical parameters, correlations between them were also an interesting field of research. The reason why the P/e correlation has been one of the most interesting to consider lies in the fact that its establishment does not need any other data for visual as for spectroscopic orbits. The only restriction lies in the need of a computed orbit.

W. DOBERCK (1878) has been the first one to bring the attention to this correlation for which he found (1898):

Mean period (years)	Mean eccentricity
30	0.37
75	0.50
206	0.62

He remarked that the real correlation is probably much stronger because "*systems with long period and large eccentricity must always be comparatively unknown as very few are at any time near periastron, where they remain a very short time, and nearly all are near apoastron*". Unfortunately this remarks does not hold for very large periods where low eccentric orbits are less easily discovered than very eccentric ones. So finally for large periods, there exists a definite selection effects in favor of large eccentricities.

Some other selection effects have been considered to reject the reality of this correlation and have been the starting point of a long historical discussion. Nevertheless, H. LUDENDORFF (1910) found the same kind of correlation for the spectroscopic binaries whereas W. W. CAMPBELL (1910) on one side and Fr. SCHLESINGER & R. H. BAKER (1910) on the other, showed that the correlation for the visual systems seems to be an extension to the larger periods of that observed for the spectroscopic ones. Many authors as R. G. AITKEN, R. E. WILSON, J. H. MOORE, K. LUNDMARK, D. BARBIER, C. K. SEYFERT, W. S. FINSEN, J. HOPMANN, R. BONNET, S. AREND, M.-L.

LABEUR, P. MULLER, J. PENSADO, St. WIERZBINSKI, etc. have then been involved in such researches with different conclusions and comments. A very complete historical sketch has been given in our research on the evolution of the double stars by mass-loss (J. DOMMANGET, 1963).

All the correlations found by these searchers are given in Figure 2. They show that in spite of the important increase of the statistical material during some 60 years, the correlation keeps a stable aspect and give evidence of a clear distribution continuity from the very close spectroscopic binaries to the very large visual ones. We thus have to count with a correlation considered by O. STRUVE (1950) as militating "*strongly against any hypothesis that would attribute a different origin to the close pairs and the wide pairs*". He added that it is "*also probable that it has a physical basis and is in some way related to the origin and evolution of binary and multiple systems*" (p. 240). We here already should mention the idea of C. M. VARSAVSKY (1962) who estimates that the semi-axis major appears as a more physically significant parameter than the period as shown by the diagram given in Figure 3 where the gap between spectroscopic and visual binaries is clearly visible.

For completeness, we shall mention that some other correlations may exist for instance between the apsidal direction or the nodal direction and some particular directions in our Galaxy.

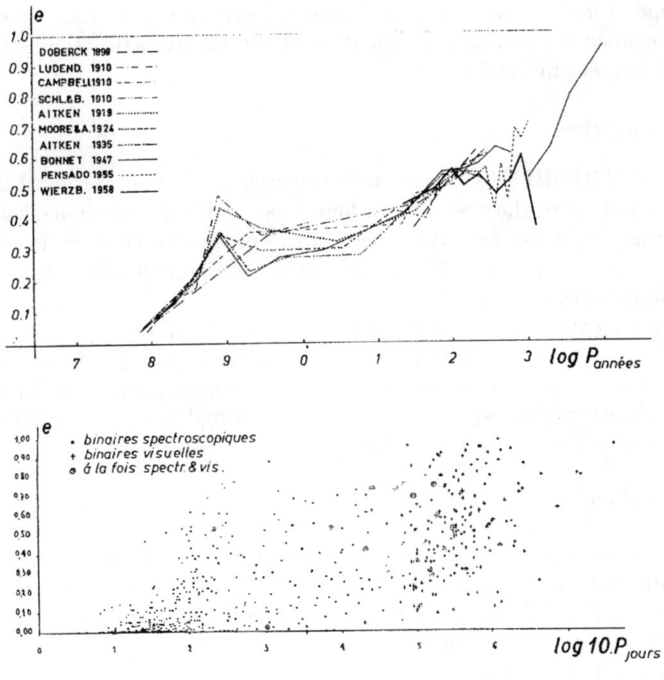

Figure 2.

To answer to this question, W. S. FINSEN (1936) made a systematic investigation between all six orbital parameters (the time of periastron transit being of course rejected). He found the correlation coefficients given in Table 1. But the author used the linear correlation technique which is not acceptable for many of them which need specific studies: the small coefficient found by Finsen for the P/e correlation, compared to some others for improbable correlations (for example: M/Ω) is thus not surprising.

	P	e	a	M	i	ω
e	+.18					
a	+.34	−.07				
M	−.25	−.07	+.09			
i	−.05	−.01	+.06	−.08		
ω	+.03	+.02	+.04	+.09	−.02	
Ω	+.10	−.26	−.04	−.25	−.09	−.16

TABLE 1.

We also have to mention the typical organisation found for the direct orbital poles (J. Dommanget, 1968, 1988) in connection with the galactic characteristics which too may be of some importance concerning the formation of the binaries. But this is not directly correlated to our present discussion.

3.3. THE MASS-RATIO

As we have seen, V. TRIMBLE by a systematic research on the mass-ratios of the spectroscopic binaries as well as of the visual ones, found particular distributions leading her to the feeling that there exist two formation mechanisms of the double stars. In particular in case of the visual pairs, she found (1986) the histogram of Figure 4 which shows a splendid maximum for the mass-ratio near 1,0.

But following an extensive research made by S. J. HOGEVEEN (1990) concerning visual pairs, all diagrams established for the mass-ratio are probably subject to important selection effects. By simulating such possible effects, the author shows that for mass-ratios larger than 0,3, observations are expected to reveal the real mass-ratio distribution but for values inferior to 0,3, selection effects begin to play a major part.

3.4. THE H.R. DIAGRAM

The distribution of the single stars in the HR diagram is of the highest interest and should also be very significant for the binaries. Many such diagrams have been established but they generally concern only medium wide and wide pairs because we generally do not know the spectra of both components for pairs having separations closer than 1 or 2", precisely those for which orbital elements may be available. Figure 5 shows a diagram established by P. MULLER (1967) on the basis of visual photometric measurements made in two colors permitting to deduce the color indices.

This contains only 51 pairs (Muller 1952) of which 60 % are closer than 3". Unfortunately such a material has insufficiently been increased since then to permit to observe

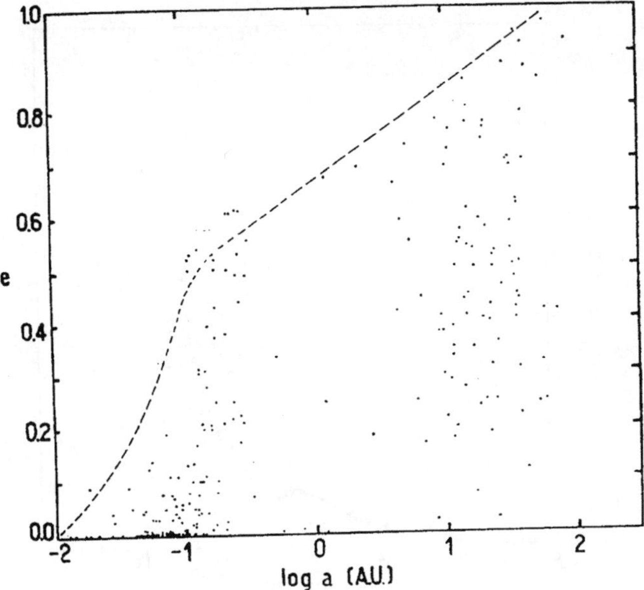

Figure 3.

some interesting particularities in the distribution of both components. Only specific visual technique (with a Muller prism for instance) and speckle technique may reach such close systems, but did not already produce its full capability.

It anyway seems that without introducing some physical or dynamical parameter, nothing special may be concluded in the case of the binaries from such a diagram.

4. THE EVOLUTION BY MASS-LOSS

As we have seen it, no consensus to-day exists regarding the mechanism of formation and evolution of the binaries. Each theory shows advantages but also offers some criticisms. The main difficulty consists to explain the evolution of the close pairs into wide ones or vice-versa. Why thus not having a look to any evolution mechanism as for example, a secular mass-loss as proposed by JEANS?

The genesis of the close and medium close binaries by the common origin theory or by the independent nuclei formation seems the most interesting to start with, especially when one remarks how much progress has been made observationally as well as theoretically in their field. To explain the genesis and evolution of all binaries, could a secular mass-loss transform a close pair into a larger one?

Mechanics with variable masses has been considered very early, first by C. DUFOUR (1866) and later by Th.von OPPOLZER, H. GYLDEN, J. MESTSCHERSKY, R. LEHMANN FILHES and many others, but generally from a theoretical point of view in very particular cases.

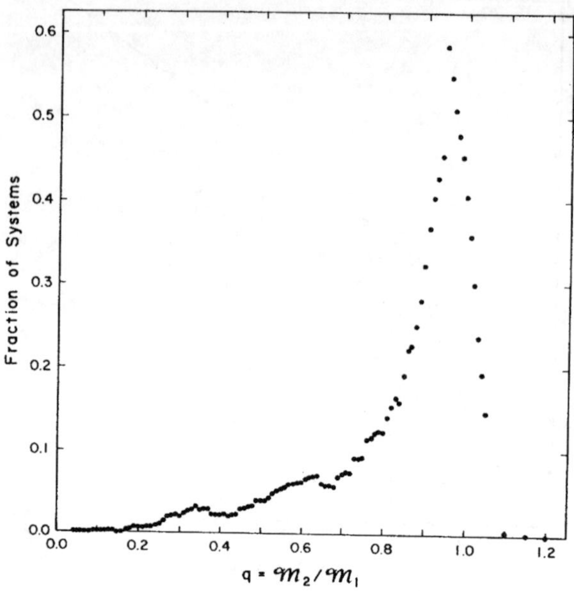

Figure 4.

The first research in the specific case of *mass-loss*, is to be credited to J. H. JEANS (1924, 1925) who based his research on the very recent theoretical explanation of the mass-luminosity relation by A. S. EDDINGTON (1923). He showed that a mass-loss expressed by the following law (α positive):

$$d\mathcal{M} = -\alpha \mathcal{M}^3 dt \tag{1}$$

is in perfect agreement with the mass-luminosity relation if one considers only stellar masses between 4 and 1 solar mass. If now the mass-loss is isotropic and of a secular nature (α very small), it appears that the eccentricity remains constant, but also that:

$$A\mathcal{M}_{AB} = \text{constant} \tag{2}$$

where A is the semi-axis major (as already shown by H. POINCARE (1911) for increasing masses with the sole condition that this increase remains constant all along a same revolution).

He also remarks that the evolution of a star from 4 to 1 solar mass under the assumption of Eddington's law, would need some:

$$7.1 \times 10^{12} \quad \textit{years} \tag{3}$$

But as we have seen, he remarks, that simultaneously each star should pass two times in the vicinity of another one, which due to the consecutive perturbations, could explain the existence of some large systems with high eccentricities. Unfortunately, this time interval

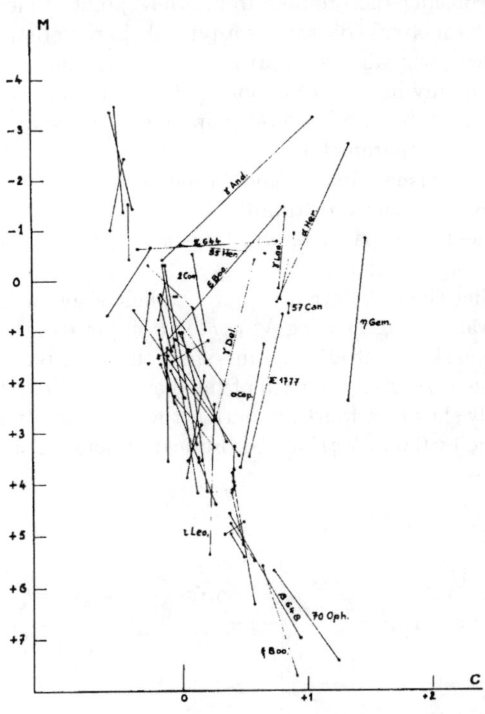

Figure 5.

is very large because it is based on a very low mass-loss rate due to electro-magnetic emission. If one considers mass-loss due to particles, this rate may be very much higher and the time interval is proportionally reduced in such a way that star encounters must be ignored and another process has to be considered for an increase of the eccentricity.

Many other researches have then been conducted on this subject. A complete historical sketch has been given in our paper of 1963 but none of them could explain an increase of the eccentricity except if one assumes that their exists a *periastron effect* as proposed by E. L. MARTIN (1934) who considers the more particular mass-loss expression:

$$d\mathcal{M}_{AB} = -\alpha \mathcal{M}_{AB}{}^{n} \frac{1}{r^2} dt \qquad (4)$$

where r is the radius vector in the true orbit. The mass-loss would thus be larger at periastron than at apoastron, which explains the increase of the eccentricity at least for the relatively close and medium wide pairs.

4.1. A MASS-ECCENTRICITY RELATION

But the hypothesis of mass-loss, whatever its expression may be, has another interesting orbital particularity: *the areal constant of the orbit is invariant if*, as one may expect, *the mass-loss is isotropic.* This incited us (J. DOMMANGET, 1963, 1964) to have a look in that direction and to consider the problem from a new point of view.

If one assumes that the areal constant is a typical characteristic of any binary during its evolution by mass-loss, why all the binaries showing the same areal constant would not be members of a same family having been more or less similarly evolving? So the idea was to compute the areal constant of all orbital pairs and to consider different categories of systems as a function of this parameter.

With a material of 212 visual binaries and 98 spectroscopic ones, 11 consecutive classes of the areal constant were considered from 0,2 to 2,4 $(a.u.)^2 y^{-1}$. Details of this research will be found in our mentioned paper of which we are here only reproducing in Figure 6 some of the most speaking diagrams.

A first remark is that the correlation is more explicit if for each given areal constant, one considers the equivalent diagrams $\log \mathcal{M}_{AB}/\log P$ in place of the original one $e/\log P$. A second important remark is that *all* diagrams show the same binary distribution. Third, the superior limit of the eccentricity in each of the diagrams *is not the result of a selection effect* as it may be easily shown. A fourth remark is that this correlation is shifted from one diagram to the next one by 0,6 in *logP* for theoretical reasons. Thus to make a better use

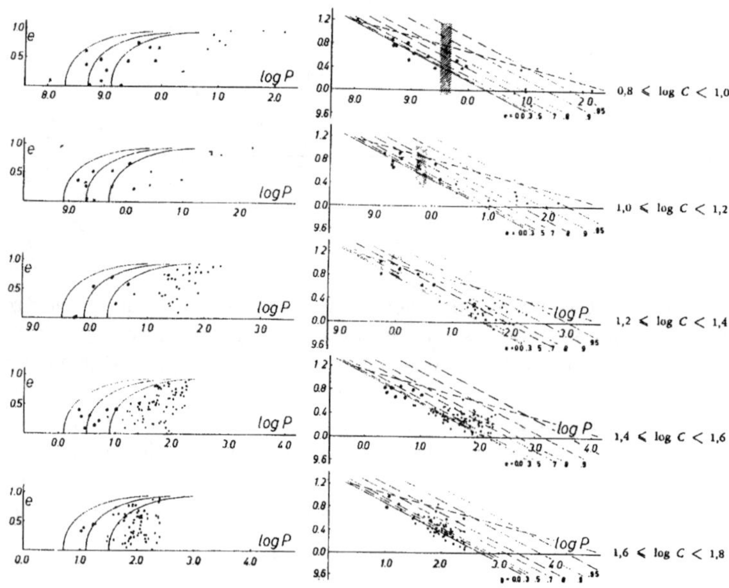

Figure 6.

of all the available data, we shifted the diagrams in such a way that the theoretical nets of straight lines representing the different values of the eccentricity superpose each other. We then found the diagram of Figure 7 where the abscissae $\log P$ have been replaced by:

$$X = \log P - 3 \log C \tag{5}$$

The superior limit of e observed in each of the individual diagrams appears better defined in the resulting one. It has later been confirmed by a second statistical research based on a more extended orbital material containing the same number of spectroscopic pairs but 551 visual systems (J. Dommanget, 1981, 1982a).

In these papers we have shown that it may be very well represented by the equation:

$$e^{2.8}\mathcal{M}_{AB} = 3.60 \tag{6}$$

We have here to remark that L. CHIARA (1955-56) showed that in case

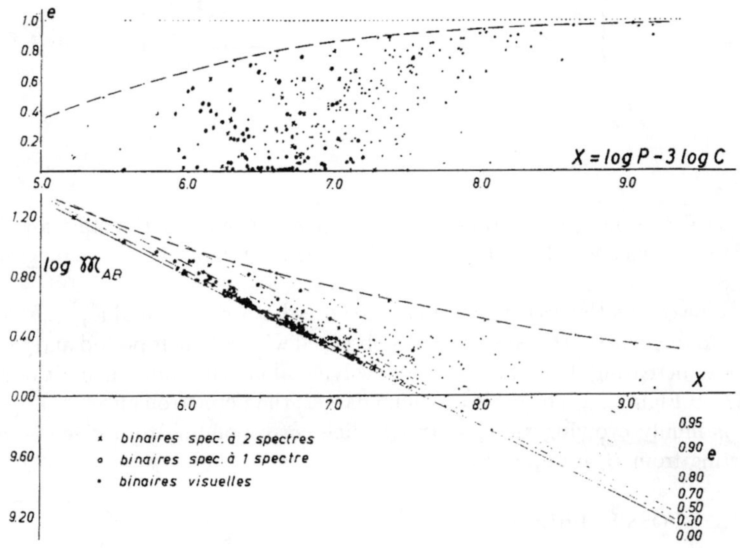

Figure 7.

of the more general mass-loss expression:

$$d\mathcal{M}_{AB} = \frac{k(t)}{r^2}dt \tag{7}$$

as the expression (4) proposed by E. L. MARTIN, the product:

$$e\mathcal{M}_{AB} \tag{8}$$

is constant, e being the mean eccentricity along one revolution and not the osculating one. A slight change with time of the expression $k(t)$ could easily lead to the expression (6).

If this is true, one could imagine the formation and evolution of a binary as starting by the splitting (or something alike that) of a heavy massive star or by the fragmentation of a protostellar cloud which due to mass-loss, mass exchange, tidal effects and any other proximity effects would evolve to produce a close binary system of two well formed stars.

The consecutive variations of the orbital elements may produce different types of close pairs having component separations spread in a somewhat extended domain so that for some of them only a mass-loss effect remains. By this effect, the period and the semi-axis major are increasing.

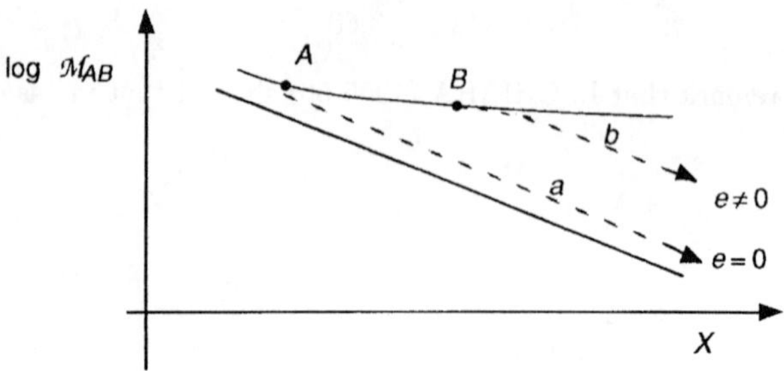

Figure 8.

For practical reasons, the orbital eccentricity must be small at the start for the great majority of the systems which are thus evolving along a straight line in the diagram (path a in Figure 8).

But for some others the eccentricity may be somewhat sensitive and a periastron effect may appear. An increase of the eccentricity is observed whereas their period and their semi-axis major are increasing: the stars are then evolving along the superior limit of equation (6) (path AB in Figure 8). The orbit becoming larger, the periastron effect disappears and the system is finally evolving along a straight line corresponding to a given eccentricity (path b starting from B in Figure 8).

4.2. THE MASS-LOSS EXPRESSION

Some information about the mass-loss expression may be deduced from the shape of the histogram of the real separation.

If one supposes that the mass-loss is given by the general expression:

$$d\mathcal{M}_{AB} = -\alpha \mathcal{M}_{AB}{}^{n+1} dt \tag{9}$$

the period and the semi-axis major increase respectively (J. DOMMANGET, 1963) by:

$$dP = 2\alpha \mathcal{M}_{AB}{}^{n} P dt \tag{10}$$

and:

$$dA = \alpha \mathcal{M}_{AB}{}^{n} A dt \tag{11}$$

This last expression may be written:

$$d \log A = \alpha \mathcal{M}_{AB}{}^{n} dt \tag{12}$$

and the density of frequency of $\log A$ is thus inversely proportional to: $\mathcal{M}_{AB}{}^{n}$ and directly proportional to A^n because, following the expressions (9) and (11) one has:

$$A\mathcal{M}_{AB} = \text{constant} \tag{13}$$

already considered above (see: equation (2)).

If we thus assume - as a working hypothesis - that the shape of the diagram has reached some stability, and that consequently in a same time interval there are as many double stars moving from each elementary interval in $\log A$ to the next one ,

- where the shape of the histogram of the real separation shows a constant frequency density, one has $\underline{n = 0}$, (in case of the law of mass-loss proposed by J. H. JEANS, one has: $n = 2$) the mass-loss expression being thus:

$$d\mathcal{M}_{AB} = -\alpha\mathcal{M}_{AB}dt \tag{14}$$

- where the histogram shows an increasing frequency, one has $\underline{n > 0}$. This seems to be in agreement with the mass-loss expression proposed by V. G. FESENKOV (1952) for the massive stars :

$$d\mathcal{M}_{AB} = -\alpha\mathcal{M}_{AB}{}^{3,90}dt \tag{15}$$

where thus: $n = 2,90$.

By mass-loss and by consecutive increase of the semi-axis major, the mass-loss expression would evolve from (15) to (14).

4.3. THE MASS-LOSS RATE

But it is not sufficient to find an acceptable law of mass-loss. One needs also to have at our disposal a sufficiently important mass-loss rate.

One knows that in case of the sun, this mass loss is of the order of $10^{-13}\mathcal{M}_{\odot}yr^{-1}$.

J. P. DE GREVE and C. DE LOORE (1972) have shown by model computations that for a star of the same spectral type and luminosity class as the sun, if the particle density at the base of the corona is of the order of $3,2 \times 10^{10}cm^{-3}$ this may amount to some $10^{-9}\mathcal{M}_{\odot}yr^{-1}$ which is needed to realise a sufficient change in mass during a star's live. Unfortunately such a mass-loss rate has never been observed but perhaps just because it is always unobservable with our present techniques.

4.4. THE MASS-RATIO

Soon after the mass-loss mechanism had been proposed by JEANS, W. M. SMART (1925) showed from a theoretical point of view that the mass-ratio should increase with a secular mass-loss and more, that this seemed to be observationally confirmed (p. 431). Somewhat later, V. M. LOSSEVA (1938) by an important discussion on the various possibilities concerning the evolution of the mass-ratio, concluded in particular, to the impossibility of a simultaneous creation of all binaries.

In a more recent research, A. JORISSEN (1984) confirmed the increase of the mass ratio by considering the difference in magnitude - in place of the mass-ratio - as a function of the parameter X defined by equation (5). By plotting Δm as a function of X (Figure 9) he first showed that there seems to exist a superior limit of Δm for values of X larger than -3.0. Admitting the secular mass-loss expression (9) he computed a table giving $d\Delta m/dX$ as a function of n and α_B/α_A. These data are in good agreement with the slope $(-1,1)$ of the observed limit.

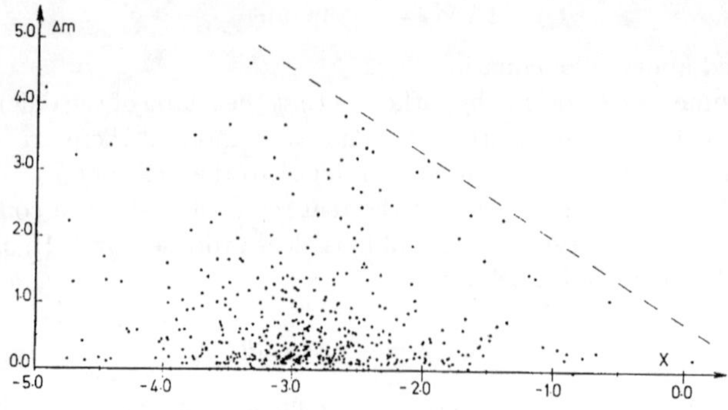

Figure 9.

By discussing such a diagram, it is useful to remember as A. H. BATTEN (1992) mentioned in his invited summary review at the IAU Colloquium $n^0 135$ (Atlanta), that the distribution functions of the mass-ratio and the separation, or the degree of multiplicity may have been seriously modified in the early stages of the life of the binaries. Therefore, at the contrary of what is sometimes believed, the mass-ratio may show a very large scatter at the time when the binaries start leaving these early stages where the proximity effects cease to be sensitive.

4.5. THE H.R. DIAGRAM

Another interesting feature concerning the areal constant, is its behavior in the HR diagram. Already at the time of our research on the evolution of the binaries by mass-loss (1963) we have shown that the distribution of the main components of the visual and spectroscopic binaries shows a light but systematic shift towards the large luminosities, with the increase of the areal constant. This was confirmed twenty years later by a more extended material (J. Dommanget, 1983a). This tendency is clearly shown (Figure 10) by the mean absolute magnitudes of these stars as a function of $\log C$ for different spectral types.

4.6. GENERAL REMARKS

Having discussed the various aspects of the theory of the evolution of the binaries by mass-loss, one must conclude that there remains some difficulties to explain the formation of very wide pairs as the product of evolved close ones. Is is still to-day not proven that stars of the main sequence loses sufficient mass to explain such an evolution. More, the main difficulty lies in the time-scale and especially in the relation (2) showing that reducing the mass of a star from ten to one solar mass yields an increase of the semi-axis major by a coefficient of 10 only, insufficient to explain the very large systems!

But this is not a reason to reject this mechanism without a more extended discussion. At least from this concept, it remains that the areal constant seems to play an important role in each of our statistical researches and that more considerations should be given to this dynamical parameter in the future. The corresponding research had also the merit of

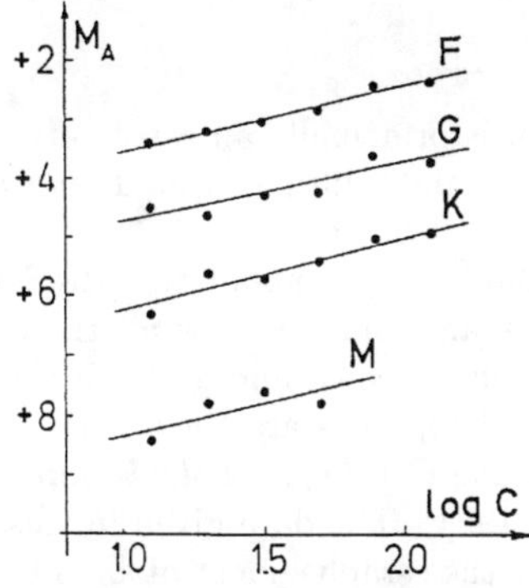

Figure 10.

showing some interesting features as for instance the "envelope" of equation (6) limiting the field covered by the binaries in the diagram $(X, log\mathcal{M}_{AB})$.

Now, from a more general point of view, we may conclude that the incompatibilities of the conclusions found by the different authors considering diagrams or correlations, have to be found in the insufficiency of our statistical material. Fortunately this may be extensively improved in the next few years, if one considers the very soon availability of the Hipparcos final catalogue.

It is thus here a good opportunity to finally discuss the "impacts" of this fundamental astrometry mission in binary evolution researches.

5. THE HIPPARCOS CONTRIBUTION

5.1. THE ANNEX OF DOUBLE AND MULTIPLE STARS

It is well know that the final Hipparcos catalogue will produce the accurate positions, the proper motions and the parallaxes as well as an accurate photometry for a sample of 118.000 stars collected in an Input Catalogue (HIC) with the aim of realising an astrometric and a photometric reference system (C.TURON & alt. 1992). The interest of this reference system realized by a unique equipment is evident for practically all researches in astronomy.

But from the double star point of view, the interest of these data shows different aspects and need a careful discussion item by item. The subset of the HIC settled for the double and multiple stars (16.000 visual systems collected in a specific Annex, published as vol 6 of the HIC and completed by recent identifications) contains systems from 22 proposals having different aims (orbital, spectroscopic, photometric, etc.). In addition to physical systems (spectroscopic, photometric and close visual pairs) there are many optical wide pairs also (to some 30") of which recognition in the Input Catalogue (HIC) was needed for observation strategy in connection, as it is well known, with the light perturbation into the IFOV due to nearby components.

5.2. THE IMPORTANCE OF THE HIPPARCOS DATA FOR BINARIES EVOLUTION

RESEARCH

5.2.1. *The accurate positions*

First of all, whereas the mission is principally an astrometric one, the positions (absolute and relative) produced for double and multiple stars do not appear as the most predominant contribution except in very restricted cases.

As a matter of fact, one should notice about the *absolute positions*, that for all the stars observed by the satellite - and thus also for all the observed double stars - accurate positions (to $\pm 1''$) have been collected and given in the Input Catalogue. The more accurate positions produced by Hipparcos are thus of no great interest for instance for the completion of catalogues as the *Catalogue of the Components of Double and Multiple stars* (J.Dommanget & O.Nys, 1994). The data given in this catalogue are very sufficient for correct identification and for the establishment of any observation program. The satellite's data may only be of some interest for proper motion determination (but these will be given by the satellite!) and in some exceptional cases for mass-ratio determination.

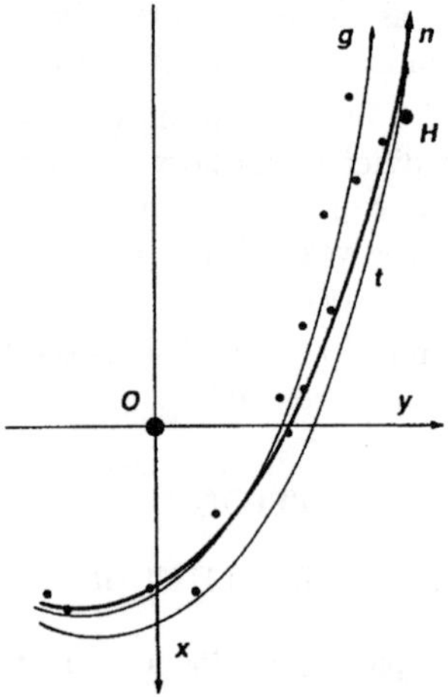

Figure 11.

Concerning the *relative positions* of the components, their interest should be seen from different points of view. As we have shown (J. Dommanget - 1995) the Hipparcos relative positions observed for classical binaries do not bring more information than any ground based observations in view of an orbit computation or recomputation, except that in principle their accuracy should be much higher. They have to be introduced in the observation material with great care because the use of a high accurate measurement together with classical ground based observations (which generally are affected by the well known existence of *systematic errors*), may lead to unacceptable or worst orbits than the ones already known. The Hipparcos position (H, in Figure 11) - if given a high weight - may put the new orbit (n) out of shape. The example given in Figure 12 (M.Froeschlé, 1995) shows such a systematic departure between ground based measurements (by speckle technique) and an Hipparcos result. If an important relative weight is given to the Hipparcos position, this last one will drag the region of the resulting orbit to it, whereas the other regions of the orbit will remain unchanged!

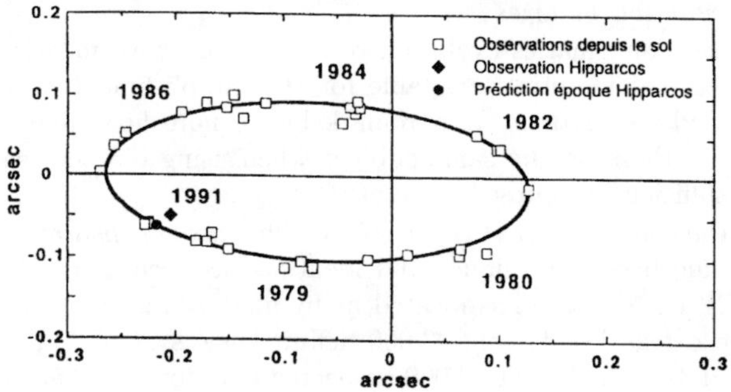

Figure 12.

Of course the situation is different when the Hipparcos measurement falls in the apparent periastron region where no other observations are available: the satellite then produces the only missing information and appears of a great help for the computation of an acceptable orbit. But this only concerns systems presently at their periastrons and showing small angular separations and thus not observable by ground based techniques in that orbital region.

Also in case of very short periods, the observations made by the satellite may cover a whole period or an important part of it, permitting the computation of an orbit of high interest because it is based on a *homogeneous material of a uniform quality.*

We also may not forget to mention an evident interest of the Hipparcos relative positions as standards for the calibration of ground based observation techniques.

5.2.2. *The parallaxes*

Nothing very much should be added concerning the parallaxes (absolute!) which is not already known. Their interest is evident for defining the real separation of the components and thus for the real separation histogram. Of course they are also fundamental for computing the absolute magnitudes. They are thus indirectly very important for the researches on binaries evolution. They also may help in recognising optical pairs when the difference in parallaxes between their components exceeds the rms attached to these data.

5.2.3. *The proper motions*

Their determination is important for galactic dynamics research. But It is not sure that they will lead to some sensible progress for binaries evolution except when they concern the components of wide pairs (cpm systems for instance). But unfortunately only in few cases both components of such systems have been observed by the satellite.

5.2.4. *The photometric data*

These data are of the greatest importance especially in case of pairs closer than say 3" which generally are not resolved by classical photometric techniques. Accurate Hipparcos measurements of magnitude differences reaching some 2,5 magnitude will thus permit to realize a very significant completion of our material to the benefit of the Hertzsprung-Russell diagram and the mass-ratio histogram researches.

5.2.5. *The discovery of new binaries*
All along this synthesis on binaries evolution researches, we have mentioned the typical poorness of the statistical materials available for the establishment of the various histograms, diagrams and correlations. It is remarkable to note how different may be the conclusions of some authors on the same subject when using different samples just because they are not sufficiently representative of the whole.

In that respect *the contribution of Hipparcos will be incommensurate*: Beside the systems known by ground-based techniques and identified as such in the Hipparcos Input Catalogue (the HIC), the satellite has located many unknown systems which may be divided in three large categories: 1) some 3.000 definite new systems, 2) some 3.000 new astrometric pairs and 3) some 8.000 to 9.000 suspected new doubles. And this on the basis of a systematic survey of the stars to a magnitude limits of the order of 8, which extends with a lesser and decreasing space density nearly to the 12-13th magnitude. To these figures should be added a few hundreds of systems showing orbital motion. All these close pairs will permit to fill - at least partially - the very well known gap between spectroscopic and visual binaries.

The stars suspected for duplicity that Hipparcos could thus not resolve for one or another reason (large difference in magnitude, insufficient or not well distributed great circle scannings), as well as the astrometric pairs, will have to be confirmed and systematically followed by sophisticated ground based techniques. The observers will just have to go to their telescopes...

Without neglecting of course the great importance of the data on parallaxes, proper motion, photometry etc. the discovery of new systems is - as we already suspected at the start of the project (J.Dommanget 1982b; 1983b) and recalled later (J. Dommanget & P. Lampens, 1993) - the most important contribution of the mission for double star astronomy. Before needing these data, one first needed of course the list of the objects to observe!

6. CONCLUSIONS

Whereas the double stars were considered at the start of the Hipparcos project as more prejudicial than productive stellar objects, the development of the mission and the need to solve many specific double star problems, yielded a particularly important contribution to double star astronomy.

With its systematic astrometric survey of the sky, the mission - including the Tycho product concerning more than one million stars to the 11.5 magnitude, ignored in the present paper - will permit the establishment of a very homogeneous sample of nearby systems with parallaxes and photometry from very large to very close pairs and thus new and decisive statistical researches. It should thus be considered as *a revolution, a fundamental "mile-stone"* on the way of a better understanding of binaries evolution.

References

ABT, H.A. & LEVY, S.G.: 1976, Multiplicity among solar-type stars; *Ap.Journal. S.S.*, **30**, pp.273-306.
ABT, H.A. & LEVY, S.G.: 1978, Binaries among B2-B5 IV,V absorption and emission stars, *Ap.Journal. S.S.*, **36**, pp.241-258.

ABT, H.A.: 1988, Visual binary separations as functions of primary types, ages and locations, *Astrophysics and Space Science*, **142**, pp.111-121.

ARCORAGI, J.P., BONNELL, I., MARTEL, H., BASTIEN, P. & BENZ, W.: 1991, Formation of binary and multiple star systems, Bulletin of the American Astronomical Society, **23**, p.1266.

BAIZE, P.: 1932, L'origine et l'évolution des étoiles doubles, *Revue Scientifique,* **70**, $n^0$14, 23 juillet, pp.418-426.

BATTEN, A.H.: 1992, Towards a comprehensive view of binary and multiple systems, *IAU Colloquium 135, Complementary approaches to double and multiple star research,* Astronomical Soc. of the Pacific, Conference Series, **32**, pp.586- 592.

BLEKSLEY, A.E.H.: 1934, Loss of mass in binary systems, *Nature,* **133**, 613.

BONEFF, N.: 1951, Sur l'origine des étoiles doubles, *Annuaire de la Faculté des Sciences, Physique et Mathématiques,* **47**, 191-199.

BOSLER, J.: 1928, *Cours d'Astronomie, III Astrophysique,* Librairie Scientifique Hermann & Cie, pp. 580-586.

BOSS, A.P.: 1992, The fragmentation mechanism, *Astronomical Society of the Pacific, Conference series,* **32**, pp.195- 205.

CAMPBELL, W.W.: 1910, Second catalogue of spectroscopic binary stars, *Lick Observatory Bulletin,* **6**, pp.17- 54.

CHIARA, L.: 1955-56, Casi in cui nel problema dei due corpi di massa decrescente l'eccentricitá varia in raggione inversa della massa, Atti dell'accademia di Scienze, Lettere e Arti di Palermo, série 4, 16, Pubbl. dell'osservatorio astron. di Palermo, N.S. **10**, $n^0$8.

CLARKE, C.J.: 1992, The formation of binary stars, Proceedings of the workshop.: Binaries as traces of Stellar Formation, Ed. Duquennoy, A. & Mayor, M., Cambridge University Press, Cambridge.

COUTEAU, P.: 1967, Relation masse-luminosité et diagram H-R dans le cas des binaires, Comptes Rendus du Colloque UAI On the Evolution of Double Stars Comm. de l'Obs. Royal. de Belg., Série B, $n^0$17, pp.181-188.

DARWIN, G.H.: 1886, On Jacobi's figure of equilibrium for a rotating mass of fluid, *Proceedings of the Royal Society* **41**, pp.319-336. Also in.: *Scientific Papers, 1910,* **3**, pp. 119-134, Cambridge University Press.

DARWIN, G.H.: 1887, On figures of equilibrium of rotating mass of fluid, *Proceedings of the Royal Society,* **42**, pp.359-362. Also in.: *Scientific Papers, 1910,* **3**, pp. 135-185, Cambridge University Press.

DARWIN, G.H.: 1900, Address delivered by the President, *Monthly Notices of the R. Astron. Soc,* **60**, pp.406-415.

DARWIN, G.H.: 1910, *Scientific papers, Vol. III, Figures of equilibrium of rotating liquid and geophysical investigations,* Cambridge University Press, Cambridge, 527 pp.

DARWIN, G.H.: 1911, *The Tides and kindred phenomena in the solar system,* Third edition, Chapter XIX, pp. 353-402, Ed. John Murray, London.

DE GREVE, J.P. & DE LOORE C.: 1972, Stellar winds and mass-loss of a rotating star, *Astrophysics and Space Science,* **18**, pp.128-134.

DOBERCK, W.: 1878, On double star orbits, *Astronomische Nachrichten,* **91**, pp.317-318.

DOBERCK, W.: 1898, On double star orbits.: their periods and eccentricities, *Astronomische Nachrichten,* **147**, pp.251-252.

DOMMANGET, J.: 1963, Recherches sur l'évolution des étoiles doubles par voie statistique et par application de la mécanique des masses variables, *Annales de l'Obs. R. de Belg.* 3éme série, **9**, 5, pp.213-304.

DOMMANGET, J.: 1964, Les étoiles doubles et l'évolution stellaire, *Ciel et Terre,* **80**, pp.315-338, Communication de l'Obs. R. de Belg. $n^0$232.

DOMMANGET, J.: 1968, La cinématique des couples stellaires visuels dans le voisinage du Soleil, *Bulletin de l'Obs. Royal de Belg.*, **6**, pp.246-249.

DOMMANGET, J.: 1970, La fréquence des étoiles doubles dans l'univers sidéral, *Ciel et Terre,* **86**, pp.463-485; *Comm. de l'Obs. Royal de Belg. série B,* $n^0$56, pp. 463-485.

DOMMANGET, J.: 1971, La distribution des distances de séparation des étoiles doubles et ses enseignements, *Sciences,* **tome II**, $n^0$3, pp.151-160; *Comm. de l'Obs. Royal de Belg.*, série B, $n^0$70.

DOMMANGET, J.: 1981, Is this diagram an argument for binary orbital evolution due to mass-loss? *Astrophysics and Space Science Library,* **89**, pp.507-513; *Comm. de l'Obs. Royal de Belg.*, série B, $n^0$119.

DOMMANGET, J.: 1982a, A mass-eccentricity correlation in spectroscopic and visual binary orbits, *Astrophysics and Space Science Library,* **98**, pp.119-122.

DOMMANGET, J.: 1982b, L'astronomie des étoiles doubles á l'heure de l'astrométrie spatiale, *L'Astronomie,* **96**, pp.15-27; *Comm. de l'Obs. Royal de Belg.*, série B, $n^0$125.

DOMMANGET, J.: 1983a, La constante des aires des binaires visuelles orbitales dans le diagram HR, *Comptes Rendus sur les Journées de Strasbourg, Les étoiles binaires dans le diagram H.R.*, 5éme réunion, pp. 39-43.

DOMMANGET, J.: 1983b, L'astronomie des étoiles doubles face á la mission du satellite astrométrique Hipparcos, *Comm. de l'Obs. Royal de Belg.*, série B, $n^0$127.
DOMMANGET, J.: 1988, The space arrangement of the orbital planes of the visual double stars, *Astrophysics and Space Science,* **142**, pp.171-176.
DOMMANGET, J. & LAMPENS, P.: 1993, How double star astronomy may develop after Hipparcos, *Astrophysics and Space Science,* **200**, pp.221-238; *Comm. de l'Obs. R. de Belg.* Série B, $n^0$161.
DOMMANGET, J. & NYS, O.: 1994, Catalogue of the Components of Double and Multiple stars (CCDM), *Communication de l'Observatoire R. de Belg.* Série A, $n^0$115, pp.26.
DOMMANGET, J.: 1995, Is this orbit really necessary ? (III) *Astronomy & Astrophysics,* (Research Note) **301**, pp. 919-921.
DUFOUR, Ch.: 1930, Sur l'accélération séculaire du mouvement de la Lune, *Comptes Rendus, Académie des Sciences de Paris,* **62**, pp. 840-842.
DUQUENNOY, A. & MAYOR, M.: 1991a, Multiplicity among solar-type stars in the solar neighbourhood I.: Coravel radial velocity observations of 291 stars, *Astronomy & Astrophysics, S.S.*, **88**, pp. 281-324.
DUQUENNOY, A. & MAYOR, M.: 1991b, Multiplicity among solar-type stars in the solar neighbourhood II.: Distribution of the orbital elements in an unbiased sample, *Astronomy & Astrophysics,* **248**, pp. 485-524.
DUQUENNOY, A. & MAYOR, M.: 1992, *Binaries as traces of Stellar Formation,* Proceedings of a workshop in Switzerland, Cambridge University Press, Cambridge.
EDDINGTON, A.S.: 1923, On the relation between the masses and luminosities of the stars, *Monthly Notices of the R. Astron. Soc.*, **84**, pp. 308-332.
FESENKOV, V.G.: 1952, Le rayonnement corpusculaire, comme facteur d'évolution du Soleil et des étoiles, *Transactions of the International Astronomical Union*, **8**, p. 702-714.
FINSEN, W.S.: 1936, The apparent statistical relation between period and eccentricity in visual binary orbits, *Monthly Notices of the R. Astron. Soc.,* **96**, pp. 862-866.
FROESCHLE, M.: 1995, La mission Hipparcos, Une belle moisson de résultats, *Bulletin de l'Association pour le Développement International de l'Observatoire de Nice*, **29,** pp.15-19.
GIANNUZZI, M.A.: 1989 -Visual binary systems in the solar neighbourhood.: a question of their separations, *Astrophysics and Space Science,* **161**, pp. 241-246.
GLIESE, W. & JAHREISS, H.: 1988, The third catalogue of nearby stars with special emphasis on wide binaries, *Astrophysics and Space Science,* **142**, pp.49-56.
HALBWACHS, J.L.: 1983, Binaries among the bright stars.: estimation of the bias and study of the main-sequence stars, *Astronomy & Astrophysics,* **128**, pp. 399-404.
HALBWACHS, J.L.: 1986, Binaries among bright stars.: systems with evolved primary components and their relation to the properties of main- sequence binaries, *Astronomy & Astrophysics,* **168**, pp. 161-158.
HEINTZ, W.: 1967, Some results of binary statistics, in.: "On the evolution of double stars" *Comm. de l'Obs. Royal de Belg.* Série B, **17,** pp. 49-52.
HEINTZ, W.: 1969, A statistical study of binary stars, *Journal of the Royal Astronomical Society of Canada,* **63**, pp.275-298.
HICKS, W.M.: 1882, Reports on the state of science, *Fluid ellipsoid and sphere under their own attractions,* Report of the fifty-second meeting of the British Association for the Advancement of Science, Ed. John Murray, London, pp.57-62.
HOGEVEEN, S.J.: 1990, The mass-ratio distribution of visual binary stars, *Astrophysics and Space Science,* **173**, pp.315-342.
HOYLE, F.: 1953, On the fragmentation of gas clouds into galaxies and stars, *Astroph. Journal,* **118**, pp.513-528.
JEANS, J.H.: 1919, *Problems of cosmogony, Stellar dynamics,*
JEANS, J.H.: 1924, Cosmogonic problems associated with a secular decrease of mass, *Monthly Notices of the R. Astron. Soc.* **85**, pp.2-11.
JEANS, J.H.: 1925, The effect of varying Mass on a binary system, *Monthly Notices of the R. Astron. Soc.* **85**, pp.912-914.
JEANS, J.H.: 1928, *Astronomy and Cosmogony,* Cambridge University Press, Cambridge, 420 pp.
JORISSEN, A.: 1984, Tendance évolutive du rapport des masses dans les binaires visuelles, *Bulletin de l'Obs. Royal de Belg.* **9**, fasc.6, pp.325-329.
KELVIN (Lord) (= Sir William THOMSON), 1891, On the Sun's heat, *Popular lectures and Addresses,* **vol I**, pp.413-417.
KUIPER, G.P.: 1935, Problems of double star astronomy II, *Publ. of the Astronomical Soc. of the Pacific,* **47**, pp.121-150.
KUIPER, G.P.: 1955, On the origin of binary stars, *Publ. of the Astronomical Soc. of the Pacific,* **67**, pp.387-396.
LIOUVILLE, J.: 1834 -Sur la figure d'une masse fluide homogéne, en équilibre, et douée d'un mouvement

de rotation, *Journal de l'Ecole Royale Polytechnique,* 23éme cahier, **14**, pp.289- 296.

LOSSEVA, V.M.: 1938, Le décroissement de la masse et le changement de la somme et du rapport des masses des composantes des systémes binaires, *Russian astronomical Journal,* **15**, pp.232-247.

LYTTLETON, R.A.: 1954, The theoretical basis of the fission theory of binary stars, *Transactions of the International Astronomical Union,* **8**, pp.717-721.

LUDENDORFF, H.: 1910, Zur Statistik des spektroskopischen Doppelsterne, *Astronomische Nachrichten,* **184**, pp.373-390.

MAC-LAURIN, C.: 1741, *De causa physica fluxus et refluxus maris,* Piéces qui ont remporté le prix de l'Académie Royal des Sciences en 1740 sur le flux et reflux de la mer, Paris 1741, pp.195-233.

MARTIN, E.L.: 1934, Sulle variazioni secolari del periastro e della eccentricitá secondo una nuova legge di irragiamento della massa per i sistemi binari, *Reale statione astronomica e geofisica di Carloforte (Cagliari),* No. **30**, 16 pp.

MAZEH, T. & GOLDBERG, D.: 1962, The mass-ratio distribution of short- period binaries, *Astronomical Society of the Pacific, Conference Series,* **32**, pp.119- 126.

MOULTON, F.R.: 1906, *An introduction to Astronomy,* pp.524-526. The Macmillan Company, London.

MOULTON, F.R.: 1909, Notes on the possibility of fission of a contracting rotating fluid of mass, (in.: *Contributions to cosmogony and the fundamental problems of Geology, The tidal and other problems,* Ed. The Carnegie Institution of Washington), pp.135-160.

MOULTON, F.R.: 1931, *Astronomy,* The Macmillan Company, New York, pp.400-403 and 440-443.

MULLER, P.: 1952, Colorimétrie de cinquante et une étoiles doubles, *Annales de l'Observatoire de Strasbourg,* **5**, fasc. 4, pp. 2-31.

MULLER, P.: 1967, Comptes Rendus du Colloque UAI *On the evolution of double stars; Comm. de l'Obs. Royal de Belg.* Série B, $n^0$17, pp.189-190.

ÖPIK, E.: 1923, On the luminosity-curve of components of double stars, *Publications de l'Observatoire Astronomique de l'Université de Tartu,* **25**, 5, 28 pp.

ÖPIK, E.: 1924, Statistical studies of double stars, *Publications de l'Observatoire Astronomique de l'Université de Tartu,* **25**, 6, 167 pp.

PLATEAU, J.: 1843, Mémoire sur les phénoménes que présente une masse liquide libre et soustraite á l'action de la pesanteur, premiére partie, *Nouveaux mémoires de l'Académie Royale des Sciences et Belles- Lettres de Bruxelles,* **16**, 34 pp. + 1 planche.

PLAVEC, M.J.: 1982, Evolution of close binary stars.: observational aspects, *Astrophysics and Space Science Library,* **98**, pp. 159-181.

POINCARE, H.: 1911, *Leçons sur les hypothéses cosmogoniques* . Librairie Scientifique Hermann et Fils, Paris, 1911, pp.184-189.

RUSSELL, H.N.: 1910, On the origin of binary stars, *Astrophysical Journal,* **31**, 3, pp.185-207.

SCHLESINGER, Fr. & BAKER, R.H.: 1910, A comparative study of spectroscopic binaries, *Publications of the Allegheny Observatory,* 1, pp.135-161.

SCHMIDT, O.J.: 1945, On the inclination of the planes of double star orbits to the plane of the Galaxy, *Comptes Rendus de l'Académie des Sciences de l'U.R.S.S.(NS),* **49**, pp.16-19.

SEE, T.J.J.: 1893, *Die Entwickelung der Doppelstern-Systeme,* Inaugural-Dissertation zur Erlangung der Doctorwürde von der Philosophischen Facultät der Königlichen Friedrich-Wilhelms-Universität zu Berlin, R.Friedländer & Sohn, Berlin.

SEE, T.J.J.: 1896, *Researches on the evolution of the stellar systems, tome I,* Thos. P. Nichols & Sons, Lynn, Mass. USA.

SEE, T.J.J.: 1910, *Researches on the evolution of the stellar systems, tome II,* Thos. P. Nichols & Sons, Lynn, Mass. USA.

SMART, W.M.: 1925, The cosmogonic time-scale and binary stars, *Monthly Notices of the R. Astron. Soc.,* **85**, pp.423-433.

STONEY, J.: 1867, On the physical constitution of the sun and stars, *Proc. R. Sc. London,* **16**, pp.25-34.

STRAND, K.Aa.: 1967, General Introductory lecture to the IAU Colloquium *On the evolution of double stars; Comm. de l'Obs. R. de Belg.,* série B, $n^0$17, pp.19-24.

STRUVE, O.: 1950, *Stellar evolution,* Princeton University Press, 266 pp.

TRIMBLE, V.: 1974, On the distribution of binary system mass ratio, *Astronomical Journal,* **79**, pp. 967-973.

TRIMBLE, V.: 1978, q (a) reconsidered, *The Observatory,* **98**, pp.163-166.

TRIMBLE, V.: 1986, The angular momentum-vs-mass relation and the distribution of mass ratios for visual binary systems, *Astrophysics and Space Science,* **126**, pp.243-253.

TURON, C & alt.: 1992, The Hipparcos Input Catalogue, *ESA SP-1136,* 6 volumes.

VARSAVSKY, C.M.: 1962, An interpretation of the period-eccentricity relation for double stars, *Symposium on Stellar Evolution,* Astronomical Observatory, Nat. Univ. of La Plata, pp.173-183.

THE $V-(B-V)$ DIAGRAM OF THE TYCHO DOUBLE STARS[1]

P. VIRELIZIER AND J.L. HALBWACHS
Observatoire Astronomique de Strasbourg, URA 1280
Equipe "Populations Stellaires et Evolution Galactique"
11 rue de l'Université, F–67 000 Strasbourg, France

Abstract. The star mapper of the Hipparcos satellite was used to derive the B_T and V_T magnitudes of about 1 million stars, within the context of the Tycho program. The separation limit of the components of double stars was about 3 arcsec.

A least square method using Tchebychev polynomial approximation is proposed, in order to build the HR diagram of the components of the field double stars from the differences of magnitude of the components and from the $(B_T - V_T)$ color indices. This method was validated with simulations, and applied to the double stars of the Tycho Catalogue.

1. Introduction

The calibration of the absolute magnitudes of the stars may be derived from several methods. Apart from the classic techniques involving trigonometric parallaxes or stellar clusters, the double stars were sometimes used for investigating some parts of the HR diagram. Thus, Stephenson (1960) checked the locations of bright evolved components, and Murphy (1969) derived the main sequence of the B stars from the visual binaries with B–type primaries. Finally, Sinachopoulos and Rakos (1983), and Rakos (1985) derived the main sequence from a sample of 147 double stars with MS components. In all these papers, one component was used to obtain the distance of the binary, and therefore the absolute magnitude of the other component.

A different approach is proposed hereafter. It consists in a variant of a method usually applied to the open clusters: each binary is considered

[1] Based on observations made with the ESA Hipparcos satellite, and on work by the Tycho Consortium in collaboration with the INCA, FAST and NDAC Consortia.

J. A. Docobo et al. (eds.), Visual Double Stars: Formation, Dynamics and Evolutionary Tracks, 429–438.

as a cluster of two stars and they are all gathered in a composite HR diagram. Since the distances of the stars are ignored, this diagram provides the absolute magnitudes *plus* a calibration constant still to be determined. The location of the main sequence is fitted by a least square computation based on Tchebychev polynomals. The giant primary components are then plotted by assuming that the secondaries are on the main sequence.

This method was applied to a selection of double stars of the Tycho Catalogue (ESA, 1997). The Tycho experiment was on board the Hipparcos satellite. A catalogue of one million star was obtained, providing astrometric and photometric data. The Tycho photometry consists in magnitudes in two passbands, called B_T and V_T. These magnitudes may easely be transformed in the B and V magnitudes of the Johnson system. A preliminary version of the Tycho Catalogue was used for this paper, and results might slightly change within the forthcoming final version.

2. The Selection of the Double Stars

The selection of a sample of double stars was performed in three steps. First at all, only the coordinates of the stars were considered. The separations between the stars were derived, and a list of double stars was set up. All the catalogue has been searched up to a separation of 30 arcsec.

The second step was the search for a separation discriminating the physical double stars and the optical pairs. For statistical reason, the separations of optical pairs obey a linear distribution. On the other hand, since physical double stars have a limited semi-major axis, they concentrate in small separation ranges. Therefore, the observed distribution of angular separations shows a minimum, separating two groups:

- close pairs with a majority of physical double stars.
- wide pairs with a majority of optical pairs.

The distribution of the separations is plotted in Figure 1. The gap between the physical binaries and the optical pairs appears clearly at about 10 arcsec. All double stars with separations wider than this limit were discarded from the sample. The other ones were assumed to be binary systems, although the proportion of optical pairs among them was evaluated as large as 27 %.

The last step of the selection consisted in discarding the double stars having a component with an unreliabe photometry: variable stars, stars with errors $\sigma_{B_T} > 0.1$ or $\sigma_{V_T} > 0.1$, and stars for which usable B_T and V_T were not available in time for the colloquium. In addition, the few pairs with separations closer than 3.5 arcsec were removed too. A sample of 1154 double stars was thus finally obtained.

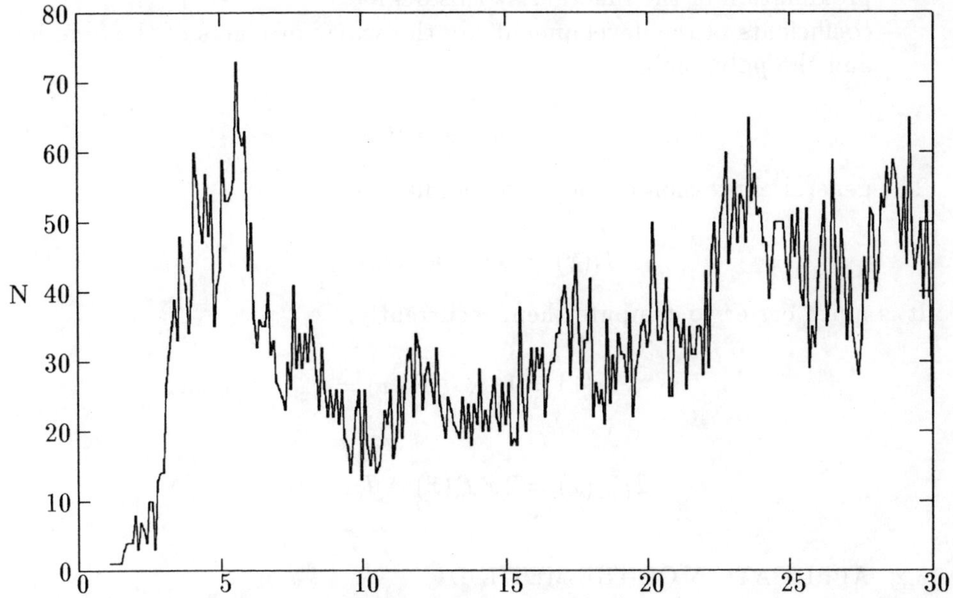

Figure 1. The distribution of the angular separations of the Tycho double stars

3. A New Method to Fit the Main Sequence

3.1. DESCRIPTION OF THE METHOD

The main sequence had to be fitted by means of a least square technique, using a polynomial approximation of the mean main sequence. Tchebychev polynomials were chosen because they offer topological qualities, as reported in Couot *et al.* (1981) and in Ciarlet (1985).

As a matter of fact Tchebychev polynomials have a very interesting property: they are an orthonormal base of the group of continuous functions on the interval [-1,1] with the scalar product given below.

$$\mathbf{f}.\mathbf{g} = \int_{-1}^{+1} f(t)\, g(t)\, dt \tag{1}$$

As a consequence:

– any continuous function on the interval may be developed as an infinite sum.

$$f(x) = \sum_{i=0}^{\infty} a_i T_i(x) \quad \text{with} \quad -1 \leq x \leq 1 \tag{2}$$

- any truncation of this sum at any degree is the best polynomial approximation of the function at this degree.
- coefficients of the development are the scalar products of the function and the polynomials:

$$a_i = \mathbf{f}.\mathbf{T_i} \tag{3}$$

The general expression of these polynomials is:

$$T_i(x) = \cos(i.\arccos(x)) \tag{4}$$

It is much easier to compute them recurrently:

$$T_0(x) = 1 \quad ; \quad T_1(x) = x$$

$$T_{i+1}(x) = 2\,x\,T_i(x) - T_{i-1}(x) \tag{5}$$

3.2. APPLICATION OF THE METHOD

In the problem we are dealing with, f represents the mean main sequence, *i.e.* the absolute magnitude M_{V_T}, expressed as a function of the color index. $(B_T - V_T)$ is not used directly, but a function c varying from -1 to 1 is taken in its place. c is defined as:

$$c = 2\frac{(B_T - V_T) - (B_T - V_T)_{min}}{(B_T - V_T)_{max} - (B_T - V_T)_{min}} - 1 \tag{6}$$

where $(B_T - V_T)_{min}$ and $(B_T - V_T)_{max}$ represent the limits of the actual range of $(B_T - V_T)$.

Then, the sum of the squares of the individual distances to the solution is derived from:

$$S = \sum_{j=1}^{N}(V_{2j} - f(c_{2j}) - V_{1j} + f(c_{1j}))^2 \tag{7}$$

Or

$$S = \sum_{j=1}^{N}\left(V_{2j} - V_{1j} + \sum_{i=0}^{\infty} a_i\ (T_i(c_{1j}) - T_i(c_{2j}))\right)^2 \tag{8}$$

where i is the order of the polynomial and j refers to the couple of stars. The zero order coefficient, a_0, can take any value and remains undetermined, since T_0 is a constant. Therefore, the diagram will provide the absolute magnitudes *plus* a calibration constant to be determined by another mean.

3.3. SEARCHING FOR THE OPTIMAL SOLUTION

It is a classic minimization problem. Partial derivatives are nulls at local extremes (minimum here):

$$dS = 0 \quad \text{is equivalent to} \quad \frac{\partial S}{\partial a_i} = 0 \ \text{ for all } \ i \tag{9}$$

Starting with any set of coefficients $(a_1, ...a_N)$, an iterative process using a gradient method is initiated. This process is accelerated if we assume that the second partial derivatives are nearly constant close to the solution. This is not true in reality, and a smoothing factor has to be used.

If k is the iteration order, λ the smoothing factor, and A_i an intermediate variable defined as:

$$S_{i1} = S(a_1, a_2, ..., a_i - \Delta a, ...a_N)$$

$$S_{i2} = S(a_1, a_2, ..., a_i, ...a_N)$$

$$S_{i3} = S(a_1, a_2, ..., a_i + \Delta a, ...a_N)$$

$$A_i = \frac{S_{i1} - S_{i3}}{S_{i3} - 2S_{i2} + S_{i1}}$$

The final iteration formula is:

$$\begin{pmatrix} a_1 \\ a_2 \\ . \\ . \\ a_i \\ . \\ a_N \end{pmatrix}_{k+1} = \begin{pmatrix} a_1 \\ a_2 \\ . \\ . \\ a_i \\ . \\ a_N \end{pmatrix}_{k} + \lambda \times \Delta a \times \begin{pmatrix} A_1 \\ A_2 \\ . \\ . \\ A_i \\ . \\ A_N \end{pmatrix} \tag{10}$$

In practice, $\Delta a = 0.1$ and $0.5 \leq \lambda \leq 0.9$ are reasonnable values in this problem.

3.4. VALIDATION

The method was validated with simulations. Synthetic double stars were generated with components with color indices randomly distributed. The absolute magnitudes of the stars were then derived assuming an input calibration function. Using samples of 500 or 1000 double stars, sequences in good agreement with the input calibration were obtained when the approximation was truncated to the degree $L = 8$. Various types of input calibration function were employed, including one very similar to the actual main sequence.

4. Derivation of the Main Sequence

It was assumed that the double stars with $\Delta(B_T - V_T) > 0$ had MS components (hereafter, $\Delta(B_T - V_T)$ is defined as $(B_T - V_T)_2 - (B_T - V_T)_1$, where the index "1" refers to the primary and the index "2" to the secondary component). 60 % of the sample were satisfying to this condition. However, the double stars with $\Delta(B_T - V_T)$ close to zero were useless to build the main sequence, and the couples with $\Delta(B_T - V_T) < 0.1$ were still discarded.

The sample contained very few stars with $(B_T - V_T)_2 \geq 1$, as expected since the red dwarf stars are not frequent in a magnitude–limited sample. Therefore, this part of the sample was contaminated by a large proportion of optical pairs involving a giant secondary, and it was necessary to use an *a priory* knowledge of the main sequence for discarding them. This was done by keeping only the double stars with:

$$\Delta V_T > 7 \times \Delta(B_T - V_T) - 2.3 \tag{11}$$

when $(B_T - V_T)_2 \geq 1$. 14 more couples were thus rejected. The 365 remaining couples constituted the sample used to fit the main sequence.

The algorithm described in the previous sections was applied. The calculation was limited to the degree $L = 8$ and $k = 30$ iterations were performed to get the polynomial coefficients. The function represented in Figure 2 was finally obtained. The calibration sequence provided by Schmidt–Kaler (1982) is also represented on this Figure. This calibration refered to Johnson photometry, but the B and V magnitudes were transformed in B_T and V_T magnitudes by inverting the linear formulae given in the Tycho Catalogue:

$$V_T = V_J + \frac{0.090}{0.850} \times (B - V)_J \tag{12}$$

$$(B_T - V_T) = \frac{(B - V)_J}{0.850} \tag{13}$$

and

(The Tycho Catalogue provides also a more accurate transformation, but this simple one was sufficent for our purpose). Moreover, the constant calibration term a_0 was chosen in order to have the same absolute magnitude when $B_T - V_T = 0$. Both sequences look in good agreement.

The components of the double stars of the sample were plotted on Figure 3. The sequence thus obtained is thinner than the one derived from the Hipparcos parallaxes (Perryman *et al.*, 1995), especially when $B_T - V_T < 0.7$. This is an artefact due to the method, however, since the components of each double star were plotted symmetrically on both sides of the sequence.

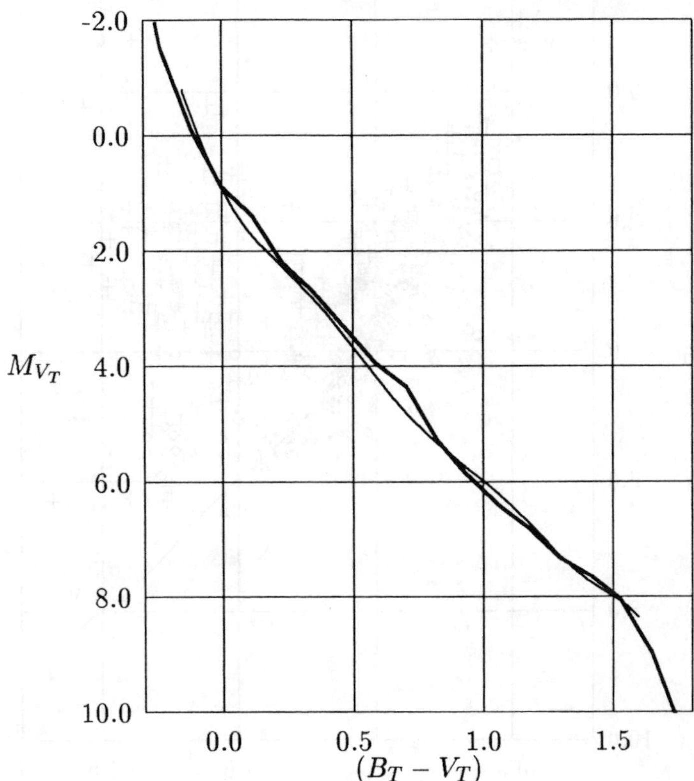

Figure 2. The main sequence derived from the Tycho double stars compared to the calibration of Schmidt-Kaler (1982). The origin of the absolute magnitude was chosen in order to have the same calibration for $(B_T - V_T) = 0$ on the main sequence. The plain line refer to the main sequence provided by Schmidt-Kaler, the thin line to the Tycho mean main sequence.

5. The Sequence of the Giants

Since, for any star, the duration of the red giant stage is much shorter than the MS stage, it was assumed that the double stars with $\Delta(B_T - V_T) < 0$ all have a giant primary component and a dwarf secondary. Moreover, in order to reduce the contamination by optical pairs, only the systems with $\Delta V_T = V_{T_2} - V_{T_1} > 1$ mag were taken into account. A sample of 271 couples was thus obtained. The primary components were added to the diagram of Figure 3 by plotting the secondaries right on the main sequence.

Two groups of stars appear in this figure. The first one is close to the

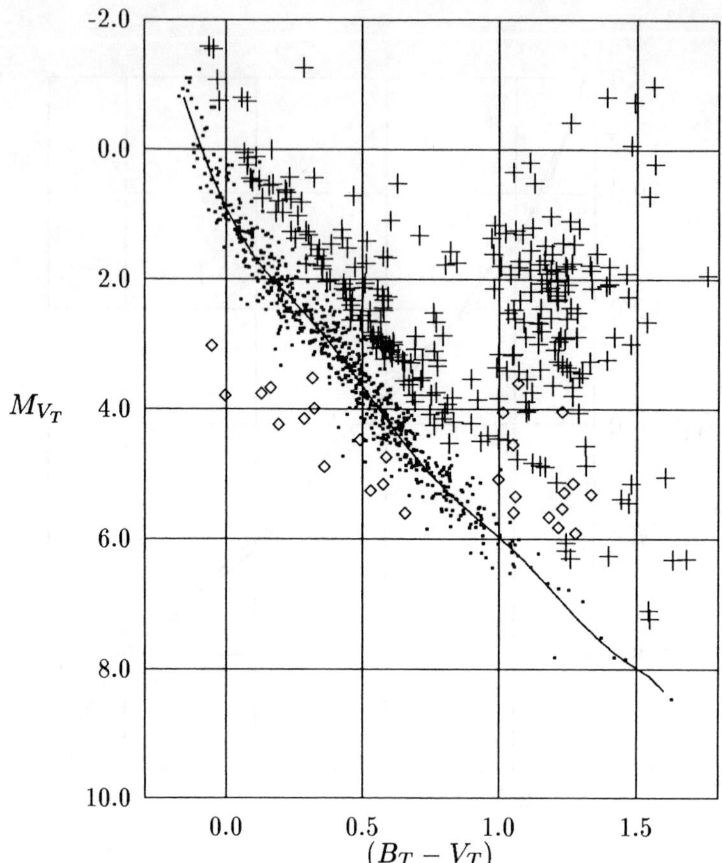

Figure 3. The HR Diagram of the Tycho double stars. The origin of the absolute magnitudes is arbitrary. The dots refer to the MS stars used to derive the calibration function represented by the full line. The diamonds represent the components of the 14 probable optical pairs that were discarded from the calculation. The crosses refer to the giants.

main sequence, just above the 1 mag limit; these stars are suspected to be components of optical pairs. The second group is in the region $(B_T - V_T) =$ 1.2 and $M_{V_T} = 2.0$. These stars are fainter than the usual clump of the giant, but they are probably evolved components of actual binary stars, since Perryman *et al.* also found a concentration of stars in this part of the HR diagram. The small density of stars at the location of the giant clump ($M_{V_T} = 0$) is probably due to a selection effect: These "normal giants" are now much brighter than when they were on the main sequence, and, therefore, their secondary components are several magnitudes fainter than the primaries (Halbwachs, 1986). As a consequence, the Tycho catalogue should contain a lot of intrinsically bright giants having a MS secondary component that is too faint for being also in the catalogue.

6. Conclusion

An original method based on the Tchebichev polynomials was developed in order to derive the main sequence of the HR diagram from a sample of double stars with dwarf components. This method was validated with simulated data, and it was applied to a sample of double stars taken from a preliminary version of the Tycho Catalogue. A reliable main sequence was obtained, although the sample was contaminated by a large proportion of optical systems.

The position of the giants in the HR diagram was also investigated, assuming that they have dwarf secondary components. The lower part of the red clump found by Perryman *et al.* was thus obtained, but the bright giant stars were not found again, due to a selection effect.

Acknowledgements

Prof. C. Jaschek has suggested the subject. This work was supported by national funding from the Centre National d'Etudes Spatiales (CNES).

References

Ciarlet, P.G. (1985) *Introduction à l'analyse numérique matricielle et à l'optimisation.* Masson, Paris

Couot, J., Gaches, J., Souhait, J. (1981) *Algorithmique Numérique, Première Partie.* Cours E.N.S.A.E., Toulouse

ESA (1997) The Tycho Catalogue, **ESA SP–1200**

Halbwachs, J.L. (1986) Binaries among bright stars : systems with evolved primary components and their relation to the properties of main–sequence binaries, *Astron. Astrophys.*, **Vol. no. 168**, pp. 161–168

Halbwachs, J.L., Høg, E., Makarov, V.V., Wagner, K., Wicenec, A. (1996) The treatment of double stars in the Hipparcos–Tycho programme, *this issue*

Murphy, R.E. (1969) A spectroscopic investigation of visual binaries with B–type primaries, *Astron. J.*, **Vol. no. 74**, pp. 1082–1094

Perryman, M.A.C., Lindegren, L., Kovalevsky, J., Turon, C., Høg, E., Grenon, M., Schrijver, H., Bernacca, P.L., Crézé, M., Donati, F., Evans, D.W., Falin, J.L., Froeschlé, M., Gómez, A., Grewing, M., van Leuuwen, F., van der Marel, H., Mignard, F., Murray, C.A., Penston M.J., Petersen, C., Le Poole, R.S., Walter, H. (1995) Parallaxes and the Hertzsprung–Russel diagram from the preliminary Hipparcos solution H30, *Astron. Astrophys.*, **Vol. no. 304**, pp. 69–81

Rakos, K.D. (1985) Visual binaries and the calibration of the main sequence, *Calibration of Fundamental Quantities*, **IAU Symposium No 111**, D.S. Hayes et al. eds., pp. 397–400

Schmidt-Kaler, T. (1982) *Numerical Data and Functionnal Relationships in Science and Technology, New series, Group VI, Vol 2, Subvol b. Edr Landolt–Börnstein.* Sringer–Verlag, Berlin–Heidelberg–New York

Sinachopoulos, D. and Rakos, K.D. (1983) M_V als Funktion von $(B-V)_0$ für die visuellen Doppelsterne der Leuchtkraftsklasse V, *Mitteilungen der Astronomische Gesellschaft*, **Vol. no 60**, pp. 315–317

Stephenson, C.B., (1960), A study of visual binaries having primaries above the main sequence, *Astron. J.*, **Vol. no. 65**, pp. 60–79

ACCURATE CCD PHOTOMETRY AND ASTROMETRY FOR HIPPARCOS VISUAL DOUBLE STARS *

P. LAMPENS AND J. CUYPERS
Koninklijke Sterrenwacht van België, Ringlaan 3, Brussels, Belgium

E. OBLAK
Observatoire de Besançon, Besançon, France

W. SEGGEWISS
Observatorium Hoher List der Universitäts-Sternwarte Bonn, Daun, Germany

AND

D. DUVAL
Koninklijke Sterrenwacht van België, Ringlaan 3, Brussels, Belgium

Abstract.

Research in double star astronomy benefits from the large-scale impacts of current astrometric (and photometric) space missions. Our aim is to complement the information for the systems observed in space by accurate and homogeneous ground-based photometric data for the individual components. Results of multi-colour CCD observations on components of some 400 visual double systems with angular separations between 1 and 15 arcseconds are presented. Standard magnitudes and colours in the V, (occasionally R) and I passbands of the Cousins system have been obtained. Mean external errors are less than a few hundredths of a magnitude. Relative positions are obtained as a by-product. Mean relative astrometric errors are better than pixel width times 0.01.

* Based on observations made at ESO, La Silla, Chile (and Key Programme 7-009-49K)

J. A. Docobo et al. (eds.), Visual Double Stars: Formation, Dynamics and Evolutionary Tracks, 439–449.

1. Introduction

Present estimates of the percentage of double versus single stars and the high number of new detections (Hubble Space Telescope, HIPPARCOS, ground-based surveys...) point toward a ratio of 60% or much higher. Nevertheless the information from the photometric point-of-view for these objects is poor: global photometry may exist for some systems but generally no high-quality data are available and component information is often lacking. Therefore, impact on theoretical models is not very strong: inclusion of double stars in models of galactic structure and evolution has, for example, not occurred yet.

We aim at obtaining homogeneous and accurate magnitudes and colours for the components of "intermediate" visual double stars of the HIPPARCOS programme (with separations in the range 1.5 to 15 arcsec). By providing complementary astrophysical information to be calibrated with respect to physical parameters for the individual components of such systems, a maximised scientific output is ensured. The combination of the high-quality space astrometry and photometry with accurate multi-colour component photometry will allow to study basic characteristics of double stars (Oblak et al., 1992). This is important for understanding the main factors governing their formation and their evolution. For the first time we should be able to get less biased double star statistics, sufficient and accurate double star data for model testing and related matters.

2. The CCD photometric programme

A CCD programme has been defined. The usefulness of CCD's for differential measurements of double stars has been shown earlier (see e.g. Sinachopoulos and Seggewiss (1990)). Our programme consists of measuring "intermediate" visual double stars of the HIPPARCOS Input Catalogue satisfying the conditions:

$1.5 <$ separation $\leq 15''$,

$0 \leq \Delta m < 3$,

lacking component photometry.

We consider that component photometry is also lacking in the case that an existing global photoelectric magnitude has to be combined with an insufficiently accurate Δm (visual estimates can be up to 0.5 mag off). Remember that, more generally, photographic magnitudes are found in double star catalogues. Accurate colour differences are almost never known and it is rare that both components have a spectral classification.

We use CCD imaging and photometry to obtain both relative (Δm, Δcolour) and standard photometry (mag_I, mag_V,(V-I)). As a by-product we obtained relative positions of sufficient accuracy to provide starting values for some double stars of the HIPPARCOS reduction process. These will eventually provide information on relative motions in combination with previous existing data. A complementary part of our programme was acquired by conventional photoelectric photometry. This part is not presented here.

Our observing programmes cover both hemispheres. We performed observations at the observatories of Calar Alto (Germany-Spain), Haute-Provence (OHP, France), La Palma (LPL, Spain), Teide (Spain) and La Silla (ESO, Chile) between October 1991 and January 1995. From April 1992 on, our programme was a part of ESO Key Programme *CCD and conventional photometry of components of visual binaries.*

Table 1 presents the status of the CCD-observations (double stars only). A total of 1600 selected double stars has been observed alltogether, taking into account the overlap between these and the La Palma data (some 4%) (Argue et al., 1992) as well as between the northern and southern samples. Thus, more than half of each sample has so far been observed.

TABLE 1. Status of CCD Observations

Sample	Major contri bution from	Number	Observed (this project)	Observed (Argue et al., 1992)	%
South	ESO	1900	720	350	52
North	LPL	1500	300	600	56
Total	All	3400	1020 ↘	↙ 950	54
Total *		3000	1600		54

*: without overlap

The results discussed here originate from 5 ESO campaigns (2 normal and 3 Key Programme missions). The 90cm Dutch telescope was used with CCD adapter and chips no. 7 (1991-92) and no. 29 (1993-94). The CCD photometry has been obtained with the filters Bessel V and Gunn i. We have applied neutral density filters when necessary. Under favourable conditions, the extinction and the standard system have been evaluated by making use of the standards lists (Menzies et al., 1989; 1991) reviewed by Grenon

(1991). Standard two-colour component photometry has thus been obtained for some 400 "intermediate" visual double stars.

3. Reduction of the CCD Frames

To improve data acquisition we decided to use the windowing option of the CCD whenever possible. Because the majority of our frames show two overlapping profiles, usually without any other isolated star in the field, we needed a profile-fitting reduction method. One of us thus developed a tool for the specific reduction of double star images. For more details on this method we refer to the poster paper by Cuypers (Cuypers, this workshop). The method is not limited to $\Delta m < 3$ but larger errors may occur for larger differences in magnitude. The same procedure also works for multiple stars in systems with up to 10 components. The tests of the method, verification of the relative results and the need for a check of the resulting standard magnitudes took a long time and delayed the presentation of our first results until now.

4. Consistency and Error Evaluation

4.1. INTERNAL CONSISTENCY

4.1.1. *Astrometry*

The fact that each observation consists of a repetitive sequence (up to 15 times for $\Delta m = 3$) allows to perform internal checks and to provide image statistics. The internal errors for an individual observation on the angular separation show a sharp increase with larger differential magnitudes (cfr. Fig. 1 in Oblak et al. (1996)). Individual errors on separation are generally below 10 to 15 mas, i.e. below pixel width $\times$ 0.04. Figure 1 shows the internal errors on the mean angular separation as a function of separation and differential magnitude. Errors on the mean separation are well below 4 mas, i.e. below pixel width $\times$ 0.01. There is a degradation of the consistency for separations smaller than 3" (due to seeing limitation) and again for separations larger than 9" (maybe due to loss of isoplanicity).

4.1.2. *Photometry*

Internal individual errors on the differential V magnitude are generally better than 0.010 to 0.015 mag. Figure 2 illustrates the internal error on the mean differential V magnitude as a function of differential magnitude and separation. The consistency is good: internal errors on the mean are better than 0.005 mag. Again, a degradation can be noticed for separations smaller than 3" but it is quasi-independent of the differential magnitude itself.

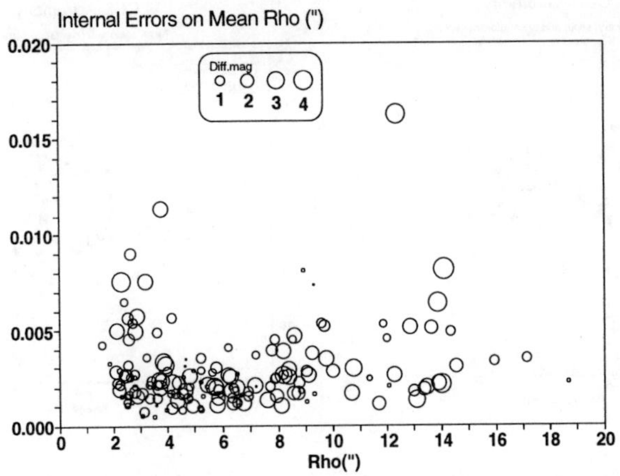

Figure 1. Internal errors on mean separation (larger symbols for larger Δm)

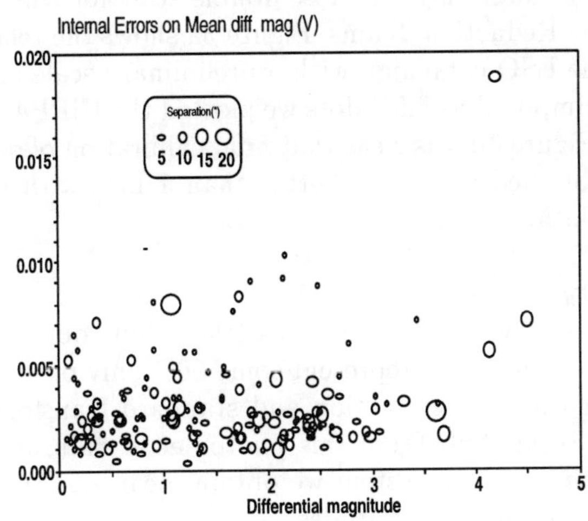

Figure 2. Internal errors on mean Δm_V (larger symbols for larger separations)

4.2. EXTERNAL CONSISTENCY AND ERROR EVALUATION

4.2.1. *Astrometry*

The comparison with existing (relative) astrometry is not easily performed in the range of separation considered (generally < 10"). It will be possible

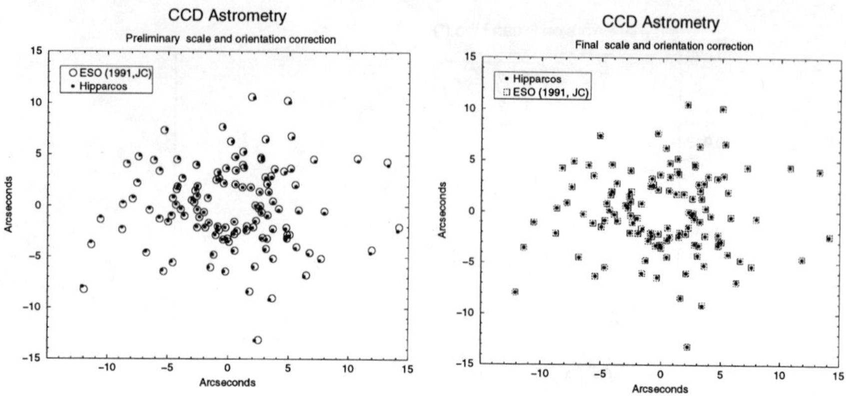

Figure 3. CCD relative positions: preliminary scale (a) and final calibration (b)

as soon as the high-quality data provided by the HIPPARCOS mission are available. For the moment being, we performed a comparison with preliminary HIPPARCOS data only for those double stars for which results were transmitted to the Reduction Teams. Figure 3a shows the relative positions obtained from one ESO campaign with a preliminary scale and orientation correction. Superimposed as filled dots we plotted the HIPPARCOS relative positions. From Figure 3b it is clear that final calibration of our astrometric results will be obtained at a level better than 1 mas with the use of the definitive space data.

4.2.2. *Photometry*

Standard stars. In favourable conditions (the results obtained under such conditions and discussed here represent some 60% only of the total amount of relative data gathered), extinction and standard transformation coefficients have been computed. From this photometric calibration and transformation into the standard system we obtain mean residuals better than 0.02 mag for both V and I magnitudes.

Photometry for individual components. Since our programme has been defined some years ago, we searched existing data bases for recent component photometry. We found individual information for thirteen components of our sample only making use of the Lausanne Photometric data base (Mermilliod et al., 1996) and the Besançon Double and Multiple Star data base (Kundera et al., 1996).

This comparison is based on too small a sample (usually only global magnitudes exist for systems with separation below 10") but a good general

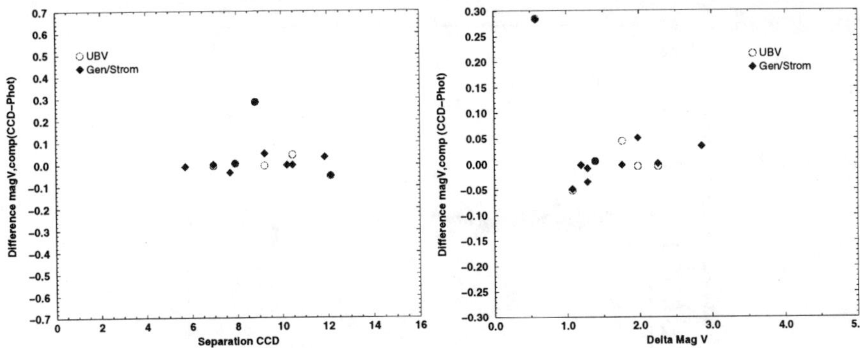

Figure 4. Difference (CCD - Photoelectric) V magnitude for one component as a function of separation (a) and differential magnitude (b)

agreement may be noticed in Figure 4a/b. An outlying data point here concerns HIC 104619 at a separation of 8.8". Both components are classified as F-type stars but their colours are very awkward and so, standard colour transformations are no longer reliable. In such comparisons, from now on, we will distinguish between existing UBV photometry and existing Geneva or Strömgren photometry where both may of course exist in parallel. With the publication of the HIPPARCOS data, it will be possible to treat and compare a much larger sample thus allowing definitive conclusions.

Photometry for global systems. For this comparison a much larger sample could be used: about 100 double stars have either UBV or Geneva or Strömgren photometry. Global magnitudes have been recomputed from standard component magnitudes. Figures 5 and 6 illustrate the differences in the sense CCD - Photoelectric respectively as a function of separation and of differential magnitude for two ranges of separation. The agreement is generally good up to a separation of about 8". In the range where the separation is of the order of the smaller diaphragm apertures ($\rho >\approx 8"$), there is a systematic effect in the sense that photoelectric global magnitudes are fainter. This separation range is critical for conventional aperture photometric techniques: the discrepancy could be due to light losses in the wings of the two profiles if both are centered in one diaphragm aperture. In some cases it may also be due to the fact that identification of the number of components in the diaphragm is uncertain. Here again, comparison with definitive HIPPARCOS data will help to give the correct interpretation and to provide an independent quality assessment of the data.

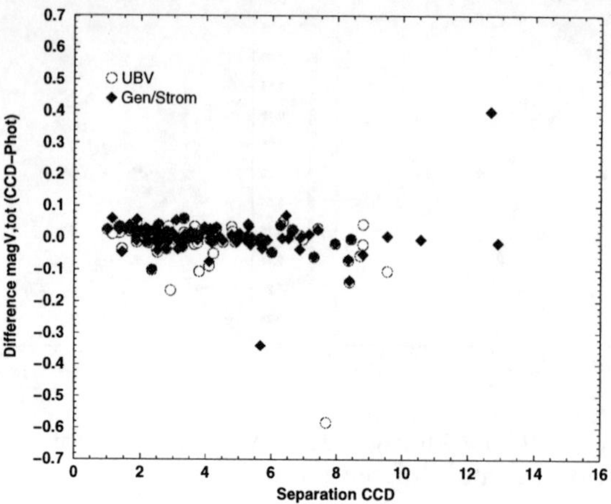

Figure 5. Difference (CCD - Photoelectric) V magnitude for the system as a function of separation

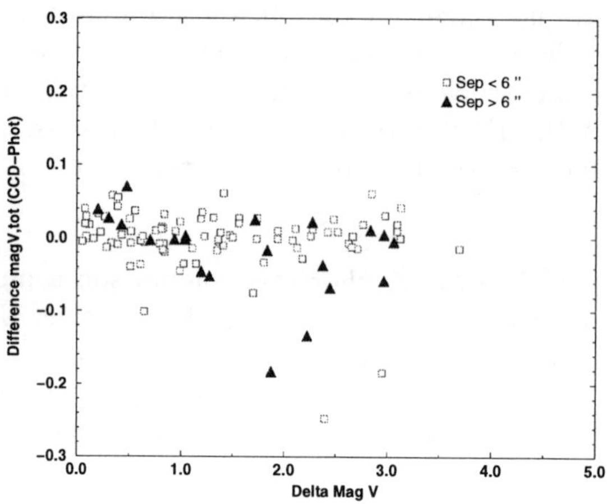

Figure 6. Difference (CCD - Photoelectric) V magnitude for the system as a function of Δm for two separation ranges (Gen/Ström only)

For illustration purpose only and based on preliminary data, we have plotted in Figure 7 the differences in the sense CCD - HIPPARCOS for all observed pairs as a function of Δm_V for two ranges of separation ($\rho <$ 6" and $\rho >$ 6"). To this purpose, we first transformed the HIPPARCOS magnitudes into V_J using our observed (V-I) values (tables 1.3.5 and 1.3.6,

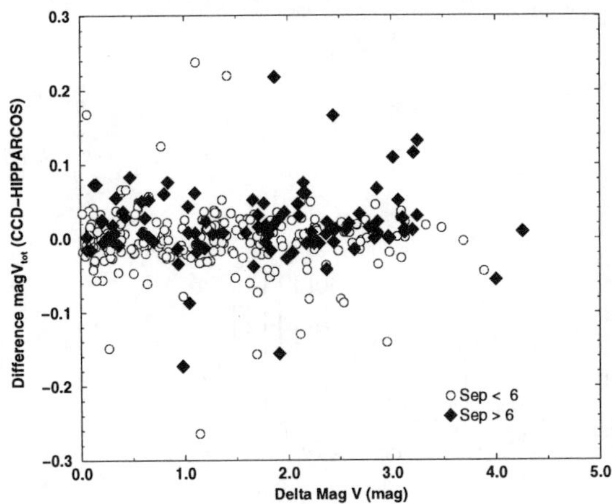

Figure 7. Difference (CCD - HIPPARCOS) V magnitude for the system as a function of differential magnitude for two separation ranges

ESA, 1997). Here no such systematic effect is found. Indeed, a very good agreement between global magnitudes of CCD observations compared to HIPPARCOS data is found once the HIPPARCOS magnitudes have been adequately transformed and colour corrected. The concordance is especially good for our data in the separation range between 6 and 10".

Figure 8 shows the histograms corresponding to the differences in global magnitude in the sense CCD - Photoelectric of Figure 5. A standard deviation of 0.036 mag is found for 90 systems having UBV photometry. For 110 systems with Geneva/Strömgren photometry, the standard deviation drops to 0.027 mag. The distribution is clearly more peaked in the latter case.

5. Conclusions

Very accurate all-sky photometry and astrometry can be achieved on modestly sized telescopes equipped with a CCD. For visual double and multiple stars with separations in the range 2 to 12" (somewhat larger than the seeing limitation), we have shown that CCD's can provide high-quality photometric and astrometric results, matching the accuracy level obtained with other ground or space-based photometric techniques. This was possible due to the application of a strict protocol and central reduction procedure. Only space projects involving astrometry and photometry can compete to acquire simultaneously both kinds of data.

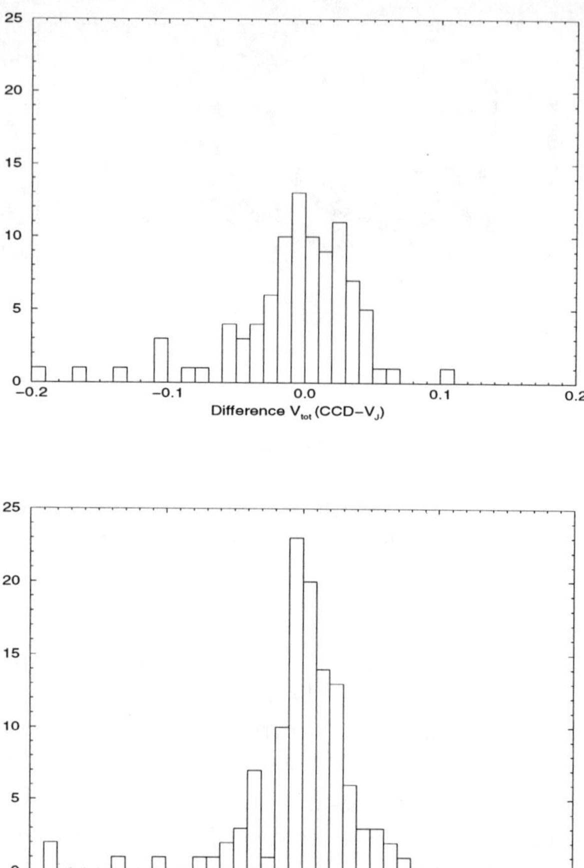

Figure 8. Histogram of the differences in global V magnitude in the sense (CCD - UBV) (top) and in the sense (CCD - Geneva/Strömgren) (bottom)

Magnitudes and colours on individual components are needed to obtain reliable and accurate standard magnitudes and colours for the whole system. It appears that CCD photometry is still reliable in the range where conventional techniques starts giving results but is imprecise and difficult ($\rho \geq 8$"). Knowledge of individual colour differences for double stars will permit the best comparison with the newest space results through adequate use of the transformations between different magnitude systems. Component colour indices are important additional astrophysical information that up to now was generally missing in many researches on visual double and multiple systems.

Acknowledgements

We thank ESO for alloting a Key Programme on small to modestly sized telescopes for this work and the Hipparcos Science Team for the preliminary data concerning those double stars for which CCD results were previously transmitted to the Reduction Teams. This research has made use of the *Besançon Double and Multiple Star* data base, operated at Besançon (France), the *Lausanne Photometric* data base, operated at Geneva (Switserland) and the *Simbad* data base, operated at CDS, Strasbourg (France). We also acknowledge financial aid from the French 'Ministère de la Recherche et de la Technologie' at the start of this project in 1992.

References

Argue, A.N., Bunclark, P.S., Irwin, M.J., Lampens, P., Sinachopoulos, D. and Wayman, P.A., 1992, 'Double Star CCD astrometry and photometry', MNRS **259**, 563-568.

ESA, 1997, in: 'The Hipparcos Catalogue', Vol.1, ESA SP-1200, to be published.

Grenon, M., 1991, private communication.

Kundera, T., Oblak, E., 1996, 'A general data base of double and multiple stars', A. A. Suppl. Ser. , to be submitted. [WWW:*dahu.obs-besancon.fr/bdb/Welcome.html*]

Menzies, J. 1989, SAAO Circ. **13**, 1-13.

Menzies, J. 1991, MNRAS **248**, 642-652.

Mermilliod, J.C., Hauck, B. and Mermilliod, M., 1996, 'General Catalogue of Photometric Data'. [WWW:*obswww.unige.ch/gcpd/gcpd.html*]

Oblak, E., Argue, A.N., Brosche, P., Cuypers, J., Dommanget, J., Duquennoy, A., Froeschlé, M., Grenon, M., Halbwachs, J.L., Jasniewicz, G., Lampens, P., Mermilliod, J.C., Mignard, F., Sinachopoulos, D., Seggewiss, W. and Van Dessel, E., 1992, 'The European Network of Laboratories: Visual Double Stars', IAU Coll. 135, ASP Conference Series **32**, eds. H.A. McAlister and W.I. Hartkopf, 454-456.

Oblak, E., Chareton, M., J. Cuypers, P. Lampens and R. Domon, 1996, Proceedings of 3rd Pacific RIM Conference on Recent Developments on Binary Star Research, 26 Oct - 1 Nov 1995, Lopburi, Thailand, in press.

Sinachopoulos, D. and Seggewiss, W., 1990, 'CCD astrometry of visual binaries. II. Differential measurements of northern double stars', A. A. Suppl. Ser.**83**, 245-250.

DETECTION, ASTROMETRY AND PHOTOMETRY OF VISUAL BINARIES WITH HIPPARCOS

C. MARTIN AND F. MIGNARD
Observatoire de la Côte d'Azur, CERGA, URA CNRS 1360
Av. N. Copernic, F-06130 Grasse, France

Abstract. We present the main properties of the Hipparcos double stars solution after the reduction of the complete dataset covering the 37 months of observation. The principle of the detection of binaries by Hipparcos is briefly stated, with statistical results concerning the discovery of 'new' binaries and a study of the completeness. We show in a second part the main results of the relative astrometry (position angle and separation) and the photometry (in the Hipparcos wide band). The last part deals with the comparisons, first, of the solutions obtained by the FAST and the NDAC consortium, followed by a comparison to the ground–based data. We have also compared the Hipparcos position of the secondary with the predicted position for the subset of 900 binaries for which at least one orbit has been computed.

1. Introduction

The ESA Hipparcos satellite was launched in August 1989, with the aim of measuring accurate positions, parallaxes, proper motions and magnitudes of a set of 118 300 stars. Both the observing mission and the data analysis have come to an end and the merging of the solutions obtained by each of the two consortiums FAST and NDAC is completed. Most of the results presented in this paper refer to the data processing by the FAST consortium, except for the comparisons to ground based observations where the merged solutions were used. Among the Hipparcos programme stars, there were approximately 11 500 double stars already known as double before the mission, ranging typically from 4 to 12.5 mag and with separations from 0.5 to 20 seconds. The relative astrometry and the photometry have been successfully determined for 90% of these stars. In addition, numerous 'new' multiple systems have been discovered (most of them being identified as double stars). A complete solution is proposed for some 7000 'probably new' binaries, with separations ranging from $0\farcs1$ to $0\farcs5$, and magnitude differences less than 3 mag. Comparisons between FAST and NDAC solutions show disagreements in both magnitude difference and separation for some 3000 common 'new' binaries, stressing the need of complementary observations from the ground. Additional comparisons (CCD, Speckle Interferometry) over small samples of known binaries show a good general agreement, consistent with the precisions obtained.

J. A. Docobo et al. (eds.), Visual Double Stars: Formation, Dynamics and Evolutionary Tracks, 451–460.

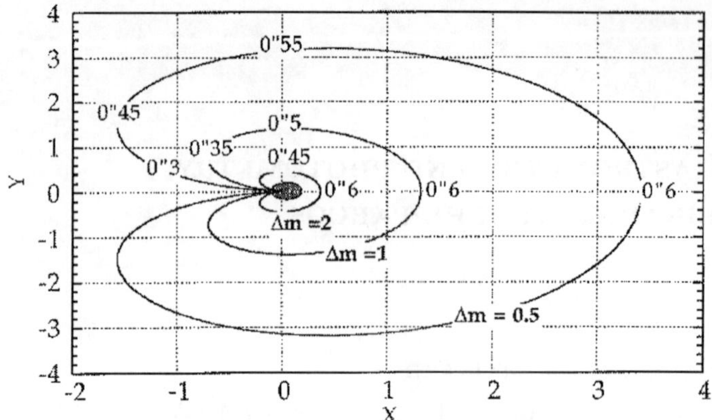

Figure 1. Configuration space for the binary stars. A point in the diagram corresponds to a double star with magnitude difference Δm and a projected separation as sampled. For a single star we have $X \approx 0$ and $Y \approx 0$. Obviously, the recognition of a double star with Δm larger than 4 and/or a separation smaller than $0\overset{''}{.}10$ was not possible.

2. The Recognition of a double star

2.1. PRINCIPLE

When a double star is observed by the satellite, two diffraction patterns are mapped on the focal plane of the telescope, each producing the two-harmonics periodic signal typical of a single star (see Eq. 1 of Martin C. *et al.*, this volume). The compound signal is simply the addition of the signal of each component, which may be modelled exactly in the same way as the single star's signal, although the modulated signal has generally a different shape. The differences concern the respective values of the modulation coefficients M and N, and the ratio of the phases ϕ/ψ. Whereas the three previous quantities are known parameters for a single star, this is no longer the case for a multiple star with two or more components. Their values depend on the magnitude differences between the components and the geometric configuration of the observation (projection on the scanning direction). A simple combination of these quantities allows to build two indicators of duplicity, namely X and Y, both equal to zero when the observed star is single, and taking non-zero values otherwise. A good indicator of the *strength* of the double star's signal may be conveniently constructed with $\sqrt{X^2+Y^2}$. This appears in Fig. 1 as the distance between the origin and the point of coordinates X and Y associated to a particular double star. The shaded area around the origin indicates the region where the signal of a multiple system cannot be distinguished from that of a single star. During the 37 months of mission, each object has been observed between 100 and 150 times at about 30 different epochs with different scanning directions on the grid, so that the probability that a strong enough binary system ($\rho > 0.1\,\mathrm{arcsec}, \Delta m < 3.5$) escapes the detection is very small.

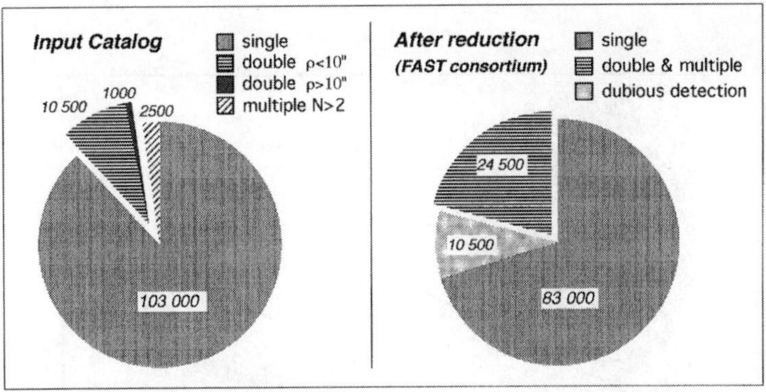

Figure 2. Statistics of detection of double and multiple stars: Comparison of the composition of the Input Catalog and the new composition obtained by FAST after reduction.

2.2. STATISTICS OF DETECTION (FAST)

The diagram of Fig. 2 shows the statistics of recognition of non single objects among the 118 300 programme stars. The content of the Input Catalog is given in order to estimate the efficiency of the detection process. It contains about 14 000 multiple systems, including some 11 500 binaries identified by their CCDM number (Catalog of Components of Double and Multiple stars). The stars flagged with $\rho > 10''$ had a particular observing mode: each component was pointed at sequentially. After the detection phase was completed, the population of non-single stars reached 24 500 objects, while the precise status of an additional set of 10 500 remained uncertain.

2.3. COMPLETENESS OF THE DISCOVERY OF BINARIES

W.D. Heintz (1969) used a diagram in $\log \rho/\Delta m$ (already introduced by Öpik in 1924) to describe the sample of binary objects up to 9th magnitude known at that epoch. He split the diagram into three zones separated by two parallel lines of equation $0.22\Delta m - \log \rho = 0.5$ and 1.0, respectively (see Fig. 3). According to Heintz, the three parts of the diagram, from the top to the bottom, corresponded respectively to the completely discovered, the half discovered, and the yet to be discovered visual pairs. Fig. 3a allows one to appreciate the evolution of the completeness between 1969 and 1982 (date of the compilation of the data included in the Input Catalog), while Fig. 3b shows the sample of 7000 binaries discovered by Hipparcos with a complete solution (see next section). As we can see, most of this new systems fall, at least statistically, in the area where we expect to find them (in the middle and the bottom zones of the diagram). For fainter stars ($9 < H_p < 12.5$), it is more difficult to draw conclusions because the content of the Input Catalog is not comprehensive in this range.

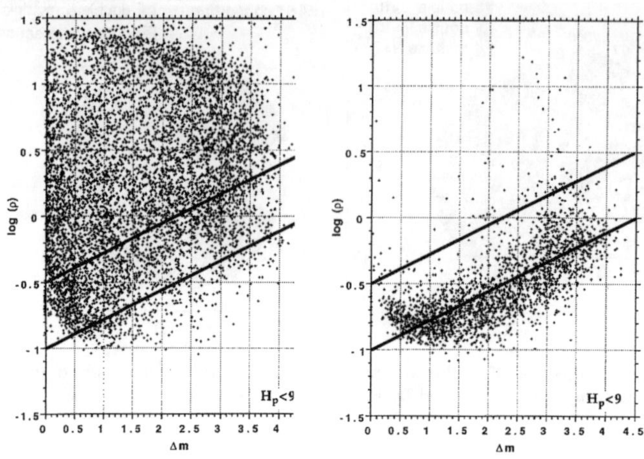

Figure 3.a and b. The completeness of the discovery of binaries up to the 9th mag for stars from the Input Catalog (left) and new pairs arising from the Hipparcos detection process (right).

3. Some Results of the Reduction

3.1. PRINCIPLE OF THE REDUCTION AND STATISTICS

Each time a double star is observed, the two components are projected on the current scanning direction (which is perfectly known and very closely perpendicular to the slits of the grid), so that the only available information on the relative astrometry is the value of the projected separation modulo one grid step, *i.e* 1.″2074. The information on the integral number of grid steps contained in this projected distance remains unknown during this step of the processing. From a set of observations with different directions of scan, the algorithm combines this truncated information and tries to retrieve the true separation and orientation of the pair. In some difficult cases of weak binaries, there is still a possibility that the solution is wrong by one or two gridsteps. If the number of observations is too small or if the angles of scan are poorly distributed, this determination may be difficult or simply impossible. In this case the use of the *a priori* separation and position angle found in the Input Catalogue proved very useful to constrain the search domain. Obviously no such information was available for the newly discovered systems and the risk of error is larger in this case.

On the other hand, there was no major problem in the photometric reduction process. The global Hipparcos magnitude H_p of the system resulted from the general processing of the signal (treated as if the star was single), while a separate algorithm split the signal into its two components in order to get the magnitude difference Δm. The individual Hipparcos magnitudes H_1 and H_2 were then straightforwardly computed. Among the 24 500 objects identified as non single without great certainty, a complete solution (relative astrometry and photometry) could be determined for some 16 800 binaries. For the others, an approximate photometric solution is proposed, but no satisfactory astrometric solution was found by the standard algorithm.

TABLE 1. The Hipparcos double stars: statistics after reduction.

	Known Binaries	New Binaries	Total
Complete solution	$\approx$ 10 000	$\approx$ 7000	17 000
Photometry only	$\approx$ 1500	$\approx$ 6000	7500
Dubious detection	$\approx$ 500	$\approx$ 10 000	10 500

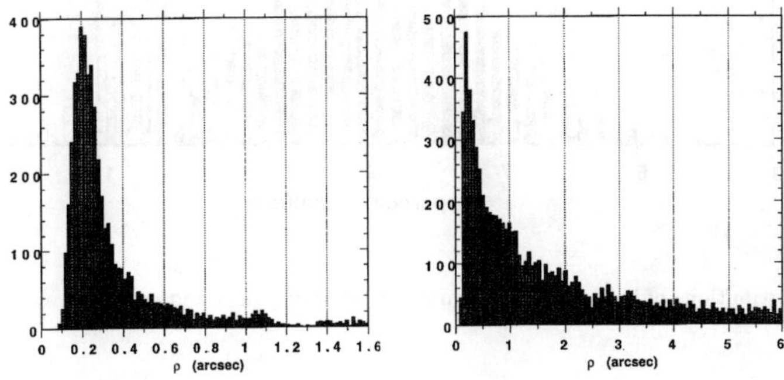

Figure 4. Distributions of separations for some 11 000 known pairs (left) and about 6000 new double stars (right). For known pairs, the wing is extending up to 25 arcsec.

3.2. RELATIVE ASTROMETRY

The distribution of the separations is shown in Fig. 4 for the known and new binaries. A maximum occurs at $\rho = 0\overset{''}{.}2$ for the new pairs. The small irregularities of the distribution beyond 1 arcsec is due to the grid step errors. The standard deviations on separations after the merging are about 10 mas for the known binaries and 30 mas for the new pairs (for a typical 9 mag star). The corresponding distribution for the set of the two pointing wide pairs ($\rho > 10''$) reveals the presence of a population of about 200 systems with very small separations ($< 1''$), showing that some of these pairs are hierarchical systems containing also one close pair.

3.3. PHOTOMETRY

The distribution of the magnitude differences is roughly flat up to $\Delta m = 3$ (however, more stars are present below $\Delta m = 1$ than beyond $\Delta m = 2$), and then decreases continuously up to Δm =4–4.2, at the Hipparcos instrumental limit. The distribution of the global magnitudes is nearly Gaussian, from $H_p \approx 4$ up to 12.5, with a peak at 8.7. The combination of Δm and H_p yields the individual magnitudes, distributed as in Fig. 5. The typical standard deviations on these quantities are 0.04 mag for Δm and respectively 0.015 and 0.04 for the magnitudes of the primary and the secondary. These values have slightly changed during the merging of the FAST and NDAC solutions. The quality of the solutions is essentially dependent upon the separation (Fig. 6), the magnitude difference, the global magnitude and the color index (lower quality for stars redder than $(V-I)_c = 1$).

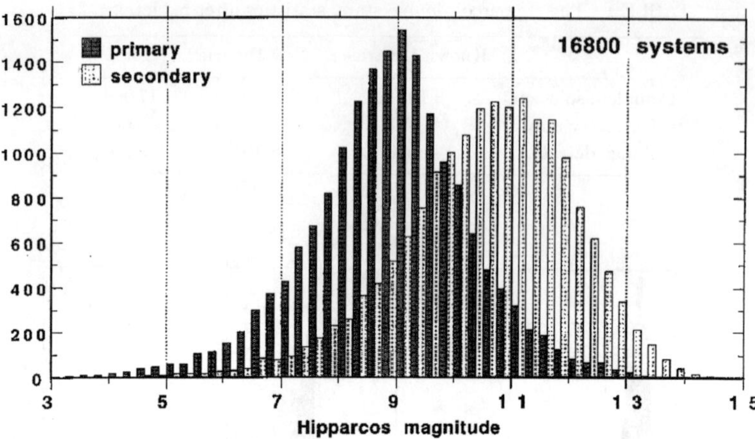

Figure 5. Distributions of individual magnitudes of primary (dark) and secondary components (light) of 16 800 double stars.

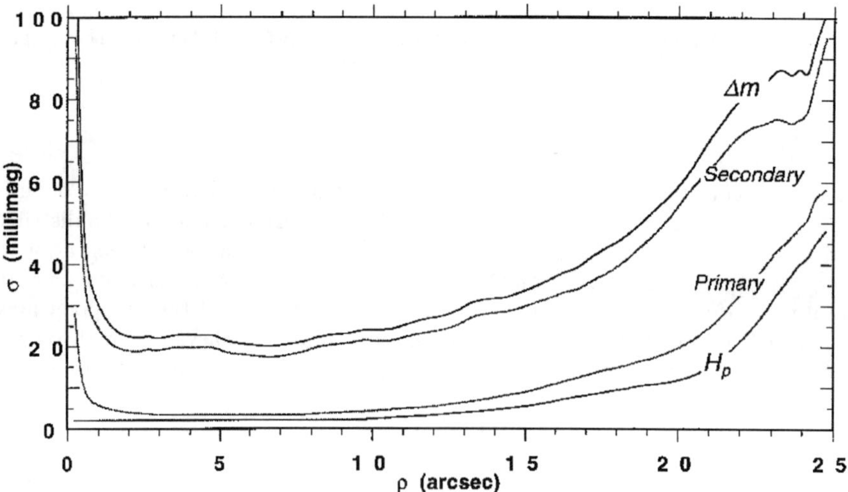

Figure 6. Standard deviation of the photometric solutions as a function of the separation. The lower quality at small separations is due to the Hipparcos instrumental limit. Beyond 15″, this is a consequence of another instrumental effect: the systematic attenuation of the component which is not in the centre of the field (there is no significant signal at distance > 30″ from the centre).

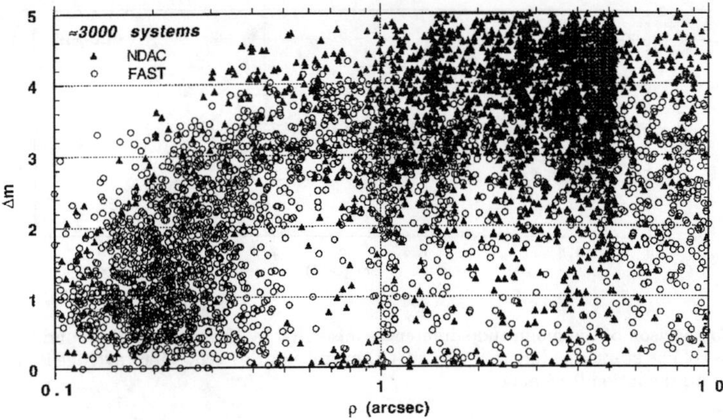

Figure 7. Distribution of the solutions in separation and Δm for 3000 new binaries common to FAST and NDAC. There are two data points per system in this plot, corresponding to the solution of either consortium. The elongated vertical structures are due to the grid step error.

4. Comparisons

4.1. COMPARISON FAST/NDAC

The comparison of the photometric results of FAST and NDAC bears upon about 13 800 systems, including ≈ 4000 new binaries. The distribution of differences in Δm is acceptable for most of the common objects (the standard deviation is $\sigma \approx 0.05$ mag), except for some 3000 new binaries. In fact, these disagreements concern both magnitude differences and separations: it is commonplace in this instance to have for FAST a small separation and Δm less than 3, while NDAC gets as a wide pair with a large Δm (Fig. 7). While such anomaly is very disturbing to a visual observer it is easily understood in the framework of the Hipparcos observations. The diagram in Fig. 1 shows that the signals produced by the two types of binaries are quite similar and hardly recognizable from each other with the Hipparcos data. As a result of this discrepancy, the merged solution to appear in the Hipparcos Catalogue contains only the 12 000 more reliable solutions. However the complete information obtained by each consortium is archived. The study of completeness presented in section 2.3 favors the FAST solution (small separations and Δm), but this has to be confirmed by complementary observations from the ground. Several such programs are being designed in Europe and in the United-States.

4.2. COMPARISON WITH GROUND BASED OBSERVATIONS

4.2.1. *CCD observations from La Palma*

A large amount of CCD observations of binary and multiple systems in V and R bands has been made available by A.N. Argue *et al.* (1992) from observations carried out at La Palma in 1986-87. More than 2300 systems were observed with separations larger than $0\farcs7$ and less than $5''$, including 1360 systems belonging to the Hipparcos programme, all detected as non single and solved for the astrometry and photometry.

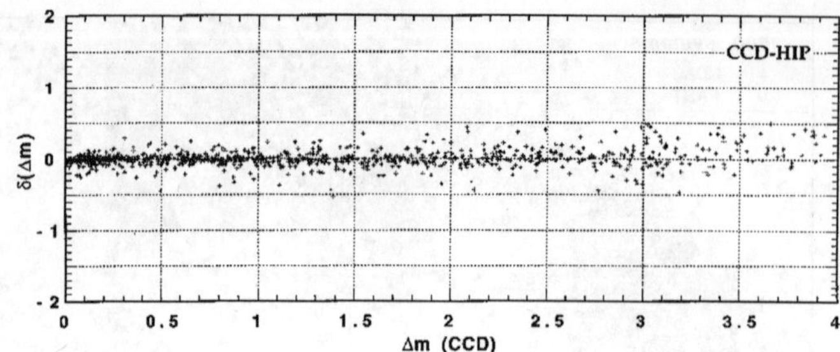

Figure 8. Comparison of the magnitude differences measured by Hipparcos and ground based CCD at La Palma for a set of 958 stars, as a function of the CCD magnitude difference. The average is $\delta(\Delta m) = -0.002$ and the standard deviation 0.13 mag.

Although the magnitude systems in Hipparcos and La Palma are different, it is possible to do a meaningful comparison for stars with similar Δm in the two bands (stars with $|\Delta m(V) - \Delta m(R)| < 0.4$), provided the $B-V$ color index is less than 1.7. Another selection based on the differences in separation allows to avoid most of the errors of identification of the objects. We finally get a sample of 958 common systems with two magnitude differences expressed in the Hipparcos photometric wide band (one from Hipparcos and the other from La Palma). The comparison has first been made for the solutions of FAST and NDAC separately. In both cases, there is no systematic difference larger than 0.02 mag, and the scatter is about 0.14 mag. The statistical distributions are close to a Gaussian in the central part but exhibit extended wings, probably due to the limited validity of the photometric transformation applied.

A comparison using the merged Hipparcos results is summarized in Fig. 8, as a function of the CCD magnitude difference. Up to $\Delta m \approx 3$ there is no systematic difference between the Hipparcos and ground based measurements. Neither is the separation a factor affecting the difference, at least for $\rho > 1''$. For smaller separations the scatter is larger and is more likely due to the low reliability of the CCD observations in this range.

4.2.2. *Speckle Interferometry (CHARA)*

The only systematic sizeable set of observations of relative astrometry of multiple systems matching the quality of the Hipparcos data are provided by the speckle observations and occultation timings compiled in the various versions of the CHARA catalogue. The following comparison is based on the third version available on the WEB (Hartkopf and McAlister 1996). Among the 6280 entries of the CHARA catalogue, about 5000 have a counterpart in the Hipparcos catalogue and 2106, out of the 5000, are associated to at least one reliable observation of separation and position angle. 1680 systems out of the 2106 have a double star solution in the Hipparcos catalogue, and after removing the stars with absolute deviations in separation larger than 50 mas ($|\Delta\rho| > 50$ mas), we get 1506 systems for the analysis. To be significant the comparison between the speckle measurements and the Hipparcos observations must be based on nearly contemporary observations. The comparison data set was eventually separated into five categories according to the reliability of the estimation of the separation at the Hipparcos epoch (T_0 = JD 1991.25):

1. Systems with at least two CHARA observations bracketing the Hipparcos epoch (an interpolation was done at T_0).

2. Systems for which the last CHARA observation was performed at a date T_C earlier than T_0 with the two sub-cases: 2a. $T_C < T_0 - 6$ months and 2b. $T_C > T_0 - 6$ months.

3. Systems for which the first CHARA observation was done at a date T_C later than T_0 with the two sub- cases: 3a. $T_C < T_0 + 6$ months ; 3b. $T_C > T_0 + 6$ months.

In case 2 (resp. 3) the last (resp. the first) observation was retained for the comparison. The content of each category is summarized in Table 2. For all the observations the possible 180 degrees ambiguity has been removed by adding 180 degrees to the position angle of the speckle data θ_C whenever $\cos(\theta_C - \theta_H) < -0.85$. The actual distribution of $\Delta\theta = \theta_C - \theta_H$ for the 949 'good' systems of categories 1, 2a and 3a has a core of 771 objects with $|\Delta\theta|$ less than 20 degrees and two small populations at ±180 degrees of about 90 objects each. The main features of the results of the comparison are summarized in Table 2. A plot

TABLE 2. Content and summary statistics of the comparison data set between Hipparcos and CHARA.

Categories	1	2a	2b	3a	3b	Total
Entries in common	906	192	866	104	38	2106
With Hipparcos DS solution	765	159	637	95	24	1680
Systems used in the analysis	710	152	540	87	17	1506
Mean of $\Delta\rho$ in mas	0.5	0.8	1.6	1.1	-1.9	
Median of $\Delta\rho$ in mas	0	2	2	1	-4	
Std. deviation of $\Delta\rho$ in mas	8.8	9.1	14.7	11.7	13.8	
Mean of $\rho\Delta\theta$ in mas	-1.1	1.1	1.7	0.1	0.6	
Median of $\rho\Delta\theta$ in mas	-0.5	1.2	2.1	0.1	1.8	
Std. deviation of $\rho\Delta\theta$ in mas	10.4	11.8	14.4	10.5	17.2	

in $\Delta\rho = \rho_H - \rho_C$ as a function of the separation ρ shows a systematic bias of +3–4 mas for $\rho < 0.2''$ and of the same amount for $\rho > 0.6''$. Its origin is still unknown. The comparison in position angle shows that there is no systematic orientation difference larger than 0.05–0.1 degree.

4.2.3. *The subset of Orbital Binaries*

The set of known binaries includes some 900 orbital pairs, mostly with periods larger than 30 years. For such long period systems, the orbital motion is unnoticed by Hipparcos, so that we can consider the position of the secondary with respect to the primary as fixed during the 3 years of the mission. The analysis reduces to the comparison between the predicted position given by the ephemerides and the normal point (1991.25) supplied by Hipparcos. The agreement is good for 80% of the objects, while an other 10% show only a 180 degrees ambiguity in the position angle. Finally, the remaining 10% are not matching the predicted position at all. New investigations are certainly required to improve the quality of the orbital elements concerned.

The other orbital binaries with shorter periods are deferred to a separate paper (Martin C.*et al.* of this volume).

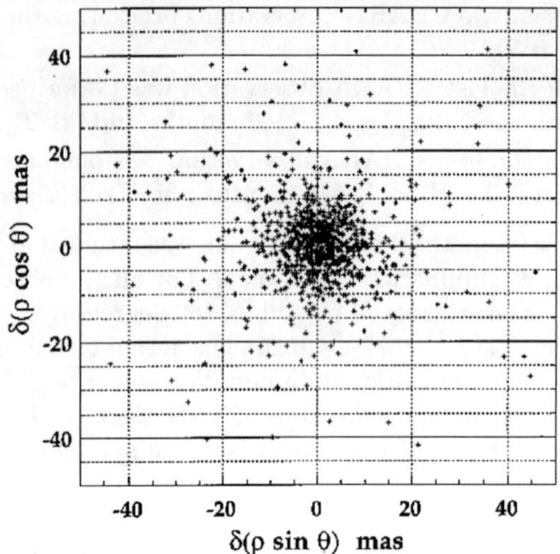

Figure 9. Relative position on the tangent plane on the sky between the position of the secondary with respect to the primary obtained by Hipparcos and CHARA. The difference is Hipparcos–Speckle.

5. Conclusion

We have shown that the Hipparcos mission is about to produce a wealth of information on visual double stars, with the detection in only three years of as many new double stars as detected during the last twenty years by visual observers. The quality of the relative astrometry is comparable, for the best solutions, to the speckle method, and the newly determined magnitude differences for thousands of binaries in the sub-arcsecond range constitutes an important astrophysical information. The comparisons have allowed us to identify minor problems in the data reduction, and to set up a precise list of a few thousands of supposedly close pairs to be carefully observed in the near future.

References

Argue, A.N., Bunclark, P.S., Irwin, M.J., Lampens, P., Sinachopoulos, D., Wayman, P.A. (1992) Double star CCD astrometry and photometry, *MNRAS*, **Vol. no. 259**, pp. 563–568.

Dommanget, J., Nys, O. (1994) CCDM (Catalog of Components of Double and Multiple stars), *Comm. of Obs. R. Belgique*, Ser. A, N. 115.

Heintz, W.D. (1969) A statistical study of Binary Stars, *J. of Royal Astr. Soc. of Canada*, **Vol. 63, No. 6**, pp. 275–298.

Hartkopf, W.I., McAlister, H.A. (1996) Third Catalog of Interferometric measurements of Binary Stars, *CHARA contribution no. 3.*

Martin, C., Mignard, F. (1996) Mass Determination of short period binaries, *this volume.*

Martin, C. (1996) Contribution à l'Étude des Étoiles Doubles Hipparcos, *thesis.*

Mignard, F., Söderhjelm, S., Bernstein, H., Pannunzio, R., Kovalevsky, J., Frœschlé, M., Falin, J.L., Lindegren, L., Martin, C., Badiali, M., Cardini, D., Emanuele, A., Spagna, A., Bernacca, P.L., Borrielo, L., Prezioso, G. (1995) Astrometry of Double Stars with Hipparcos, *Astron. and Astroph.*, **Vol. no. 304**, pp. 94.

THE TREATMENT OF DOUBLE STARS IN THE HIPPARCOS–TYCHO PROGRAMME[1]

J.L. HALBWACHS
Observatoire Astronomique de Strasbourg, URA 1280
Equipe "Populations Stellaires et Evolution Galactique"
11 rue de l'Université, F–67 000 Strasbourg, France
E. HØG AND V.V. MAKAROV
Copenhagen University Observatory
Juliane Maries Vej 30, DK 2100 Copenhagen OE, Denmark
AND
K. WAGNER AND A. WICENEC
Institut für Astronomie und Astrophysik Tübingen
Waldhäuser Straße 64, D–72 076 Tübingen, Germany

Abstract. The Tycho Catalogue[1] contains astrometric data and (B, V) photometry for more than one million stars. The double stars with separations wider than about 3 arcsec appear as two separate entries as if they were two single stars. Between 1.5 and 3 arcsec, the two components are possibly also separate entries, but the photometric data is somewhat less accurate. Double stars with separations as close as 0.4 arcsec were also detected in the Tycho observations, and they are flagged in the Catalogue.

1. The Tycho project

Tycho is a survey using the star mapper of the Hipparcos satellite for deriving astrometric data and photometric data in a two–colour system. The Tycho photometric magnitudes, referred to as B_T and V_T are close to the B and V magnitudes of the Johnson system. The star mapper was constituted by two slit groups. Each slit group consisted of four slits separated by multiple of a basic step of 5.625 arcsec. The total width of a slit group was 33.75 arcsec. A description of the instrument and an overview of the data analysis was presented by Høg *et al.*.

The selection of the stars of the Tycho Catalogue (ESA, 1997) was done in three steps. The first one was the preparation of the Tycho Input Catalogue (Egret *et al.*, 1992): 3 million stars were taken from the INCA database and from the Guide Star Catalog of the Hubble Space Telescope (GSC). The INCA database was created for the preparation of the

[1]Based on observations made with the ESA Hipparcos satellite, and on work by the Tycho Consortium in collaboration with the INCA, FAST and NDAC Consortia.

J. A. Docobo et al. (eds.), Visual Double Stars: Formation, Dynamics and Evolutionary Tracks, 461–467.

Hipparcos mission. It contains double stars with separate components, but, unfortunately, the number of stars (200 000) was only a small part of the Tycho Input Catalogue. The vast majority were coming from the GSC, that often provided only the brightest components of visual double stars. The incompleteness of the Tycho Input Catalogue was corrected in the one–year data processing, sometimes refers as the *recognition* (Halbwachs *et al.*, 1992). The missing components with separations between 3 arcsec and 20 arcsec were then added to the programme (Halbwachs, 1992). In the same time, about 2 million stars appeared to be too faint for the Tycho detection, and they were discarded from the programme. The remaining 1 million stars were finally included in the Tycho Catalogue, but only when their astrometric data were of good quality. This condition was adapted to the origin of the star in the Catalogue, in order to avoid false stars: The criterion was more severe for the stars added in the one–year data processing than for the stars that already had a correct position in the Tycho Input Catalogue.

The reduction of the Tycho data is now just finished. The Tycho Catalogue contains 1 058 000 stars, but only 1 034 000 received positions and complete (B_T, V_T) photometry coming from the Tycho reduction. The median errors of the astrometric data are 25 mas for the coordinates, 31 mas for the parallaxes, and 30 mas/y for the proper motions. The median V_T magnitude is 10.5 mag, with the median error 0.06 mag, and the median B_T magnitude is 11.3 mag, with the median error 0.07 mag. The Catalogue will be published in early April, 1997.

2. The double stars with wide separations

In the present paper, "wide separations" refer to separations sufficent for treating the double star components as single stars in the routine data reduction. As explained above, the minimum separation here is about 3 arcsec. Double stars were searched among all the 1 038 000 stars with complete Tycho data, and about 32 000 pairs with separations closer than 1 arcmin were found. The distribution of the separations is in Figure 1. Several relevant points appear in this figure:

- The abundance of separations smaller than 10 arcsec is obviously due to physical double stars. About half of these stars were added during the one–year data processing.
- The distribution of optical pairs seems to obey two linear relations at different separations. It is expected that, for statistical reasons, the optical pairs obey a linear relation. It is obvious in Figure 1 that the slope of the distribution is steeper for separations closer than about 35 arcsec than for separations wider than this limit. On the other hand, the width of the slit groups of the star mapper was 33.75 arcsec. This suggests that the detection of a double star component was easier when the other component was crossing the star mapper at the same time.
- The distribution of the separations closer than 35 arcsec exhibits peaks with a 5.6 arcsec periodicity. This matches the basic step of the distance between the slits of the star mapper. When two stars separated by $n\times$ this step, or a bit more, were crossing the star mapper, there was some probability that they could appear in two different slits simultaneously. Therefore, in the integrated photon count, each star could receive as much as $\frac{1}{4}$ of the photons of the other one. The effects of this contamination were restricted by the non–linear filter used in the detection process (Bässgen *et al.*, 1992), but they did not disappear entirely.

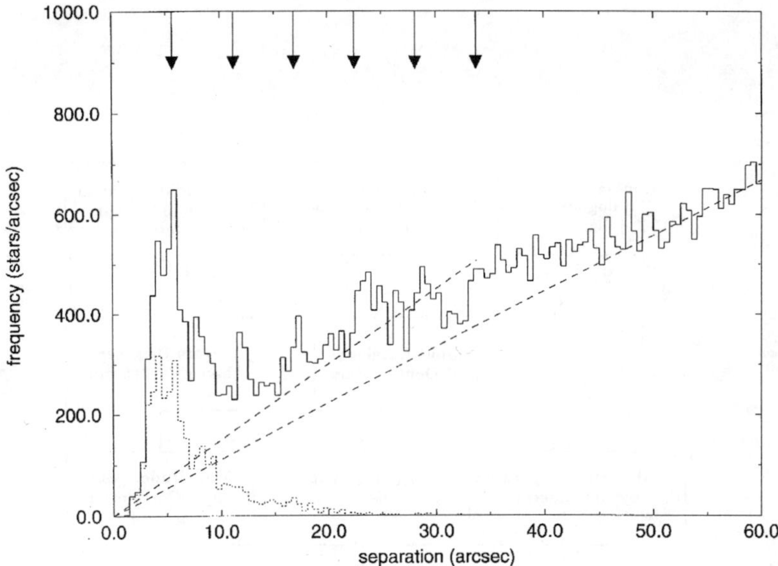

Figure 1. The distribution of the separations for the 32 000 Tycho pairs of stars with separations up to 1 arcmin. The dotted line refers to the pairs with one components added in the one–year data processing. The dashed lines fit the distribution of the optical pairs, and the arrows indicate all the distances between the slits of a star mapper slit group (see explanations in text).

Therefore, although the Tycho Catalogue provides an important amount of data about double stars with separations wider than about 3 arcsec, it must be used with caution in any attempt to derive the intrinsic statistical properties of wide binaries, and especially the distribution of the semi–major axes.

3. Search for close double stars

3.1. THREE METHODS FOR FINDING CLOSE DOUBLE STARS

The double stars with separations closer than 3 arcsec were usually not separated in the routine Tycho reduction, due to several reasons: First at all, the width of the slits of the star mapper was 0.9 arcsec, and the bins of the photon count sampling was 0.28 arcsec wide; since the detection algorithm was adapted to single stars, it was never possible to separate stars closer than about 1 arcsec. The second reason is that the 1–year data processing was based on a preliminary determination of the attitude of the satellite, with errors sometimes as large as 1 arcsec.

It is possible to separate close double stars from the Tycho data however, with two independent methods: an astrometric reduction dedicated to double stars, and a search for double peak in the photon counts, that is fully described in another paper of this issue (Wagner & Halbwachs, 1996). Unfortunately, these processes are time–consuming, and they were applied only to an input catalogue of close double stars, presented in the next subsection.

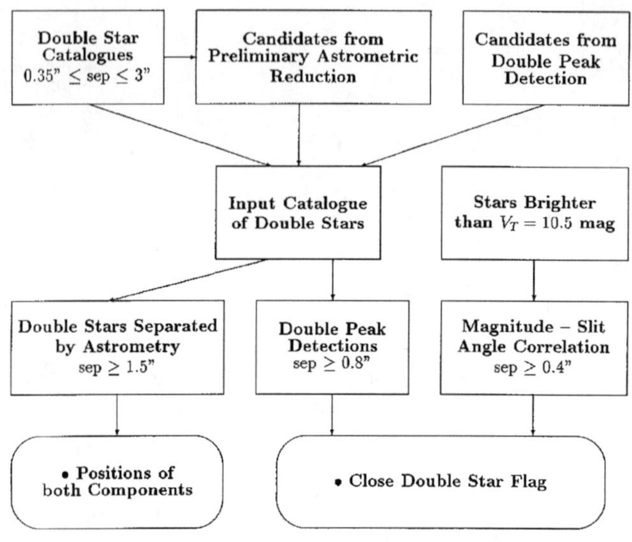

Figure 2. The treatment of close double stars in the preparation of the Tycho Catalogue

Close double stars were also detected with a photometric method. The application of this technique was not restricted to a short selection, but it was applied to about half a million stars.

These three methods and their related data flows are summarized in Figure 2.

3.2. THE INPUT CATALOGUE OF CLOSE DOUBLE STARS

The aim of this input catalogue was to select the stars deserving some computing time for confirming duplicity, and, if possible, for deriving the positions and some photometric data of the components. It contained:

- The known double stars with separations between 0.35 and 3 arcsec, components brighter than 12 mag, and $\Delta m < 3$ mag for the pairs closer than 1.5 arcsec. These stars were taken from the following catalogues: the Washington Double Star (Worley & Douglass, 1984), called WDS hereafter, the part of the Catalogue of Components of Double and Multiple stars included in the Hipparcos Input Catalogue, (Dommanget, 1992), and the COU catalogue (Couteau, 1990). When possible, the positions in WDS and COU were superseded by positions from the PPM (Röser & Bastian, 1991, Bastian *et al*, 1993). 13 234 known double stars were matching 14 237 Tycho stars.

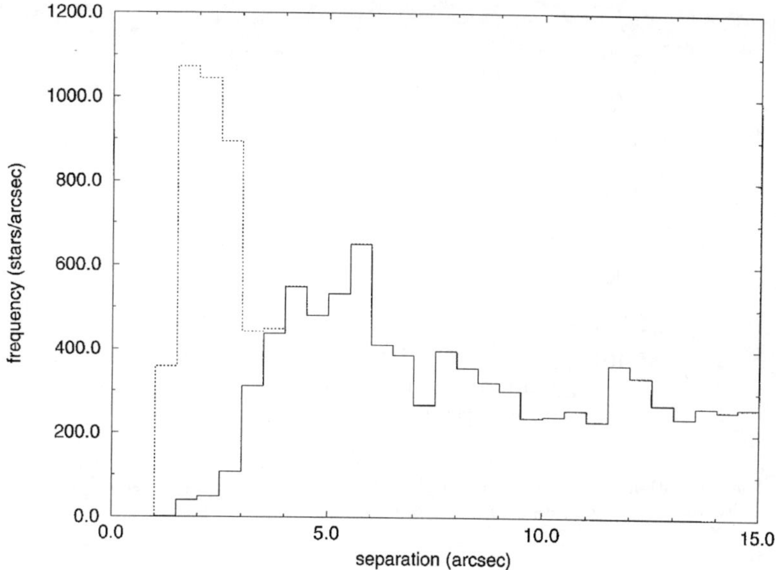

Figure 3. The distribution of the separations of the Tycho double stars when the close pairs found by the astrometric reduction are taken into account. The full line represents the double stars coming from the standard treatment, and the dotted line the pairs separated by astrometry.

- Candidate new double stars, coming from two sources. The first was a list of stars separated by the double–peak detection, when this method was applied to a subset of the observations. 8 687 stars, including 1 608 'known double stars' were thus selected. A second set was provided by a preliminary astrometric reduction. It contained 1 217 stars, including 420 'known double stars', that all had large residuals in their astrometric solutions.

The input catalogue of close double stars finally contained 22 083 stars from the selection performed during the 1–year data processing, but only 17 255 of them are in the final Tycho catalogue.

3.3. THE CLOSE DOUBLE STARS SEPARATED BY THE ASTROMETRIC REDUCTION

A dedicated routine was used in order to separate the components of the input catalogue of double stars in the astrometric reduction. 1657 pairs with separations between 1 and 3 arcsec were thus separated. The distribution of separations of the Tycho double stars, completed with these additional pairs, is in Figure 3.

Each component became an entry in the Tycho Catalogue. However, they received only approximate (B_T, V_T) photometry.

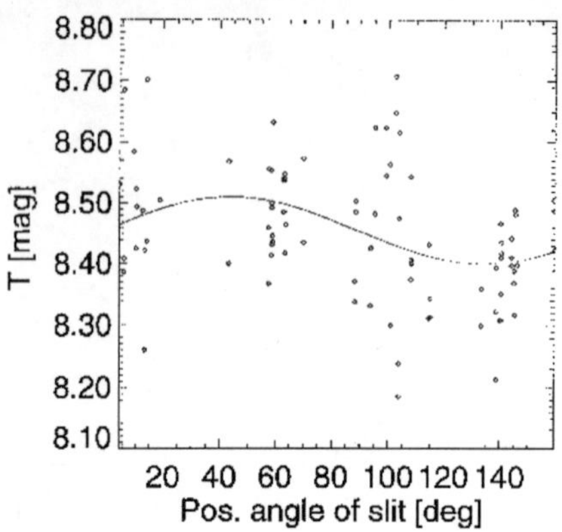

Figure 4. The correlation between the estimated Tycho magnitude of a double star and the orientation of the slits of the star mapper of Hipparcos.

3.4. THE DOUBLE–PEAK DETECTION

A double–peak detection algorithm was used to distinguish the transits of the components of the double stars of the input catalogue, in order to determine their relative positions. This process is fully described in another paper of this issue (Wagner & Halbwachs, 1996). The calculation of the positions and the derivation of the photometric parameters of the pairs was not included in the preparation of the Tycho Catalogue, but it was treated as an off–line task. However, when a double star was separated, it received a "double star flag" in the Tycho catalogue.

3.5. THE PHOTOMETRIC DETECTION OF CLOSE DOUBLE STARS

When a double star is crossing the slits of the star mapper, the estimation of its magnitude depends on the geometric configuration of the transit. If the components are parallel to the slits, the transit appears in the photon counts exactly as if the star were single. At the opposite, when the radius vector of the double star is perpendicular to the slit, the peak in the photon count is wider, although the total number of photons in the signal is the same as previously. Since the amplitude of the signal is estimated by fitting a slit response function to the photon count, the signal seems smaller, and the star look fainter. The consequence is that double stars seem to be variable stars, but it is possible to distinguish them from true variables, since their magnitude is correlated to the orientation of the slit. This is illustrated in Figure 4.

This method is efficient for separations down to 0.4 arcsec. Since it is not expensive in computing time, it was applied to all the 480 000 stars of the Tycho Catalogue that are brighter than $V_T = 10.5$ mag. As the double stars separated by the double–peak detection, the close double stars found by this process received the "double star flag" in the Tycho catalogue.

3.6. THE DUPLICITY FLAG

The double–peak detection and the photometric detection of close double stars concluded to non–duplicity for 449 078 stars. In the same time, 7011 stars received the "certain duplicity" flag, and 22 485 were flagged as possible double stars (the components of the close pairs separated by astrometry were not counted in this statistic). Among these stars, 5000 "certain double stars" and 20 000 "possible double stars" were not in the selection of known double stars, and are possibly new double stars.

4. Conclusions

The Tycho Catalogue provides accurate positions, parallaxes, proper motions, and (B_T, V_T) photometry for components of double stars wider than 3 arcsec. 3314 components of pairs with separations between 1 and 3 arcsec were separated by the astrometric reduction, and became independant entries in the catalogue, with an approximate (B_T, V_T) photometry. 7000 other stars were found to be certainly double, including 5000 stars that were not already included in the published double star catalogues.

Acknowledgements

This work was supported by national funding from Centre National d'Etudes Spatiales (CNES), the Danish Space Board, and Deutsches Agentur für Raumfahrtangelegenheiten (DARA).

References

Bässgen G., Wicenec A., Andreasen G.K., Høg, E., Wagner K., Wesselius P. (1992) Tycho transit detection, *Astron. Astrophys.*, **Vol. no. 258**, pp. 186–192

Bastian, U., Röser, S. *et al.* (1993) *PPM Star Catalogue, Vols. III and IV, positions and proper motions of 197179 stars south of -2.5 degrees declination.* Astronomisches Rechen-Institut, Heidelberg (Spektrum Akademischer Verlag, Heidelberg, Berlin, New York, 1993)

Couteau, P. (1990) *Catalogue de 2550 étoiles COU*, Observatoire de la Côte d'Azur

Dommanget, J. (1992) *The Hipparcos Input Catalogue, Annex 1: Double and Multiple Stars* ESA SP-1136, **Vol. 6**

Egret, D., Didelon, P., McLean, B.J., Russel, J.L., Turon, C. (1992) The Tycho Input Catalogue. Cross–matching the Guide Star Catalog with the Hipparcos INCA Data Base, *Astron. Astrophys.*, **Vol. no. 258**, pp. 217–222

ESA, 1997, The Tycho Catalogue, **ESA SP–1200**

Halbwachs, J.L., Høg, E., Bastian, U., Hansen, P.C., Schwekendiek P. (1992) Tycho star recognition, *Astron. Astrophys.*, **Vol. no. 258**, pp. 193–200

Halbwachs, J.L. (1992) First double star observations in the Tycho project of the Hipparcos satellite, *ASP Conference Series*, **Vol. no. 32**, pp. 429–434

Høg E., Bastian, U., Egret, D., Grewing, M., Halbwachs, J.L., Wicenec, A., Bässgen, G., Bernacca, P.L., Donati, F., Kovalevsky, J., van Leeuwen, F., Lindegren, L., Pedersen, H., Perryman, M.A.C., Petersen, C., Scales, D., Snijders, M.A.J., Wesselius, P.R. (1992) Tycho data analysis : Overview of the adopted reduction software and first results, *Astron. Astrophys.*, **Vol. no. 258**, pp. 177–185

Röser, S., Bastian, U. (1991) *PPM Star Catalogue, Vols. I and II, positions and proper motions of 181731 stars north of -2.5 degrees declination.* Astronomisches Rechen-Institut, Heidelberg (Spektrum Akademischer Verlag, Heidelberg, Berlin, New York, 1991)

Wagner, K., Halbwachs, J.L. (1996) Double Star Detection in the Tycho Photon Counts, *this issue*

Worley, C.E. Douglass, G.G. (1984) *The Washington Double Star Catalog*, U.S. Naval Obs.

DOUBLE STAR DETECTION IN THE TYCHO PHOTON COUNTS[1]

K. WAGNER
Institut für Astronomie und Astrophysik Tübingen
Waldhäuserstraße 64, D-72076 Tübingen, Germany

AND

J.L. HALBWACHS
Observatoire Astronomique de Strasbourg, URA 1280
Equipe "Populations Stellaires et Evolution Galactique"
11 rue de l' Université, F-67000 Strasbourg, France

Abstract. During data reductions of the Tycho project, the single measurements of double stars with separations less than 3 arcsec were not resolved due to several reasons. In an off-line task the single measurements of 22083 stars contained in a double star input catalogue were treated by a special double star algorithm. If the measurements of a double star could be separated under different scanning angles, its relative position was obtained. Comparison of these results with preliminary results of the NDAC double star working group led to the selection of 5816 reliable positions.

1. Double stars in the Tycho project

In the routine Tycho data reduction, double stars with separations less than 3 arcsec were usually not separated for different reasons which were already listed in another paper of this issue (Halbwachs J.L. *et al.*, 1996). As the detection algorithm (Bässgen G. *et al.*, 1992) was adapted to single stars, it often produced only one measurement (called "transit") for the crossing of a double star over the slits of the star mapper, especially when the star distance projected to the slits got below 2 arcsec.

2. Selection of candidate double stars

During the routine Tycho detection process, a goodness-of-fit value was calculated for all transits. For double stars these numbers are significantly higher than for the average of all stars and could thus be used as a double star indicator. The sensitivity of the goodness-of-fit parameter has a maximum for projected separations of the two photon count peaks around 1 arcsec. All double stars with separations above 1 arcsec have projected separations around 1 arcsec several times during the mission. But the sensitivity was good enough to detect pairs down to 0.7 arcsec and $\Delta m = 2$.

[1]Based on observations made with the ESA Hipparcos satellite, and on work by the Tycho Consortium in collaboration with the NDAC consortium

J. A. Docobo et al. (eds.), Visual Double Stars: Formation, Dynamics and Evolutionary Tracks, 469–474.

In a first step all transits with significantly enlarged goodness-of-fit values were selected. A double peak test fit, which will be described in the next section, was applied to the photon counts of the selected transits. For all stars having more than 3 successful fits the correlation between projected peak distance and projection angle was verified. The 8687 stars with the best correlation were selected in the end, because the number of candidate double stars had been limited to about 10000 for project internal reasons. These stars were incorporated in the input catalogue of double stars as described in another paper of this issue (Halbwachs J.L. *et al.*, 1996). As time for the preparation of the list of candidate double stars was limited, the goodness-of-fit check and the test fits were applied only to a part of the mission data corresponding to 247 full days of observations.

3. Double peak fit and angle correlation

The photon count rates of all transits of the stars of the input catalogue of double stars were extracted from the Tycho telemetry data together with some data from the so-called "transit prediction", giving e.g. the star ID, the actual scanning velocity and the position angle of the scanning direction. This time the data of the whole mission were used. A double peak test was applied to the extracted photon counts. Figure 1 shows the different steps of this test. It started with a single peak fit at the position of the maximum count rate. If this was successful, the fitted peak was subtracted from the count rates and a second maximum was searched in an interval with a full width corresponding to 5.6 arcsec centred on the initial transit position. If a maximum was found, a final simultaneous fit of both positions and amplitudes was made. The test was considered to be successful if the signal-to-noise ratio of both peaks was larger than 1.5 and if the separation of the peaks was between 0.45 and 4.2 arcsec.

Taking all successful double peak fits of one double star together, one gets the projected distances of the two components for a number of different projection angles. As the position angles of the scanning direction were known, it was possible to derive the relative position of the secondary component. An example of a double star with a separation of 1.54 arcsec and a position angle of 256 degrees is shown in Figure 2. Because the two components could not be distinguished (especially for pairs with small magnitude differences), in the first run the position angle was determined with a uncertainty of π. In a second run each peak could be assigned to one of the both components and a decision between the two possible position angles was made.

In the above Figure a number of peak distance measurements can be seen at scanning angles for which the projected peak distance should be almost zero. These measurements are due to either noise detections or another star, the two components of the binary system are not resolved in these cases. In cases of small separations and/or faint magnitudes it is possible that their number exceeds the number of correct detections. Two different methods were applied to discard as many as possible of them before the calculation of separation and position angle. The first method is based on the correlation between the amplitude of the single peak fit described above and the scanning angle. The behaviour of this correlation is described in section 3.5 of another paper of this issue (Halbwachs J.L. *et al.*, 1996). The second method uses the fact that the amplitudes of the separated components should not show this correlation.

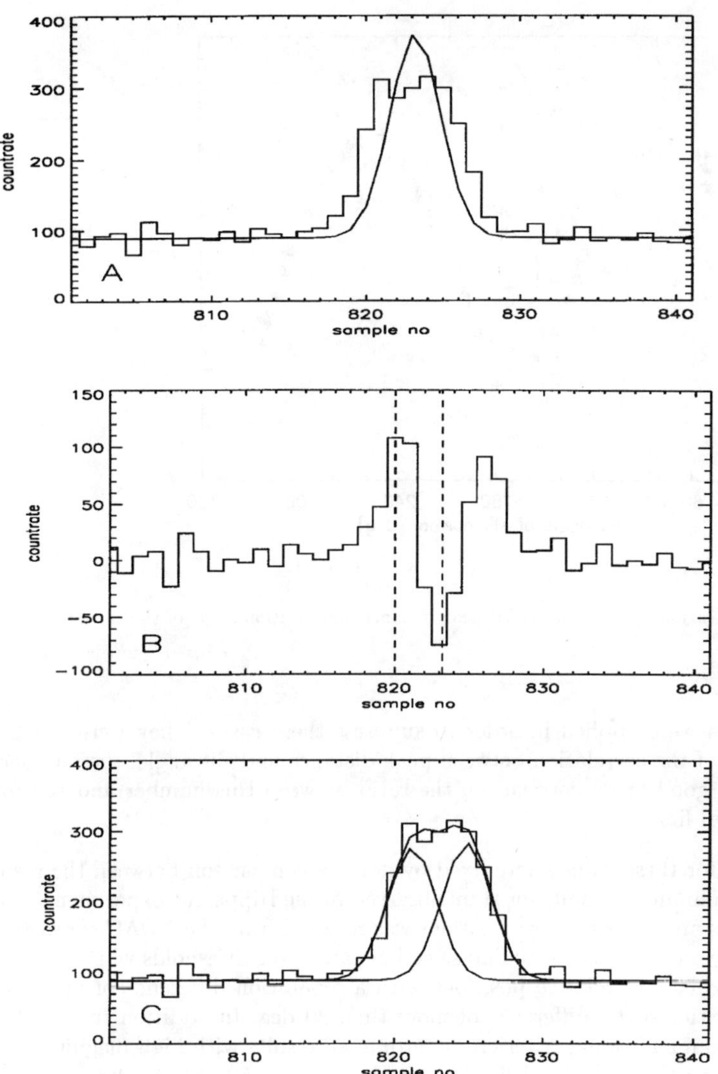

Figure 1. Resolving a photon countrate peak into two components. A: countrates with single peak fit, B: single peak subtracted, dashed lines are on positions of single peak and of new maximum, C: simultaneous fit of two peak position and amplitudes

4. Comparison with preliminary Hipparcos results

The selection of single measurements was not able however to discard all wrong measurements. Relative positions were obtained for almost all of the 22083 stars of the double star input catalogue. Especially if the correlation between peak distances and slit angle is only weak, it may be created by pure chance from a random combination of noise peaks.

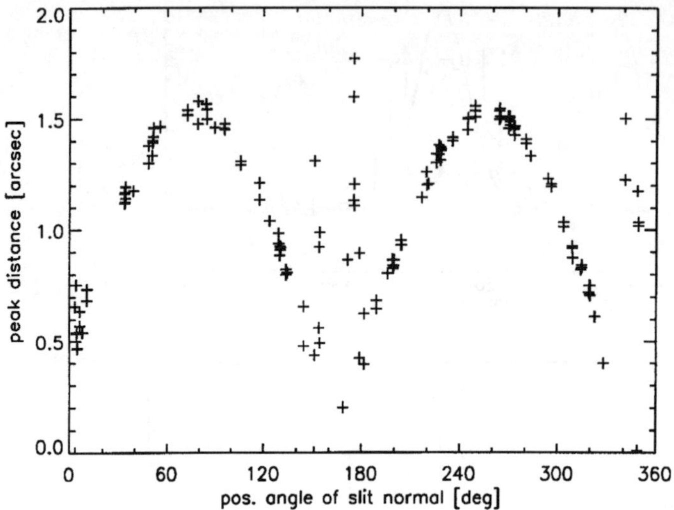

Figure 2. Correlation between projected peak distance and position angle of the slit normal

So selection criteria were applied in order to suppress these cases. They were based on the goodness-of-fit of the correlation between peak distance and slit angle, the number of successful fits with good angle correlation, the ratio between this number and the total number of successful fits.

The thresholds for these values were fixed by doing a comparison between the results of this work and preliminary results for stars observed in the Hipparcos experiment. A list of 2357 double stars and their relative positions was received from the NDAC consortium. During the comparison, the aim was to find selection parameter thresholds which minimise "false" solutions, which are relative positions with a separation difference of more than 0.2 arcsec or a position angle difference of more than 20 deg. In addition to the above mentioned criteria, an additional criterion discarding all results with high magnitude differences and small separations was defined, as it showed up that many solutions in this range were false.

Two sets of selection parameters were defined from the results of the comparison. Among the comparison sample of 2357 stars, applying the first set leads to the selection of 1339 double stars with an error rate of false solutions clearly below 0.5%. The median differences between their solutions and the NDAC reference are 0.021 arcsec in separation and 1.1 deg in position angle. Applying the second set leads to the additional selection of 471 double stars. The error rate in this combined group is 5.5%, the median differences are 0.028 arcsec and 1.4 deg respectively. Figures 3 and 4 show the median differences for the first group of stars, Figure 5 shows the fractions of resolved doubles for different separations and magnitude differences.

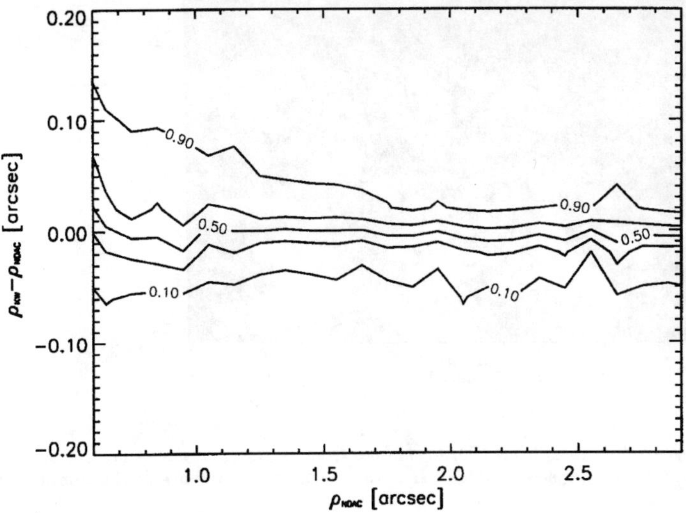

Figure 3. Separation differences from comparison with NDAC results, shown with 0.1, 0.33, 0.5, 0.67 and 0.9 quantiles

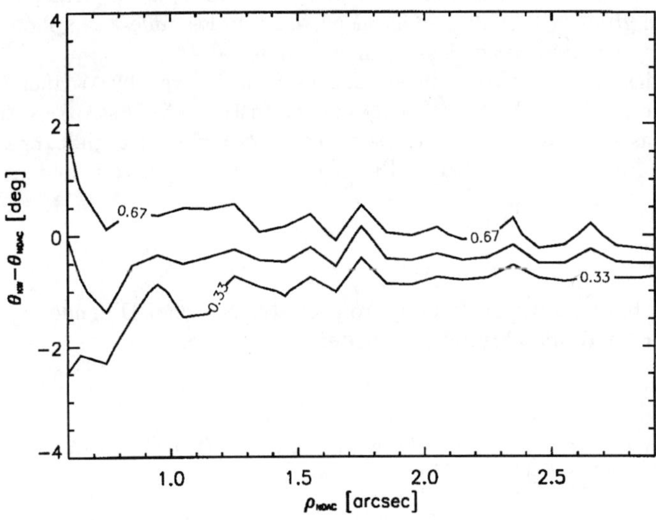

Figure 4. Position angle differences from comparison with NDAC results, shown with 0.33, 0.5, and 0.67 quantiles

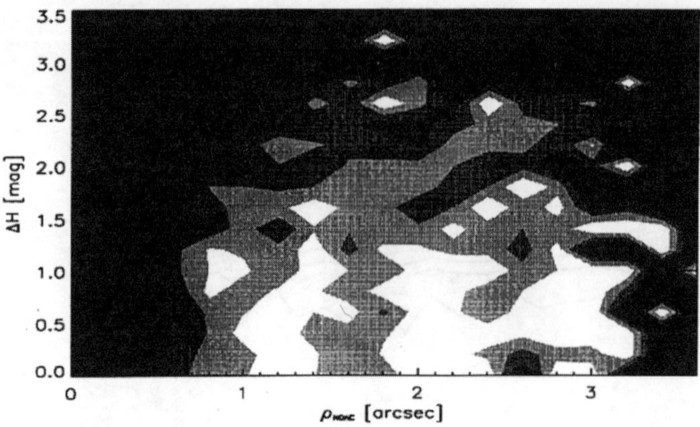

Figure 5. Fraction of resolved doubles from NDAC sample. Colours refer to more than 20, 40, 60 and 80% of the stars of the NDAC comparison sample. ΔH is the magnitude difference of the components in the Hipparcos magnitude.

5. Results for all stars

Applying the two selection criteria sets to all 22083 candidate stars of the double star input catalogue leads to the selection of 3239 and 2577 stars respectively. The stars of the first group will be flagged in the Tycho Catalogue (ESA, 1997) as confirmed doubles, the stars of the second group will be flagged as suspected doubles, unless the photometric method of double star search could confirm them as double.

A more detailed description of these investigations will be given by Wagner (1996). The results will be accessible on the WWW pages of the Institute for Astronony and Astrophysics Tübingen as soon as they are released. (`http://astro.uni-tuebingen.de/`, follow links "working groups" and "Tycho"). Publication at other places is under consideration.

Acknowledgements

This work was supported by national funding from Centre National D'Etudes Spatiales and Deutsche Agentur für Raumfahrtangelegenheiten.

References

Bässgen G., Wicenec A., Andreasen G.K., Hog E., Wagner K., Wesselius P. (1992) Tycho transit detection, *Astron. Astrophys.*, **Vol. no. 258**, pp. 186-192

ESA, (1997) The Tycho Catalogue, **ESA SP-1200**

Halbwachs J.L, Hog E., Makarov V.V., Wagner K., Wicenec A. (1996) The treatment of double stars in the Hipparcos-Tycho programme, *this issue*

Wagner K. (1996) Bestimmung relativer Positionen von 5816 visuellen Doppelsternen mit Hilfe des Tycho Satellitenexperiments, *Universität Tübingen*, doctoral thesis, to be published

MASS DETERMINATION OF ASTROMETRIC BINARIES WITH HIPPARCOS

C. MARTIN, F. MIGNARD AND M. FRŒSCHLÉ
Observatoire de la Côte d'Azur, Dept. CERGA
Av. N. Copernic, F-06130 Grasse, France

Abstract. Among the 11 500 known double stars observed by the ESA satellite Hipparcos, there are some 900 binaries for which orbits have been published since the beginning of the century. When the apparent displacement of the components due to the orbital motion is significant over the 37 months of the mission, the five astrometric parameters introduced in the general processing (position, parallax and proper motion), are not sufficient to allow for the motion of the components around the centre of mass of the system. It is therefore necessary to correct each observation in order to recover the astrometric properties of the centre of mass. We show that the knowledge of the orbital motion of a binary, when combined with the Hipparcos observations, may lead to an estimation of the mass and intensity ratios of the components, provided that the period is less than 30 years and the semi-major axis is larger than 0.2 arcsec. For smaller separations, we only get a linear combination of these two quantities.

1. Introduction

The aim of the general astrometric processing of the observations carried out by Hipparcos was to produce, for each of the 118 300 program stars, the five astrometric parameters representing the position, the parallax and the proper motion. When the program star was single, the previous quantities were associated to the center of light of the star, that is to say its astrometric direction. In the case of a double star without detectable orbital motion, they corresponded to a point which was specific to the Hipparcos observations, called 'Hippacentre', which was closely related to the photocentre of the pair, but not always identical. In these two cases, the absolute motion

J. A. Docobo et al. (eds.), Visual Double Stars: Formation, Dynamics and Evolutionary Tracks, 475–481.

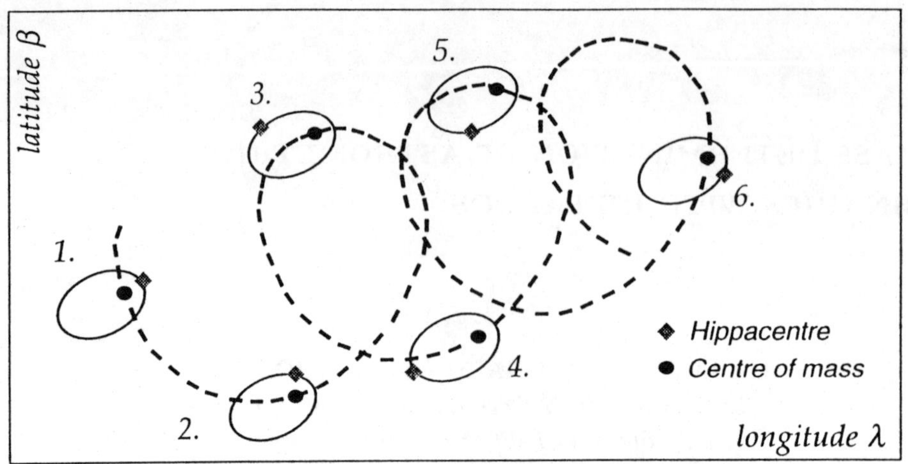

Figure 1. The combination of the parallactic effect and the rectilinear proper motion of a star during the mission may be represented on the sky by the dotted curve above. When a short-period binary is observed, the corresponding point, or Hippacentre, oscillates around the path of the centre of mass from one observation to another.

on the sky, either of the photocentre or of the Hippacentre, was completely described by the general five parameters model (see the dotted curve of Fig. 1).

A different situation occurs when the pair presented an orbital motion with a period not larger than few times the mission length, in which case the hippacentre had a wavy motion on the sky superimposed to that of the centre of mass of the system, and might no longer be modelled by only five parameters. The aim of the work reported hereafter is precisely to correct the sample of observations of such short–period pairs for the parasitic orbital motion and then retrieve the classical shape of the absolute motion, followed by the centre of mass. The knowledge of the relative orbital motion of a component around the other is absolutely necessary to apply this correction, and, as explained in the next sections, sometimes also sufficient.

2. The astrometric information

2.1. DEFINITION OF THE HIPPACENTRE ON THE GRID

For the sake of simplicity, we consider here that the line joining the two components is perpendicular to the the grid–slits. The basic Hipparcos record during the transit of a star on the grid (single or not) was modelled as (Murray et al. 1989).

$$S(t) = I + B + IM\cos(\omega t + \phi) + IN\cos(2\omega t + \psi)\ \ \mathrm{s}^{-1} \tag{1}$$

where I is the total intensity, B the unmodulated background, M and N the modulation coefficients of the first and second harmonic and ϕ and ψ the corresponding phases. For a double star, the positional information contained in the phase ϕ refers to a point called 'Hippacentre' and depends on the relative position and relative luminosity of the components, via the expression :

$$\phi \approx \beta\xi + \left(-\frac{\beta}{6} + \frac{\beta^2}{2} - \frac{\beta^3}{3}\right)\xi^3 \tag{2}$$

where $\beta = \frac{I_2}{I_1+I_2}$ is the normalized intensity of the secondary component. $\xi = 2\pi\rho/s$ is the phase difference between the components with $s = 1''.2074$ for the grid step. When the separation is smaller than $0''.2 - 0''.3$, the second term of Eq. 2 disappears and we get $\phi \approx \beta\xi$, which is nothing but the phase of the photocentre. In other cases, the additive nonlinear term is larger than 10 mas and cannot be neglected in the Hipparcos observations.

2.2. ABSOLUTE ASTROMETRY OF A DOUBLE STAR

We consider now the observation of a close binary in the absence of orbital motion, when the orientation with respect to the grid is arbitrary. The position of the Hippacentre H on the grid with respect to the projections of the Primary P, the Secondary S, the centre of mass G and the photocentre F is represented in Fig. 2, in a case where $\rho > 0''.3$ (so that H and F are distinct). In addition, the relative position of H with respect to P, S and F changes from one observation to another as a result of the scanning direction and the non–linearity of Eq. 2 (which is not true for the photocentre).

Let $\boldsymbol{\rho}$ be the vector **PS** oriented from the primary to the secondary and $B = \frac{\mathcal{M}_2}{\mathcal{M}_1+\mathcal{M}_2}$ the mass fraction of the components. The position of the photocentre on the sky and on the grid with respect to the centre of mass are given by :

$$\mathbf{GF} = (\beta - B)\,\boldsymbol{\rho} \quad , \quad \mathbf{G'F'} = (\beta - B)\,\boldsymbol{\rho}' \tag{3}$$

which shows that the photocentric orbit has the same shape as the relative orbit with a scale defined by $\beta - B$. The corresponding equations for the hippacentre H read:

$$\mathbf{GH} = \left(\frac{\phi}{\xi} - B\right)\boldsymbol{\rho} \quad , \quad \mathbf{G'H'} = \left(\frac{\phi}{\xi} - B\right)\boldsymbol{\rho}' \tag{4}$$

For small separations $\frac{\phi}{\xi} \approx \beta$, (Eq. 2) and there is no difference between an Hipparcos observation and the classical observation of the photocentre of an unresolved system. The same conclusions may be drawn when Δm goes to zero. For larger separations the actual position of H depends separately

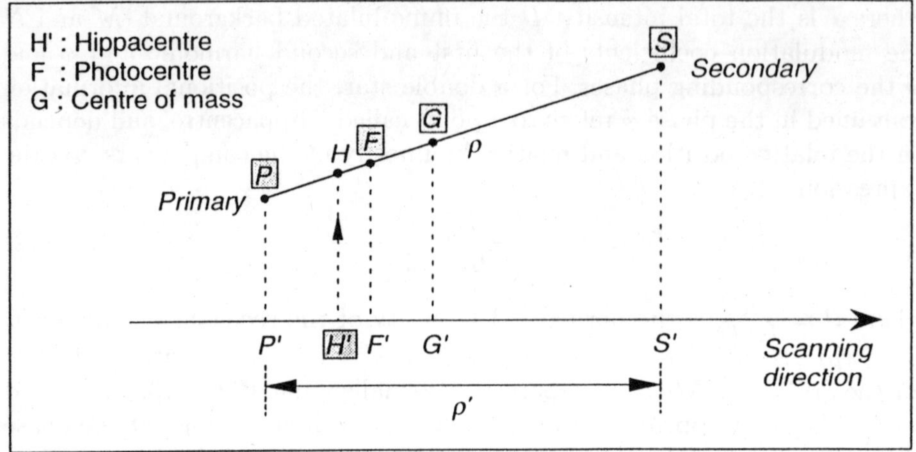

Figure 2. Projection of the points characterizing a double star on the scanning direction. The Hippacentre H' is defined on the grid but can be back-projected to a point H on the sky.

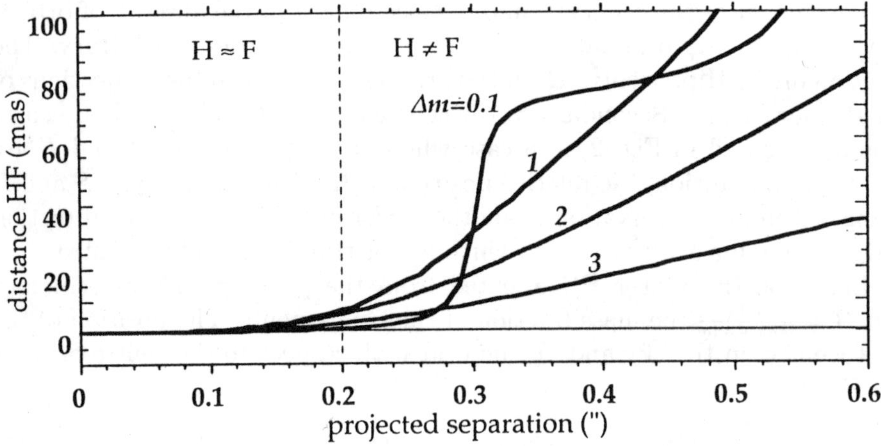

Figure 3. Angular distance between H and F as a function of the projected separation and magnitude difference. The two points can be distinguished beyond a threshold of about 0.″2 – 0.″3, depending on the object's global magnitude. These values will be made more precise in Sec. 4 on the basis of an extensive simulation.

on the mass and intensity ratio and on the scanning direction, a parameter which was perfectly known for every Hipparcos observations. The absolute path of the Hippacentre on the sky during one orbital period is thus generally non-elliptical. Fig. 3 shows the theoretical behaviour of the distance between the hippacentre and the photocentre as a function of the projected separation.

3. Schematic principle of the specific processing

As mentioned in the Introduction, the reduction consisted of producing the absolute astrometry of the centre of mass by correcting each observation by the quantity **GH** (Eq. 4). If we assume that the relative orbit is perfectly known (so that the phase difference ξ could be determined for each configuration), we just need to know β and B (or the difference $\beta - B$ when H and F merge) to apply the correction and solve an overdetermined system with five unknowns. But the two previous quantities are generally not adequately known and it is better to introduce them as two additional unknowns, yielding, in addition to the 5 astrometric parameters of the centre of mass, an estimation of the mass and intensity fractions or only of their difference for very close pairs.

This ideal program worked only when the signature of the absolute orbital motion during the mission was significant and could be easily separated from the two other contributions (proper motion and parallax). Thus, the orbital period appears as a key parameter: the curvature of the part of orbit sampled by the Hipparcos observations had to be large enough so that the orbital motion and the proper motion could be disentangled. An extensive simulation was set up to determine the range of periods, separations, magnitude differences and mass–ratios in which there was some chance of success.

4. The simulation

In the simulation we have generated few relatively short period binaries of various separations and computed several series of observations, based on the time sampling of the Hipparcos data sets. According to the theoretical approach, we assumed that the quality of the determination of β and B or of their difference depended primarily on the three following parameters : the semi-major axis a, the orbital period P and the Hipparcos global magnitude H_p. Each final output of the simulation was a median value of 30 realistic cases, each resulting from a random drawing of the five remaining orbital elements.

The main goal of the simulation was to determine in the space semi-major axis – period, the regions where a separate determination of the mass– and intensity–ratio was achievable and where only the scale of the photocentric orbit will be obtained. A summary of the main results appears in Fig. 4 as contour plots. As expected, there is no hope to measure the mass ratio for separations less than $0\overset{\prime\prime}{.}2$ and periods larger than about 25 years.

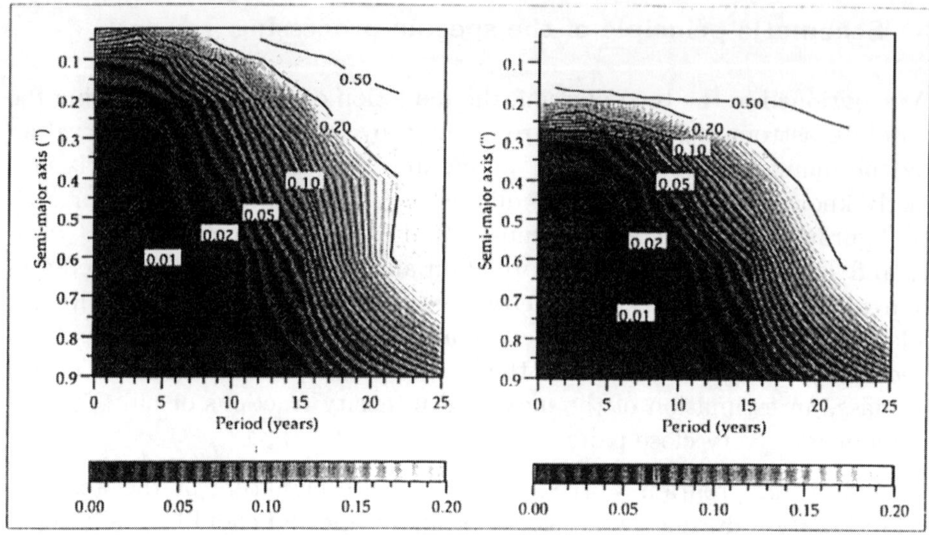

Figure 4. Typical standard deviation of $\beta - B$ (left) and of the mass fraction B alone (right) as a function of the period and semi-major axis, for a tenth magnitude binary star. The light area in the upper right of each diagram represents the domain where no valuable information on β and/or B can be extracted from the observations.

5. Selection of objects and first results

According to the results of the simulation, it is possible to set up a list of Hipparcos stars which are likely to be good candidates for such an analysis, assuming the relative orbits are sufficiently well known. From an extensive search in different files of orbits and catalogues as in recent publications, a selection of 277 orbits for 173 objects has been done. Some of the most interesting stars are listed in Table 1. Among these 173 candidates, significant results have only been obtained for some 50 stars, probably because of the low quality of most of the orbital elements found in the literature (only

TABLE 1. Some good candidates for a direct determination of the mass ratio.

HIP	a''	P(yr)	HIP	a''	P(yr)	HIP	a''	P(yr)
171	0.83	26.3	44248	0.65	21.8	86032	0.48	8.5
2762	0.24	6.9	46706	0.55	18.3	93506	0.53	21.1
2941	0.67	25.0	51986	0.34	16.3	103655	0.73	28.2
7918	0.56	19.5	64241	0.67	25.8	104771	0.23	5.7
11452	0.54	25.3	80346	0.27	3.7	104858	0.26	5.7
38382	0.57	23.3	84140	0.76	12.9	107522	0.31	12.5

20% of the known orbits are considered as definitive or very good). For 8 of these stars, the mass fraction B could be determined separately from β. For the others, one may obtain in some cases the magnitude difference from ground-based observations, so that the mass fraction B can be also derived with a good accuracy. For example, Algol belongs to this category. Then, by using the corrected parallax produced by the same processing, individual masses have been calculated as well, together with their uncertainty.

For six stars out of the eight mentioned above, the standard error of the mass ratio is better than 0.03, and for half of the whole set of 50 stars, $\sigma_{\beta-B} \leq 0.02$. The best case is that of Algol AB-C, with $B = 0.260 \pm 0.008$, and the worst values are close to 0.08. Concerning the individual masses, the mass of the primary component is known with an absolute error less than 10% for 8 systems and less than 20% for 22 systems (resp. 7 and 18 systems for the secondary component).

6. Conclusion

In this paper we have identified a new category of astrometric binaries very specific to the Hipparcos measurement procedures. It appears that for pairs with orbital period less than 25 years and with angular separation larger than $0\farcs3$, it is possible to derive the mass ratio between the two components and their magnitude difference at the same time as the five astrometric parameters of the centre of mass. An extensive simulation has allowed to locate accurately in the separation–period–magnitude space various regions where the processing might end up with the two physical parameters or only with their difference. In some favorable situations an accuracy of few hundreths for the mass ratio can be achieved. The number and the quality of the results will increase as new accurate orbits are determined in the next few years, particularly from the speckle interferometry technique.

References

Martin, C., Mignard, F., Froeschlé, M. (1996) Mass Determination of Astrometric Binaries with Hipparcos, *Astron. and Astroph. Suppl. Ser.*, in press.

Martin, C., Mignard, F. (1996) Detection, Astrometry and Photometry of Visual Double Stars with Hipparcos, *this volume.*

Martin, C. (1996) Contribution à l'Étude des Étoiles Doubles Hipparcos, *thesis.*

Mignard, F., Söderhjelm, S., Bernstein, H., Pannunzio, R., Kovalevsky, J., Froeschlé, M., Falin, J.L., Lindegren, L., Martin, C., Badiali, M., Cardini, D., Emanuele, A., Spagna, A., Bernacca, P.L., Borrielo, L., Prezioso, G. (1995) Astrometry of Double Stars with Hipparcos, *Astron. and Astroph.*, **Vol. no. 304**, pp. 94.

Murray et al. (1989) The Hipparcos mission, ESA report SP-1111, **Vol. III**, pp. 33–49.

20% of the known orbits are considered as definitive or very good. For 24 % of the systems, the mass fraction B could be determined separately from the other elements; [illegible] in some cases the [illegible] ground-based observations, so that the mass fraction B can be also derived with a good accuracy. For example, [illegible] belongs to this category. Then, by using the corrected parallax provided by the same processing, individual masses have been calculated as well, together with their uncertainty.

[illegible] out of the [illegible] mentioned above, the standard error of the mass [illegible] 0.08 and for half of the whole set of the stars [illegible] 0.12. The best case is that of [illegible] with B = 0.260 ± 0.008 [illegible] values are [illegible] 0.08. [illegible] the mass of the primary component is known with [illegible] less than 5% or 8 [illegible] than 20% [illegible] for the secondary component.

[illegible]

In this paper we [illegible] the Hipparcos measurement process [illegible] separation [illegible] the magnitude difference of the [illegible] components [illegible] the extensive [illegible] the quality of the results will increase [illegible]

SECTION V

CATALOGUES AND DATABASES

THIRTY YEARS OF WORK: CONSTRUCTION OF THE WASHINGTON VISUAL DOUBLE STAR CATALOG AND ITS FUTURE

CHARLES E. WORLEY AND GEOFFREY G. DOUGLASS
U.S. Naval Observatory Washington, D.C. 20392-5420. U.S.A.

Abstract. A new edition of the Washington Visual Double Star Catalog (1996.0) has now been completed and issued. This report discusses some of the back-ground concerning such compilations, their contents and arrangement of the data, as well as prospects for future enhancements.

1. Introduction

Although the first double star discovered was announced more than three centuries ago by Riccioli, little attention was paid to any systematic study for over 100 years. Much of this neglect stemmed from the belief that such objects represented chance alignments of physically unrelated objects, but also must have been influenced by the poor quality of early telescopic equipment. Almost simultaneously, in 1781 and 1782, C. Mayer and W. Herschel published short lists of double stars which, in fact, constitute the first catalogs of such objects. Herschel also seemed to haw a suspicion that some of the double stars were true binaries, and, after long observation, announced that fact in 1803.

Following some experiments with small instruments, F.G.W Struve secured a 24cm refractor at Dorpat. Here, for the first time, was a equatorially- mounted, clock-driven telescope with superior optics, equipped with an accurate filar micrometer. With this telescope Struve began the first systematic survey aimed at the discovery of new double stars, of which nearly 3000 were detected. Following determination of their positions with a meridian circle, Struve published his great catalog "Mensurae Micrometricae" in 1837.

Curiously enough, little survey work was done in most of the rest of the Nineteenth Century, until Burnham began to discover considerable numbers of new pairs. This work inspired Aitken and Hussey to resume systematic work in the northern skies, and later Innes, van den Bos, Rossiter, and others to begin southern surveys. There resulted three important catalogs: A General Catalogue of Double Stars (Burhan, 1906); Southern Double Star Catalogue (Innes, 1927); and New General Catalogue of Double Stars (Aitken, 1932). These catalogs presented a summary of known pairs within their declination, separation, and magnitude limits (if any), as well as extensive lists of measures, notes, references, and astrometric and astrophysical data.

J. A. Docobo et al. (eds.), Visual Double Stars: Formation, Dynamics and Evolutionary Tracks, 485–487.

Following the initial systematic visual surveys, many thousands of additional pairs were found in various special surveys, as well as on the plates of the Astrographic Catalogue. And, of course, observations of known pairs continued to accumulate. Fortunately, from the time of Burnham, a comprehensive data base has been maintained by several generations of astronomers. As technology changed, so did the form of the data base, with the manuscript data transferred first to punch-cards, and later to magnetic disk storage.

In the 1950's, the northern and southern data bases were merged, and the first all-sky catalog was issued in 1963 (Jeffers, van den Bos, and Greeby). In 1965 the data base was transferred to the U.S. Naval Observatory, and in 1984 we produced The Washington Visual Double Star Catalog 1984.0, (Worley and Douglass, 1984.0). We have now produced a considerably augmented and revised version, which will be described below.

2. The Washington Visual Double Star Catalog, 1996.0

Survey work for new double stars has continued, with concentration on the northern hemisphere. Much of this work has been visual, but substantial numbers of objects have been found by techniques involving speckle interferometry and CCD imaging. The new catalog contains 78100 objects versus the 3000 in Struve's work, and rests on a observational data base of 452000 measures, versus perhaps 10000 at the time of Struve.

Positions are nominally J2000, though we hasten to add that there are many pairs whose positions are poorly established. Corrections for proper motions have been made where warranted. All Durchmusterung stars have been checked with the ACRS and PPM for modern proper motions. Many discoverer designations have received numbers.

Perhaps the most inhomogeneous aspect of the double star data base remains the magnitudes. Very many are rather crude visual estimates, while the Astrographic Catalog stars have photographic values. For the 1996 double star catalog, we have used V magnitudes whenever available. For blended pairs, we have determined mean magnitude differences as estimated by reliable observers, and derived the individual values. We hope the large new surveys, such as TYCHO, will considerably alleviate this problem.

Whenever available, we attempt to list MK spectral types, many of which come from the new Michigan Spectral Survey. We have also updated the dates of observation, number of observations, position angles and separations, and the Notes and References. In particular, the Notes now include expanded information on orbital motion, spectroscopic and photometric duplicity, and variability. All new pairs and known corrections are included. Because of the many changes, we have altered the format to exclude some less important or obsolete data.

3. Distribution of Data

The 1996 WDS is now available on the U.S. Naval Observatory World Wide Web page at http://aries.usno.navy.mil/ad/wds/. It is also obtainable from the astronomical data centers. The Fourth Catalog of Orbits of Visual Binary Stars (Worley and Heintz, 1983) is also included at this address. It now has been expanded to include ephemerides for the orbits.

Further, we maintain the Catalog of Observations, upon which both the WDS and Orbit Catalogs ultimately depend. This catalog is complete for data since 1927, but not before. (As time permits, we strive to add the earlier material). Requests for data from this data base for individual objects are honored to the extent that our limited resources can support

4. Future of the Double Star Data Bases

Any project of this nature requires long-term attention by knowlegeable, dedicated persons. Furthermore, an extensive library is a necessity if completeness and correctness is to be achieved. Decisions as to the formats, contents, and distribution of the data have to be made, bearing in mind the principal mission-oriented interest of the Naval Observatory in a project such as this. With the development of new observing techniques the former distinction between visual and spectroscopic binaries is becoming a thing of the past, and one can envision the time when a single catalog embodying a more complete description of binary systems will become necessary.

References

Aitken, R.G.: 1932, *New General Catalogue of Double Stars within 120 degrees of the North Pole,* (Carnegie Institute of Washington)

Burnham S.W.: 1906, *A General Catalogue of Double Stars within 121 degrees of the North Pole,* (Carnegie Institute of Washington)

Innes, R.T.A. 1927, *Southern Double Star Catalogue, -19 degrees to -90 degrees,* (Union Observatory, Johannesburg, South Africa)

Jeffers, H.M., van den Bos, W.H., and Greeby, F.M.: 1963, "Index Catalogue of Visual Double Stars, 1961.0," *Pub. Lick Observatory,* **XXI**

Worley, C.E. and Douglass, G.G.: 1984, *The Washington Visual Double Star Catalog, 1984.0,* (available on tape and CD-ROM)

Worley, C.E. and Heintz, W.D.: 1983, *Pub. U.S. Naval Observatory,* Second Series, **XXIV,** Pt. VII

RESEARCH INTO DOUBLE STARS AND CELESTIAL MECHANICS AT THE R.M. ALLER OBSERVATORY: A NEW EPHEMERIS CATALOGUE

J. A. DOCOBO[1], J. F. LING[1], C. PRIETO[2], P. MAGDALENA[1]
Observatorio Astronómico "Ramón María Aller"
P.O. Box 197; Universidade de Santiago de Compostela, Spain
[1]*Dpto. Matemática Aplicada. Universidade de Santiago. Spain.*
[2]*Dpto. Matemática Aplicada. Universidade de Vigo. Spain.*

1. Introduction

Since it was founded in 1943, the Ramón María Aller Observatory at the University of Santiago de Compostela (Spain) has centred its research on double stars.

Even before this date, Dr. Aller had introduced this branch of astronomy into Spain and had been carrying out not only micrometric measurements (Aller 1930, 1932, 1934, 1936, 1941) but also calculating certain orbits in his own private observatory in Lalín (50km from Santiago). More specifically, the orbit of the system STT 77 was the first to be determined in Spain (Aller 1935).

In the years which followed the most outstanding studies were of a theoretical nature, in particular Professor E. Vidal's orbital calculations and a new geometrical method of passing from an apparent to a relative orbit (Vidal 1946) and the determination of the parabolic orbit using three normal sites $(\theta_i, \rho_i; t_i)$, (Vidal 1948).

The so-called "orbigraph", an instrument that facilitated the automatic calculation of the curves $\rho(t), \theta(t)$ which verified the law of areas, was based on the ideas of A. Docobo (J. A. Docobo's uncle) and E. Vidal himself (Docobo 1946; Vidal 1952, 1957). Besides a new series of observations carried out by Professor Aller (Aller 1947, 1951, 1952, 1960) the following contributions by his coworkers are also worthy of mention: E. Vidal (Vidal 1944, 1945, 1947, 1953), J. A. Zaera (Zaera 1963, 1966a, 1966b) and in particu-

J. A. Docobo et al. (eds.), Visual Double Stars: Formation, Dynamics and Evolutionary Tracks, 489–495.

lar the work of R. Cid (Cid 1946, 1950 , 1953), who was first to solve the so-called "fundamental problem" of calculating elliptical orbits, which consisted of analytically determining the seven elements of relative orbit using seven observation data or normal sites: $3(\theta_i, \rho_i; t_i), i = 1, 2, 3$ and $(\theta_4; t_4)$ (Cid, 1958).

J. L. Dopico (1961) later discovered another solution to the same problem.

Professor Cid moved to the University of Zaragoza where he created an important research group to investigate celestial mechanics.

Although numerous orbits for double stars had already been calculated it was J.M. Costa, and at a later date C. Morales, who initiated the systematic determination of these calculations in the Observatory at the University of Santiago de Compostela in the 60s (Costa 1966, 1970, 1972; Costa & Morales 1975, 1976).

In 1983, Dr. J. A. Docobo, up until then a collaborator of Professor R. Cid at the University of Zaragoza, was appointed the new director of the Observatory.

From this moment on and thanks to the opportunities Spanish universities were now offering, more staff was appointed and diverse research projects went underway which not only enabled the Centre to promote its traditional studies but also opened up new lines of work.

The orbits for double stars calulated by members of the Observatory have been recorded both in Information circulars published by Commission 26 of the I.A.U. and in different international journals (Docobo & Costa 1986, 1990, 1992; Ling 1992; Docobo et al. 1994b).

All of these orbits were obtained using an original analytical method designed by Professor J. A. Docobo (Docobo 1985) which is based on the fact that 3 normal sites $(\theta_i, \rho_i; t_i)$ can generate a set of relative orbits whose apparent orbits pass through the three points selected. The orbit is finally chosen from within this set as that which best adjusts to the rest of the observations available bearing in mind their weight.

One the main advantages of this method is that it is not necessary to calculate the areal constant.

It is worth noting that this method also solves the "fundamental problem" in a simple manner. Moreover it can be generalized when radial velocity data are provided (Docobo et al 1988, 1992a)

The development of the aforementioned research projects has permitted the calculation of the relative positions of visual double stars not only in Santiago de Compostela but also in larger observatories such as those at Nice and Pic du Midi (France), Calar Alto and Fabra (Spain) (Docobo et al. 1983, 1984, 1991; Docobo, 1986, 1989; Ling, 1987; Couteau & Ling, 1988;

Couteau et al. 1989, 1992; Ling & Couteau 1992; Docobo & Prieto 1993; Ling & Lanchares 1993; Docobo & Ling 1995) [1]

At the present moment, the relationships which have been established with other groups (CHARA Geogia State University, U.S.A.; Mérida Observatory, Venezuela; Astronomic Centre in Yebes, Spain) has led to collaboration in techniques such as speckle interferometry (Hartkopf et al 1993, 1996) and CCD registers.

In 1993, Dr. Paul Couteau (Couteau 1993) passed the responsibility of editing the Information Bulletin of Commission 26 (double and multiple stars) of the I.A.U. onto Dr. Docobo. From then on Professors J. A. Docobo and J. F. Ling have aimed to boost this important vehicle of information which was first begun in by Dr. Paul Muller (Muller 1954).

As far as celestial mechanics as such is concerned, J. A. Docobo has worked on the three-body stellar problem (Docobo 1977a, 1977b; Docobo & Prieto 1988; Docobo et al 1992b has worked on the three-body stellar problem) in which it is treated using the Hamiltonian form and later analytically integrated by methods based on Lie transformations (Deprit 1969). Professors J. A. Docobo and A. J. Abad supervised J. F. Ling's doctoral thesis which dealt with the same problem. In Dr. Ling's study the Langrange equations are integrated using the so-called stroboscopic method (Roth 1973, 1979). The most relevant results of this study were later published (Ling 1991; Ling et al. 1995).

More recently J. A. Docobo suggested that C. Prieto should discuss the theme of the two-body problem with slowly decreasing mass and its application to double stars as the title of his doctoral thesis.

The results of this work (Prieto 1995) is an exhaustive analytical study containing comparative calculations with the results obtained both from the exact solution of Mestsersky as well numerical methods.

Lastly and thanks to the recent arrival of Professor V. Tamazian, our Observatory has opened a new line of work in relation to double stars with variable components.

Within this context and with the experience our group has gathered over the years, we have decided to publish a new ephemeris catalogue of the orbits of visual double stars which is briefly described below.

2. New ephemeris catalogue of visual double stars:

Both the observer and calculators of the orbits of visual double and multiple stars are aware of the importance of an up-to-date ephemeris catalogue.

[1]Many of these studies have been made possible thanks to the invitations of Dr. Paul Couteau with whom we have been collaborating for over ten years.

For this reason we think it would be interesting to present the catalogue we are preparing at this Workshop. It aims to be a continuation of that published by the Nice Observatory in 1986 (Couteau et al 1986) which is still valid today.

We hope to continue the tradition which began when Dr. P. Muller published the first catalogue (Muller 1954). Subsequently, the same P. Muller elaborated other catalogues in collaboration with C. Meyer and P. Couteau (Muller & Meyer 1964, 1969; Muller & Couteau 1979).

Our catalogue will be different to the last one published by P. Couteau in the sense that all the orbits will be on the same list which will cover the decade which spans from 1995 to 2004, inclusive.

However for short period stars, we will give the values of ρ and θ each half year.

The catalogue is organized in the following manner: the W.D.S. number of each star, that is the 2000.0 coordinates, its A.D.S. number, should this be the case, the magnitudes and the spectral type which figures in the W.D.S., the author or authors of the orbit and the date of the last observation used in the calculations will be given in the top square (see Tables 1 and 2). The year of publication of the orbit will appear in brackets.

	01409N1117	
ADS 1321	A 2320	
9.7	9.7	G5
BAIZE	(1987)	
1982.85		
P=37.2	a=0.183	
Epoch	θ	ρ
1995	226.3	0.13
1996	232.5	0.14
1997	237.8	0.15
1998	242.5	0.16
1999	246.7	0.17
2000	250.5	0.18
2001	254.0	0.19
2002	257.2	0.19
2003	260.2	0.20
2004	263.1	0.21

TABLE 1.

	01376S0924	
	KUI 7	
7.0	7.4	F5
CHARA	(1996)	
1996		
P=29.05	a=0.3179	
Epoch	θ	ρ
1994.0	155.6	0.28
1994.5	155.0	0.29
1995.0	154.5	0.29
1995.5	153.9	0.30
1996.0	153.4	0.30
1996.5	152.9	0.30
1997.0	152.4	0.30
1997.5	151.9	0.30
1998.0	151.4	0.29
1998.5	150.9	0.29
1999.0	150.3	0.29
1999.5	149.8	0.28
2000.0	149.2	0.27
2000.5	148.6	0.27
2001.0	147.9	0.26
2001.5	147.2	0.25
2002.0	146.5	0.24
2002.5	145.7	0.23
2003.0	144.9	0.23
2003.5	143.9	0.22

TABLE 2.

Lastly, we will include the values of the orbital period (in years) and the semimajor axis (in arcsec). The bottom square will show the ephemeris values.

The orbits have been taken from our up-to-date files although they have almost all been included in the catalogue published by C.E. Worley and W.D. Heintz (Worley & Heintz 1983) and the most recent in different astronomic journals or in the Information circulars published by the I.A.U.

At any event, the catalogue will include old and new orbits but invariably what we believe are the best.

In many cases different orbits will appear for the same star, in particular when a real difference has yet to be established between them. Nevertheless our criteria is that if a new orbit does not clearly improve an already existing orbit then only the earlier will appear.

The catalogue will be edited on paper, C.D. Rom etcetera with a view to satisfying different needs.

We hope that it will be ready for publication and available to the international community within a few months and that our esteemed friend and colleague Dr. Paul Couteau will honour us with a prologue.

Acknowledgements

This study was carried out as part of the research project XUGA 24301B96 financed by XUNTA DE GALICIA (Spain).

References

Aller, R. M.:(1930) *Astron. Nach.*, **238**, 71
Aller, R. M.:(1932) *Astron. Nach.*, **245**, 213
Aller, R. M.:(1934) *Astron. Nach.*, **251**, 273
Aller, R. M.:(1935) *Astron. Nach.*, **256**, NR 6133, 245
Aller, R. M.:(1936) *Astron. Nach.*, **259**, 133
Aller, R. M.:(1941) *La Ciencias*, **VI**, 259. Spain
Aller, R. M.: (1946) *Rev. Geofísica*, **Año V**. Spain
Aller, R. M.: (1947) *Rev. Geofísica*, **24**, 603
Aller, R. M.: (1951) *Urania*, **225**, 7. Spain
Aller, R. M.: (1952) *Urania*, **229**, 1
Aller, R. M.: (1960) *Urania*, **252**, 138
Cid, R.: (1946) *Rev. Geofísica*, **Año V**
Cid, R.: (1950) *Rev. Geofísica*, **35, 36**
Cid, R.: (1953) *Urania*, **232**
Cid, R.: (1958) *Astron. J.*, **63**, 395
Costa, J. M.: (1966) *Urania*, **263**, 71
Costa, J. M.: (1970) *Urania*, **271**, 126
Costa, J. M.: (1972) *Urania*, **276**, 145
Costa, J. M. and Morales, C.: (1975) *Urania*, **263**, 21
Costa, J. M. and Morales, C.: (1976) *Urania*, **286**, 175
Couteau, P.; Morel, P.J. and Fulconis, M.: (1986) *Pub. Obs. de Nice*
Couteau, P. and Ling, J. F.: (1988) *Astron. Astrophys. Suppl. Ser.*, **73**, 449
Couteau, P.; Docobo, J. A.; Elipe, A. and Ling, J. F.: (1989) *Astron. Astrophys. Suppl. Ser.*, **78**, 483
Couteau, P.; Docobo, J. A. and Ling, J. F.: (1992) *Astron. Astrophys. Suppl. Ser.*, **100**, 305
Couteau, P.: (1993) *Circulaire d'Information*, **120**, I.A.U. Com.26
Docobo, A.: (1946) *Rev. Geofísica*, **Año IV**
Docobo, J. A.:(1977a) Aplicación de la Teoría de Perturbaciones al Estudio de Sistemas Estelares Triples. *Ph. D.*, University of Zaragoza. Spain
Docobo, J. A.:(1977b) *Actas II Asamblea Nacional de Astronomía y Astrofísica*, **Vol 1**, 265. San Fernando, Cadiz. Spain
Docobo, J. A.; Costa, J. M. and Ling, J. F.:(1983) *Bol. Academia Galega de Ciencias*, **Vo. II**, 131. Spain
Docobo, J. A.; Costa, J. M. and Ling, J. F.:(1984) *Astron. Astrophys. Suppl. Ser.*, **58**, 287
Docobo, J. A.: (1985) *Cel. Mech.*, **36**, 143
Docobo, J. A.: (1986) *Acta Astron.*, **36**, 175

Docobo, J. A. and Costa, J. M.:(1986) *Astrophys. J. Suppl. Ser.*, **60**, 945
Docobo, J. A. and Prieto, C.:(1988) *Pub. Obs. Astron. R. M. Aller, 2nd epoch*, No. 46. Spain
Docobo, J. A.; Elipe, A. C. and Abad, A. J.:(1988) *Astrophys. Space Science*, **142**, 195
Docobo, J. A.: (1989) *Pub. Astron. Society of the Pacific*, **101**, 274
Docobo, J. A. and Costa, J. M.:(1990) *Pub. Astron. Society of the Pacific*, **102**, No. 658
Docobo, J. A.; Ling, J. F.; Prieto, C. and Lanchares, V.: (1991) *Astron. Astrophys. Suppl. Ser.*, **91**, 229
Docobo, J. A. and Costa, J. M.:(1992) *Astrophys. J. Suppl. Ser.*, **82**, 323
Docobo, J. A.; Ling, J. F. and Prieto, C.:(1992a) *I.A.U. Col. 135. A.S.P. Conf. Ser*, **32**, 220
Docobo, J. A.; Prieto, C. and Ling, J. F.:(1992b) *I.A.U. Col. 135. A.S.P. Conf. Ser*, **32**, 223
Docobo, J. A. and Prieto, C.: (1993) *Astron. Astrophys. Suppl. Ser.*, **100**, 641
Docobo, J. A. and Ling, J. F.: (1994) *Astron. Astrophys. Suppl. Ser.*, **105**, 337
Docobo, J. A.; Ling, J. F. and Prieto, C.:(1994) *Astrophys. J. Suppl. Ser.*, **91**, 793
Dopico, J. L.: (1961) *Urania*, **253**, 3
Hartkopf, W. I.; Mason, B. D.; Barry, D. J.; McAlister, H. A. and Prieto, C.: (1993) *Astron. J.*, **106**, (1), 352
Hartkopf, W. I.; Mason, B. D.; McAlister, H. A.; Turner, N. H.; Franz, O. G. and Prieto, C.: (1996) *Astron. J.*, **111**, (2), 936
Ling, J. F.:(1987) *Astron. Astrophys. Suppl. Ser.*, **71**, 115
Ling, J. F.:(1991) *Astrophys. Space Science*, **185**, 51
Ling, J. F.:(1992) *Astron. Nach.*, **313**, 2
Ling, J. F. and Couteau, P.: (1992) *Astron. Astrophys. Suppl. Ser.*, **95**, 423
Ling, J. F. and Lanchares, V.: (1993) *Astron. Nach.*, **314**, (2), 303
Ling, J. F.; Docobo, J. A. and Abad, A. J.: (1995) *Astron. J.*, **110**, (2) 875
Muller, P.: (1953) Catalogue de 304 éphémérides d'étoiles doubles visuelles *Jour. Observ.*, **36**, 61
Muller, P.: (1954) *Circulaire d'Information*, **1**, I.A.U. Com.26
Muller, P.: (1955) Supplément au Catalogue d'Ephémérides d'Etoiles Doubles Visuelles *Jour. Observ.*, **37**, 153
Muller, P. and Meyer, C.: (1964) Second Catalogue d'Ephémérides d'Etoiles Doubles Visuelles *Notes et Informations Obs. de Paris*, **22**, 1
Muller, P. and Meyer, C.: (1969) Troisiéme Catalogue d'Ephémérides d'Etoiles Doubles Visuelles *Pub. Obs. de Paris*
Muller, P. and Couteau, P.: (1979) Quatriéme Catalogue d'Ephémérides d'Etoiles Doubles Visuelles *Pub. Obs. de Paris*
Prieto, C.: (1995) *PhD. Pub. Dep. Matemática Aplicada*, **1**, Univ. de Santiago. Spain
Roth, E. A.: (1973) *Celes. Mech.*, **8**, 245
Roth, E. A.: (1979) *J. Appl. Math. Phys.*, **30**, 315
Vidal, E.: (1944) *Pubs. Obs. Santiago II*, **6**, Univ. de Santiago. Spain
Vidal, E.: (1945) *Rev. Geofísica*, **Año IV**,
Vidal, E.: (1946) *Rev. Geofísica*, **Año V**
Vidal, E.: (1947) *Rev. Geofísica*, **Año VI**
Vidal, E.: (1948) *Astron. J.*, **54**
Vidal, E.: (1952) *Urania*, **230**
Vidal, E.: (1953) Cálculo de órbitas de estrellas dobles visuales *C.S.I.C., Mon. de Astron. y Ciencias Afines*, **1**. Spain
Vidal, E.: (1957) *Urania*, **245**, 45
Worley, C. E. and Heintz, W. D.: (1983) Fourth Catalog of orbits of Visual Binary Stars *Pub. U.S.N.O.*, **VOL XXIV**, PART VII
Zaera de Toledo, J. A.:(1963) *Urania*, **257**, 3
Zaera de Toledo, J. A.:(1966a) *Urania*, **263**, 3
Zaera de Toledo, J. A.:(1966b) *Urania*, **264**, 129

MASS-RATIO DISTRIBUTIONS FROM ΔM-STATISTICS FOR NEARBY HIPPARCOS BINARIES[1]

S.SÖDERHJELM
Lund Observatory
Box 43, S-22100 Lund, Sweden

Abstract. The Hipparcos satellite has produced a large body of double-star data that can now begin to be exploited. For an almost complete sample of nearby binaries with linear separations 30-500 a.u., the observed magnitude differences are translated to mass-ratios, and their distribution are studied as a function of separation and/or primary mass. An excess of equal-mass pairs is found especially for the smaller (linear) separations, being possibly a signature of different formation processes for wider and closer binaries.

1. INTRODUCTION

The Hipparcos Catalogue reductions have recently been completed, and the data are available for use within the reduction consortia before the general release of the catalogue around June 1, 1997. Numerous papers describing the complicated reductions are available in vol:s **258** (May 1992) and **304** (December 1995) of Astronomy&Astrophysics, with the double-star processing described e.g. by Mignard *et al.* (1992), Söderhjelm *et al.* (1992) and Mignard *et al.* (1995).

There are especially three features of the Hipparcos double star data that make them ideally suited to tackle the difficult question about double-star mass-ratio distributions. Firstly, the Hipparcos data give an almost complete sampling of the bright ($V < 9$) visual double stars with separations in the range 0.3-10 arcsec and with magnitude-differences smaller than 3.5. Secondly, each double has a trigonometric parallax, allowing reasonably unbiased distance-limited sampling. Finally and most crucially, each dou-

[1]Based on observations made with the ESA Hipparcos astrometry satellite

J. A. Docobo et al. (eds.), Visual Double Stars: Formation, Dynamics and Evolutionary Tracks, 497–508.

ble has a well-determined magnitude-difference which for the main-sequence pairs is closely related to the mass-ratio.

In practice, there are many problems that make the analysis less straightforward, but even the preliminary results presented below show features in the q-distribution that have not previously been described.

2. MASS-RATIOS

The mass-ratio ($q \equiv M_{sec}/M_{pr} < 1$) is a key parameter characterizing a binary star, and before any large-scale mass-loss or mass-transfer processes it remains at its original value. The distribution of mass-ratios for a group of binaries will thus tell something about the birth-processes for this group, and much work has been spent in deriving such statistics. One of the key difficulties is that fundamental mass-ratios can best be derived from double-lined spectroscopic binaries, which by definition are strongly biased towards $q \simeq 1$. A large sample of single-lined spectroscopic binaries may give a statistical estimate of the underlying q-distribution, but still with some model-dependence for observational selection-effects. For visual binaries, systems with known absolute orbits (giving individual component masses) are still few. The standard alternative procedure (for systems at a known evolutionary state) is to translate the luminosity-ratio to a mass-ratio, but here the lack of good magnitude-differences for close systems has been a key bottleneck.

2.1. PREVIOUS RESULTS FOR THE Q-DISTRIBUTION

A very large number of investigations have been published, and rather different results have been obtained, both for different samples of binaries and for different methods to treat the many selection-effects in the observations. In later years, there has been a consensus towards a generally decreasing $f(q)$, with large numbers of small-q binaries and the 'observed' peak at $q = 1$ explained away as due to observational selection. For main-sequence binaries, see e.g. papers by Duquennoy and Mayor(1991), Halbwachs(1987) and Trimble (1990).

2.2. RELATION TO THE ΔM:S

A key assumption in the present work is that there is a rather well-defined relation between q and Δm for main-sequence binaries, as a natural consequence of a 'mass-luminosity-relation'. In the light of theoretical stellar evolution, the matter is more complicated, and it is instructive to look first at typical isochrones in the observed $Hp/(V - I)$ plane. Fig. 1 has been derived with the useful program iso.f (made available on the WWW

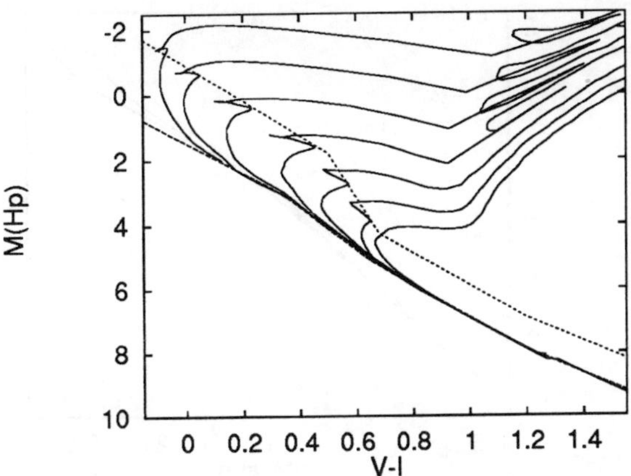

Figure 1. Theoretical stellar isochrones for standard Z=0.02 models with lg age from 8.2 to 10.0 in steps of 0.3 dex

by G. Meynet) interpolating the results of standard-composition (Z=0.02) stellar models (Schaller *et al.*, 1992). The $M_V/(B-V)$ coordinates were only slightly transformed to $Hp/(V-I)$, using the calibrations given in the preliminary Introduction to the Hipparcos Catalogue. (For the broad-band Hp-magnitude, $Hp - V \equiv f_{\mathrm{HpV}}$ is absolutely less than 0.2 mag for all except the reddest stars). Also plotted in the figure are the schematical 'MS-limits' used below in the selection of stars.

Using only the 'main-sequence' part of the data, one obtains the theoretical mass-luminosity diagram of Fig. 2. In the 1-2 M_{sun} interval, there is a reasonably tight relation, but at lower and higher masses, the age of the system starts to influence the results critically, and more detailed modelling is ideally required. For the present study, most of the data are in the 'easy' interval, and no individual age-determinations are attempted. Statistically, however, most isochrones have a steeper slope than the global mean relation, and this has to be taken into account.

3. THE HIPPARCOS DATA

Hipparcos data are available for an almost complete (magnitude-limited) sample with $B-V < 0.8$ and $V < 7.9 + 1.1\sin(|b|)$. (This so-called 'Survey' is described in the Hipparcos Input Catalogue, Turon et al., 1992). The Survey-limit transformed to Hp-magnitude is then

$$Hp_{\mathrm{lim}} = 7.9 + f_{\mathrm{HpV}}(V-I) + 1.1\sin(|b|) \quad (1)$$

and using a mean main sequence, this corresponds to typical distance-limits around 300/500 pc at $V-I=0$, 130/210 pc at $V-I=0.4$ and

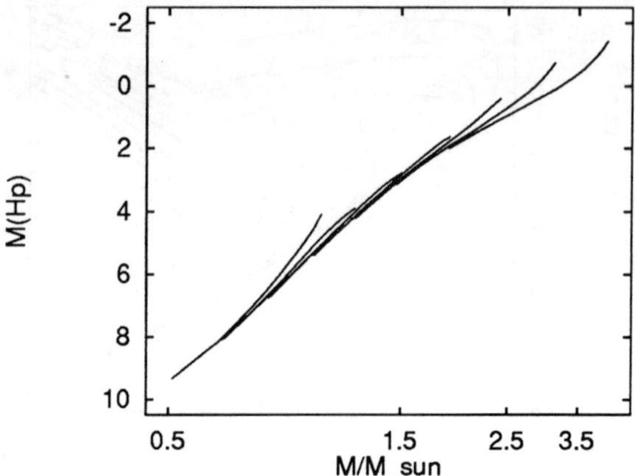

Figure 2. Theoretical Mass-Luminosity diagram for the main-sequence part of the isochrones in Fig.1

40/70 pc at $V - I = 0.7$ (the larger numbers at the galactic poles). With a typical parallax-error around 1.2 mas, the absolute magnitude will be given with 0.5 mag mean error at 200 pc, and this is taken as the 'absolute' upper distance-limit. An almost complete (distance-limited) sample of main-sequence binaries can then be obtained by letting the distance-limit decrease with $V - I$ in such a way that all main-sequence primaries are within the above magnitude-limit.

3.1. MAIN-SEQUENCE SAMPLES

An observed close double has mostly only a single 'combined' $(V - I)_c$. Assuming the two components to be on a 'typical' main sequence

$$Hp = C + \alpha(V - I) \tag{2}$$

($\alpha \simeq 6$), it is easy to show that the approximate colours of the components are

$$\begin{aligned} (V - I)_{pr} &= (V - I)_c - \alpha^{-1}\Delta Hp/(1 + 10^{0.4\Delta Hp}) \\ (V - I)_{sec} &= (V - I)_{pr} + \alpha^{-1}\Delta Hp \end{aligned} \tag{3}$$

An improved model is obtained by schematically fitting the components to the particular isochrone where the primary is at the 'turnoff' with minimal $V - I$. (Although this isochrone is selected quite ad hoc, the fit should be statistically better than with the first model, because all isochrones are steeper than the mean relation $Hp(V - I)$. From the observed parallax, the primary absolute magnitude is derived, and if it is within 3σ (or at most 0.75 mag) of the upper or lower edges of the schematical MS-band in Fig. 1, the system is accepted as a main-sequence pair. A red-giant primary in a

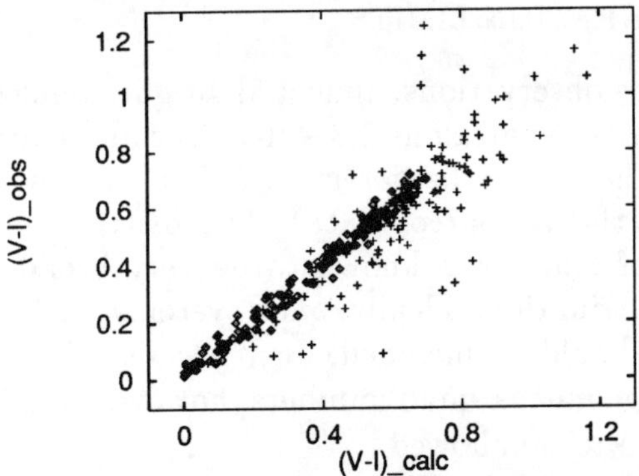

Figure 3. Observed vs. computed colours for systems where the components were observed individually in the Tycho experiment. The diamonds are for primary components, the crosses for the secondaries.

physical pair with a blue main-sequence star will give a too bright absolute magnitude even for the combined (erroneous) $V-I$, and these systems will therefore be automatically excluded.

For a number of systems with angular separation above some 2 arcsec, individual $(B_T - V_T)$ colours have been derived from Tycho 'starmapper' observations, and these observed colours (transformed to $V - I$) can be compared with the 'computed' ones according to the above prescription. About 150 'accepted MS'-systems can be used for the comparison, giving the qualitatively satisfactory plot shown in Fig. 3. There are clearly some remaining systematic effects, generally in the sense that the real isochrones are steeper than the modelled ones. The important output from the fitting is the primary colour, however, which is given with almost no bias (The mean and standard deviation are about -0.004/0.028).

To have an almost complete sample, the distance-limit must be 'inside' the volume defined by the magnitude-limit above. This is realized by adopting a distance-limit $r_{\rm lim}(V-I,|b|)$ corresponding to a ZAMS primary (absolute magnitude M_Z) at the magnitude-limit, viz.

$$r_{\rm lim} = 10^{1+0.2[Hp_{\rm lim}(V-I,|b|)-M_Z(V-I)]} \tag{4}$$

All systems in the magnitude-limited sample have a brighter absolute magnitude both because of evolution away from ZAMS and because the selection was for the combined (primary+secondary) system, and thus all will be within this conservative distance-limit.

The space-volume surveyed is continously variable with $V-I$ (and $|b|$), but the final q-distributions are conveniently studied by defining an 'A-star' group with $0.00 < V-I < 0.40$, and an 'F-star' group with $0.40 < V-I < 0.70$.

3.2. LINEAR SEPARATION LIMITS

After the Hipparcos observations, almost all bright doubles (primary magnitude less than 9) with separations $0.3-10$ arcsec and magnitude-differences less than 3.5 are thought to be detected. (This claim will be elucidated in further studies of the Hipparcos data). The distribution-function for the semi-major axes of binaries is known to be very broad, and for studying the mass-ratio distribution, ideally only systems within small ranges of linear separation should be included. Just because the a-distribution is so broad, a small range means small numbers, however, and in practice, rather large intervals have to be allowed.

The key advantage of the Hipparcos data is that each angular separation can be directly converted to a linear one via the known parallax. The physically significant parameter is of course the semi-major axis a, but in a statistical context, the distribution of (instantaneous) linear separation is very similar to the a-distribution. The 0.3 arcsec limit corresponds to 60 a.u. at 200 pc, and this becomes then a natural lower limit of linear separations for the 'A'-group. Putting the upper limit at 480 a.u. necessitates also (at least in the forthcoming studies of the binary frequencies) that all systems *closer* than 48 pc be excluded, because otherwise some angular separations may go above 10 arcsec. For the 'F'-group, $r_{\rm lim}$ is mostly below 100 pc, and a lower separation-limit around 30 a.u. is marginally acceptable. An upper separation-limit 240 a.u. makes for an excluded 'hole' 24 pc in radius, again with a rather small loss of systems. The final sampling has then three partly overlapping separation-intervals: 60-120/120-240/240-480 a.u. for the 'A'-group and 30-60/60-120/120-240 a.u. for the 'F'-group.

3.3. PHOTOMETRIC MASS-RATIOS

With close to complete sampling of the main-sequence binaries in these distance and separation-intervals, the idea is to use each observed Δm to derive a corresponding mass-ratio q. This can be done in more or less detail, and to see the influence on the final results, two different calibrations were used independently.

The first method adopts a rough 'mean' mass-luminosity relation through the isochrones in Fig. 2:

$$\lg m = \begin{cases} 0.48 - 0.150 M_{Hp} + 0.0125 M_{Hp}^2 & M_{Hp} < 2.7 \\ 0.40 + 0.095 M_{Hp} + 0.0030 M_{Hp}^2 & 2.7 < M_{Hp} < 5.4 \\ 0.28 + 0.050 M_{Hp} - 0.0011 M_{Hp}^2 & M_{Hp} > 5.4 \end{cases} \qquad (5)$$

From the observed absolute magnitudes of the two components, the masses and thus the mass-ratio is directly obtained.

TABLE 1. The observed number of binaries in q-bins defined by the more rigorous M/L-model. In parentheses, for comparison, the numbers obtained with the mean model in eq.(5) are given

	A (linear sep in a.u.)			F (linear sep in a.u.)		
q	60-120	120-240	240-480	30-60	60-120	120-240
1.00-0.95	16(19)	9(9)	6(7)	15(15)	15(15)	5(5)
0.95-0.90	13(13)	6(4)	5(4)	8(8)	8(7)	8(7)
0.90-0.80	8(7)	7(11)	10(9)	9(8)	13(13)	7(8)
0.80-0.70	16(16)	16(17)	16(21)	8(9)	16(17)	15(15)
0.70-0.60	18(20)	23(22)	29(32)	16(15)	14(15)	14(14)
0.60-0.55	16(13)	8(10)	13(9)			

The second model assumes instead that the two components are on the 'special' isochrone (primary at minimum colour), and the mass-ratio is obtained from the stellar-model results for this particular age. (The ultimate refinement not yet tested would be to use the known absolute magnitude to put the system on the correct isochrone, but observational parallax errors would make application difficult beyond 50 or 100 pc distance).

4. RESULTS

For each of the two stellar samples (A or F), there are three ranges of linear separation, making six 'observed' q-distributions for each of the two M/L-models above. The raw counts in five or six q-bins are given in Table 1. (Note that the first two bins are half the size of the others, which means the normalized frequencies have to be doubled). The steep $q(\Delta Hp)$-relation for the 'F'-stars puts the 3.5 mag-limit in ΔHp at about $q = 0.60$, thus the complete sampling covers only the limited range $q = 0.60 - 1.0$. (For the 'A'-group, an extra, narrow bin to $q = 0.55$ is added). Although there are obvious differences between the two M/L-models (especially for the 'A'-group), the main features of the distributions seem to be rather stable. For the reasons given above, the 'isochrone'-model is deemed more reliable, and the 'model-error' σ_m is taken formally as half the difference between the two numbers in each bin.

The 'random' error in the numbers can be estimated by varying the input-data. One may simply input a number of 'alternative' sets of Hipparcos data, obtained by adding random corrections (with the estimated variances) to each catalogue datum, and looking at the variations in the output q-distributions. After 100 such trials, the mean and standard deviation (σ_r) in each bin is obtained as shown in Table 2. In the folllowing

TABLE 2. Mean and standard deviations for the number of binaries in the above q-bins after testing with an ensamble of 100 'randomized' Hipparcos catalogues

	A (linear sep in a.u.)			F (linear sep in a.u.)		
q	60-120	120-240	240-480	30-60	60-120	120-240
1.00-0.95	14.4(2.4)	8.3(2.1)	7.2(1.8)	16.3(2.4)	16.0(3.0)	8.1(2.0)
0.95-0.90	12.0(2.6)	8.3(1.9)	4.6(1.4)	9.4(1.9)	10.2(1.8)	7.0(1.7)
0.90-0.80	11.4(3.0)	8.8(2.0)	14.2(2.7)	11.9(2.3)	15.1(2.1)	7.6(2.0)
0.80-0.70	18.4(2.6)	18.8(3.5)	16.9(2.6)	10.8(2.0)	15.4(2.4)	16.6(2.1)
0.70-0.60	20.8(3.4)	24.1(3.5)	30.6(3.8)	16.6(2.5)	17.5(2.7)	13.7(2.1)
0.60-0.65	15.9(2.9)	11.0(2.7)	14.1(2.7)			

tables, the two sigmas are quadratically combined to a 'total'

$$\sigma_t = \sqrt{\sigma_m^2 + \sigma_r^2} \tag{6}$$

The 'completeness-distance' at each $V - I$ was so far calculated so as to make a single ZAMS-star just make the magnitude-limit for completeness. This can be checked by counting how many of the selected stars are actually flagged as 'S' in the Hipparcos Input Catalogue. The fractions (in different bins of q and linear separation) are mostly above 95%, and one may experiment with some increase of the distance-limits. With the standard limits corresponding to ZAMS, an increase or decrease is conveniently parametrized by a magnitude-offset $\Delta m_{\rm off}$ to the value M_Z in eq. (4), a 10% increase in the distances corresponding to about $\Delta m_{\rm off} = -0.2$.

Table 3 gives the total number of 'S' and 'non-S' stars when the distance-limits are changed in this way, and as expected, the numbers increase rapidly with decreasing $\Delta m_{\rm off}$. A priori, it seems like a 92% mean completeness even at 'ZAMS-0.75 mag' would be enough to preserve the statistical distribution of q, with the advantage of doubled counts in each bin. This is potentially a dangerous argument, however, because it reintroduces the classical (Öpik, 1923) selection effect that a magnitude-limited sample contains a disproportionate number of large-q binaries. (Each individual component is too faint, but the combined magnitude puts it in the sample). It is necessary therefore to check carefully how the q-distributions vary with $\Delta m_{\rm off}$, as has been done in Fig. 4 for one of the q bins. One would expect 'random' variations for $\Delta m_{\rm off} > 0$ and then possibly some systematic trend when the distances are increased, but the variations are usually well within the calculated mean errors. We can thus safely use the runs with $\Delta m_{\rm off} = -0.25$ as listed in Table 4 and plotted in Fig. 5. (The

TABLE 3. Total number of binaries(n_t) in the 'A' and 'F' categories, and the number(n_x) and fraction that are not in the Input Catalogue 'Survey'. The runs have distance-limits corresponding to ZAMS plus or minus a constant magnitude-offset.

	A			F		
$\Delta m_{\rm off}$	n_t	n_x	(%)	n_t	n_x	(%)
1.00	85	0	0	52	0	0
0.75	108	0	0	68	0	0
0.50	157	2	1	95	0	0
0.25	194	6	3	138	0	0
0.0	235	13	6	182	3	2
-0.25	280	20	7	258	10	4
-0.50	317	26	8	347	21	6
-0.75	346	32	9	450	36	8
-1.00	369	36	10	556	58	10

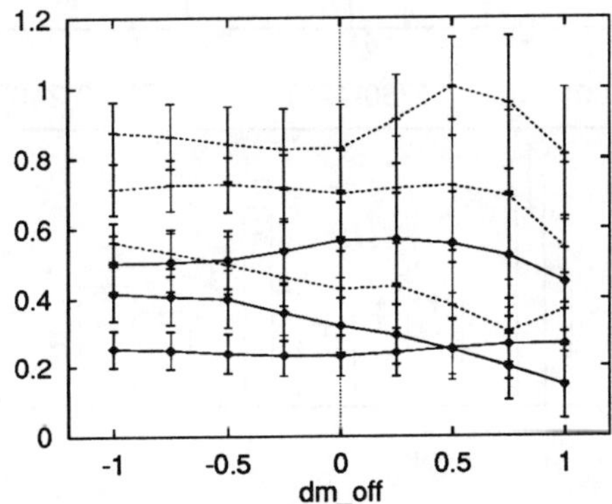

Figure 4. Histogram frequencies for the bin q=0.95-1.0 in all the six star/separation groups plotted as a function of $\Delta m_{\rm off}$.

ad hoc normalization puts the mean frequency in the interval $q = 0.6 - 0.9$ to unity).

In most of the diagrams, it is immediately apparent that there is a 'peak' near $q = 1$ that deviates strongly from the otherwise smooth trend of smaller numbers towards $q = 1$. The problem when one tries to be more specific is that there seems to be systematic trends both with respect to mass ('A' vs 'F') and with respect to separation. A simplistic χ^2 test for the 'A' vs 'F' differences (in the two overlapping separation-intervals) gives

TABLE 4. Arbitrarily normalized q-distributions with estimated total mean errors

q	A (linear sep in a.u.) 60-120	120-240	240-480	F (linear sep in a.u.) 30-60	60-120	120-240
1.00-0.95	0.54(0.09)	0.36(0.08)	0.23(0.06)	0.83(0.12)	0.72(0.10)	0.46(0.08)
0.95-0.90	0.45(0.08)	0.34(0.08)	0.15(0.04)	0.49(0.09)	0.42(0.07)	0.34(0.07)
0.90-0.80	0.23(0.05)	0.18(0.05)	0.23(0.04)	0.34(0.05)	0.29(0.04)	0.22(0.04)
0.80-0.70	0.36(0.05)	0.36(0.07)	0.27(0.06)	0.28(0.05)	0.33(0.05)	0.44(0.05)
0.70-0.60	0.42(0.07)	0.46(0.06)	0.50(0.06)	0.39(0.06)	0.38(0.05)	0.34(0.04)
0.60-0.65	0.58(0.06)	0.42(0.05)	0.46(0.06)			

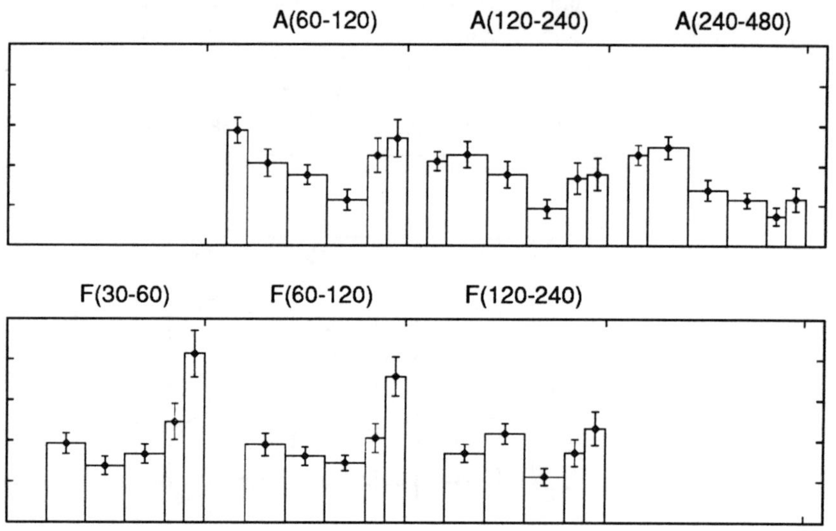

Figure 5. Normalized q-distributions for the six different groups of binaries. The mean errors are thought to be reasonable estimates of the total uncertainties.

however χ^2 less than about 5 for 5 d.o.f., that is, these differences are probably not significant.

The variation of the q-distribution with separation is more interesting. For a long time, and from many talks at this conference also, there has been a suspicion that closer pairs have a different dominating formation mechanism (involving disk evolution) than the wider ones formed by protostellar fragmentation. The limiting separation is not very well-determined, but values between 10 and 100 a.u. are often suggested. Unfortunately, the present study does not go to sufficiently small separations to be conclusive, but it

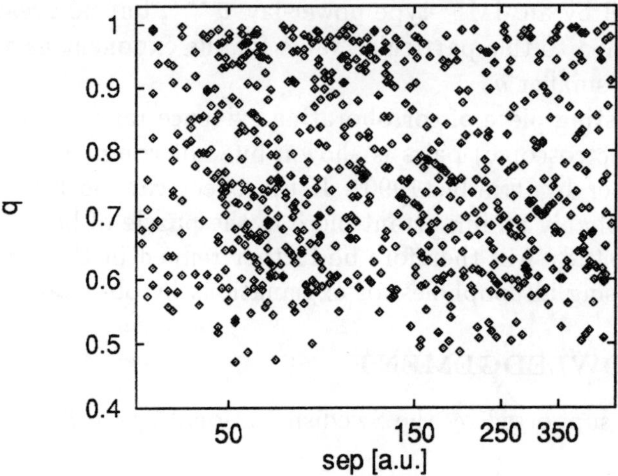

Figure 6. Distribution of the selected doubles in the q/lg sep plane.

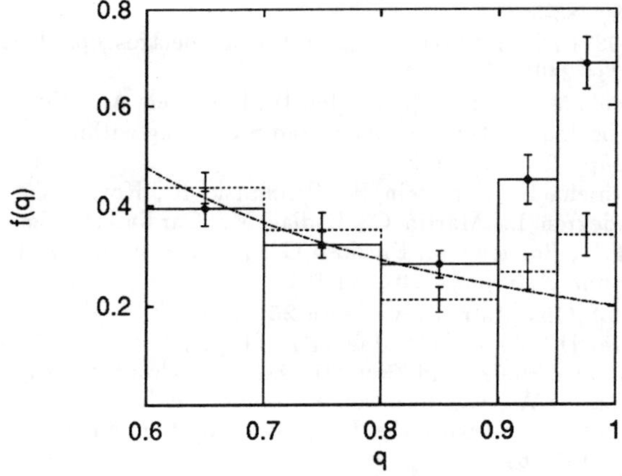

Figure 7. The final relative frequency distributions for q for separations below (full line) and above (dotted) 120 a.u. The curve is a schematic $q^{-1.7}$ relation.

starts to give some definite hints.

If one neglects the A/F-differences, and looks only at the total distribution in the lg sep/q-plane, a plot like the one in Fig 6 is obtained. One can definitely see that there are more pairs with q around 1 for smaller separations, and a division of the data into 'close' (30-120 a.u.) and 'wide' (120-480 a.u.) gives the frequency histograms in Fig. 7. The χ^2 difference is now around 40, and the conclusion is that there is indeed observational evidence for a process producing a high proportion of equal-mass pairs at separations below some 100 a.u. In the interval $q = 0.6 - 0.9$, the histograms are fitted well by an 'IMF'-type power-law $q^{-1.7}$, but no great significance can be attached to this particular value for the exponent as a guide to the behaviour at smaller q.

An interesting piece of corroborating evidence for an equal-mass excess for close (spectroscopic) pairs is shown, but not commented upon, in Fig. 5 in the paper by Trimble(1990). If the effect can be further substantiated, it is probably an important piece in the puzzle of binary star origins. The present study will therefore be further refined in the coming months, hopefully closing all loopholes for explanations by observational bias.

5. ACKNOWLEDGEMENT

This work is supported by the Swedish National Space Board.

References

Duquennoy A., Mayor M. (1991) Multiplicity among solar-type stars in the solar neighborhood: II. Distribution of the orbital elements in an unbiased sample, *Astron. Astrophys.* **248**, 485-524

Halbwachs J.L. (1987) Distribution of mass-ratios in spectroscopic binaries, *Astron. Astrophys.* **183**, 234-240

Mignard F., Froeschlé M., Badiali M., Cardini D., Emanuele A., Falin J.L., Kovalevsky J. (1992) Hipparcos double star recognition and processing within the FAST consortium, *Astron. Astrophys.* **258**, 165-172

Mignard F., Söderhjelm S., Bernstein H., Pannunzio R., Kovalevsky J., Froeschlé M., Falin J.L., Lindegren L., Martin C., Badiali M., Cardini D., Emanuele A., Spagna A., Bernacca P.L., Borriello L., Prezioso G. (1995) Astrometry of double stars with Hipparcos, *Astron. Astrophys.* **304**, 94-104

Öpik, E (1923) Publ. Obs. Astr. Univ. Tartu **25**, No 6

Schaller G., Schaerer D., Meynet G., Maeder A. (1992) New grids of stellar models from 0.8 to 120 Msun at Z=0.020 and Z=0.001, *Astron. Astrophys. Suppl. Ser.* **96**, 269-331

Söderhjelm S., Evans D.W., van Leeuwen F., Lindegren L. (1992) Detection and measurement of double stars with the Hipparcos satellite: NDAC reductions, *Astron. Astrophys.* **258**, 157-164

Trimble V. (1990) The distributions of binary system mass ratios: a less biased sample, *Mon. Not. R. astr. Soc.* **242**, 79-87

Turon, C. et al. (1992) The Hipparcos Input Catalogue, ESA SP-1136